# Promising Antiviral Herbal and Medicinal Plants

With the prevalence of viral diseases causing significant health challenges and the emergence of highly infectious and difficult-to-treat conditions, such as Ebola, HIV/AIDS, and Hepatitis B, there is a pressing need for effective antiviral treatments. Throughout history, viral infections have caused devastating pandemics, leading to widespread illness and mortality. Notable examples include the Black Death caused by *Yersinia pestis* and the Spanish flu pandemic of 1918. To address this ongoing threat, exploring the potential of herbal and medicinal plants as a source of antiviral compounds is crucial. *Promising Antiviral Herbal and Medicinal Plants* provides a comprehensive overview of the latest research in the field of antiviral properties of herbal and medicinal plants.

Features:

- Compiles cutting-edge research on the antiviral properties of various plants, highlighting their potential therapeutic applications.
- Discusses various chemical components of the medicinal and herbal plants with the known bioactive constituents and their antiviral applications.
- Elaborates on medicinal and herbal plants' opportunities and future direction for drug synthesis against various viral diseases.

A volume in the *Exploring Medicinal Plants* series, this book is an essential resource for researchers, healthcare professionals, and those interested in viral infections and their management.

## Exploring Medicinal Plants

*Series Editor: Azamal Husen, Department of Biotechnology, Smt. S. S. Patel Nootan Science & Commerce College, Sankalchand Patel University, Visnagar - 384315, Gujarat, India*

Medicinal plants render a rich source of bioactive compounds used in drug formulation and development; they play a key role in traditional or indigenous health systems. As the demand for herbal medicines increases worldwide, supply is declining as most of the harvest is derived from naturally growing vegetation. Considering global interests and covering several important aspects associated with medicinal plants, the Exploring Medicinal Plants series comprises volumes valuable to academia, practitioners, and researchers interested in medicinal plants. Topics provide information on a range of subjects including diversity, conservation, propagation, cultivation, physiology, molecular biology, growth response under extreme environment, handling, storage, bioactive compounds, secondary metabolites, extraction, therapeutics, mode of action, and healthcare practices.

Led by Azamal Husen, PhD, this series is directed to a broad range of researchers and professionals consisting of topical books exploring information related to medicinal plants. It includes edited volumes, references, and textbooks available for individual print and electronic purchases.

*Medicinal Plant Responses to Stressful Conditions*
Edited by Arafat Abdel Hamed Abdel Latef

*Aromatic and Medicinal Plants of Drylands and Deserts*
*Ecology, Ethnobiology and Potential Uses*
Edited by David Ramiro Aguillón Gutiérrez, Cristian Torres León, and Jorge Alejandro Aguirre Joya

*Secondary Metabolites from Medicinal Plants*
*Nanoparticles Synthesis and their Applications*
Edited by Rakesh Kumar Bachheti, Archana Bachheti

*Aquatic Medicinal Plants*
Edited by Archana Bachheti, Rakesh Kumar Bachheti, and Azamal Husen

*Antidiabetic Medicinal Plants and Herbal Treatments*
Edited by Azamal Husen

*Ethnobotany and Ethnopharmacology of Medicinal and Aromatic Plants*
*Steps Towards Drugs Discovery*
Edited by Mohd Adnan, Mitesh Patel and Mejdi Snoussi

*Wild Mushrooms and Health*
*Diversity, Phytochemistry, Medicinal Benefits, and Cultivation*
Edited by Kamal Ch. Semwal, Steve L. Stephenson, and Azamal Husen

*Medicinal Roots and Tubers for Pharmaceutical and Commercial Applications*
Edited by Rakesh Kumar Bachheti and Archana Bachheti

*Pharmacological Aspects of Essential Oils*
*Current and Future Trends*
Edited by Sunita Singh, Pankaj Kumar Chaurasia and Shashi Lata Bharati

*Promising Antiviral Herbal and Medicinal Plants*
Edited by Nadeem Akhtar, Azamal Husen, Vagish Dwibedi, and Santosh Kumar Rath

**For more information about this series, please visit:** www.routledge.com/Exploring-Medicinal-Plants/book-series/CRCEMP

# Promising Antiviral Herbal and Medicinal Plants

Edited by
Nadeem Akhtar, Azamal Husen, Vagish Dwibedi,
and Santosh Kumar Rath

CRC Press
Taylor & Francis Group
Boca Raton London New York

CRC Press is an imprint of the
Taylor & Francis Group, an **informa** business

Designed cover image: © Shutterstock

First edition published 2024
by CRC Press
2385 NW Executive Center Drive, Suite 320, Boca Raton FL 33431

and by CRC Press
4 Park Square, Milton Park, Abingdon, Oxon, OX14 4RN

*CRC Press is an imprint of Taylor & Francis Group, LLC*



*Library of Congress Cataloging-in-Publication Data*
Names: Akhtar, Nadeem, editor.
Title: Promising antiviral herbal and medicinal plants /
edited by Dr. Nadeem Akhtar, Dr. Azamal Husen, Dr. Vagish Dwibedi,
and Dr. Santosh Kumar Rath.
Description: First edition. | Boca Raton, FL : CRC Press, 2024. |
Series: Exploring medicinal plants ; vol. 18 | Includes bibliographical
references and index. |
Identifiers: LCCN 2023036427 (print) | LCCN 2023036428 (ebook) |
ISBN 9781032358802 (hbk) | ISBN 9781032358819 (pbk) |
ISBN 9781003329169 (ebk)
Subjects: LCSH: Herbs—Therapeutic use. | Medicinal plants.
Classification: LCC RM666.H33 P767 2024 (print) | LCC RM666.H33 (ebook) |
DDC 615.3/21—dc23/eng/20231214
LC record available at https://lccn.loc.gov/2023036427
LC ebook record available at https://lccn.loc.gov/2023036428

ISBN: 9781032358802 (hbk)
ISBN: 9781032358819 (pbk)
ISBN: 9781003329169 (ebk)

DOI: 10.1201/9781003329169

Typeset in Times
by codeMantra

# Contents

# About the Editors

**Dr. Nadeem Akhtar**, a scientist, is currently making significant strides in the animal health and longevity sector for a leading Canadian company, CBS Bio Platforms Inc. His impressive background demonstrates that he has held a variety of important positions in academia and research throughout his career. Dr. Akhtar was a Research Associate at the University of Guelph, Canada, where the objective of his work was to improve companion animal's health in order to promote a superior and more health-conscious lifestyle. Prior to assuming this position, he excelled as a postdoctoral researcher at the Department of Chemical Engineering, University of Waterloo, Canada (2019–2021) and the Department of Animal Biosciences, University of Guelph, Canada (2015–2019). His accomplishments in the National Eligibility Test (NET-2010) and Graduate Aptitude Test (GATE) in Life Sciences (2010) and Biotechnology (2011) further emphasize his academic excellence. He earned his Ph.D. from the esteemed Department of Biotechnology at the Thapar Institute of Engineering and Technology in India. Dr. Akhtar's scientific contributions are nothing short of remarkable, with over 27 articles published in peer-reviewed international journals as well as 12 book chapters and two edited books. His abilities extend to patent accomplishments, as he has one patent to his name. In addition to his research, he has shown his dedication to knowledge advancement by serving as a reviewer for prestigious journals such as *RSC Advances,* the *Journal of Applied Microbiology, Scientific Reports,* and *Poultry Science.* Dr. Akhtar's career exemplifies dedication and excellence, demonstrating his invaluable contributions to the fields of animal science and biotechnology. His journey continues to motivate and shape the global scientific landscape.

**Professor Azamal Husen** (MSc from Hamdard University, New Delhi, and PhD from Forest Research Institute, Dehra Dun, India) is a Foreign Delegate at Wolaita Sodo University, Wolaita, Ethiopia. He has served the University of Gondar, Ethiopia, as a Full Professor of Biology, and worked there as the Coordinator of the MSc Programme and as the Head of the Department of Biology. He was a Visiting Faculty member of the Forest Research Institute, and the Doon College of Agriculture and Forest at Dehra Dun, India. He has more than 20 years' experience of teaching, research, and administration. Dr Husen specializes in biogenic nanomaterial fabrication and application, plant responses to nanomaterials, plant adaptation to harsh environments at the physiological, biochemical, and molecular levels, herbal medicine, and clonal propagation for improvement of tree species. He has conducted several research projects sponsored by various funding agencies, including the World Bank, the Indian Council of Agriculture Research (ICAR), the Indian Council of Forest Research Education (ICFRE), and the Japan Bank for International Cooperation (JBIC). He has published over 150 research papers, review articles, and book chapters, edited books of international repute, presented papers at several conferences, and produced over a dozen manuals and monographs. He has received four fellowships from India and a recognition award from the University of Gondar, Ethiopia, for excellent teaching, research, and community service. An active organizer of seminars/conferences and an efficient evaluator of research projects and book proposals, Dr Husen has been on the Editorial board and the panel of reviewers of several reputed journals published by Elsevier, Frontiers Media SA, Taylor & Francis, Springer Nature, RSC, Oxford University Press, Sciendo, The Royal Society, CSIRO, PLOS, and John Wiley & Sons. He is on the advisory board of Cambridge Scholars Publishing, UK. He is a Fellow of the Plantae group of the American Society of Plant Biologists, and a Member of the International Society of Root Research, Asian Council of Science Editors, and INPST. Dr Husen is Editor-in-Chief of the *American Journal of Plant Physiology.* He is also working as Series Editor of Exploring *Medicinal Plants,* published by Taylor & Francis Group, USA; Plant Biology, Sustainability, and Climate Change, published by Elsevier Inc., MA 02139, USA; and *Smart Nanomaterials Technology*, published by Springer Nature Singapore Pte Ltd., Singapore.

**Dr. Vagish Dwibedi** completed his B.Sc. from Lucknow University, Lucknow, Uttar Pradesh, and Ph.D. from Thapar Institute of Engineering and Technology, Patiala, India. Dr. Dwibedi is an academician and researcher with more than 8 years of experience in biotech research and development. He has conducted research projects and consultancy work in the areas of plant-microbe interaction, bioassay-guided drug discovery and development, food security, sustainable agriculture and waste water treatment. Presently, he is working as Assistant Professor at University Institute of Biotechnology, Chandigarh University, Gharuan Mohali, Punjab. Formerly, he worked as Research Scientist at Agpharm Bioinnovations LLP incubated at Thapar Institute of Engineering & Technology, Patiala, India. He was also the winner in the Down select competition—DST—Lockheed Martin—Tata Trust _IIGP 2.0 (2k18) (India Innovation Growth Programme 2.0) Innovation ID: IIGPUIBSUNB. (The award carries financial support up to 11 Lac INR.) Dr. Dwibedi's work is directed towards the development of screening platforms for different biological activities such as anti-microbial, antioxidant, anti-cancer, or finding novel molecules that interfere in the mechanism of development of diseases such as Alzheimer's dementia (AD), Parkinson's disease (PD), obesity, anti-gout (arthritis), and type 2 diabetes. He is also interested in food security, which predominantly involves the exploitation of plant-microbe interaction to combat abiotic stress and post-harvest preservation to enhance the shelf life of fresh crop/horticulture produces. He has published more than 25 research papers in various national and international journals and holds two Indian patents.

**Dr. Santosh Kumar Rath** is a medicinal and natural product chemist with more than 10 years of experience in research, academia, and industry. Dr. Rath received his doctorate from AcSIR at CSIR-Institute of Minerals and Materials Technology (CSIR-IMMT), Bhubaneswar, India. Presently, he is working as Assistant Professor at Department of SoPPHI, Faculty of Pharmacy, DIT University, Dehradun, Uttarakhand, India. Formerly, he worked as an Associate Professor and Research Head at the Department of Pharmaceutical Chemistry, Danteshwari College of Pharmacy, Jagdalpur, Chhattisgarh, India. He also worked as Research Associate (CSIR-RA) in the School of Chemistry & Biochemistry at Thapar Institute of Engineering & Technology, Patiala, India. Dr. Rath has received Research Associateship from CSIR and Senior Research Fellowship (ICMR-SRF) from ICMR, New Delhi, India. He did research at CSIR-Indian Institute of Integrative Medicine (CSIR-IIIM), Jammu, India. Dr. Rath has many publications to his credit, published more than 25 research papers in highly reputed international journals, 4 book chapters and 1 Indian patent. His area of research are Natural Product Chemistry and Organic Synthesis, in which he has expertise in isolation, identification, and characterization of bioactive secondary metabolites, structural modification of major chemical constituents from plants as well as fungal sources. His research areas involve multistep synthesis of biologically active synthetic and/or natural product-based hybrid scaffolds for lead identification in special targets to neurological disease, cancer, HIV, COVID-19, and infectious diseases. His research is mainly focused on the synthesis of novel P-gp and bacterial efflux pump inhibitors, besides synthetic modifications of the bioactive natural products for better activity/minimize toxicity profiles. Currently, he is developing new synthetic methodologies for C-H functionalization of heterocycles and other medicinally relevant molecules.

# Contributors

**Mandeep Kumar Arora**
School of Pharmaceutical and Population Health Informatics
DIT University
Dehradun, India

**S. R. Assumi**
Scientist, ICAR Research Complex for NEH Region
Umiam, Meghalaya

**Aadil Bathla**
Department of Civil and Environmental Engineering
Hanyang University
Seoul, Republic of Korea

**Pooja Bhadrecha**
Department of Biotechnology
University Institute of Biotechnology, Chandigarh University
Mohali, India

**Aleena James Chirayimmel**
School of Bio Sciences and Technology
Vellore Institute of Technology
Vellore, India

**Michael Onyegbulam Chukwuma**
Department of Pharmacognosy and Environmental Medicines, Faculty of Pharmaceutical Sciences
University of Nigeria
Nsukka, Nigeria

**Pritha Dalapati**
School of Pharmaceutical and Population Health Informatics
DIT University
Dehradun, India

**Abdulkadir Mohammed Danyaro**
Department of Biochemistry, Natural Product Research Laboratory
Bauchi State University
Gadau, Nigeria

**M. Bilashini Devi**
Scientist, ICAR Research Complex for NEH Region
Umiam, Meghalaya

**Amit Kumar Dubey**
Parasite Immunology Lab, Division of Microbiology
Indian Council of Medical Research (ICMR)-Rajendra Memorial Research Institute of Medical Sciences (RMRIMS)
Patna, India
School of Pharmaceutical and Population
Health Informatics
DIT University
Dehradun, India

**Vagish Dwibedi**
Department of Biotechnology
University Institute of Biotechnology, Chandigarh University
Mohali, India
Institute of Soil, Water and Environmental Sciences
Volcani Resaerch Center, Agricultural Research Organization
Rishon Lezion, 7528809, Israel

**Odoh Uchenna Estella**
Department of Pharmacognosy and Environmental Medicines, Faculty of Pharmaceutical Sciences
University of Nigeria
Nsukka, Nigeria

**Manoj Gadewar**
School of Medical and Allied Sciences
KR Mangalam University
Gurgaon, India

**Nancy George**
Department of Biotechnology
University Institute of Biotechnology, Chandigarh University
Mohali, India

**Sahil Jain**
Department of Biochemistry and Molecular Biology, Faculty of Life Sciences
Tel-Aviv University
Tel-Aviv, Israel

**Gurleen Kaur**
Department of Biotechnology
University Institute of Biotechnology, Chandigarh University
Mohali, India

**Gurpreet Kaur**
Department of Biotechnology
Thapar Institute of Engineering and Technology
Patiala, India

**Gursharan Kaur**
Department of Biotechnology
University Institute of Biotechnology, Chandigarh University
Mohali, India

**Rajinder Kaur**
Department of Plant Sciences
University of Idaho
Moscow, USA

**Sukhminderjit Kaur**
Department of Biotechnology
University Institute of Biotechnology, Chandigarh University
Mohali, India

**Yadveer Kaur**
Department of Biotechnology
Thapar Institute of Engineering and Technology
Patiala, India

**Zaved Ahmed Khan**
Faculty of Sciences, Baba Farid College
Baba Farid Group of Institutions
Bathinda, India

**Abhijit Kumar**
Department of Biotechnology
University Institute of Biotechnology, Chandigarh University
Mohali, India

**Priyanshu Kumar**
School of Pharmaceutical and Population Health Informatics
DIT University
Dehradun, India

**Jyoti Kumari**
Parasite Immunology Lab, Division of Microbiology
Indian Council of Medical Research (ICMR)-Rajendra Memorial Research Institute of Medical Sciences (RMRIMS)
Patna, India

**Reena Kumari**
Department of Chemistry
Indian Institute of Technology
Roorkee, India

**Shweta Kumari**
Department of Biotechnology
University Institute of Biotechnology, Chandigarh University
Mohali, India

**Tanisha Mahajan**
Department of Biotechnology
University Institute of Biotechnology, Chandigarh University
Mohali, India

**Avijit Malakar**
Department of Biotechnology
University Institute of Biotechnology, Chandigarh University
Mohali, India

**Chaitali Mallick**
Department of Pharmaceutical Technology, School of Medical Sciences
Adamas University
Kolkata, India

**Manisha**
Department of Biotechnology
University Institute of Biotechnology, Chandigarh University
Mohali, India

**Shashank Shekher Mishra**
School of Pharmaceutical and Population Health Informatics
DIT University
Dehradun, India

**Pritam Mukherjee**
Department of Biotechnology
University Institute of Biotechnology, Chandigarh University
Mohali, India

**Shaon Mukherjee**
Department of Biotechnology
University Institute of Biotechnology, Chandigarh University
Mohali, India

**Srilekha Das Nag**
Department of Pharmaceutical Technology, School of Medical Sciences
Adamas University
Kolkata, India

**Nitu**
Department of Biotechnology
University Institute of Biotechnology, Chandigarh University
Mohali, India

**Khadija Aliyu Nuru**
Department of Biochemistry, Natural Product Research Laboratory
Bauchi State University
Gadau, Nigeria

**Palak Rana**
Department of Biotechnology
University Institute of Biotechnology, Chandigarh University
Mohali, India

**Sneha Rani**
Parasite Immunology Lab, Division of Microbiology
Indian Council of Medical Research (ICMR)-Rajendra Memorial Research Institute of Medical Sciences (RMRIMS)
Patna, India

**Santosh Kumar Rath**
School of Pharmaceutical and Population Health Informatics
DIT University
Dehradun, India

**Saumaya Rustogi**
School of Pharmaceutical and Population Health Informatics
DIT University
Dehradun, India

**Nandan Sarkar**
Department of Pharmaceutical Technology, School of Medical Sciences
Adamas University
Kolkata, India

**Puja Shah**
School of Pharmaceutical and Population Health Informatics
DIT University
Dehradun, India

**Swati Sharma**
Department of Biotechnology
Shoolini Institute of Life Science and Business Management
Solan, India

**Davinder Singh**
Faculty of Sciences, Baba Farid College
Baba Farid Group of Institutions
Bathinda, India

**Lakhveer Singh**
School of Pharmaceutical and Population Health Informatics
DIT University
Dehradun, India

**Manpreet Singh**
Faculty of Sciences, Baba Farid College
Baba Farid Group of Institutions
Bathinda, India

**Niva Singh**
Parasite Immunology Lab, Division of Microbiology
Indian Council of Medical Research (ICMR)-Rajendra Memorial Research Institute of Medical Sciences (RMRIMS)
Patna, India

**Shubhankar Kumar Singh**
Parasite Immunology Lab, Division of Microbiology
Indian Council of Medical Research (ICMR)-Rajendra Memorial Research Institute of Medical Sciences (RMRIMS)
Patna, India

**H.D. Talang**
Scientist, ICAR Research Complex for NEH Region
Umiam, Meghalaya

**Babita Thakur**
Department of Biotechnology
University Institute of Biotechnology, Chandigarh University
Mohali, India

**Monika Thakur**
Department of Biotechnology
Shoolini Institute of Life Science and Business Management
Solan, India

**Habibu Tijjani**
Department of Biochemistry, Natural Product Research Laboratory
Bauchi State University
Gadau, Nigeria

**Prachi Vaid**
Department of Biotechnology, School of Sciences
AP Goyal Shimla University
Shimla, India

**Vanlalruati**
Scientist, ICAR Research Complex for NEH Region
Umiam, Meghalaya

**Mohini Yadav**
Department of Biotechnology
University Institute of Biotechnology, Chandigarh University
Mohali, India

# Preface

The endeavour of this book, *Promising Antiviral Herbal and Medicinal Plants*, is to provide comprehensive information on the latest research in the field of antiviral properties of herbal and medicinal plants. Viruses are responsible for a wide range of human diseases, including highly infectious and often difficult-to-treat conditions such as Ebola, HIV/AIDS, and Hepatitis B. Throughout history, viral infections have caused devastating pandemics. The Black Death, caused by the bacterium *Yersinia pestis*, is known as one of the deadliest pandemics ever recorded. Another notable pandemic was the H1N1 influenza pandemic of 1918, commonly referred to as the "Spanish flu," which infected a significant portion of the global population and resulted in millions of deaths. The book *Promising Antiviral Herbal and Medicinal Plants* seeks to address the urgent need for effective antiviral treatments by exploring the potential of herbal and medicinal plants. By compiling cutting-edge research in this field, this book aims to contribute to the development of novel therapeutic approaches and expand our understanding of antiviral properties present in nature.

It is hoped that this book will serve as a valuable resource for researchers, healthcare professionals, and individuals interested in combating viral infections and improving public health.

# Acknowledgements

Viruses cause various human diseases, of which several deadly communicable diseases such as Ebola, HIV/AIDS, and Hepatitis B are challenging to treat. Viral infections have triggered many pandemic outbreaks throughout world history. Herbal medicines and plant-derived preparations have been used to combat many communicable diseases for centuries. Over the past few decades, natural phytochemicals have been tested for antiviral properties more specifically. However, viral infections are becoming a significant threat to humans; further research is still needed to learn about viral infections or their mutations. In addition, medicinal plants and plant-derived bioactive secondary metabolites are increasingly being suggested as suitable alternative sources of antiviral agents. The development of suitable, precise *in vitro* and *in vivo* screening techniques may contribute to identifying biochemical and physiological effects of compounds. This advancement in bioassay evaluation techniques helps to identify potential bioactive plants rapidly. The edited book presented here serves as a valuable resource, providing up-to-date knowledge to researchers in this field. Compiling diverse studies within a single publication poses a considerable challenge, but it offers an opportunity to consolidate and disseminate important information in the related field. First of all, we would like to acknowledge all the contributor authors of chapters in this book for bringing such an exhaustive compilation for a wider readership. All the editors gratefully acknowledge their parent organization for all the support and encouragement rendered. Thanks are also due to all those researchers whose original work has made the basis of this compilation, without the endless efforts of the researchers, science and society could not progress. We are also thankful to the team Taylor & Francis for bringing this compilation to the scientific fraternity.

# Abbreviations

| S. NO. | ABBREVIATIONS | FULL FORM |
|---|---|---|
| 1. | **amiRNA** | Artificial MicroRNA |
| 2. | **ARDS** | Acute Respiratory Distress Syndrome |
| 3. | **CCR** | C-C Chemokine Receptor |
| 4. | **CCR5** | C-C Chemokine Receptor Type 5 |
| 5. | **CHIKV** | Chikungunya virus |
| 6. | **COVID-19** | Coronavirus Disease 2019 |
| 7. | **CXCR4** | C-X-C Chemokine Receptor Type 4 |
| 8. | **DENVax** | Dengue vaccine candidate |
| 9. | **EBOV** | Ebola virus |
| 10. | **eIF-2α** | Eukaryotic translation initiation factor 2-alpha |
| 11. | **GC-MS** | Gas Chromatography-Mass Spectrometry |
| 12. | **H1N1** | Influenza A subtype H1N1 |
| 13. | **H7N9** | Influenza A subtype H7N9 |
| 14. | **HAART** | Highly Active Antiretroviral Therapy |
| 15. | **HBeAg** | Hepatitis B e antigen |
| 16. | **HBsAg** | Hepatitis B surface antigen |
| 17. | **HBV** | Hepatitis B virus |
| 18. | **HCC** | Hepatocellular carcinoma |
| 19. | **HESNs** | HIV-Exposed Seronegative Individuals |
| 20. | **HIV** | Human immunodeficiency virus |
| 21. | **HSV** | Herpes simplex virus |
| 22. | **HSV 2** | Herpes simplex virus type 2 |
| 23. | **IC50** | Half-maximal Inhibitory Concentration |
| 24. | **IL-6** | Interleukin 6 |
| 25. | **MDR** | Multidrug-Resistant |
| 26. | **MERS** | Middle East respiratory syndrome |
| 27. | **MHC II** | Major Histocompatibility Complex Class II |
| 28. | **MIP-1α** | Macrophage Inflammatory Protein 1-alpha |
| 29. | **MPLC** | Medium Pressure Liquid Chromatography |
| 30. | **MTT** | 3-(4,5-Dimethylthiazol-2-yl)-2,5-Diphenyltetrazolium Bromide |
| 31. | **NF-κB** | Nuclear Factor kappa B |
| 32. | **NIH** | National Institute of Health, USA |
| 33. | **NMR** | Nuclear Magnetic Resonance |
| 34. | **NRTIs** | Nucleoside Reverse Transcriptase Inhibitors |
| 35. | **NS3/4A** | Non-structural protein 3/4A |
| 36. | **PEDV** | Porcine Epidemic Diarrhea Virus |
| 37. | **PPCPs** | Pharmaceuticals and personal care products |
| 38. | **PR8** | Influenza A Virus Strain PR8 |
| 39. | **qPCR** | Quantitative Polymerase Chain Reaction |
| 40. | **RdRp** | RNA-dependent RNA polymerase |

| S. NO. | ABBREVIATIONS | FULL FORM |
|---|---|---|
| 41. | **RISC** | RNA-Induced Silencing Complex |
| 42. | **ROS** | Reactive Oxygen Species |
| 43. | **RSV** | Respiratory syncytial virus |
| 44. | **SARS** | Severe Acute Respiratory Syndrome |
| 45. | **SARS-CoV-2** | Severe acute respiratory syndrome coronavirus 2 |
| 46. | **SDS–PAGE** | Sodium dodecyl sulphate–Polyacrylamide gel electrophoresis |
| 47. | **SSCP** | Single-Strand Conformation Polymorphism |
| 48. | **STDs** | Sexually Transmitted Diseases |
| 49. | **TAE** | Tris acetate EDTA |
| 50. | **TCA** | Trichloro acetic acid |
| 51. | **TCM** | Traditional Chinese Medicine |
| 52. | **TLC** | Thin-Layer Chromatography |
| 53. | **TMV** | Tobacco Mosaic Virus |
| 54. | **TRM** | Lung-resident memory T cells |
| 55. | **TSLP** | Thymic stromal lymphopoietin |
| 56. | **UHPLC** | Ultra-high-performance liquid chromatography |
| 57. | **USFDA** | United States Food and Drug Administrations |
| 58. | **Vero cells** | African green monkey kidney cells |
| 59. | **VOCs** | Volatile organic compounds |
| 60. | **WHO** | World Health Organization |

# *Section A*

## *Emerging Viral Diseases and Bioactive Compounds*

# 1 Emerging Viral Diseases
## *A Global Challenge and Herbal Medicines*

*Odoh Uchenna Estella*
*and Michael Onyegbulam Chukwuma*
University of Nigeria

*Habibu Tijjani, Abdulkadir Mohammed Danyaro,*
*and Khadija Aliyu Nuru*
Bauchi State University

## 1.1 INTRODUCTION

Viruses are primarily responsible for causing human diseases. Numerous studies have reported associations between viral infections and a variety of incurable diseases and complicated disorders, including hepatocellular carcinoma, Alzheimer's disease and type 1 diabetes, among others (Hober et al., 2012; Ball et al., 2013; Morgan et al., 2013). Newly emerging viral infectious diseases, as well as those that have existed previously but are rapidly becoming prevalent, continue to pose a significant threat to global health and stability (Excler et al., 2021). The emergence and spread of viral diseases in humans are becoming increasingly alarming due to technological advancements, uncontrolled population growth, changes in population structure and movement that encourage mass gatherings, shifts in human behavioural changes in response to globalization and climate/environmental changes, and inadequate public health infrastructures and absorptive capacity (Excler et al., 2021; Huttly, 1990).

Studies have illustrated the ongoing resurgence of viruses such as dengue, measles, influenza and severe acute respiratory syndrome (SARS) viruses (Louten, 2016). The recent emergence of the 2019 coronavirus disease (COVID-19), caused by the outbreak of an unfamiliar virus known as SARS-CoV-2 (severe acute respiratory syndrome coronavirus 2) (Huang et al., 2020; Zhu et al., 2020; Wang et al., 2020), overwhelmed the globe in less than 6 months, leading to economic problems and millions of deaths, especially among the elderly and individuals with pre-existing medical conditions. Governments and stakeholders were compelled to take various measures, including imposition of social distancing, travel restrictions and total lockdowns, as well as the use of personal protective equipment (PPE) such as facemasks, to mitigate the crisis before vaccines and drugs were later developed by researchers. Thereupon, attention has turned towards the surveillance and management of emerging infectious disease to prevent the occurrence of another pandemic (Gully, 2020; David et al., 2021).

DOI: 10.1201/9781003329169-2

## 1.2 EMERGING VIRAL DISEASES

### 1.2.1 COVID-19 (SARS-CoV-2)

The occurrence of SARS-CoV-2 in December 2019 resulted in COVID-19, which has emerged as a significant global health challenge (WHO, 2020a; Lim et al., 2021). As of 23 September 2020, about 960,000 cases were reported out of over 31 million infections recorded (WHO, 2020b). The crisis prompted researchers and clinicians to study the pathology and pathogenesis of the virus, leading to the adoption of the initial preventive measures such as hand hygiene, use of PPE, social distancing, travel restrictions and lockdowns to curb its transmission. Consequently, vaccines and drugs were later developed, which helped to mitigate the impact of the pandemic. Resurgence do, however, occur in some countries that relapse the COVID-19 preventive measures either because of the availability of preventive vaccines and therapeutic drugs or because of its impact on their economies (Bonaccorsi et al., 2020; Chu et al., 2020; Sardar et al., 2020; UNDP, 2020). The high rate of mutation of SARS.CoV-2 makes it possible for new strains to emerge, potentially overwhelming acquired immunity in humans and leading to possibly cause another pandemic (Konishi, 2022).

### 1.2.2 H1N1; H7N9 'Swine Flu'

Swine Influenza (commonly known as Swine Flu or pig flu) is a type of influenza A subtype H1N1 viral infection that causes respiratory tract infection in pigs (Mujoriya et al., 2011; Jilani et al., 2022). The term "influenza" was first coined by Italians in 1357 AD, as they believed that the disease was caused by the influence of the stars. The 1918 influenza pandemic, caused by avian H1N1 virus, is considered the deadliest influenza pandemic in recent times. H1N1 resurfaced to cause the 1977 Russian Flu and the 2009 Swine Flu through mutation of the viral strains from human, birds and pigs (Donaldson et al., 2009). On the other hand, H7N9 is the causative agent of Avian Influenza (commonly known as Bird Flu) that affects human populations. The first human cases occurred on 31 March 2013, in Shanghai and Anhui, China, following recombination of genes from poultry and wild birds in Asia (Koopmans and De Jong, 2013). The virus also reemerged severally to cause the fifth epidemic in 2016/2017. Novel influenza virus strains appear yearly, causing seasonal flu epidemics (Mujoriya et al., 2011).

### 1.2.3 Chikungunya Virus

Chikungunya is a disease characterized by fever, rashes, headache, fatigue, severe joints and muscles pains, as well as swelling of the joints, and it is transmitted to humans by infected mosquitoes (Robinson, 1955). The causative agent is the chikungunya virus (CHIKV), which led to an outbreak in Southern Tanzania and was first identified in 1952. The virus derived its name, 'chikungunya,' from the Kimakonde language of the Makonde people of Southern Tanzania and Northern Mozambique, referring to the bent nature of patients due to excruciating joint pain (Schwartz and Albert, 2010; Thiberville et al., 2013). Since its identification, the virus has occasionally caused outbreaks in Africa and Asia until 2002, when it rapidly spread to over 60 countries in Africa, Asia, Europe, as well as North and South America (Rodrigues et al., 2016), causing millions of deaths including the recent Sudan, Yemen and Cambodia outbreaks in 2018, 2019 and 2020, respectively (WHO, 2022b).

CHIKV is a 12 kb long +ssRNA virus that possesses a 5′ open reading frames (5′ORFs). This genomic RNA is translated to encode non-structural proteins 1–4 (termed nsP1, nsP2, nsP3 and nsP4). Additionally, there is a 3′ORF, translated from subgenomic RNA, which encodes the polyprotein used for the formation of capsid (C), envelope 1 (E1) and envelope 2 (E2), and two peptides (E3 and 6K)] (Schwartz and Albert, 2010). Three genotypes of CHIKV have been identified, including the West African genotype; the East/Central/Southern Africa (ECSA) genotype, which reemerges in Africa, Asia, Europe and some parts of Brazil; and the Asian genotype, which evolves from the ECSA genotype and affects Latin America (Rodrigues et al., 2016; Bettis et al., 2020). It has become one of the major public health challenges due to increased invasive nature of its vector, *Aedes albopictus* mosquito, and the E1-A226V mutation in the ECSA genotype, which enhances vector transmission (Zeller et al., 2016).

### 1.2.4 Zika Virus

Zika virus (ZIKV) is one of the mosquito-borne flaviviruses that causes Zika fever, which is characterized by fever, headache, rashes, joints pain and myalgia (muscles aches and pains), similar to that of Dengue fever (Noorbakhsh et al., 2019; Gorshkov et al., 2019). The virus was first isolated from a febrile rhesus macaque monkey and *Aedes africanus* mosquitoes found in Zika forest of Uganda in 1945 (Gorshkov et al., 2019). The first recorded outbreak of ZIKV infection occurred on Yap Island in the Federated States of Micronesia, with subsequent outbreaks in Europe, North America and South America (Duffy et al., 2009). ZIKV is a +ssRNA virus belonging to the Flaviviridae family that has the possibility of recombination, which is rare in flaviviruses. Hence, its propensity to infect a large number of people poses economic and health challenges (Plourde and Bloch, 2016). The foregoing and the ability of the virus to cause microcephaly and other neuro-developmental abnormalities in maternal infection has raised serious concerns within the international medical community, leading to its declaration as a Public Health Emergency of International Concern (PHEIC) by the World Health Organization in early 2016 (Heymann et al., 2016; Sikka et al., 2016; WHO, 2016). Since diagnosis is low and no vaccines and/or drugs for its treatment are currently available, prevention of mosquito bites and elimination of vectors remain the surest means of preventing its recurrence (Plourde and Bloch, 2016).

### 1.2.5 Ebola Haemorrhagic Fever

Ebola virus (EBOV) is a filamentous enveloped virus (filovirus) transmitted to humans through contact with the tissue, blood, body fluid or secretions from an infected host (human or animals). This transmission causes Ebola haemorrhagic fever or simply Ebola virus disease, a highly infectious disease characterized by fever, headache, anorexia, diarrhoea, vomiting, unexplained bleeding, etc. (Johnson et al., 1977; Murphy and Peters, 1998; Bermejo et al., 2006; Francesconi et al., 2003). The virus was first identified near Ebola River in Democratic Republic of Congo (DRC) in 1976, from which it derived its name, during two concurrent outbreaks in DRC and what is now South Sudan (WHO, 1976, 1978; Dixon and Schafer, 2014; Breman et al., 2016). EBOV is a –ssRNA virus with a genome of about 18 kb. It encodes several proteins, including nucleoprotein (NP), matrix protein (VP40), polymerase (L), polymerase cofactor (VP35), secondary matrix protein (VP24), glycoprotein (GP) and transcriptional activator (VP30). This genetic complexity makes the virus a challenging target for drug design, prevents humans from acquiring immunity against it and classifies it as a potent bioterrorism agent (Ferman et al., 2020).

From 1976, the year of its first epidemic, to 2012, there were 24 outbreaks of the disease resulting in the death of over 1,590 people (Dixon and Schafer, 2014). The most fatal epidemic occurred in Liberia between 2014 and 2016, resulting in 4,809 deaths (Statista, 2022).

### 1.2.6 Severe Acute Respiratory Syndrome

Severe Acute Respiratory Syndrome coronavirus (SARS-CoV) is a zoonotic +ssRNA virus of the *Coronaviridae* family that causes SARS, a severe viral respiratory disease in humans characterized by fatal flu-like symptoms, fever and pneumonia (Peiris et al., 2003). It is transmitted from its animal reservoir, believed to be horseshoe bats, to humans through contact with the respiratory droplets from its host or any contaminated inanimate object (Peiris et al., 2003; Hui and Zumla, 2019;

Zumla and Hui, 2019). SARs-CoV was first discovered in 2002 during an outbreak in Guangdong, China. The epidemic later spread to over 30 countries and became the first pandemic of the 21st century, causing several deaths (LuDec and Barry, 2004). Its treatment is primarily through supplementing oxygen ventilation (noninvasive and invasive ventilation) and other supportive care, but no antiviral drug has been formulated yet (David et al., 2021; Lau et al., 2004). It is contained through clinical isolation of infected persons, quarantine, hand hygiene, use of PPE, etc. (David et al., 2021; Hui and Zumla, 2019; Zumla and Hui, 2019).

### 1.2.7 Middle East Respiratory Syndrome

The MERS coronavirus (MERS-CoV) is another zoonotic +ssRNAβ-coronavirus regularly found in Dromedary Camels of East Africa and the Arabian Peninsula. It invades host cells by binding to dipeptidyl peptidase 4 (DPP4) receptor, to cause a severe respiratory infection symptomized by mild to severe fever, cough, shortness of breath and diarrhoea (Zaki et al., 2012). MERS-CoV was first isolated in June 2012 from the sputum of a Saudi Arabian patient, who suffered from pneumonia and acute kidney injury (Ramadan and Shaib, 2019). This marked the beginning of the first outbreak in the Arabian Peninsula, which later spread to over 25 countries and resulted in high number of fatalities (WHO, 2022a). Subsequently, numerous outbreaks occurred resulting from human-to-human transmissions, with the latest reported in Qatar in 2022, bringing the total number of cases to 2,591 and 94 deaths as of May 2022 (WHO, 2022a).

## 1.3 TREATMENT OF VIRAL INFECTIONS

Viral infections are treated with drugs, many of which are semi-synthetic or synthetic. There are numerous antiviral drugs available to treat infections, some specific for a particular virus, while others have the potential to affect multiple viruses (Table 1.1) (Arankalle et al., 2007). Several antiviral drugs currently in use are associated with significant adverse effects. This has necessitated the need for alternative agents to be used as antiviral drugs.

## 1.4 HERBAL MEDICINE AND THEIR EFFICACY ON EMERGING VIRAL DISEASE

### 1.4.1 Coronavirus

Studies conducted by Li et al. (2005) demonstrated that extracts prepared from *Pyrrosia lingua*, *Lycoris radiata*, *Lindera aggregata* and *Artemisia annua* exhibited potent antiviral activity against SARS-CoV. Alternatively, a water extract from *Houttuynia cordata* demonstrated potency to inhibit the virus by targeting 3-chymotrypsin-like cysteine (3CL) protease and blocking its RNA-dependent RNA polymerase activity (Lin et al., 2014).

### 1.4.2 Coxsackievirus

An 80% ethanolic extract derived from *Macaranga kilimandscharica* Pax (Euphorbiaceae) leaves has expressed potency to block the replication of Coxsackieviruses, which are non-enveloped enteroviruses similar to hepatitis A and polioviruses) (Cos et al., 2002a). The scholars also demonstrated that extracts from the leaves of *Guizotiascabra* (Vis.) Chiov. (Asteraceae), *Pavetta ternifolia* Hiern (Rubiaceae) and *Eriosema montanum* Baker f. (Fabaceae) have significant pharmacological activities against Coxsackieviruses. Badia-Boungou et al. (2019) also showed that the aqueous extract of Pellegr. (Myrtaceae) and *Syzygium brazzavillense* Aubrév inhibits the replication of Coxsackie B4 virus (CVB4) in human larynx epithelioma cancer (HEp-2) cells culture and lowers its cytopathic effect signifying the structural changes that occur in a host cell caused by viral infection, as well as that of CVB2 and CVB3.

**TABLE 1.1**
**Drugs for Treatment of Viral Diseases**

| Primarily Active Against | Generic Name (Abbreviation) Trade Name | Nature | Mechanism of Viral Activity | Manufacturer |
|---|---|---|---|---|
| Hepadnaviruses (HBV) | Lamuvidine (LMV or 3TC) Zeffixs, Hetovir | Nucleoside analogue | Polymerase inhibitor | GSK |
| | Adefovir dipivxil (ADV) Hepseras® | Prodrug of adefovir nucleotide analogue | Polymerase inhibitor | Gilead |
| | Entecavir Baracludes® | Nucleoside analogue | Polymerase | BMS |
| RNA viruses | Zanamivir Relenzas® | Sialic acid analogue | Neuraminidase inhibitor, Influenza A and B | GSK |
| | Oseltamivir Tamiflus® | Sialic acid analogue | Neuraminidase inhibitor | Roche |
| | Ribavirin Virazole® | Nucleoside analogue | Possibly no direct viral target | Schering-Plough |
| Herpes viruses | Valaciclovir (VAC) Valitrex® | | Polymerase inhibitor | GSK |
| | Famiciclovir (FCV) Famivir® | | Activated by viral Thymidine Kinase, inhibits viral polymerase | Novatis |
| | Valganiciclovir (VGCV) Valcyte® | Prodrug of vaniclovir | Activated by kinase encoded UL 97, polymerase inhibitor | CMV |
| Retroviruses (HIV) | Zidovudine/Lamuvidine (AZT/3TC) Combivirs ® | | | |
| | Efuvritide (T20) Fuzeon® | Polypeptide | Envelope protein g41 | Roche |
| | Maraviroc Selzentry® | CCR5 ligand mimc | Blocks receptor on host cell membrane | Pfizer |
| Influenza viruses | Peramivir Rapivab® | Influenza virus neuraminidase inhibitor | Competitively binds to neuraminidase | BioCryst |
| | Zanamivir Relenza® | Influenza virus neuraminidase inhibitor | Alteration of virus particle aggregation and release | GSK |
| | Oseltamivir phosphate Tamiflu® | Influenza virus neuraminidase inhibitor | Alteration of virus particle aggregation and release | Roche |
| | baloxavirmarboxil Xofluza® | Polymerase acidic (PA) endonuclease inhibitor | | Genentech |

### 1.4.3 Dengue Virus

Extract of *Terminalia chebula* has been reported as a broad-spectrum antiviral agent (Lin et al., 2013, 2014). This extract is effective in halting Dengue virus infection through irreversible interaction with its virions and by interfering with glycosaminoglycans (GAGs)-assisted viral binding, fusion and early entry spread. The extract of Marine Seaweed also possesses significant anti-Dengue virus activity (Low et al., 2011; Zandi et al., 2011; Koishi et al., 2012).

### 1.4.4 Enterovirus 71

Non-fermented *Camellia sinensis* leaves and buds (green tea) restrict enterovirus 71 replication by curcumin suppression of protein kinase C-δ (PKCδ) activity (Ho et al., 2009; Low et al., 2011). Infusions of the bioactive components of *O. basilicum* and *Woodfordia fruticosa* flowers also possess anti-replication activities against EV71, effectively preventing infection (Chiang et al., 2005; Choi et al., 2010).

### 1.4.5 Hepatitis B Virus

Curcumin, the active ingredient found in the spice Turmeric, serves as a potent therapeutic agent in anti-hepatitis B virus (HBV) drugs. It can induce HBV gene expression and replication by suppressing the acetylation of histones 3 and 4 (H3/H4) that are bound to covalently closed circular DNA of HBV. Additionally, it can inhibit peroxisome proliferator-activated receptor gamma coactivator (PGC) 1alpha (an HBV transcriptional coactivator) (Rechtman et al., 2010; Wei et al., 2017). The ethanolic extract of *Polygonum cuspidatum* and extracts of *Corydalis saxicola*, *Piper longum* and *Laggeraalata* possess anti-HBV activities (Chiang et al., 2003; Zeng et al., 2013). *Prunus africana* Kalkman (*Rosaceae*), *Acacia mellifera* Benth (*Fabaceae*) and *Carissa edulis* (*Apocynaceae*) also have anti-HBV activities (Beressa et al., 2021).

### 1.4.6 Hepatitis C Virus

*Silybum marianum* exhibits *in vitro* anti-HCV activity (Polyak et al., 2007; Polyak et al., 2010) and demonstrates promising potency of reducing the number of HCV viral particles capable of invading host cells (viral load) (Ferenci et al., 2008; Neumann et al., 2010; Marino et al., 2013).

### 1.4.7 Herpes Simplex Virus

Medicinal plants are used in Sub-Saharan Africa as alternatives for the treatment of infections from herpes simplex virus types 1 and 2 (HSV and HSV 2) (Beressa et al., 2021). For instance, the aqueous extract prepared from the root bark of *C. edulis* inhibits HSV both *in vitro* and *in vivo* (Omino and Kokwaro, 1993). Extracts of *Ficus sycomorus*, *Combretum adenogonium*, *Terminalia mollis* and *Lannea schweinfurthi* possess significant anti-HSV1. Other plants that possess strong antiviral activities against HSV include *Ephedra alata* and *Moringa peregrine* (Beressa et al., 2021), *M. lanceolata* (Neumann et al., 2010; Sindambiwe et al., 1999), *Rhododendron ferrugineum L.* and *Myrothamnus flabellifolia* Welw (Gescher et al., 2011a, 2011b; Danaher et al., 2011).

### 1.4.8 Human Immunodeficiency Virus

Butanol in the aqueous solution demonstrated anti-reverse transcriptase activities against HIV, while the terpenes fractions demonstrated proteolytic processing of polyproteins used for assembly of virus particles (Hussein et al., 1999: Huang and Chen, 2002; Yu et al., 2006). The ethanolic extracts of some Rwandan plants have also demonstrated anti-HIV activities. These plants include *Rumex nepalensis* Spreng. (*Polygonaceae*), which enhances cell protection against HIV-induced cytopathic effect, *Aspilia pluriseta* Schweinf. ex Engl. (*Asteraceae*), which induces absolute cell-resilience against HIV-1-provoked cytopathic effect (Cos et al., 2002b) and *Tithonia diversifolia* (Hemsl.) A. gray (*Asteraceae*), which inhibits HIV-1 (Cos et al., 2002b; Hussein et al., 1999). Furthermore, Ethiopian medicinal plants (*Combretum paniculatum* Vent (*Combretaceae*) and *Bersama abyssinica* Fresen. (*Francoaceae*)) demonstrated antiviral activities against both HIV-1 and HIV-2 (Attah et al., 2021).

### 1.4.9 Influenza Virus

Research on Ethiopian medicinal plants has established that *Acokanthera schimperi*, *Inula conferttiflora* and *Euclea schimperi* inhibit Influenza A virus (IAV) replication (Gebre-Mariam et al., 2006). The hot and cold aqueous extracts obtained from *Pittosporum viridiflorum* and *Rapanea melanophloeos* have been effective against IAV (Cos et al., 2002b; Mehrbod et al., 2018a). *Melilotus elegans*, *Cussonia spicata*, *Clerodendrum glabrum* and *Tabernaemontana ventricosa* have anti-IAV effects (Ma et al., 2003; Huang et al., 2010).

### 1.4.10 Measles Virus

The leaves of *Macaranga kilimandscharica* Pax (Euphorbiaceae) and *Eriosema montanum* Baker f. (*Fabaceae*) demonstrated significant activity against measles virus *in vitro* (Cos et al., 2002b).

### 1.4.11 Respiratory Syncytial Virus

Extracts prepared from the leaves of *Radix wikstroemiae*, *Lophatherum gracile* and *Selaginella uncinata* exhibited potent antiviral effects against RSV infection (Lin et al., 2014; Wang et al., 2012a; Berit et al., 2020; Goyal, 2020).

### 1.4.12 Chikungunya

*Azadirachta indica* A. Juss (Neem) leaves juice or bark powder, as well as powders of Guduchi and Sudrashana, prevent chikungunya virus infection (Kappor et al., 2017; Rex et al., 2018) (Table 1.2).

## 1.5 ROLE OF SECONDARY METABOLITES IN VIRAL TREATMENTS

Secondary metabolites and bioactive compounds obtained from medicinal plants, including alkaloids, flavonoids, terpenoids, tannins, glycosides, steroids, polypeptides, lignans, coumarins, antimicrobial peptides, phenylpropanoids and defensins, are reported to possess vast antiviral activity (Kapoor et al., 2017; Rex et al., 2018; Younas et al., 2018; Ghildiyal et al., 2020; He et al., 2020). These bioactivities exert a controlling influence on how the body recognizes and defends itself against viruses and other pathogens (Han et al., 1998; Zhang et al., 2002; Shivaprasad et al., 2006; Cruz et al., 2007; Fialho et al., 2016; Sodagari et al., 2018) and correspondingly reduce the amount of virus or parasites in infected persons' blood (Attah et al., 2021; Omoregie and Pal, 2016; Michelini et al., 2018; Jasso-Miranda et al., 2019; Salinas et al., 2019; Afolayan et al., 2020).

For instance, Quercetin-3-O-α-L-rhamnopyranoside, a glycoside flavone from *R. melanophloeos*, reduces the number of IAV particles capable of invading host cells. Quercetin-3-O-α-L rhamnopyranoside was reported to decrease the virus titer of IAV at 150 µg/mL by inhibiting replication of influenza A virus and inducing cytokine production (Mehrbod et al., 2018a; ColungaBiancatelli et al., 2020). Triterpene glycosides from *Scrophularia scorodonia*, *Heteromorpha* spp. and *Bupleurum* spp. exhibit anti-HCoV 22E9 effects (Cheng et al., 2006) by thwarting HCoV 22E9 binding, fusion and incubation that occur during the early stages of infection (Lin et al., 2014). Another metabolite called Ent-epiafzelechin-(4α→8)-epiafzelechin, extracted from *Cassia javanica*, blocks herpes simplex virus (HSV 2) replication, while 1,3,4,6-tetra-O-galloyl beta d glucose and excoecarianin from *Phyllanthus urinaria* potently impede HSV infection (Lin et al., 2014). Hydrolysable tannins like chebulagic acid and punicalagin inhibit HSV1 entry and cell-to-cell spread by binding to cell surface (glycosaminoglycans, GAGs) (Lin et al., 2011). These tannins also inactivate free HCV particles, avert their binding and entry into the host cells and thwart cell-to-cell spread (Lin et al., 2013).

**TABLE 1.2**
**Herbal Medicine and Their Efficacy on Emerging Viral Disease**

| Medicinal Plant | Role Viral Disease | References |
|---|---|---|
| *Artemisia annua, Lindera aggregate, Lycoris radiata, Pyrrosia lingua* | Inhibits SARS-CoV 3-chymotrypsin-like cysteine protease and blocks the RNA-dependent RNA polymerase activity. | Li et al. (2005) |
| *Macaranga kilimandscharica Pax, Guizotia scabra, Pavetta ternifolia Hiern, Eriosema montanum Baker, Syzygium brazzavillense* | Blocks the replication of Coxsackievirus B4 in human larynx endothelioma cancer cell cultures and limits its cytopathic effect, as well as that of CVB2 and CVB3. | Cos et al. (2002a); Badia-Boungou et al. (2019) |
| *Terminalia chebula* | Directly inactivates free Dengue virions and interferes with glycosaminoglycans-assisted viral binding, fusion and early entry spread. | Lin et al. (2013) |
| *O. basilicum, Woodfordia fruticosa* | Blocks EV71 replication by curcumin suppression of protein kinase C-δ activity. | Chang et al. (2005); Choi et al. (2010) |
| *Curcumin, Corydalis saxicola, Laggera alata, Piper longum, Polygonum cuspidatum* | Inhibits HBV gene expression and replication by down-regulating cccDNa-bound histones or PGC-1α. | Hao et al. (2012); Jiang et al. (2013); Zeng et al. (2013) |
| *Silybum marianum* | Can reduce the number of HCV viral particles capable of invading host cells. | Polyak et al. (2007); Polyak et al. (2010); Marino et al. (2013) |
| *M. lanceolata, Myrothamnus flabellifolia* | Reduces the amount of HSV in the blood of infected individuals and diminishes the ability of HSV 2 to establish an infection. | Sindambiwe et al. (1999); Gescher et al. (2011b) |
| *Aspilia pluriseta, Rumex nepalensis, Tithonia diversifolia, Bersama abyssinica, Combretum paniculatum* | Induces anti-reverse transcriptase activities against HIV, while its terpenes constituents inhibit proteolytic processing of polyproteins used for the assembly of virus particles. | Hussein et al. (1999); Huang and Chen (2002); Yu et al. (2006) |
| *Pittosporum viridiflorum, Rapanea melanophloeos* | Inhibits IAV replication. | Gebre-Mariam et al. (2006); Mehrbod et al. (2018a) |
| *Macaranga kilimandscharica, Eriosema montanum* | Demonstrates *in vitro* effect against measles | Cos et al. (2002a) |
| *Selaginella uncinata, Radix wikstroemiae, Lophatherum gracile* | Expresses potent antiviral effects against RSV infection | Wang et al. (2012b); Lin et al. (2014); Berit et al. (2020); Goyal et al. (2020) |
| *Azadirachta indica* | Prevent chikungunya virus infection | Kapoor et al. (2017) |
| *Bupleurum* spp., *Heteromorpha* spp., *Scrophularia scorodonia* | Prevents binding, entry and incubation of HCoV-22E9. | Lin et al. (2014) |
| *Cassia javanica, Phyllanthus urinaria, Houttuynia cordata* | Inhibits HSV replication and impedes infection. | Lin et al. (2014) |

Houttuynoids A-E, the flavonoids from *Houttuynia cordata*, are effective against HSV 1, while the biflavonoids of *Radix wikstroemiae* such as genkwanol B, genkwanol C and stelleranol are effective against RSV (Berit et al., 2020; Chen et al., 2012).

Researchers have revealed a series of compounds isolated, such as amide alkaloid from isochlorogenic acid A of *Laggeraalata*, *Piper longum* and dehydrocheilanthifoline of *Corydalis saxicola*, to have positive activities against HBV (Hao et al., 2012; Jiang et al., 2013; Zeng et al., 2013).

The two chromone glycosides of *Selaginella uncinata*, uncinosides A and B, effectively hinder RSV infection (Wang et al., 2012b), much like a number of flavone 6-C monoglycosides from *Lophatherum gracile* do in a cytopathic effect reduction assay (Goyal et al., 2020) (Table 1.3).

## 1.6 FUTURE PROSPECTIVE

Infectious viral diseases present a never-ending challenge that can emerge or re-emerge in unpredictable regions and at unpredictable times (Nii-Trebi, 2017). An emerging viral disease is one that is newly evolved, newly recognized or has not been observed previously within a population or geographic location (Nii-Trebi, 2017; Salah et al., 2019). Reemerging viral diseases are characterized by resurgence or recurring outbreaks affecting a significant part of the population after a previous decline in incidence (Asselah et al., 2021). Over the past 50 years, dozens of viruses have emerged or reemerged within the human population.

The physical and biochemical properties of the virus pose challenges for the design of antiviral drugs. Few targets and platforms are available for antiviral drug design due to the simple nature of the viruses (encapsulated nucleic acid), the limited number of targets in the viruses, the rapid evolution of viral genes, the quick emergence of drug-resistant pathways and the development of new viral strains through mutations (Salah et al., 2019). Viruses have shown resistance to therapy or prophylaxis longer than any other life forms due to their nature that they totally depend on the cells they infect for their multiplication and survival. The success rate of the current antiviral strategies is not 100% guaranteed. Mutation, resistance and the fact that the vaccines are not developed to cover all viruses create opportunities for failure when excluded viral strains are involved in infections. Though the burden of some viruses has been significantly reduced by the use of vaccines, there is still ongoing research needed. Viruses like hepatitis B, hepatitis C and HIV are currently resistant to vaccines and are managed with antiviral drugs. Understanding cellular recognition receptors is important for the development of intervention strategies towards viral infection. Inhibitors of viral entry by targeting ACE2 receptors, the transmembrane serine protease 2, an inhibitor of viral membrane fusion or inhibitor of RNA-dependent RNA polymerase are being developed (Musarra-Pizzo et al., 2021).

Characterizing natural extracts has the potential to lead to the development of remedies or agents that can be used as adjuvant therapy or prophylactic treatment against adenoviral infection (Hsieh et al., 2020). Phytochemical compounds exert antiviral activity by directly targeting virus-host receptor interactions, particularly by blocking the attachment of enterovirus A71 to cellular receptors such as p-selectin glycoprotein ligand-1 (PSGL1). Some compounds interact with cellular molecules, rather than viral molecules, thereby inhibiting RNA replication. Others exhibit their antiviral activity by stimulating the immune response and developing a protective state in uninfected cells. The stimulation of IFN-α and IFN-1β protein expression reduces viral infectivity in a dose-dependent manner (Hsieh et al., 2020). Viral infection can be counteracted by inhibiting viral genome replication by using specific inhibitors of replication organelles, which are membranous structures that support the replication of viral genome (Hsu et al., 2010). The proteolytic cleavage of polyprotein at specific sites by proteinase-chymotrypsin-like protease (3CLpro) is essential for the replication of SARS-CoV and is used to assess the anti-enzymatic activity of compounds.

## 1.7 CONCLUSION

Emerging viral diseases present a concern for researchers to provide remedies for combatting these diseases. The use of herbal medicines has emerged as an alternative in their treatments, as several secondary metabolites have demonstrated activity for their prevention and treatment. An even more challenging aspect of viral infections is the reemergence of infectious viral diseases, which continue to threaten humanity. Most emerging or reemerging viral diseases lack preventive vaccines and efficient treatments, underscoring the importance of prioritizing viral infection research within

**TABLE 1.3**
**Some Plant-Derived Compounds with Antiviral Activities**

| S. No. | Compound | Chemical Structure | Plant | Virus | Mechanism | Ref. |
|---|---|---|---|---|---|---|
| 1 | Saikosaponins A, B2 | 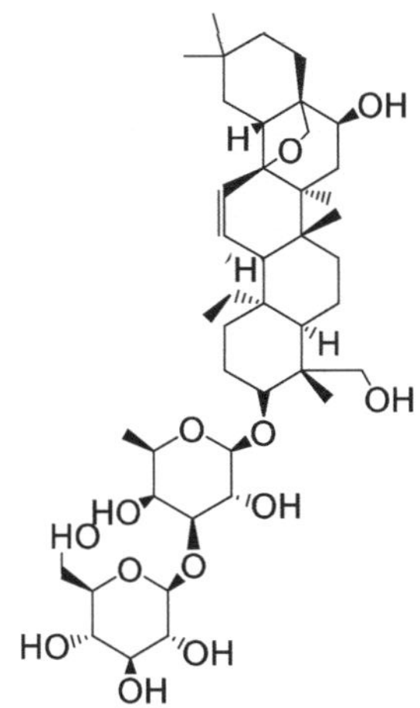 | *Bupleurum chinense* DC (Apiaceae) | Coronavirus | Inhibits viral attachment and penetration stages | Cheng et al. (2006) |

(*Continued*)

**TABLE 1.3** (*Continued*)
**Some Plant-Derived Compounds with Antiviral Activities**

| S. No. | Compound | Chemical Structure | Plant | Virus | Mechanism | Ref. |
|---|---|---|---|---|---|---|
| | | 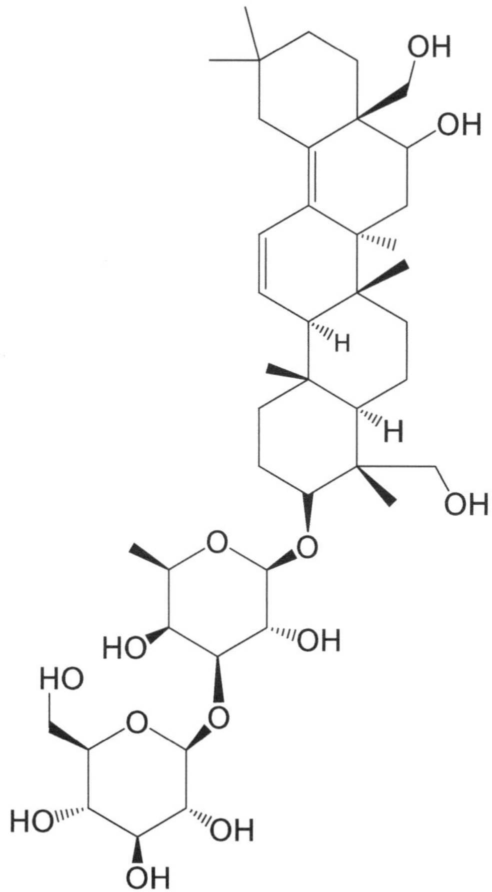 | | | | |

(*Continued*)

**TABLE 1.3 (*Continued*)**
**Some Plant-Derived Compounds with Antiviral Activities**

| S. No. | Compound | Chemical Structure | Plant | Virus | Mechanism | Ref. |
|---|---|---|---|---|---|---|
| 2 | Lycorine | 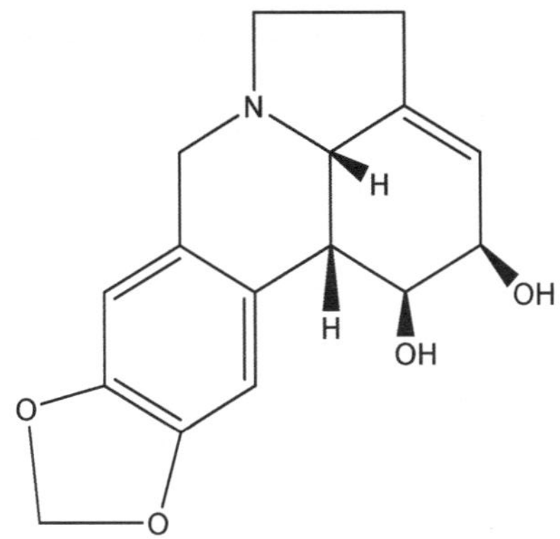 | *Lycoris radiata* (L'Hér) Herb (Amarylidaceae) | Coronavirus | | Li et al. (2005) |
| 3 | Amentoflavone | 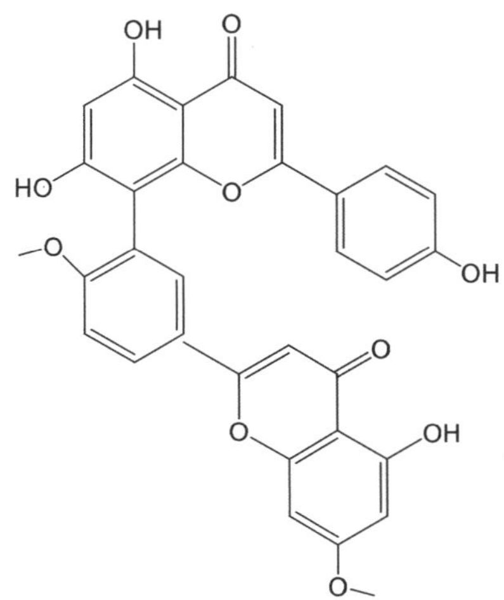 | *Torreya nucifera* (L) Siebold and Zucc (Taxaceae) | SARS-CoV | Inhibits SARS-CoV 3CL protease | Ryu et al. (2010) |

(*Continued*)

**TABLE 1.3 (*Continued*)**
**Some Plant-Derived Compounds with Antiviral Activities**

| S. No. | Compound | Chemical Structure | Plant | Virus | Mechanism | Ref. |
|---|---|---|---|---|---|---|
| 4 | Myricetin and scutellarein | 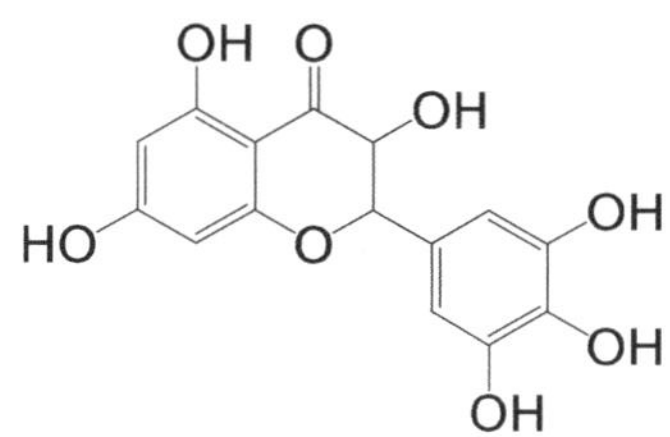 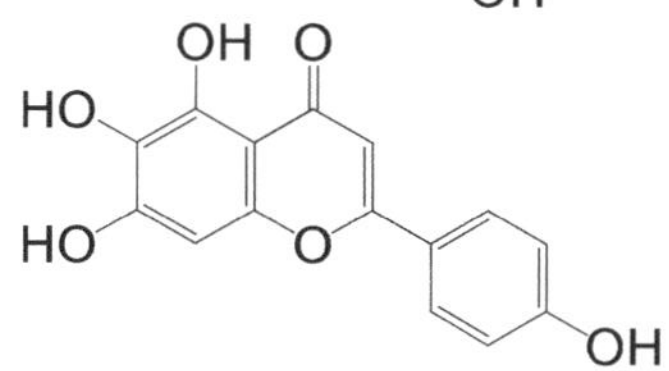 | Myricetin–vegetable fruits, nuts, beverages (e.g. Tea), berries. Scutellarein–*Scutellaria laterifflora* L (Lamiaceae) and other members of the genus *Scutellaria* | SARS-CoV | Inhibits SARS-CoV helicase | Yu et al. (2012) |
| 5 | Linalool | 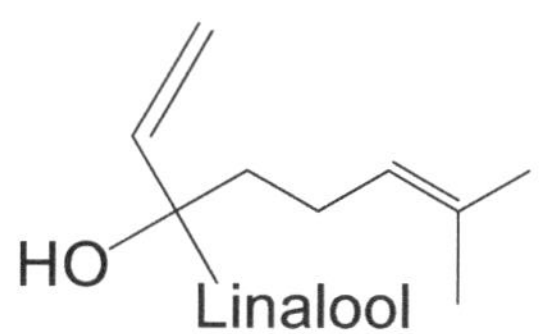  | *Ocimum basilicum* L (Lamiaceae) | Coronavirus | Interference with viral replication | Chiang et al. (2005) |
| 6 | Baicalein | 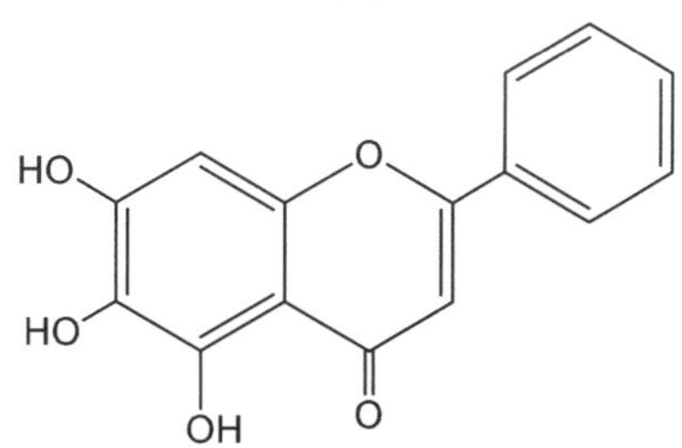 | *Scutellaria baicalensis* Georgi (Lamiaceae) | DENV-2 | Impedes viral adsorption onto host cell; viricidal activity against extracellular viruses; inhibits viral post entry replication | Zandi et al. (2012) |

(*Continued*)

**TABLE 1.3 (*Continued*)**
**Some Plant-Derived Compounds with Antiviral Activities**

| S. No. | Compound | Chemical Structure | Plant | Virus | Mechanism | Ref. |
|---|---|---|---|---|---|---|
| 7 | Quercetin | 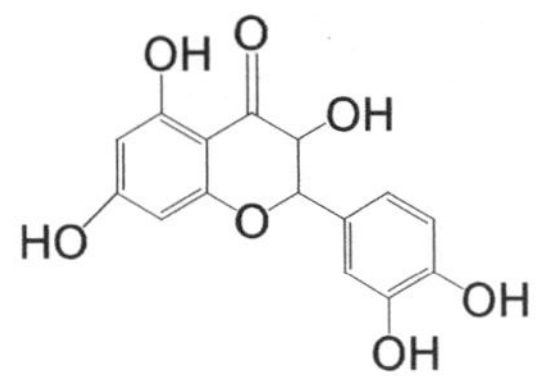  | *Allium cepa* L. (Amaryllidaceae), *Citrus* species (Rutaceae) | DENV-2 | Inhibits viral replication | Zandi et al. (2011) |
| 8 | Gallic acid | 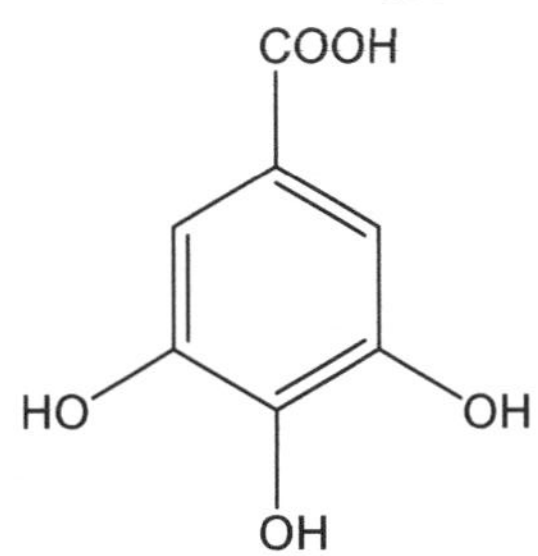  | *Woodfordia fruticosa* (L) Kurtz (Lythraceae) | Eterovirus 71 | Unclear | Choi et al. (2010) |
| 9 | (-)-Epigallocatechin gallate | 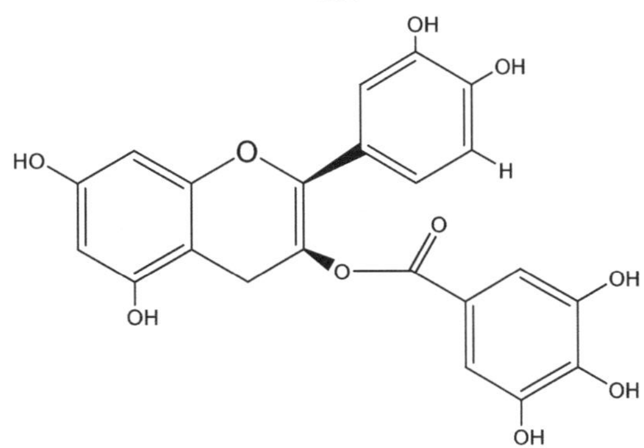  | *Camellia sinensis* (L.) Kuntze (Theaceae) Green tea | Eterovirus 71 | Modulation of cellular redox environment, thereby interfering with viral replication | Ho et al. (2009) |
| | | | | Hepatitis B virus | Viral entry inhibition | Calland et al. (2012) |

(*Continued*)

**TABLE 1.3 (*Continued*)**
**Some Plant-Derived Compounds with Antiviral Activities**

| S. No. | Compound | Chemical Structure | Plant | Virus | Mechanism | Ref. |
|---|---|---|---|---|---|---|
| 10 | Saikosaponins C | 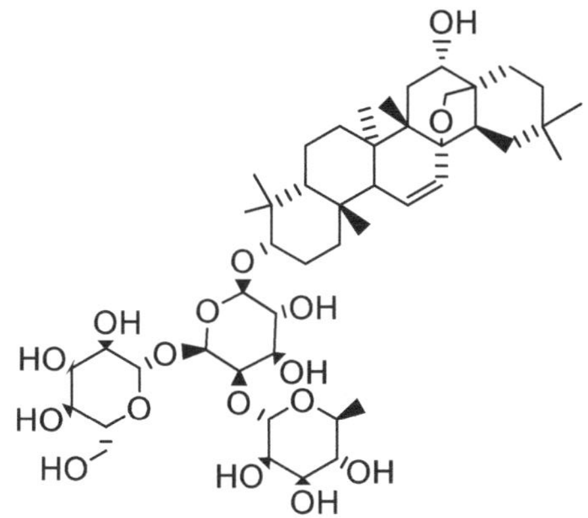 | *Bupleurum* species | Hepatitis B Virus | Saiksaponnin C inhibits HBeAg expression and HBV DNA replication | Rechtman et al. (2010) |
| 11 | Tellimagrandin I | 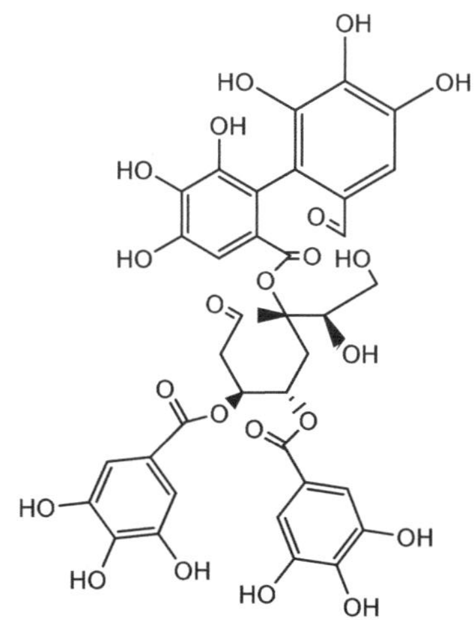 | *Rosa rugosa* (Rosaceae) | Hepatitis B Virus | HCV invasion inhibition | Haid et al. (2012) |

(*Continued*)

**TABLE 1.3 (*Continued*)**
**Some Plant-Derived Compounds with Antiviral Activities**

| S. No. | Compound | Chemical Structure | Plant | Virus | Mechanism | Ref. |
|---|---|---|---|---|---|---|
| 12 | Curcumin | 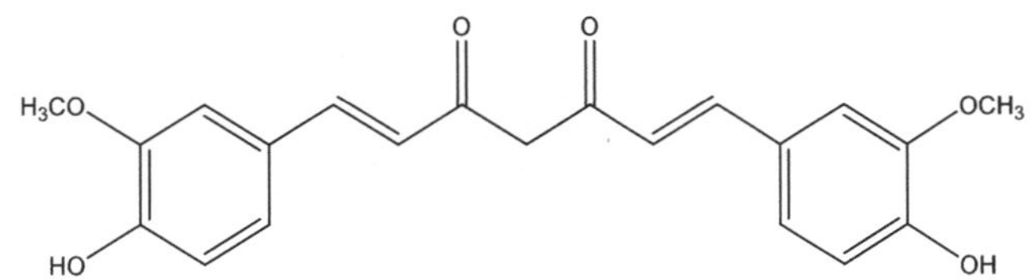 | *Curcuma longa* L. (Zingiberaceae) | Hepatitis B virus | Suppresses viral transcription | Polyak et al. (2007) |
| | | | | | Inhibits HCV replication | Anggakusuma et al. (2013) |
| 13 | Meliacine | 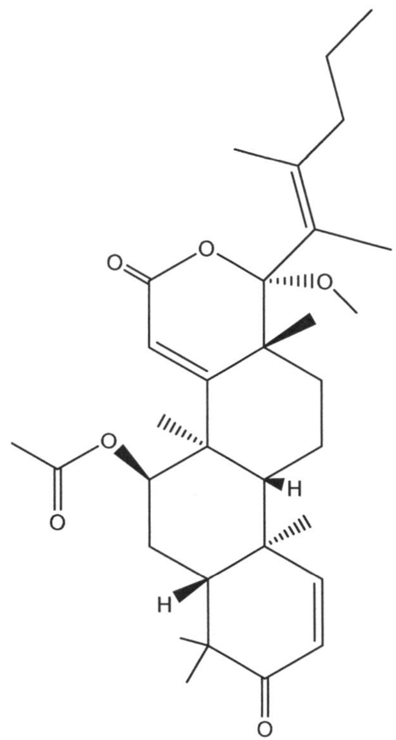 | *Melia azedarach* L. (Meliaceae) | HSV.2 infection in a mouse model | TNF-α and IFN-α production induction | Petrera and Coto (2009) |

(*Continued*)

**TABLE 1.3** (*Continued*)
**Some Plant-Derived Compounds with Antiviral Activities**

| S. No. | Compound | Chemical Structure | Plant | Virus | Mechanism | Ref. |
|---|---|---|---|---|---|---|
| 14 | Andrographolide | 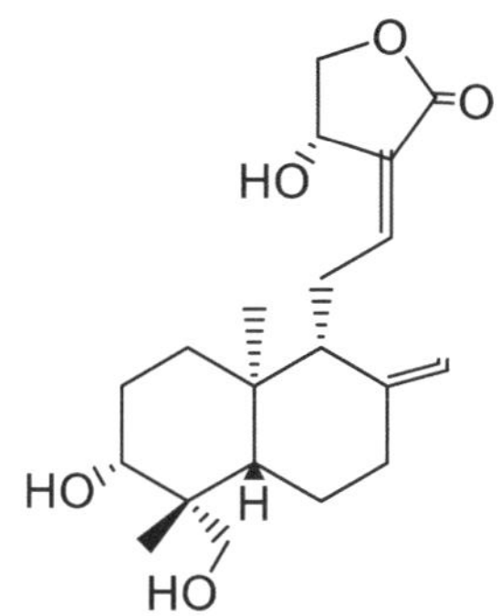 | *Andrgraphis paniculate* (Burm.f.) Nees | HSV-I, HIV and EBV | a) Inhibit replication of HSV-I and II<br>b) Inhibit multiplication of HCV | Jayakumar et al. (2013) |
| 15 | Hesperidin | 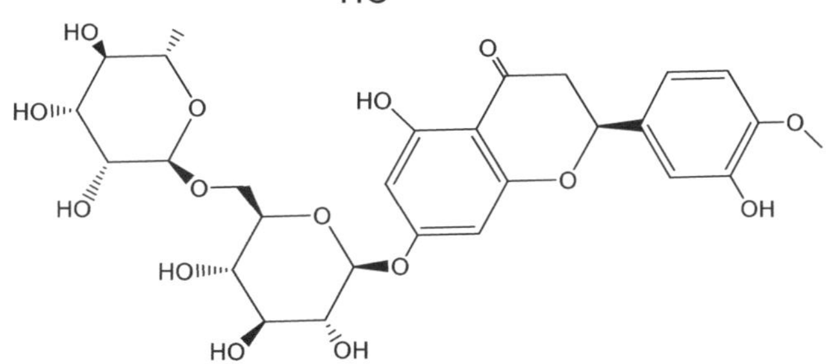 | *Citrus sinensi* (L) Osbeck (Rutaceae) | SARS-CoV | SARS-CoV-2 receptor inhibition | Bellavite and Donzelli (2020) |

(*Continued*)

**TABLE 1.3 (*Continued*)**
**Some Plant-Derived Compounds with Antiviral Activities**

| S. No. | Compound | Chemical Structure | Plant | Virus | Mechanism | Ref. |
|---|---|---|---|---|---|---|
| 16 | Anolignan A, Anolignan B | 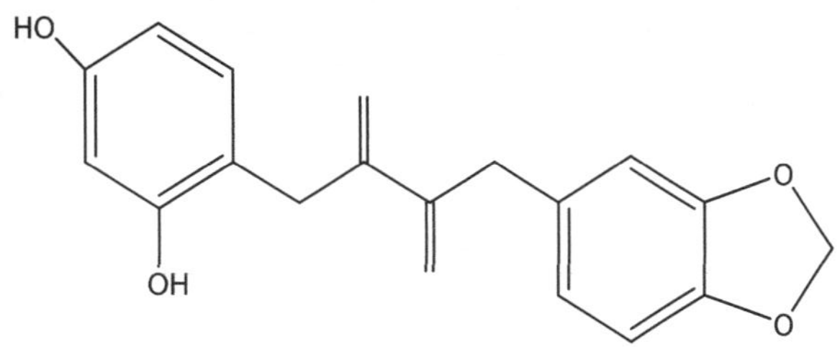 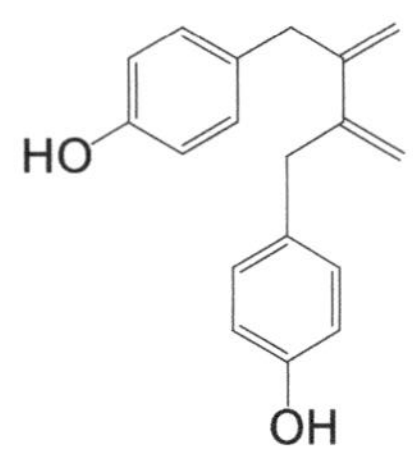 | *Anogeissus acuminata* (Roxb. ex DC) Wall, ex Guillem&Perr (Combretaceae) | HIV | Inhibits reverse transcriptase in HIV-1 | El-Ansari et al. (2020) |
| 17 | Humulene epoxide | 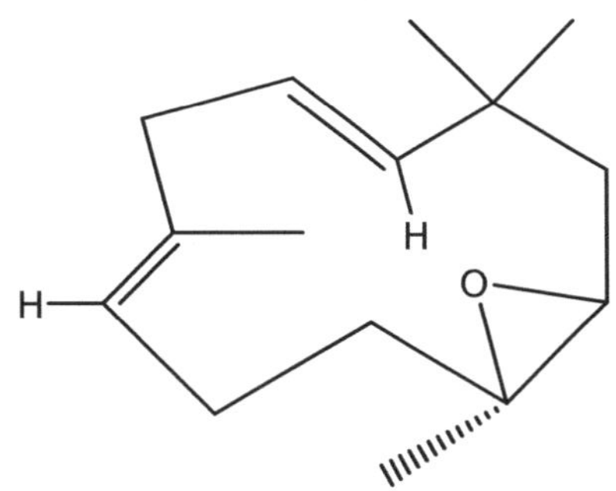 | *Cyperus rotundus* L (Cyperaceae) | SARS-CoV | SARS-CoV-2 target protein | Amparo et al. (2021) |

(*Continued*)

**TABLE 1.3 (*Continued*)**
**Some Plant-Derived Compounds with Antiviral Activities**

| S. No. | Compound | Chemical Structure | Plant | Virus | Mechanism | Ref. |
|---|---|---|---|---|---|---|
| 18 | Quercetin and kaempferol | 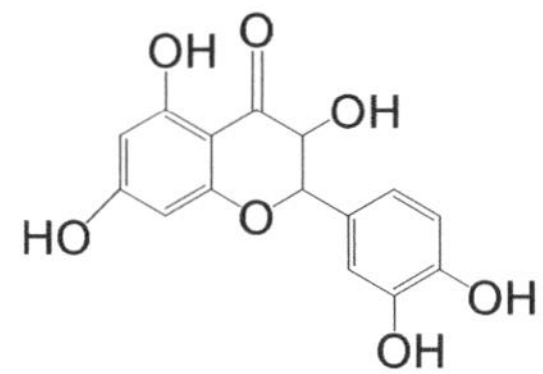 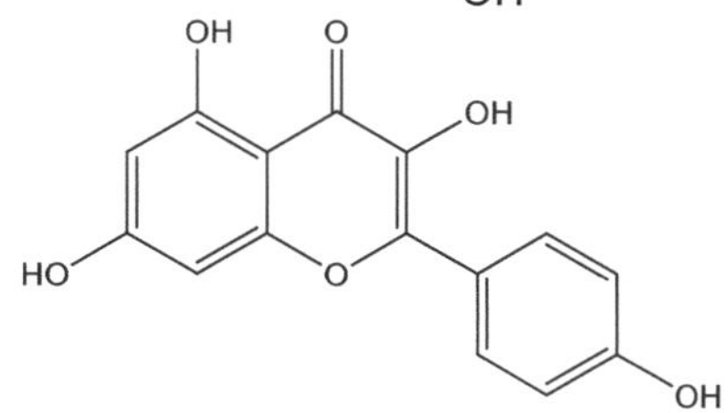 | *Aloe vera* (L) Burm.f (Asphodelaceae) | SARS-CoV | Autophagy and viral M2 RNA synthesis inhibition and M2 protein expression, SARS-CoV-2 Mpro and ACE2 receptor inhibition | Khaerunnisa et al. (2020) |
| 19 | Phyllanthin and hypophyllantin | 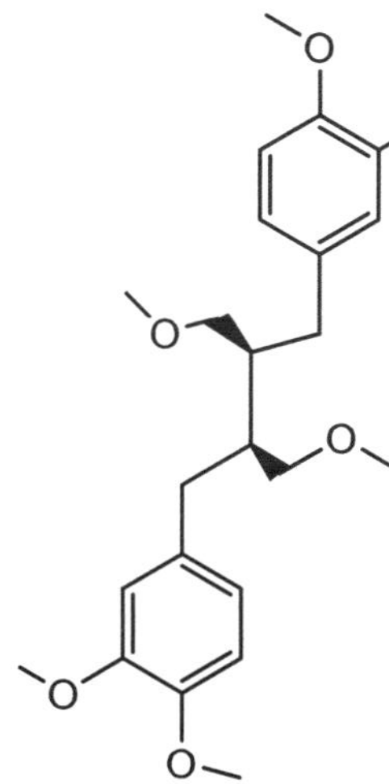 | *Phyllanthus niruri* L. (Phyllanthaceae) | Coronavirus | Interference with viral entry into host cells | Wahyuni et al. (2019) |

(*Continued*)

**TABLE 1.3 (*Continued*)**
**Some Plant-Derived Compounds with Antiviral Activities**

| S. No. | Compound | Chemical Structure | Plant | Virus | Mechanism | Ref. |
|---|---|---|---|---|---|---|
| | | 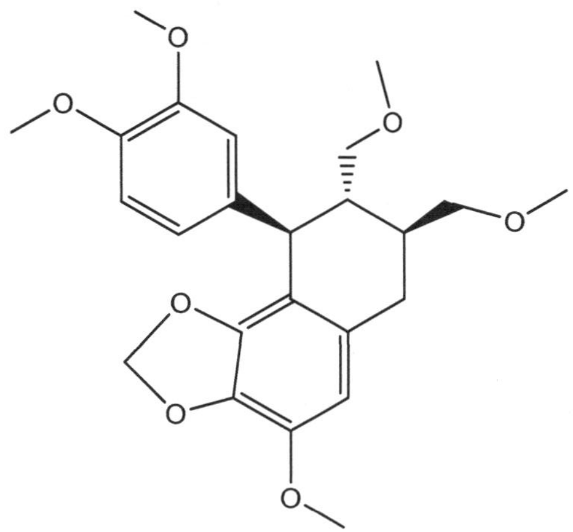 | | | | |
| 20 | Epigallocatechin-3-gallate (EGCG) | 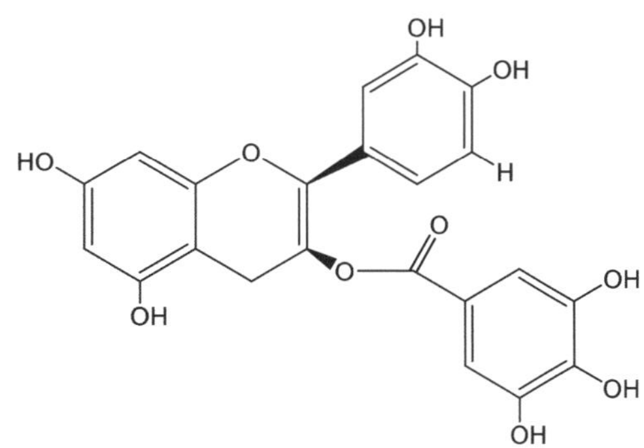 | *Camellia sinensis* (L.) Kuntze (Theaceae) | SARS-CoV | Inhibition of SARS-CoV-2 Mpro enzyme and S-protein receptor interaction<br>Inhibition of HIV reverse transcriptase<br>RNA and DNA synthesis and antigen expression inhibition in HBV<br>Blocking of HSV1 and HIV-1 attachment to host cells | Goyal et al. (2020) |

(*Continued*)

**TABLE 1.3 (*Continued*)**
**Some Plant-Derived Compounds with Antiviral Activities**

| S. No. | Compound | Chemical Structure | Plant | Virus | Mechanism | Ref. |
|---|---|---|---|---|---|---|
| 21 | Apigenin | 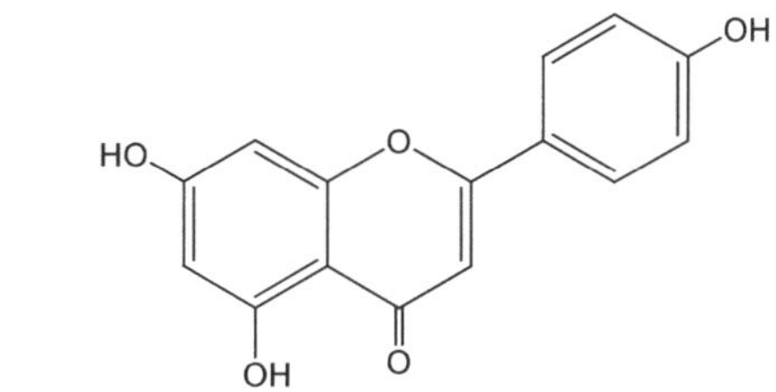 | *Ocimum basilicum* L. (Lamiaceae) | SARS-CoV | a. Inhibits ACE2 receptor and 3CLpro of SARS-CoV-2 Benencia<br>b. Inhibits replication of HBV | Tshilanda et al. (2020) |
| 22 | Ursolic acid | 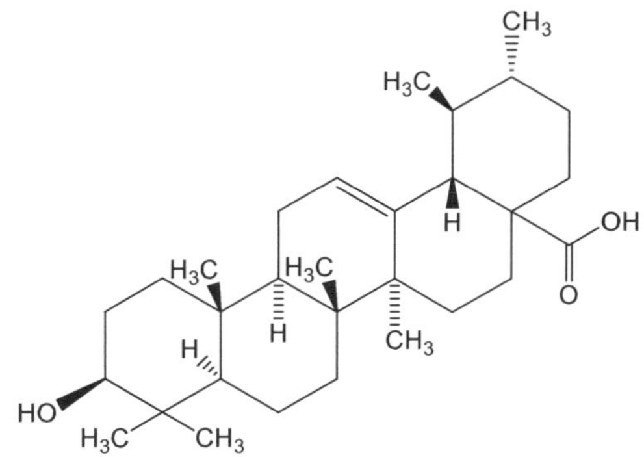 | Many fruits and herbs Including *Mirabilis jalapa* L. (Myctaginaceae) | HSV-I | a. Inhibits replication of HSV-I and II<br>b. Inhibits multiplicatio n of HCV | Ghosh et al. (2021) |
| 23 | Anisotine | 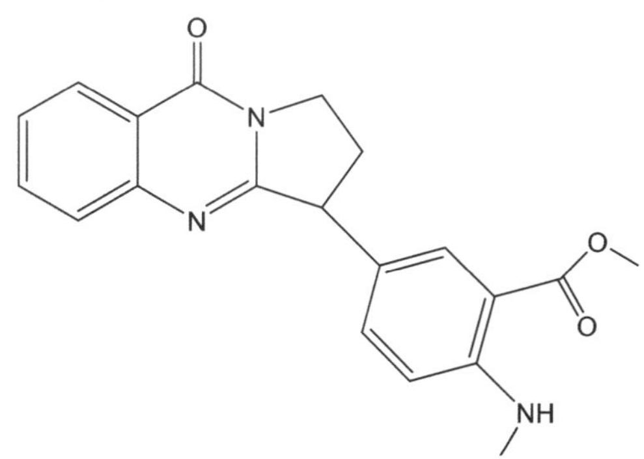 | *Justica adhatoda* L. (Acanthaseae) | SARS-CoV-2 Influenza virus and HSV | Inhibits Mpro of SARS-CoV-2 which mediates the cleavage of polyprotein to get matured and acquire infectivity | Rouf et al. (2020) |

(*Continued*)

**TABLE 1.3 (*Continued*)**
**Some Plant-Derived Compounds with Antiviral Activities**

| S. No. | Compound | Chemical Structure | Plant | Virus | Mechanism | Ref. |
|---|---|---|---|---|---|---|
| 24 | Ajoene | 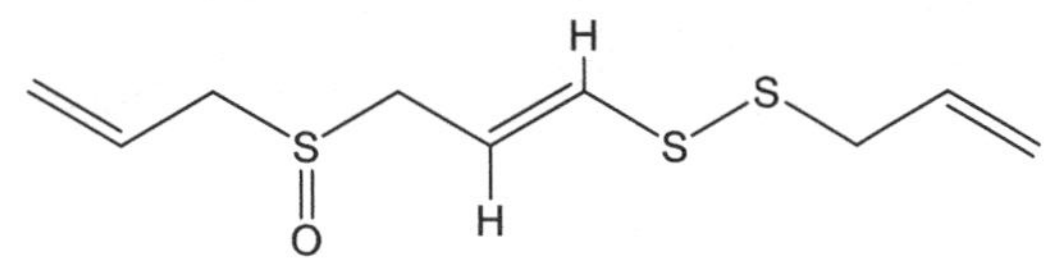 | *Allium sativum* L (Alliaceae) | HIV | a. Prevents HIV-induced destruction of CD+cells<br>b. Enhances cellular immunity<br>c. Inhibits virus-cell attachment and viral reverse transcriptase of HIV-I<br>d. Induces apoptosis of Human cytomegalovirus (HCMV)-infected cells | Kaushik et al. (2020) |
| 25 | 6-Gingerol and gingeronone A | 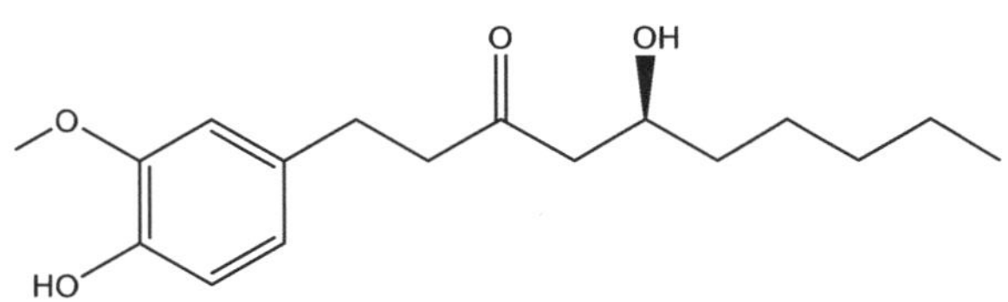<br>6-gingerol<br>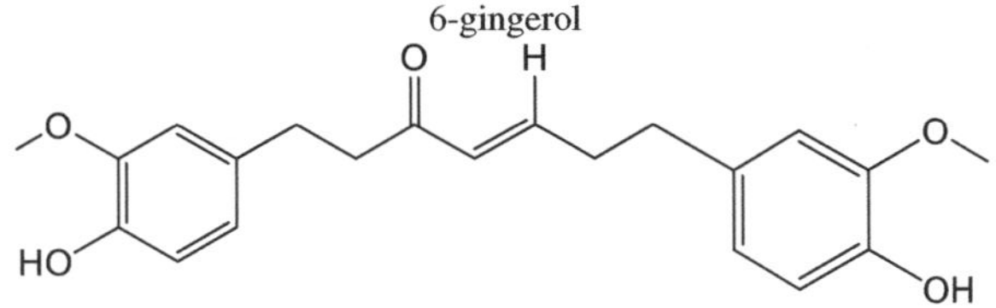<br>Gingeronone A | *Zingiber officinale* Roscoe (Zingiberaceae) | CHIKV, HCV and SARS-CoV-2 | Inhibits HCV NS5B protein leading to RNA replication<br>Inhibits ACE2 receptor and 3CLpro of SARS-CoV-2 Benencia | Rathinavel et al. (2020); Tshilanda et al. (2020) |

(*Continued*)

**TABLE 1.3 (*Continued*)**
**Some Plant-Derived Compounds with Antiviral Activities**

| S. No. | Compound | Chemical Structure | Plant | Virus | Mechanism | Ref. |
|---|---|---|---|---|---|---|
| 26 | Rosmarinic acid | 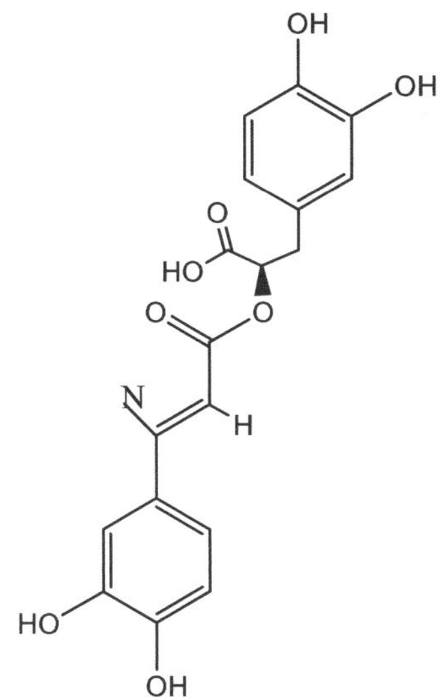 | *Ocimum tenuiflorum* L.<br>*Ocimum americanum* L.<br>(Lamiaceae) | HSV-1 | a. Inhibits replication of HSV-I and II<br>b. Inhibits protease enzyme of HSV-I and I1 | Ren et al. (2020) |
| 27 | Coumestan | 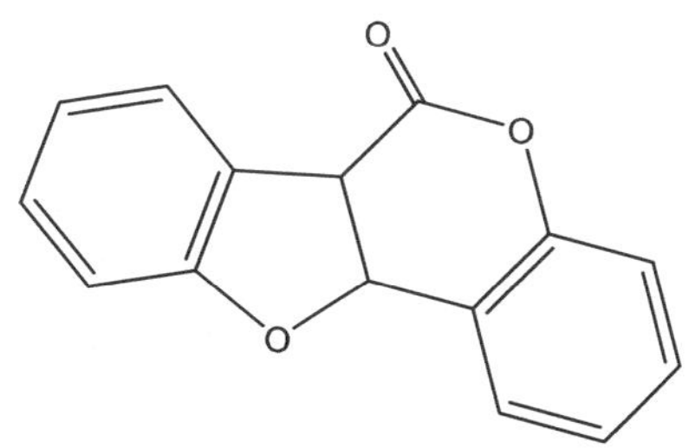 | *Eclipta prostrata* L. (Asteraceae) | HCV | Inhibits HCV NS5B protein leading to RNA replication | Wang et al. (2012b) |

(*Continued*)

**TABLE 1.3 (*Continued*)**
**Some Plant-Derived Compounds with Antiviral Activities**

| S. No. | Compound | Chemical Structure | Plant | Virus | Mechanism | Ref. |
|---|---|---|---|---|---|---|
| 28 | Kaempferol-3-O-(6″-O-Ep-coumaroyl)-β-D-glucopyranoside | 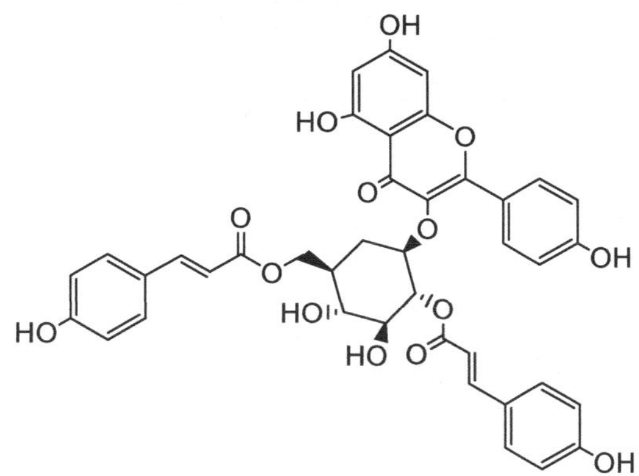 | *Bombax ceiba* L. (Malvaceae) | RSV, SARS-CoV-2 | a. Inhibits cytopathic effect of RSV<br>b. Inhibits ORF 3a protein of SARS-CoV-2 leading to interference of virus release mechanism and reduce poptosis | Lai et al. (2012) |
| 29 | Cimicifugin | 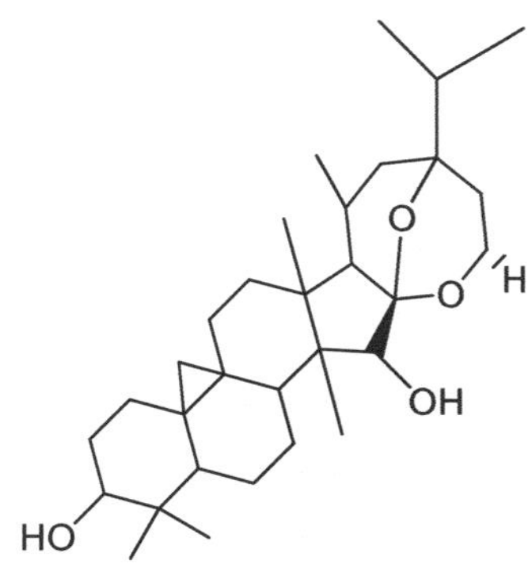 | *Cimicifuga foetida* L (Ranunculaceae) | Coronavirus | Inhibits viral attachment and internalization steps; stimulates IFN *f*À secretion | Nesari et al. (2021) |

(*Continued*)

**TABLE 1.3 (*Continued*)**
**Some Plant-Derived Compounds with Antiviral Activities**

| S. No. | Compound | Chemical Structure | Plant | Virus | Mechanism | Ref. |
|---|---|---|---|---|---|---|
| 30 | Ursolic acid, eugenol, 1,8,-cineole, | 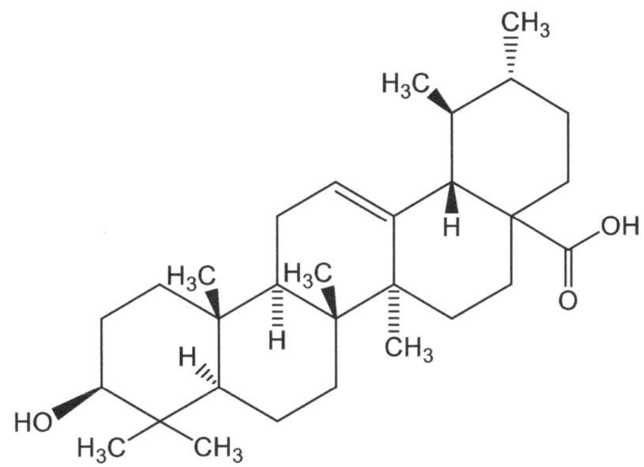 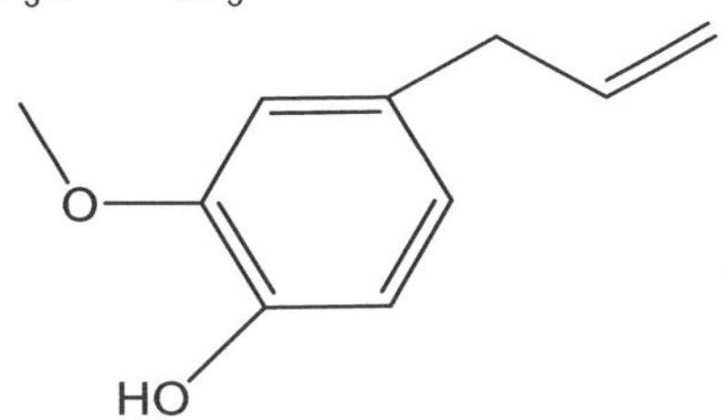 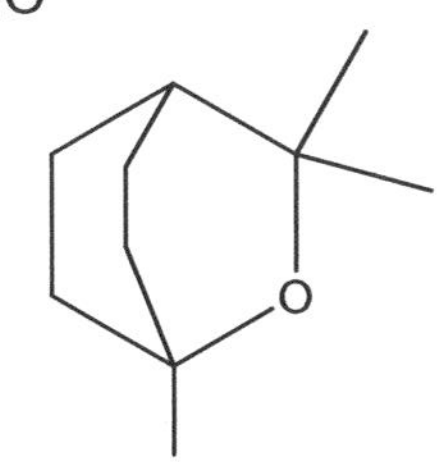 | *Ocimum tenuiflorum* L. (Lamiaceae) | HSV-1 and II | | Wahyuni et al. (2019) |

(*Continued*)

**TABLE 1.3** (*Continued*)
**Some Plant-Derived Compounds with Antiviral Activities**

| S. No. | Compound | Chemical Structure | Plant | Virus | Mechanism | Ref. |
|---|---|---|---|---|---|---|
| 31 | Eugenol and thymol | 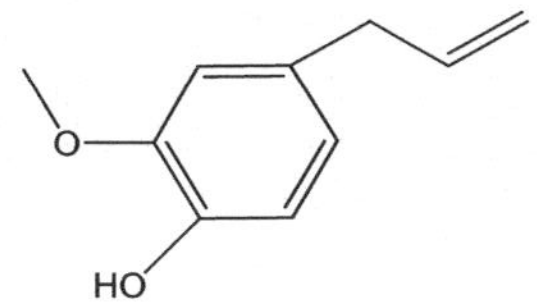 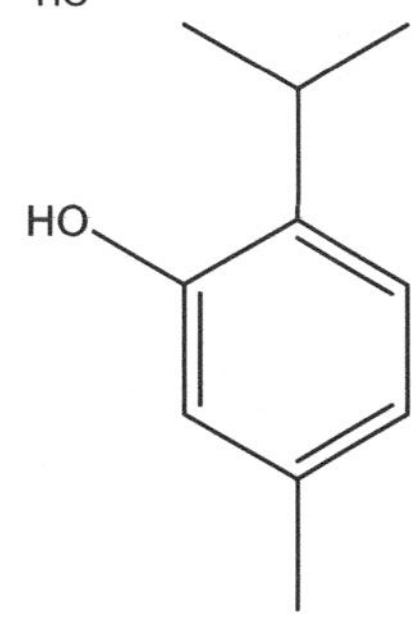 | *Ocimum campechianum* Mill (Lamiaceae) | HSV-1 and II | a. Inhibits replication of HSV-I and II<br>b. Inhibits protease enzyme of HSV-I and I1 | Jaijyan et al. (2018) |
| 32 | Gedunin, Pongamol | 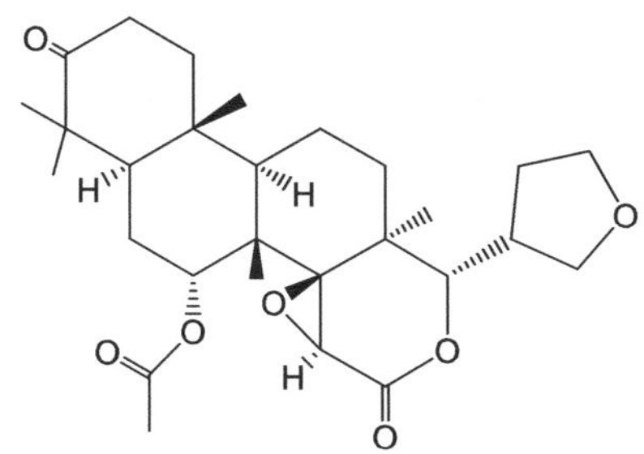 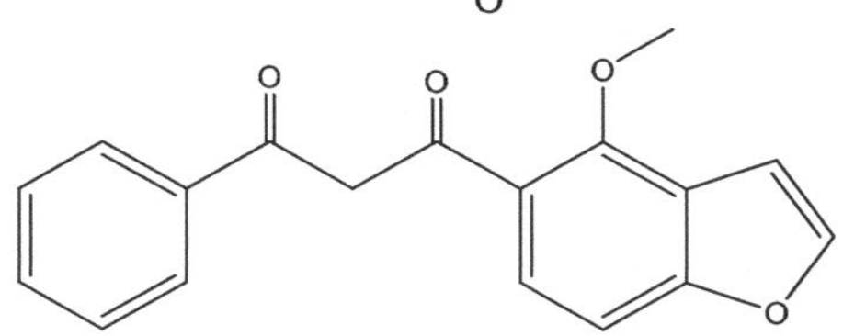 | *Azadirachta indica* A. Juss (Meliaceae) | HSV-I, CVB-B4, HBV and SARS-CoV-2 | a. Inhibits replication of HSV-I<br>b. Inhibits ACE2 receptor and 3CLpro of SARS-CoV-2 Benencia | Martins et al. (2016) |

the scientific community. Medical plants hold promise for their management and treatment of these diseases. The active phytochemical found in these plants should be further explored as potential alternatives for effectively managing emerging viral diseases.

## LIST OF ABBREVIATIONS

**CHIKV** – Chikungunya Virus
**COVID-19** – Corona Virus
**DENV** – Dengue Virus
**EBV** – Epstein-Barr Virus
**GAG** – Glycosaminoglycan
**H1N1; H7N9** – Swine Flu
**HBV** – Hepatitis B Virus
**HIV** – Human Immunodeficiency Virus
**HSV** – Herpes Simplex Virus
**MERS** – Middle East Respiratory Syndrome coronavirus
**RSV** – Respiratory Syncytial Virus
**SARS** – Severe Acute Respiratory Syndrome
**SARS-CoV-2** – Severe Acute Respiratory Syndrome Coronavirus 2

## REFERENCES

Afolayan, F.I.D., Adegbolagun, O., Mwikwabe, N.N., Orwa, J., and Anumudu, C. (2020). Cytokine modulation during malaria infections by some medicinal plants. *Scientific African* 8, e00428. https://doi.org/10.1016/j.sciaf.2020.e00428

Amparo, T.R., Seibert, J.B., Silveira, B.M., Costa, F.S.F., Almeida, T.C., Braga, S.F.P., et al. (2021). Brazilian essential oils as source for the discovery of new anti-COVID-19 drug: a review guided by in silico study. *Phytochemistry Reviews*, 1–20. https://doi.org/10.1007/s11101-021-09754-4.

Anggakusuma, C.C., Schang, L.M., Rachmawati, H., Frentzen, A., Pfaender, S., Behrendt, P., et al. (2013). Turmeric curcumin inhibits entry of all hepatitis C virus genotypes into human liver cells. *Gut*. https://doi.org/10.1136/gutjnl-2012-304299.

Arankalle, V.A., Shrivastava, S., Cherian, S., Gunjikar, R.S., Walimbe, A.M., and Jadhav,S.M. (2007). Genetic divergence of Chikungunya viruses in India (1963–2006) with special reference to the 2005-2006 explosive epidemic. *Journal of General Virology*. 88(Pt 7), 1967–1976.

Asselah, T., Durantel, D., Pasmant, E., Lau, G., Schinazi, R.F., Asselah, T., Durantel, D., Pasmant, E., Lau, G., and Schinazi, R.F. (2021). COVID-19: discovery, diagnostics and drug development. *Journal of Hepatology* 74, 168–184.

Attah, A.F., Fagbemi, A.A., Olubiyi, O., Dada-Adegbola, H., Oluwadotun, A., Elujoba, A., and Babalola, C.P. (2021). Therapeutic Potentials of antiviral plants used in traditional African medicine with COVID-19 in focus: a Nigerian perspective. *Frontiers in Pharmacology* 12, 596855. https://doi.org/10.3389/fphar.2021.596855.

Badia-Boungou, F., Sane, F., Alidjinou, E.K., Hennebelle, T., Roumy, V., Ngakegni-Limbili, A. C., et al. (2019). Aqueous extracts of Syzygiumbrazzavillense can inhibit the infection with coxsackievirus B4 in vitro. *Journal of Medical Virology* 91(7), 1210–1216. https://doi.org/10.1002/jmv.25436.

Ball, M.J., Lukiw, W.J, Kammerman, E.M., and Hill, J.M. (2013). Intracerebral propagation of Alzheimer's disease: strengthening evidence of a herpes simplex virus etiology. *Alzheimers Dement* 9, 169–175.

Bellavite, P. and Donzelli, A. (2020). Hesperidin and SARS-CoV-2: new light on the healthy function of citrus fruits. *Antioxidants (Basel)* 9, 742. https://doi.org/10.3390/antiox9080742.

Beressa, T.B., Deyno, S., Mtewa, A.G., Aidah, N., Tuyiringire, N., Lukbye, B., Weisheit, A., Tolo, C.U., and Ogwang, P.E. (2021). Potential benefits of antiviral African medicinal plants in the management of viral infections: systematic review. *Frontiers in Phamacology* 12, 682794. https://doi.org/10.3389/fphar.2021.682794.

Berit, T., Mulder, L.M., Mayra, D.T., and Smit, J.M. (2020). Tomatidine, a natural steroidal alkaloid shows antiviral activity towards chikungunya virus in vitro. *Scientific Reports (Nature Publ Group)* 10(1), 6364. https://doi.org/10.1038/s41598-020-63397-7.

Bermejo, M, Rodriguez-Teijeiro, J.D., Illera, G., Barroso, A., Vilà, C., and Walsh, P.D. (2006). Ebola outbreak killed 5000 gorillas. *Science* 314, 1564.

Bettis, A.A., L'Azou Jackson, M., Yoon, I.-K., Breugelmans, J.G., Goios, A., Gubler, D.J., and Powers, A.M. (2020). The global epidemiology of chikungunya from 1999 to 2020: a systematic literature review to inform the development and introduction of vaccines. *Public Library of Sciences (PLoS) Neglected Tropical Diseases* 16(1), e0010069. https://doi.org/10.1371/pntd.0010069.s001.

Bonaccorsi, G., Pierri, F., Cinelli, M., Flori, A., Galeazzi, A., Porcelli, F., et al. (2020). Economic and social consequences of human mobility restrictions under COVID-19. *PNAS* 117(27), 15530–15535. https://doi.org/10.1073/pnas.2007658117.

Breman, J.G., Haymann, D.L., Lloyd, G., McCormick, J.B., Miatudila, M, Murphy, F.A., Muyembé-Tamfun, J.J., Piot, P., Ruppol, J.F., Sureau, P., Van Der groen, G., and Johnson, K.M. (2016). Discovery and description of Ebola Zaire virus in 1976 and relevance to the West African epidemic during 2013–2016. *The Journal of Infectious Diseases* 214(suppl 3), S93–S101. https://doi.org/10.1093/infdis/jiw207.

Calland, N., Albecka, A., Belouzard, S., Wychowski, C., Duverlie, G., Descamps, V., et al. (2012). (-)-Epigallocatechin-3-gallate is a new inhibitor of hepatitis C virus entry. *Hepatology* 55, 720.

Chang, J.S., Liu, H.W., Wang, K.C., Chen, M.C., Chiang, L.C., Hua, Y.C., et al. (2005). Ethanol extract of Polygonum cuspidatum inhibits hepatitis B virus in a stable HBV-producing cell line. *Antiviral Research* 66, 29–34.

Cheng, P.W., Ng, L.T., Chiang, L.C., and Lin, C.C. (2006). Antiviral effects of saikosaponins on human coronavirus 229E in vitro. *Clinical and Experimental Pharmacology and Physiology* 33, 612–616.

Chiang, L.C., Ng, L.T., Liu, L.T., Shieh, D.E., and Lin, C.C. (2003). Cytotoxicity and anti-hepatitis B virus activities of saikosaponins from Bupleurum species. *Planta Medica* 69, 705–709.

Chiang, L.C., Ng, L.T., Cheng, P.W., Chiang, W., and Lin C.C. (2005). Antiviral activities of extracts and selected pure constituents of Ocimumbasilicum. *Clinical and Experimental Pharmacology and Physiology* 32, 811–816.

Choi, H.J., Song, J.H., Park, K.S., and Baek, S.H. (2010). In vitro anti-enterovirus 71 activity of gallic acid from Woodfordiafruticosa flowers. *Letters in Applied Microbiology* 50, 438–440.

Chu, D.K., Akl, E.A., Duda, S., Solo, K., Yaacoub, S., Schünemann, H. J., et al. (2020). Physical distancing, face masks, and eye protection to prevent personto- person transmission of SARS-CoV-2 and COVID-19: a systematic review and meta-analysis. *Lancet* 395(10242), 1973–1987. https://doi.org/10.1016/S0140-6736(20)31142-9.

ColungaBiancatelli, R.M.L., Berrill, M., Catravas, J.D., and Marik, P.E. (2020). Quercetin and vitamin C: an experimental, synergistic therapy for the prevention and treatment of SARS-CoV-2 related disease (COVID-19). *Frontiers in Immunology* 11, 11. https://doi.org/10.3389/fimmu.2020.01451.

Cos, P., Hermans, N., De Bruyne, T., Apers, S., Sindambiwe, J.B., Vanden Berghe, D., et al. (2002a). Further evaluation of Rwandan medicinal plant extracts for their antimicrobial and antiviral activities. *Journal of Ethnopharmacology* 79(2), 155–163. https://doi.org/10.1016/S0378-8741(01)00362-2.

Cos, P., Hermans, N., De Bruyne, T., Apers, S., Sindambiwe, J. B., Witvrouw, M., et al. (2002b). Antiviral activity of Rwandan medicinal plants against human immunodeficiency virus type-1 (HIV-1). *Phytomedicine* 9(1), 62–68. https://doi.org/10.1078/0944-7113-00083.

Cruz, G.V.B., Pereira, P.V.S., Patrício, F.J., Costa, G.C., Sousa, S.M., Frazão, J.B., et al. (2007). Increase of cellular recruitment, phagocytosis ability and nitric oxide production induced by hydroalcoholic extract from Chenopodium ambrosioides leaves. *Journal of Ethnopharmacology* 111(1), 148–154. https://doi.org/10.1016/j.jep.2006.11.006.

Danaher, R.J., Wang, C., Dai, J., Mumper, R.J., and Miller, C.S. (2011). Antiviral effects of blackberry extract against herpes simplex virus type 1. *Oral Surgery, Oral Medicine, Oral Pathology, Oral Radiology, and Endodontology* 112, e31–5.

Dixon, M.G. and Schafer, I.J. (2014). Ebola viral disease outbreak-West Africa, 2014. *Moorbidity and Mortality Weekly Report* 63(25), 548–551.

Donaldson, L.L.J., Rutter, P.D., Ellis, B.M., Greaves, F.E., Mytton, O.T., Pebody, R.G., and Yardley, I.E. (2009). Mortality from pandemic A/H1N1 influenza in England: public health surveillance study. *British Medical Journal* 339, b5214. https://doi.org/10.1136/bmj.b5213.

Duffy, M.R., Chen, T.H., Hancook, W.T., Powera, A.M., Kool, J.L., Lanciotti, R.S., Pretrick, M., Marfel, M., Holzbauer, S., Dubray, C., Guillaumot, L., Griggs, A., Bel, M., Lambert, A.J., Laven, J., Kosoy, O., Panella, A., Biggerstaff, B.J., Fischer, M., and Hayes, E.B. (2009). Zika virus outbreak on Yap Island, Federated States of Micronesia. *The New England Journal of Medicine* 360(24), 2536–2543. https://doi.org/10.1056/NEJMoa08057.

El-Ansari, M.A., Ibrahim, L.F., and Sharaf, M. (2020). Anti-HIV activity of some natural phenolics. *Herba Polonica* 66, 34–43. https://doi.org/10.2478/hepo-2020-0010.

Excler, J.L., Saville, M., Berkley, S., and Kim, J.H. (2021). Vaccine development for emerging infectious diseases. *Nature Medicine* 27, 591–600. https://doi.org10.1038/s41591-021-01301-0.

Ferenci, P., Scherzer, T.M., Kerschner, H., Rutter, K., Beinhardt, S., Hofer, H., et al. (2008). Silibinin is a potent antiviral agent in patients with chronic hepatitis C not responding to pegylated interferon/ribavirin therapy. *Gastroenterology* 135, 1561–1567.

Ferman, A., Badshah, S.L., Khan, K., Ahmad, N., and Naeem, A. (2020). Ebola, the negative stranded RNA. *IntechOpen*. https://doi.org/10.5772/intechopen.91776.

Fialho, L.G., da Silva, V.P., Reis, S.R.N.I., Azeredo, E.L., Kaplan, M.A.C., Figueiredo, M.R., et al. (2016). Antiviral and immunomodulatory effects of noranteabrasiliensischoisy on dengue virus-2. *Intervirology* 59(4), 217–227. https://doi.org/10.1159/000455855.

Francesconi, P., Yoti, Z., Declich, S., Onek, P.A., Fabiani, M., Olango, J., Andraghetti, R., Rollin, P.E., Opira, C., Greco, D., and Salmaso, S. (2003). Ebola haemorrhagic fever transmission and risk factors of conntats, Uganda. *Emerging Infectious Diseases* 9(11), 1430–1437. https://doi.org/10.3201/eid0911.030339.

Friedler, A. (2021). Sociocultural, behavioural and political factors shaping the COVID-19 pandemic: the need for a biocultural approach to understanding pandemics and (re)emerging pathogens. *Global Public Health* 16, 17–35.

Gebre-Mariam, T., Neubert, R., Schmidt, P.C., Wutzler, P., and Schmidtke, M. (2006). Antiviral activities of some Ethopian medicinal plants used for the treatment of dermatological disorders. *Journal of Ethnopharmacology* 104(1–2), 182–187. https://doi.org/10.016/j.jep.2005.08,071.

Gescher, K., Kuhn, J., Hafezi, W., Louis, A., Derksen, A., Deters, A., et al. (2011a). Inhibition of viral adsorption and penetration by an aqueous extract from Rhododendron ferrugineum L. as antiviral principle against herpes simplex virus type-1. *Fitoterapia* 82, 408–413.

Gescher, K., Kuhn, J., Lorentzen, E., Hafezi, W., Derksen, A., Deters, A., et al. (2011b). Proanthocyanidin-enriched extract from MyrothamnusflabellifoliaWelw. exerts antiviral activity against herpes simplex virus type 1 by inhibition of viral adsorption and penetration. *Journal of Ethnopharmacology* 134, 468–474

Ghildiyal, R., Prakash, V., Chaudhary, V.K., Gupta, V., and Gabrani, R. (2020). Phytochemicals as antiviral agents: recent updates. Plant-Derived Bioactives 279–295.

Ghosh, R., Chakraborty, A., Biswas, A., and Chowdhuri, S. (2021). Identification of alkaloids from justicia adhatoda as potent SARS CoV-2 main protease inhibitors: an in silico perspective. *Journal of Molecular Structure* 1229, 129489. https://doi.org/10.1016/j.molstruc.2020.129489.

Gorshkov, K., Shiryaev, S.A., Fertel, S., Lin, Y.-W., Huang, C.-T., Pinto, A., Farhy, C., Strongin, A.Y., Zheng, W., and Terskikh, A.V. (2019). Zika virus: origins, pathological action and treatment. *Frontiers Microbiology* 0, 3252. https://doi.org/10.2289/fmicb.2018.03252.

Goyal, M. (2020). Threats and challenges of emerging viral diseases and scope of Ayurveda in its prevention. *AYU* 40, 67–68.

Goyal, R.K., Majeed, J., Tonk, R., Dhobi, M., Patel, B., Sharma, K., et al. Current targets and drug candidates for prevention and treatment of SARSCoV-2 (COVID-19) infection. *Reviews in Cardiovascular Medicine* 21, 365–384. https://doi.org/10.31083/j.rcm.2020.03.118.

Gully, P.R. (2020). Pandemics, regional outbreaks, and sudden-onset disasters. *Healthcare Management Forum* 33, 164–169.

Haid, S., Novodomska, A., Gentzsch, J., Grethe, C., Geuenich, S., Bankwitz, D., et al. (2012). A plant-derived flavonoid inhibits entry of all HCV genotypes into human hepatocytes. *Gastroenterology* 143213–22-25.

Han, S.B., Kim, Y.H., Lee, C.W., Park, S.M., Lee, H.Y., Ahn, K.S., et al. (1998). Characteristic immunostimulation by angelan isolated from Angelica gigas Nakai. *Immunopharmacology* 40(1), 39–48. https://doi.org/10.1016/S0162-3109(98)00026-5.

Hao, B.J., Wu, Y.H., Wang, J.G., Hu, S.Q., Keil, D.J., Hu, H.J., et al. (2012). Hepatoprotective and antiviral properties of isochlorogenic acid A from Laggeraalata against hepatitis B virus infection. *Journal of Ethnopharmacology* 144, 190–194.

He, X., Fang, J., Guo, Q., Wang, M., Li, Y., Meng, Y., et al. (2020). Advances in antiviral polysaccharides derived from edible and medicinal plants and mushrooms. *Carbohydrate Polymers* 229, 115548. https://doi.org/10.1016/j.carbpol.2019.115548.

Heymann, D.L., Hodgson, A., Sall, A.A., Freedman, D.O., Erin Staples, J., Althabe, F., et al (2016). Zika virus and microcephaly: why is this situation a PHEIC? *Lancet* 387(10020), 719–721.

Ho, H.Y., Cheng, M.L., Weng, S.F., Leu, Y.L., and Chiu D.T. (2009). Antiviral effect of epigallocatechin gallate on enterovirus 71. *Journal of Agricultural and Food Chemistry* 57, 6140–6147.

Hober, D., Sane, F., Jaidane, H., Riedweg, K., Goffard, A., and Desailloud R. (2012). Immunology in the clinic review series; focus on type 1 diabetes and viruses: role of antibodies enhancing the infection with Coxsackievirus B in the pathogenesis of type 1 diabetes. *Clinical and Experimental Immunology* 168, 47–51.

Hsieh, C.-F., Jheng, J.-R., Lin, G.-H., Chen, Y.-L., Ho, J.-Y., Liu, C.-J., et al. (2020). Rosmarinic acid exhibits broad anti-enterovirus A71 activity by inhibiting the interaction between the five-fold axis of capsid VP1 and cognate sulfated receptors. *Emerging Microbes & Infections* 9, 1194–1205. https://doi.org/10.1080/22221751.2020.1767512.

Hsu, N.Y., Ilnytska, O., Belov, G., Santiana, M., Chen, Y.H., Takvorian, P.M., et al. (2010). Viral reorganization of the secretory pathway generates distinct organelles for RNA replication. *Cell* 141, 799–811. https://doi.org/10.1016/j.cell.2010.03.050.

https://www.fda.gov/drugs/information-drug-class/influenza-flu-antiviral-drugs-and-related-information

Huang, L., and Chen, C.H. (2002). Molecular targets of anti-HIV-1 Triterpenes. Current drug Targets. *Infectious Disorders* 2(1), 33–36. https://doi.org/10.2174/1568005024605936

Huang, W., Zhang, X., Wang, Y., Ye, W., Ooi, V.E., Chung, H.Y., et al. (2010). Antiviral biflavonoids from Radix wikstroemiae (Liaogewanggen). *Chinese Medicine* 5, 23.

Huang, C., Wang, Y., Li, X., Ren, L., Zhao, J., Hu, Y., et al. (2020) Clinical features of patients infected with 2019 novel coronavirus in Wuhan, China. *Lancet* 395, 497–506.

Hui, D.S.C., and Zumla, A. (2019). Severe acute respiratory syndrome: Historical, epidemiologic, and clinical features. *Infectious Disease Clinics of North America* 33(8), 869–889. https://doi.org/10.1016/j.idc.2019.07.001

Hussein, G., Miyashiro, H., Nakamura, N., Hattori, M., Kawahata, T., Otake, T., Kakiuchi, N., and Shimotohno, K. (1999). Inhibitory effects of Sudanese plant extracts on HIV-1 replication and HIV-1 protease. *Phytotherapy Research* 13(1), 31–36. https://doi.org/10.1002/(SICI)1099-1573(199902)13:1<31:AID-PTR381>3.0.CO;2-C

Huttly, S.R. (1990). The impact of inadequate sanitary conditions on health in developing countries. *World Health statistics Quarterly. Rapport Trimestriel de StatistiquesSanitairesMondiales* 43(3), 118–126.

Jaijyan, D.K., Liu, J., Hai, R., and Zhu, H. (2018). Emerging and Reemerging Human Viral Diseases. *Annals of Biological Research* 2, 31–44.

Jasso-Miranda, C., Herrera-Camacho, I., Flores-Mendoza, L.K., Dominguez, F., Vallejo-Ruiz, V., Sanchez-Burgos, G.G., et al. (2019). Antiviral and immunomodulatory effects of polyphenols on macrophages infected with dengue virus serotypes 2 and 3 enhanced or not with antibodies. *Infection and Drug Resistance* 12, 1833–1852. https://doi.org/10.2147/idr.s210890

Jayakumar, T., Hsieh, C.Y., Lee, J.J., and Sheu, J.R. (2013). Experimental and clinical pharmacology of andrographis paniculata and its major bioactive phytoconstituent andrographolide. *Evidence-Based Complementary and Alternative Medicine*, 846740. https://doi.org/10.1155/2013/846740.

Jiang, Z.Y., Liu, W.F., Zhang, X.M., Luo, J., Ma, Y.B., and Chen, J.J. (2013). Anti-HBV active constituents from Piper longum. *Bioorganic & Medicinal Chemistry Letters* 23, 2123–2127.

Jilani, T.N., Jamil, R.T., and Siddiqui, A.H. (2022). H1N1 Influenza. 2022 Jul 18. *StatPearls* [Internet]. Treasure Island (FL): StatPearls Publishing. PMID: 30020613.

Johnson, K.M., Webb, P.A., Lange, J.V., and Murphy, F.A. (1977). Isolation and characterization of a new virus (Ebola virus) causing acute hemorrhagic fever in Zaire. *Lancet* 1, 569–571.

Kapoor, R., Sharma, B., and Kanwar, S.S. (2017). Antiviral phytochemicals: an overview. *Biochemistry and Physiology* 6 (2), 7.

Kaushik, S., Jangra, G., Kundu, V., Yadav, J.P., and Kaushik, S. (2020). Anti-viral activity of *Zingiber officinale* (ginger) ingredients against the Chikungunya virus. *Virus Disease* 31, 1–7. https://doi.org/10.1007/s13337-020-00584-0.

Khaerunnisa, S., Kurniawan, H., Awaluddin, R., Suhartati, S., and Soetjipto, S. (2020). Potential inhibitor of COVID-19 main protease (Mpro) from several medicinal plant compounds by molecular docking study. *Preprints*, 1–14. https://doi.org/10.20944/preprints202003.0226.v1.

Koishi, A.C., Zanello, P.R., Bianco, E.M., Bordignon, J., and Nunes Duarte dos Santos, C. (2012). Screening of Dengue virus antiviral activity of marine seaweeds by an in situ enzyme-linked immunosorbent assay. *PLoS One* 7, e51089.

Konishi, T. (2022). Mutations in SARS-CoV-2 are on the increase against the acquired immunity. *PLoS One* 17(7), e0271305. https://doi.org/10.1371/journal.pone.0271305.

Koopmans, M., and De Jong, M.D. (2013). Avian influenza A H7N9 in Zhejiang, China. *The Lancet* 381(9881), 1882–1883. https://doi.org/10.1016/S-140-6736(13)60936-8.

Lai, W.L., Chuang, H.S., Lee, M.H., Wei, C.L., Lin, C.F., and Tsai, Y.C. (2012). Inhibition of herpes simplex virus type 1 by thymol-related monoterpenoids. *Planta Medica* 78, 1636–1638. https://doi.org/10.1055/s-0032-1315208.

Lau, A.C., Yam, L.Y., and So, L.K. (2004). Management of critically Ill patients with severe acute respiratory syndrome (SARS). *International Journal of Medical Sciences* 1(1), 1–10. https://doi.org/10.7150/ijms.1.1. Epub 2004 Mar 10. PMID: 15912185; PMCID: PMC1074505.

Li, S.Y., Chen, C., Zhang, H.Q., Guo, H.Y., Wang, H., Wang, L., et al. (2005). Identification of natural compounds with antiviral activities against SARS-associated coronavirus. *Antiviral Research* 67, 18–23.

Lim, X.Y., Teh, B.P., and Tan, T.Y.C. (2021). Medicinal plants in COVID-19: Potential and limitations. *Frontiers in Pharmacology* 12, 611408. https://doi.org/10.3389/fphar.2021.611408.

Lin, L.T., Chen, T.Y., Lin, S.C., Chung, C.Y., Lin, T.C., Wang, G.H., et al. (2013). Broad-spectrum antiviral activity of chebulagic acid and punicalagin against viruses that use glycosaminoglycans for entry. *BMC Microbiology* 13, 187.

Lin, L.T., Hsu, W.C., and Lin, C.C. (2014). Antiviral natural products and herbal medicines. *Journal of Traditional and Complementary Medicine* 4(1), 24–35.

Louten, J. (2016). *Essential Human Virology*. London: Academic Press, 291–310. https://doi.org/10.1016/B978-0-12-800947-5.00014-8.

Low, J.S., Wu, K.X., Chen, K.C., Ng, M.M., and Chu J.J. (2011). Narasin, a novel antiviralcompound that blocks dengue virus protein expression. *AntivirTher* 16, 1203–1218.

LuDec, J.W., and Barry, M.A. (2004). SARS. The first pandemic of the 21st century. *Emerging Infectious Diseases* 10(11), e26. https://doi.org/3201/eid1011.040797_02.

Ma, L.Y., Ma, S.C., Wei, F., Lin, R.C., But, P.P., Lee, S.H., et al. (2003). Uncinoside A and B, two new antiviral chromone glycosides from Selaginella uncinata. *Chem Pharm Bull (Tokyo)* 51, 1264–1267.

Marino, Z., Crespo, G., D'Amato, M., Brambilla, N., Giacovelli, G., Rovati, L., et al. (2013). Intravenous silibinin monotherapy shows significant antiviral activity in HCV-infected patients in the peri-transplantation period. *Journal of Hepatology* 58, 415–420.

Martins, N., Imler, J.-L., and Meignin, C. (2016). Discovery of novel targets for antivirals: Learning from flies. *Current Opinion in Virology* 20, 64–70. https://doi.org/10.1016/j. coviro.2016.09.005.

Mehrbod, P., Abdalla, M.A., Njoya, E.M., Ahmed, A.S., Fotouhi, F., Farahmand, B., et al. (2018a). South African medicinal plant extracts active against influenza A virus. *BMC Complementary and Alternative Medicine* 18 (1), 1–10. https://doi.org/10.1186/s12906-018-2184-y.

Michelini, F.M., Alché, L.E., and Bueno, C.A. (2018). Virucidal, antiviral and immunomodulatory activities of β-escin and Aesculus hippocastanum extract. *Journal of Pharmacy and Pharmacology* 70 (11), 1561–1571. https://doi.org/10.1111/jphp.13002.

Morgan, R.L., Baack, B., Smith, B.D., Yartel, A., Pitasi, M., and Falck Ytter, Y. (2013). Eradication of hepatitis C virus infection and the development of hepatocellular carcinoma: A meta-analysis of observational studies. *Annals of Internal Medicine* 158, 329–337.

Mujoriya, R.Z., Dhamande, K., and Bodla, R. (2011). A review on study of swine flu. *Indo-Global Research Journal of Pharmaceutical Sciences* 06, 1.

Murphy, F.A., and Peters, C.J. (1998). Ebola virus: Where does it come from and where is it going? In: Krause, R.M., ed. *Emerging Infections*. New York: Academic Press; 375–410.

Musarra-Pizzo, M., Pennisi, R., Ben-Amor, I., Mandalari, G., and Sciortino, M.T. (2021). Antiviral activity exerted by natural products against human viruses. *Viruses* 13, 828. https://doi.org/10.3390/v13050828.

Nesari, T.M., Bhardwaj, A., Shrikrishna, R., Ruknuddin, G., Ghildiyal, S., Das, A., et al. (2021). Anti-HIV activity of some natural phenolics. *Alternative Therapies in Health and Medicine* 23, 1–8.

Neumann, U.P., Biermer, M., Eurich, D., Neuhaus, P., and Berg T. (2010). Successful prevention of hepatitis C virus (HCV) liver graft reinfection by silibinin mono-therapy. *Journal of Hepatology* 52, 951–952.

Nii-Trebi, N.I. (2017). Emerging and neglected infectious diseases: Insights, advances, and challenges. *BioMed Research International* 2017, 5245021.

Noorbakhsh, F., Abdolmohammadi, K., Fatahi, Y., Dalili, H., Rasoolinejad, M., Rezaei, F., Salehi-Vaziri, M., Shafiei-Jandaghi, N.Z., Gooshki, E.S., Zaim, M., and Nicknam, M.H. (2019). Zika virus infection, basic and clinical aspects: A review article. *Iranian Journal of Public Health* 48 (1), 20–31.

Omino, E.A., and Kokwaro, J.O. (1993). Ethnobotany of Apocynaceae species in Kenya. *Journal of Ethnopharmacology* 40(3), 167–180. https://doi.org/10.1016/0378-8741(93)90065-d.

Omoregie, E.S., and Pal, A. (2016). Antiplasmodial, antioxidant and immunomodulatory activities of ethanol extract of Vernonia amygdalina del. Leaf in Swiss mice. *Avicenna Journal of Phytomedicine* 6 (2), 236.

Peiris, J.S., Chu, C.M., Cheng, V.C., Chan, K.S., Hung, I.F., Poon, L.L., Law, K.I., Tang, B.S., Hon, T.Y., Chan, C.S., Chan, K.H., Ng, J.S., Zheng, B.J., Ng, W.L., Lai, R.W., Guan, Y., Yuen, K.Y., and KHU/UCH SARS study Group (2003). Clinical progression and viral load in community outbreak of coronavirus-associated SARS pneumonia: A prospective study. *Lancet (London, England)* 361 (9371), 1767–1772. https://doi.org/10.1016/s0140-6736(03)13412-5.

Petrera, E., and Coto, C.E. (2009). Therapeutic effect of meliacine, an antiviral derived from Melia azedarach L., in mice genital herpetic infection. *Phytotherapy Research* 23 (12), 1771–1777.

Plourde, AR., and Bloch, E.M. (2016). A literature review of zika virus. *Emerging Infectious Diseases* 22 (7), 1185–1192. https://doi.org/10.3201/eid2207.151990.

Polyak, S.J., Morishima, C., Shuhart, M.C., Wang, C.C., Liu, Y., and Lee, D.Y. (2007). Inhibition of T-cell inflammatory cytokines, hepatocyte NF-kappaBsignaling, and HCV infection by standardized Silymarin. *Gastroenterology* 132, 1925–1936.

Polyak, S.J., Morishima, C., Lohmann, V., Pal, S., Lee, D.Y., Liu, Y., et al. (2010). Identification of hepatoprotective flavonolignans from silymarin. *Proceedings of the National Academy of Sciences of the United States of America* 107, 5995–5999.

Ramadan, N., and Shaib, H. (2019). Middle East respiratory syndrome coronavirus (MERS-CoV): A review. *Germs* 9(1), 35–42. https://doi.org/10.18683/germs.2019.1155. PMID: 31119115; PMCID: PMC6446491.

Rathinavel, T., Palanisamy, M.; Srinivasan, P.; Subramanian, A.; Thangaswamy, S Phytochemical 6-Gingerol -A Promising Drug of Choice for COVID-19.; *Int. J. Adv. Sci. Eng.*2020, 06, 1482.

Rechtman, M.M., Har-Noy, O., Bar-Yishay, I., Fishman, S., Adamovich, Y., Shaul, Y., et al. (2010). Curcumin inhibits hepatitis B virus via down-regulation of the metabolic coactivator PGC-1alpha. *FEBS Letters* 584, 2485–2490.

Ren, Y., Shu, T., Wu, D., Mu, J., Wang, C., Huang, M., et al. The ORF3a protein of SARS-CoV-2 induces apoptosis in cells. *Cellular & Molecular Immunolog* 17, 881–883. https://doi.org/10.1038/s41423-020-0485-9.

Rex, J.R.S., Muthukumar, N.M.S.A., and Selvakumar, P.M. (2018). Phytochemicals as a potential source for antimicrobial, anti-oxidant and wound healing-A review. *MOJ Bioorganic & Organic Chemistry* 2 (2), 61–70.

Robinson, M.C. (1955). An epidemic of virus disease in Southern Province, Tanganyika Territory, in 1952-53. I. Clinical features. *Transactions of the Royal Society of Tropical Medicine and Hygiene* 49 (1), 28–32.

Rodrigues, F.N., Lourenço, J., Cerqueira, E.M., Lima, M.M., Pybus, O., Alcantara, L.C. (2016). Epidemiology of chikungunya virus in Bahia, Brazil, 2014-2015. *PLoS Currents* 1 (8), pii: ecurrents.outbreaks.c97507e3e48efb946401755d468c28b2.

Rouf, R., Uddin, S.J., Sarker, D.K., Islam, M.T., Ali, E.S., Shilpi, J.A., et al. (2020). Antiviral Potential of Garlic (Allium Sativum) and its Organosulfur compounds: A systematic update of pre-clinical and clinical data. *Trends in Food Science and Technology* 104, 219–234. https://doi.org/10.1016/j.tifs.2020.08.006.

Ryu, Y.B., Jeong, H.J., Kim, J.H., Kim, Y.M., Park, J.Y., Kim, D., et al. (2010). Biflavonoids from Torreya nucifera displaying SARS-CoV 3CL (pro) inhibition. *Bioorganic & Medicinal Chemistry* 18, 7940–7947.

Salah, T., Awaidy, A., and Khamis, F. (2019). Middle East respiratory syndrome coronavirus (MERS-COV) in Oman: Current situation and going forward. *Oman Medical Journal* 34 (3), 181–183. https://doi.org/10.5001/omj.2019.36.

Salinas, F.M., Vázquez, L., Gentilini, M.V., O´Donohoe, A., Regueira, E., NabaesJodar, M.S., et al. (2019). Aesculus hippocastanum L. seed extract shows virucidal and antiviral activities against respiratory syncytial virus (RSV) and reduces lung inflammation in vivo. *Antiviral Research* 164, 1–11. https://doi.org/10.1016/j.antiviral.2019.01.018.

Sardar, T., Nadim, S.S., Rana, S., and Chattopadhyay, J. (2020). Assessment of lockdown effect in some states and overall India: A predictive mathematical study on COVID-19 outbreak. *Chaos, Solitons & Fractals* 139, 110078. https://doi.org/10.1016/j.chaos.2020.110078.

Schwartz, O. and Albert, M.L. (2010). Biology and pathogenesis of chikungunya virus. *Nature Reviews Microbiology* 8, 491–500.

Shivaprasad, H.N., Kharya, M.D., Rana, A.C., and Mohan, S. (2006). Preliminary immunomodulatory activities of the aqueous extract of Terminaliachebula. *Pharmaceutical Biology* 44 (1), 32–34. https://doi.org/10.1080/13880200500530542.

Sikka, V., Chattu, V.K., Popli, R.K., Galwankar, S.C., Kelkar, D., Sawicki, S.G., et al. (2016). The emergence of Zika virus as a global health security threat: A review and a consensus statement of the INDUSEM joint working group (JWG). *Journal of Global Infectious Diseases* 8 (1), 3–15.

Sindambiwe, J.B., Calomme, M., Cos, P., Totté, J., Pieters, L., Vlietinck, A., et al. (1999). Screening of seven selected Rwandan medicinal plants for antimicrobial and antiviral activities. *Journal of Ethnopharmacology* 65 (1), 71–77. https://doi.org/10.1016/S0378-8741(98)00154-8.

Sodagari, H.R., Aryan, Z., Abdolghaffari, A.H., Rezaei, N., and Sahebkar, A. (2018). Immunomodulatory and anti-inflammatory phytochemicals for the treatment of inflammatory bowel disease (IBD): Turning strong rationale into strong evidence? *Journal of Pharmacopuncture* 21 (4), 294–295. https://doi.org/10.3831/KPI.2018.21.033.

Statista (2022). Number of cases and deaths from ebola virus outbreaks between 1976 and 2020, by country. Statistics on ebola virus in West Africa. Accessed on 18th October 2022, from https://www.statista.com/statistics/328962/ebola-virus-disease-outbreaks -by-country-deaths-case-fatality/

Thiberville, S.-D., Moyen, N., Dupuis-Maguiraga, L., Nougairede, A., Gould, E.A., Roques, P., et al. (2013). Chikungunya fever: epidemiology, clinical syndrome,pathogenesis and therapy. *Antiviral Research* 99(3), 345–370.

Tshilanda, D.D., Ngoyi, E.M., Kabengele, C.N., Matondo, A., Bongo, G.N., Inkoto, C.L., et al. (2020). Ocimum species as potential bioresources against COVID-19: A review of their phytochemistry and antiviral activity. *International Journal of Pathogen Research* 42–54. https://doi.org/10.9734/ijpr/2020/v5i430143.

United Nations Development Programme (UNDP) (2020). Addressing the COVID-19 economic crisis in Asia through social protection, United States of America.

Wahyuni, T., Azmi, D., Permanasari, A., Adianti, M., Tumewu, L., Widiandani, T., et al. (2019). Anti-viral activity of Phyllanthus Niruri against hepatitis C virus. *Malaysian Applied Biology* 48, 105–111.

Wang, K.C., Chang, J.S., Lin, L.T., Chiang, L.C., and Lin, C.C. (2012a). Antiviral effect of cimicifugin from Cimicifuga foetida against human respiratory syncytial virus. *The American Journal of Chinese Medicine* 40, 1033–1045.

Wang, Y., Chen, M., Zhang, J., Zhang, X.L., Huang, X.J., Wu, X., et al. (2012b). Flavone C-glycosides from the leaves of Lophatherum gracile and their in vitro antiviral activity. *Planta Medica* 78, 46–51.

Wang, D., Hu, B., Hu, C., Zhu, F., Liu, X., Zhang, J., et al. (2020). Clinical characteristics of 138 hospitalized patients with 2019 novel coronavirus-infected pneumonia in Wuhan, China. *JAMA* 323, 1061–1069.

Wei, Z.Q., Zhang, Y.H., Ke, C.Z., Chen, H.X., Ren, P., He, Y.L., Hu, P., Ma, D.Q., Luo, J., and Meng, Z.J. (2017). Curcumin inhibits hepatitis B virus infection by down-regulating cccDNA-bound histonw acetylation. *World Journal of Gastroenterology* 23 (34), 6252–6260. https://doi.org/10.374/wjg.v23.i34.6252

World Health Organization (1976). Ebola haemorrhagic fever in Zaire, 1976. Report of an International Commission. *Bull WHO* 56, 271–293.

World Health Organization (1978). Ebola haemorrhagic fever in Sudan, 1976. Report of a World Health Organization International Study Team. *Bull WHO* 1978 (56), 247–270.

World Health Organization (2016). WHO Director-General summarizes the outcome of the Emergency Committee regarding clusters of microcephaly and Guillain-Barré syndrome.

World Health Organization (2020a). Naming the coronavirus disease (COVID-19) and the virus that causes it. Accessed on 18th August 2020, from https://www.who.int/emergencies/diseases/novel-coronavirus-2019/technical-guidance/naming-thecoronavirusdisease-(covid-2019)-and-the-virus-that-causes-it

World Health Organization (2020b). Timeline: WHO's COVID-19 response. Accessed on 23rd September 2020, from https://www.who.int/emergencies/diseases/novel-coronavirus-2019/interactive-timeline

World Health Organization (2022a). Middle East respiratory syndrome coronavirus (MERS-CoV) - Qatar. Accessed on 18th October 2022, from https://www.who.int/emergencies/disease-outbreak-news/item/2022-DON3q70

World Health Organization (2022b). Chikungunya. Accessed on 17th October 2022, from https://www.who.intnews-room/facts-sheets/detail/chikungunya

Younas, M., Hano, C., Giglioli-Guivarc'h, N., and Abbasi, B.H. (2018). Mechanistic evaluation of phytochemicals in breast cancer remedy: Current understanding and future perspectives. *RSC Advances* 8 (52), 29714–29744. https://doi.org/10.1039/c8ra04879g.

Yu, D., Sakurai, Y., Chen, C.H., Chang, F.R., Huang, L., Kashiwada, Y., and Lee, K-H. (2006). Anti-AIDS agents 69. Moronicacid and other Triterpen derivatives as novel potent anti-HIV. *Journal of Medicinal Chemistry* 49(18), 5462–5469. https://doi.org/10.1021/jm0601912.

Yu, M.S., Lee, J., Lee, J.M., Kim, Y., Chin, Y.W., Jee, J.G., Keum, Y.S., and Jeong, Y.J. (2012). Identification of myricetin and scutellarein as novel chemical inhibitors of the SARS coronavirus helicase, nsP13. *Bioorganic & Medicinal Chemistry Letters* 22, 4049–4054.

Zaki, A.M., Van Boheemwn, S., Besteroer, T.M., Osterhaus, A.D., and Foucher, R.A. (2012). Isolation of a novel coronavirus from a man with pneumonia in Saudi Arabia. *The New England Journal of Medicine* 367 (19), 1814–1820. https://doi.org/10.1056/NEJMoa1211721.

Zandi, K., Teoh, B.T., Sam, S.S., Wong, P.F., Mustafa, M.R., and Abubakar, S. (2011). Antiviral activity of four types of bioflavonoid against dengue virus type-2. *Journal of Virology* 8, 560.

Zandi, K., Teoh, B.T., Sam, S.S., Wong, P.F., Mustafa, M.R., and Abubakar, S. (2012). Novel antiviral activity of baicalein against dengue virus. *BMC Complementary and Alternative Medicine* 12, 214.

Zeller, H., Van Bortel, W., and Saudre, B. (2016). Chikungunya: Its history in Africa and Asia and its spread to new regions in 2013-2914. *The Journal of Infectious Diseases* 214 (5), 436–440. https://doi.org/10.1093/infdis/jiw391

Zeng, F.L., Xiang, Y.F., Liang, Z.R., Wang, X., Huang, D.E., Zhu, S.N., et al. (2013). Anti-hepatitis B virus effects of dehydrocheilanthifoline from Corydalis saxicola. *The American Journal of Chinese Medicine* 41, 119–130.

Zhang, J., Tang, Q., Zimmerman-Kordmann, M., Reutter, W., and Fan, H. (2002). Activation of B lymphocytes by GLIS, a bioactive proteoglycan from Ganoderma lucidum. *Life Science* 71 (6), 623–638. https://doi.org/10.1016/S0024-3205(02)01690-9

Zhu, N., Zhang, D., Wang, W., Li, X., Yang, B., Song, J., Zhao, X., Huang, B., Shi, W., Lu, R., Niu, P., Zhan, F., Ma, X., Wang, D., Xu, W., Wu, G., Gao, G. F., Tan, W., and China Novel Coronavirus Investigating and Research Team (2020). A novel coronavirus from patients with pneumonia in China, 19. The New England Journal of Medicine 382, 727–733.

Zumla, A., and Hui, D.S.C. (2019). Emerging and re-emerging infectious diseases. *Infectious Disease Clinics of North America* 33 (4), 869–1158. https://doi.org/10.S0891-5520(19)30071-6.

# 2 Medicinal Plants and Herbs
## *Fortune for Antiviral Drug Discovery*

*M. Bilashini Devi, S.R. Assumi, H.D. Talang, and Vanlalruati*
ICAR Research Complex for NEH Region

*Pooja Bhadrecha*
University Institute of Biotechnology, Chandigarh University

## 2.1 INTRODUCTION

Insufficient therapeutics for efficient antiviral potentials, the emergence of resistant mutants, along with expensive therapies, have made it a need of the hour to find alternative approaches like 'natural antiviral compounds' to combat viral infections. Natural compounds from plants are extensively explored as treasures of natural drugs against several diseases and pathogens. They are potential therapeutic alternatives for various diseases because of being better tolerated innately in the human body. Thousands of plants are documented as 'medicinal plants' and hence utilized in the form of 'novel drugs' (Lemoine et al., 2013). Plants and herbs are studied invariably and well known for their high antioxidant as well as antimicrobial activity. Major phytocompounds exclusively essential in providing therapeutic properties include flavonoids, tannins, phenols, sulfuric compounds, alkaloids, and phenolic compounds (Panpatil et al., 2013; Devi et al., 2012). About 20,000 Indian plants are well evaluated for their medicinal potentials, but not even 1,000 have been procured to extract therapeutic compounds from them.

In recent years, plant-derived phytochemicals as nutraceutical agents have attracted researchers because of their eminent therapeutic power against numerous infectious agents. Similarly, several medicinal herbs and plants are well known to possess antiviral properties, and since they act as immunity boosters, they can be explored for utilization in the drug industry. Hence, in this chapter, some of the potential medicinal plants and herbs are comprehensively highlighted for their curative values against numerous viruses, mentioning the mode of action induced by these compounds. Thirteen such medicinal plants are shortlisted, namely turmeric, ginger, black pepper, black cumin, capsicum, black turmeric, clove, cinnamon, garlic, basil, neem, giloy, and gotu kola. Figure 2.1 depicts the viruses these plants are exclusively effective against, and their essential compounds and mechanisms of action are explained later in the chapter.

## 2.2 TURMERIC (*CURCUMA LONGA* L.)

*Curcuma longa* L., commonly known as 'turmeric', belongs to the family *Zingiberaceae*, is indigenous to India, and is natively cultivated in North Eastern Region of India. Among the turmeric cultivated in India, turmeric from the North Eastern Region has observed an increasing demand from pharmaceutical companies since the farmers of the region cultivate the crop as organic by default through traditional practice. The major bioactive substance, namely curcuminoids, sesquiterpenes, steroids, and polyphenol present in rhizomes, are potential secondary metabolite for exploitation in pharmaceutical industries (Omosa et al., 2017). Three major curcuminoids, namely curcumin,

DOI: 10.1201/9781003329169-3

| | |
|---|---|
| Turmeric | – Dengue, herpes, hepatitis, HIV, SARS Coronavirus, Zika virus |
| Ginger | – Herpes simplex virus, influenza, SARS Coronavirus |
| Black Pepper | – Dengue, Ebola, hepatitis B |
| Black Cumin | – SARS Coronavirus, herpes simplex virus, hepatitis C |
| Capiscum | – Herpes simplex virus, Lissa virus |
| Black Turmeric | – HIV, Epstien-Barr Virus |
| Clove | – Herpes simplex, Influenza A virus |
| Cinnamon | – Human respiratory syntial virus, SARS Coronavirus |
| Garlic | – Rhinovirus, herpes simplex, HIV, Influenza B |
| Basil | – H9N2 virus, SARS Coronavirus |
| Neem | – Dengue, hepatitis B & C, Influenza, polio, SARS Coronavirus |
| Giloy | – HIV, SARS Coronavirus |
| Gotu Kola | – Herpes simplex virus, Pseudorabies virus |

**FIGURE 2.1** Medicinally important plants with antiviral activity against several viruses.

bisdemethoxy curcumin and demethoxy curcumin present in the rhizome primarily attributed to turmeric's therapeutic properties. Curcumin has been the most abundant bioactive curcuminoid in turmeric rhizome, and is known for its anti-inflammatory, antibacterial, antioxidant, antiviral as well as immunomodulatory activity (Aggarwal and Sung, 2009). These curcuminoids have been procured for traditional medicines since a very long time, to treat various diseases and disorders especially in Asian countries. Gupta et al. (2013) conducted clinical trials and reported that consumption of curcumin (12 g/day) is safe for humans. Numerous studies conducted worldwide have documented antiviral potential of curcumin against SARS-CoV, HSV (herpes simplex virus), hepatitis, HIV, dengue and zika virus. The dynamic antiviral activities of curcumin are due to its ability to reduce viral replication, contributing to numerous cellular events such as transcription through regulating various molecular targets and by activating pathways like cellular signaling (Aggarwal and Sung, 2009). Several studies have highlighted curcumin's strong mode of action against SARS Coronavirus by neutralizing the virus through inhibition of virus replication as well as the activity of proteases. Wen et al. (2007) conducted research in 2007 and reported the inhibition of SARS-CoV through curcumin administration in the range of 3–10 μM. They also reported the specific inhibitory effect of curcumin on 3CL protease at IC50 of 40 μM. Similarly, in the case of HSV-1, curcumin was found to reduce its infectivity and IE gene expression (Kutluay et al., 2008). Curcumin is also reported to have potent antiviral activities against Hepatitis B and C virus by targeting and thus, inhibiting various stages of viral cycle, including entry (Hepatitis C), replication and cccDNA stability (Hepatitis B). Curcumin has also been documented for inhibiting protease and integrase enzymes and promotes Tat protein degradation of human immunodeficiency virus. The antiviral effects of curcumin against dengue and zika viruses through inhibition of entry and particle production are also documented. In addition to its potent effect against the above viruses, curcumin also acts against avian influenza virus by interrupting the early stage of infection such as inhibition of viral uptake, replication and production of viral particles. It also acts against zika virus at early stage of infection by inhibiting the virus entry (Singh et al., 2021). Thus, curcumin is a noble candidate as an antiviral drug.

### 2.2.1 Curcumin against SARS Coronavirus

Curcumin has been reported to act against SARS Coronavirus by inhibiting viral replication and protease activity. SARS-associated coronavirus-2 (SARS-CoV-2) is the causative agent of noble respiratory illness and is devastating in nature, with serious complications such as acute pneumonia, failure of several organs, as well as, the acute respiratory distress syndrome (ARDS). Several lines of studies have suggested that curcumin can be considered as an efficient prophylactic and therapeutic candidate against SARS-CoV-2. Curcumin elicits antiviral activity through multiple mechanisms, so it can be promising against several enveloped viruses, including the deadly SARS-CoV-2. It does so by directly interacting with the viral proteins, resulting in protease inhibition, disruption of the viral envelope or inducing the antiviral response in the host. Curcumin can target NF-κB, inflammasome, HMGB1 and IL-6 trans signal pathways, thereby hindering lethal pneumonia and ARDS (Thimmulappa et al., 2021). Thus, there is a potent scope for curcumin in prophylactic and therapeutic treatment of SARS-CoV-2. Molecular docking studies have indicated that curcumin successfully binds to and prohibits SARS-CoV-2 invasion by inhibiting target receptors such as protease, spike glycoprotein-RBD and PD-ACE2 of SARS-CoV-2.

### 2.2.2 Curcumin as a Potential Candidate against Herpes Simplex Virus

Several studies have demonstrated curcumin to be a phytocompound having broader approaches as antiviral agents against other viruses. This has been documented in their study on bioconjugates of curcumin (*viz.*, di-*O*-decanoyl curcumin, di-*O*-pamitoyl curcumin, di-*O*-tryptophanylphenylalanine curcumin, di-*O*-bis-($\gamma,\gamma$) folyl curcumin, 4-*O*-ethyl-*O*-$\gamma$-folyl curcumin and $C^4$-ethyl-*O*-$\gamma$-folyl curcumin) against HSV. This evaluation was conducted via the MTT test, which showed efficient activity against the virus (Moghadamtousi et al., 2014).

### 2.2.3 Curcumin as a Potential Candidate against Human Immunodeficiency Virus

Curcumin is reported as an efficient phytocompound that can effectively suppress gene expression directed by HIV-1 LTR without effecting cell viability. Balasubramanyam et al. (2004) evaluated curcumin's antiviral potential against HIV and reported that curcumin targets the p300/CREB-binding protein (CBP) acetyl transferase. They stated that it deserves to be considered efficient against the HIV virus. Ungphaiboon et al. (2005) reported the reduction of wound infection in HIV patients, as well as complete disappearance of itching, using a 0.5% w/v ethanol extract in the form of liquid soap. They are observed up to 78.6% conversion of abscess to dry scabs within a duration of 2 weeks.

### 2.2.4 Curcumin as a Potential Candidate against Influenza Viruses

Chen et al. (2010) evaluated the antiviral potential offered by curcumin against PR8, H1N1, as well as H6N1 influenza viruses. They observed a reduction of viral yield in cell culture up to 90% when using 30 μM of curcumin.

## 2.3 GINGER (*ZINGIBER OFFICINALE*)

*Zingiber officinale*, which is also referred as ginger, belongs to the family *Zingiberaceae* and India cultivates it on a very large scale. Since it is famously known for its potent and diverse therapeutic properties, it has been used in Ayurveda since time immemorial. Ginger is reported to have antifungal, antimicrobial, anticancer, respiratory protective, immunomodulatory and cardiovascular protective properties, among others. It was used in preparation of several popular plant-derived drugs as traditional medicines for different ailments in ancient times (Bashir and Afrin, 2019). Ginger reportedly has antiviral potential against several viruses (Hayakawa et al., 2011). Ginger's economic part, *i.e.,* the rhizome is well known as a library of potent bioactive phenolic compounds like zingerones, gingerols, paradols, etc. (Anand et al., 2007). Among others, 6-gingerol, a phytophenolic compound, is a promising candidate for antiviral drug. Several lines of evidence have reported ginger's effective antiviral activity, as well as the activity of its bioactive compounds against many viruses.

### 2.3.1 Ginger as a Potential Candidate against SARS Coronavirus

Ahkam et al. (2020) conducted a study on the bioactive phytocompounds of ginger against SARS-CoV-2 and revealed that the compounds such as gingerol, gingerenone A, zingiberene, shogaol, zingiberenol and zingerone act against coronavirus by blocking the spike protein with receptor ACE2 or inhibiting MPro. At the initiation of infection, the spike protein binds with receptor ACE2 from host cell, creating a congenial environment that favors the replication of the virus (Walls et al., 2020). During replication, poly-proteins pp1a and pp1ab are processed by Main Protease (MPro) (Hilgenfeld et al., 2014). Rathinavel et al. (2020) also studied 6- gingerol through an in silico approach against the SARS-CoV-2 proteins, revealing it as an excellent parameter of drug likeliness and ADME pharmacokinetic properties of 6-Gingerol. Their study identified 6-gingerol as a potent and promising antiviral candidate against SARS-CoV-2. This compound exhibited maximum binding affinity as well as exclusive interactions with the spike protein, RNA binding protein and viral proteases.

### 2.3.2 Ginger as a Potential Candidate against Other Viruses

A study on the virucidal effect of ginger oil with an inhibitor concentration (IC50) at 0.004% against HSV-2 revealed a significant reduction in the viral plaque formation via interactions with the viral envelope (Koch et al., 2008). Singh et al. (2021) also reported ginger-derived essential oil against HSV through disruption of the viral envelope. Similarly, ginger's viricidal effect against the human respiratory syncytial virus was studied and it was found that fresh ginger at a concentration of 300 μg/mL, given before viral incubation results in reduction of plaque counts to 19.7% and 27.0% in human upper (A549) and lower (HEp-2) respiratory tract line, respectively. This antiviral effect may result from secretion of IFN-β from mucosal cells, which counteracts viral infection (Chang et al., 2013). Ginger-induced conditioned medium was also reported to possess viricidal effect against influenza A virus. The mode of action involved the activation of macrophage for the production of TNF-α (Singh et al., 2021).

## 2.4 BLACK PEPPER (*PIPER NIGRUM*)

*Piper nigrum*, commonly referred to as "black pepper," belongs to the *Piperaceae* family and is famously termed as the king of spices due to the strong fragrance. Indian states in the northeastern region, like Assam and Arunachal Pradesh, are also documented as a promising belt for black pepper cultivation. Black pepper possesses healing properties because of its bioactive phytocompound piperine (1-peperoyl piperidine). It is commercially used in medicine, perfumery and preservative. This alkaloid is known for its antimicrobial, antioxidant, antidepressant, antihypertensive, anti-Alzheimer's, anti-inflammatory properties (Damanhouri and Ahmad, 2014).

### 2.4.1 Black Pepper as a Potential Candidate against SARS Coronavirus

A recent docking-based study by Rajagopal et al. (2020) reported that compounds like piperdardiin and piperanine in black pepper enable it to be efficiently active against SARS coronavirus.

### 2.4.2 Black Pepper as a Potential Candidate against Other Viruses

Nag and Chowdhury (2020) evaluated the antiviral activity of piperine compared to ribavirin, an antiviral drug, through a molecular docking-based study and revealed that piperine is a promising candidate against Dengue and Ebola viruses. It inhibits methyltransferase in Dengue virus, along with inhibiting VP35 interferon domain in Ebola virus. Similarly, Rajagopal et al. (2020) also reported piperdardiine and piperanine from black pepper as a promising antiviral agents against SARS-CoV-2. Another study showed that the piperine elicits antiviral activity against hepatitis B virus by inhibiting viral surface antigen (HBsAg) and e antigen (HBeAg) secretion from the virus. Srinivasan and Venkatesan (2022), in their study on viricidal efficacy of phytochemicals derived from black pepper against zika virus, revealed that among the 26 bioactive compounds evaluated, piperine and isoscutellarein were found to elicit effective inhibitory effect against the virus compare to antiviral drugs presently used for treatment of zika virus.

## 2.5 BLACK CUMIN (*NIGELLA SATIVA*)

*Nigella sativa*, commonly referred to as 'black cumin', belongs to the *Ranunculacea* family and is also known as kalonji. There is a popular saying attributed to Prophet Muhammad that reads, "In the black cumin, there is a cure for every disease except death." Black cumin has been utilized in traditional medicines since time immemorial because of its therapeutic properties against various ailments such as rheumatic diseases, common cold, headache, warts, etc. (Anand et al., 2007). More recently, black cumin has been reported as a potent herb in the treatment of hypertension, cardiovascular diseases, cancer, diabetes, etc. (Ammayappan and Moses, 2009). The therapeutic properties of black cumin are well presented in traditional medicines such as Ayurveda and Unani (Han and Yang, 2005). Black cumin is a rich source of diverse bioactive phytocompounds, among which 'thymoquinone (TQ)' is one of the most prominent (Varaprasad et al., 2011).

### 2.5.1 Black Cumin as a Potential Candidate against SARS-CoV-2

Black cumin, with its reported antiviral, antidiabetic, antihypertensive, bronchodilatory, anticoagulant and immunomodulatory properties, could be a potential herb for the treatment of SARS-CoV-2 patients with comorbidity (Maideen, 2021). Besides, nigellidine and α-hederin, another bioactive components of black cumin, have been recorded to possess suppressing effects against SARS-CoV-2 (Salim and Noureddine, 2020). The possible efficacy of thymoquinone (TQ) against SARS-CoV-2 was reviewed by Khazdair and Sadegi, who reported extensive to moderate affinity of TQ with A5 shock protein's substrate-binding domain β (Khazdair et al., 2021). Thus, this bioactive compound may be effective in treatment of SARS-CoV-2 patient with comorbidities (Elfiky, 2020). In another study, the efficacy of thymoquinone in binding to the E protein of SARS-Cov-2, as well as its diminishing effects on the ion channel of E protein, were also taken into account for this molecular docking study (Mohideen, 2021).

### 2.5.2 Black Cumin as a Potential Candidate against Other Viruses

Black cumin possesses antimicrobial, antifungal, as well as antiviral potential (Alijabre et al., 2005). In in vitro studies, replication inhibition in hepatitis C virus Alpha-zam was reported in a selective manner using a herbal formulation of *Nigella sativa* seed. In case of HIV virus, it has been reported that the viral load (HIV-RNA) of stage I patients reduced from the average of 19,000 copies/mL to undetectable level and of stage IV patients reduced from the average of 51,000 copies/mL to (Maideen, 2021).

## 2.6 CAPSICUM SPECIES

*Capsicum* is a genus comprising of five cultivated species, which are *annuum*, *baccatum*, *chinense*, *frutescens* and *pubenscens*. The genus belongs to the *Solanaceae* family and is economically important worldwide. Capsicum species contain bioactive compounds like violaxanthin, capsaicinoids, lutein, β-cryptoxanthin and β-carotene (Szolcsanyi, 2004). It also contains capsaicinoids, namely capsaicin, dihydrocapsaicin and nordihydrocapsaicin, which are pungency principles, in *Capsicum annuum*. Capsicum species is reported to elicit antimicrobial activity against Gram-positive as well as Gram-negative bacteria, including *Staphylococcus* species and *E. coli*.

### 2.6.1 Capsaicin as Potent Candidate against SARS-COV-2

Gharaibeh et al. (2020) pointed out a hypothetic relation between SARS-CoV-2 and high capsaicin dietary consumption. According to their study, they correlated the low incidence of SARCoV-2 in various Asian and American countries with their high capsaicin intake in daily cuisine. Their study reveals that capsaicin may have a stimulating effect on immune responses against the virus.

### 2.6.2 Capsaicin as Potent Candidate against Herpes Simplex Virus 1 and 2

Hafiz et al. (2017) conducted a study to evaluate the methalonic extract from *Capsicum annuum* against HSV-1 and HSV-2 for its potential as an antimicrobial agent. They revealed that the extract totally inhibits the HSV-1 growth in vitro. Similarly, the extract inhibited HSV-2 growth in vitro. They also observed low cytotoxicity of the extract, ensuring its potentiality to treat infections caused by herpes virus. Similarly, Carole et al. (2015) examined the antiviral activity of methanolican dethanolic extracts from *Capsicum* species against herpes simplex virus type1. Among the six species of *Capsicum* studied, *Capsicum annuum* and *Capsicum chinense* at the concentration of 0.09 mg/mL (methanol extract) and 0.1 mg/mL (ethanol extract) were the most potential against HSV-1.

### 2.6.3 Capsaicin as Potent Candidate against Lassa Virus

Lassa virus causes severe viral infections, with symptoms manifesting in the form of severe hemorrhagic fever in humans, which can be fatal. According to the World Health Organization (WHO), Lassa virus is categorized as a priority pathogen that requires urgent diagnostic tests, antiviral drugs and vaccines. Tang et al. (2020) reported in their study that capsaicin, a capsaicinoids present in *Capsicum* species, inhibits entry of Lassa virus (LASV) with an $EC_{50}$ of 6.9–10.0 μmol/L employing an HIV-based pseudovirus platform.

## 2.7 BLACK TURMERIC (*CURCUMA CAESIA* ROXB.)

*Curcuma caesia* Roxb., commonly known as "Black turmeric," belongs to the *Zingiberaceae* family. The plants are perennial with a rhizome of bluish-black color which is different from yellow rhizome of common turmeric. Due to its numerous medicinal properties, this species is gradually gaining popularity in the pharmaceutical industry. As discussed in the earlier part of the chapter, turmeric (i.e., *Curcuma longa*) is exclusively utilized by traditional medicines for treating several types of ailments. However, black turmeric (*Curcuma caesia* Roxb.) is a lesser-known species that remains almost untouched and might possess bioactive compounds for herbal drugs. The rhizomes of black turmeric possess antioxidant, antibacterial and antifungal activity, muscle relaxant and CNS depressant activities, among others. The rhizome of black turmeric is reported to contain active phytochemical compound such as curcumin in high concentration. As mention in the earlier part of chapter regarding the efficacy of curcumin and its antiviral activity, black turmeric can also be categorized as a potential plant species for development of antiviral drugs.

### 2.7.1 Black Turmeric as a Potential Candidate against Viruses

Curcumin is the key bioactive phytochemical compound that attributes to the highly acclaimed medicinal properties of genus *Curcuma*. It is a phenolic compound found naturally in different species of genus *Curcuma*. Curcumin consists of one molecule of two 4-hydroxy-3-methoxyphenyl rings tightened via α, β-unsaturated carbonyl groups (Hosny et al., 2011). The efficiency of curcumin in treating diverse viral infections, such as HIV, hepatitis B and C, influenza A, etc., has been extensively studied and proven to be beneficial (Kaur et al., 2010). Thus, black turmeric can be considered as effective medicinal plant for future exploration in modern medicine against several viruses.

## 2.8 CLOVE (*SYZYGIUM AROMATICUM*)

*Syzygium aromaticum*, also referred to as "clove," is a popular spice and from the family *Myrtaceae*. Clove-derived essential oil has been used in traditional medicines for a long time. Clove is employed in the treatment of several illness. Among other compounds, eugenol, a bioactive compound, is known to attribute to the anesthetic, stimulating, antimicrobial, antifungal, antiviral and antiseptic properties of clove.

### 2.8.1 Clove as a Potent Candidate against SARCoV-2

Vicidomini et al. (2021) highlighted clove's potential against the SARS Coronavirus, keeping in view its therapeutic properties for treating respiratory illnesses, various viruses as well as its anti-inflammatory, antithrombotic and immunostimulatory properties. Computational studies on virucidal effect of clove and its phytocompounds on SARS-CoV-2 reported its potent antiviral activities against the virus. Rehman et al. (2020), in their in silico study on kaempferol, a phytocompound extracted from clove, reported that the phytocompound binds SARS-CoV-2 protease with high specificity at active site residues (Cys145 and His41). Their study suggests that clove-derived phytocompounds may prove to be valuable agents in SARS-CoV-2 drugs. Similarly, they also revealed that other clove-derived phytocompounds such as bicornin and biflorin were also recorded to possess inhibitory effects against the virus in molecular docking studies.

### 2.8.2 Clove as a Potent Candidate against Other Viruses

Kurokawa et al. (2001) evaluated the viricidal potential of euginiin, a bioactive compound in clove, reported its antiviral activity against HSV-1. Their evaluation suggested that eugeniin reduced skin lesions development in mice infected with HSV-1. Similarly, Rojas et al. (2014) also reported eugeniin at 5 μg/mL targets the viral DNA synthesis of HSV by inhibiting the DNA polymerase enzyme. They also reported the viricidal potential of clove's aqueous extracts, daikonso, sumac as well as chebulic myrobalan combined with acyclovir against the HSV-1. The ethanolic extract from clove flower buds showed a reduction in plague upon treatment by inhibiting virus replication. Various researchers have examined the inhibitory effect of eugenol, another bioactive compound from clove, against the replication mechanisms of HSV-1 and HSV-2. Eugenol has also been reported for eliciting antiviral efficiency against influenza A and Ebola viruses by inhibiting replication process (Dai et al., 2013). Clove extract has been proven to possess virucidal effect against feline calcivirus (Aboubakr et al., 2016).

## 2.9 CINNAMON (*CINNAMOMUM CASSIA*)

Cinnamon, which belongs to the *Lauraceae* family, is a perennial tree species. It has been prominently utilized in traditional systems of medicine since time immemorial and is also known to possess antiviral property. Cinnamon consists of various bioactive compounds whose antiviral activities are well-documented. The procyanidin type A compound has been reported as an active agent against SARS-CoV-2 (Ghosh, 2020). Aqueous extracts of cinnamon are active against the human respiratory syncytial virus (Yeh et al., 2013). Silver nano particles of cinnamon bark have been reported to be active against Avian influenza virus subtype H7N3 (Goldstein and Shumaker, 2019). Cinnamaldehyde is active against T2 bacteriophage, and cinnamon extracts have also been recorded as active against hepatitis C virus (Ghosh, 2020).

## 2.10 GARLIC (*ALLIUM SATIVUM* L.)

*Allium sativum* L., which is also referred to as "garlic," belongs to the *Alliaceae* family, and is efficiently consumed worldwide as both a spice and functional food. This herb has been extensively employed in traditional medicines as a prevention for variety of ailments since time immemorial. Garlic contains bioactive organosulfur compounds that are known to be potent against viral infections (Rauf et al., 2020). Weber et al. (1992), in their research on VERO and HeLa cell, reported that garlic extracts inhibit entry of HSV 1 and 2 by damaging the viral envelope and cell membrane. Besides, the sulfur-containing compounds of garlic are also known to possess antiviral activities against the same virus. Diallyl disulfide, a garlic-derived bioactive compound, is reported to inhibit proliferation of HIV-1 infected cells. Singh et al. (2021) also documented its antiviral role against HIV through inhibition of integrin dependent process. Garlic-derived essential oil and allicin, another bioactive compound, were also reported to have a significant inhibitory effect against SARCoV-2 virus. Mosbauer et al. (2021) reported that allicin inhibits viral replication in the Vero E6 and Calu-3 lung cells infected by SARCoV-2. Allitridin, a garlic-derived bioactive compound, is reported to suppress gene transcription of human cytomegalovirus in vitro (Zhang et al., 2013). The antiviral properties of garlic have been well explored and documented by several researchers worldwide. Garlic has been reported efficiently active against adenovirus-3 and adenovirus-41 (Khanal et al., 2018), SARS Coronavirus (Mehrbod et al., 2013), dengue virus (Alejandria, 2015), HSV-1 and 2, Human cytomegalovirus (HCMV), Influenza A virus H9N2 and H1N1, Influenza B virus (Straface et al., 2012) and HIV (Wang et al., 2017).

## 2.11 BASIL (*OCIMUM SANCTUM* LINN.)

Basil, a member of the *Lamiaceae* family, is an herb with incredible medicinal properties and has been utilized since time immemorial in traditional medicine. The plant and its extract are potent against viral, fungal and bacterial pathogens. It has been demonstrated to be effective against Coxsackie virus, Bovine viral diarrhea virus, H9N2 virus SARS-CoV-2 and HIV. The antiviral potential of basil is attributed to the novel compounds present in the plant, which have been well studied and documented. Ursolic acid in basil makes it effective against coxsackievirus and enterovirus 71; crude extracts and terpenoid in basil make it efficiently effective against influenza H9N2 virus (Ghoke et al., 2018). Additionally, compounds such as rosmarinic acid, oleanolic SARS-CoV-2acid, ursolic acid and methyl eugenol make basil very effective against SARS Coronavirus (Kumar, 2020).

## 2.12 NEEM (*AZARDIRACHATA INDICA*)

Neem, belonging to the family *Meliaceae*, is extensively explored in traditional systems of medicine. Various studies have reported potent bioactive compounds in neem with significant antiviral activity. The plant and its extract obtained from different parts are reported to be antifungal, antibacterial, antiviral, anti-malarial, antioxidant and anticarcinogenic. Many studies have reported phytocompounds with potential antiviral activity in neem such as NIM-76 against polio virus; azadirachtin against hepatitis B virus (Parvez et al., 2019); terpenoids against SARCoV-2 (Borkotoky and Banerjee, 2020); rutin against the influenza virus (Ahmad et al., 2016); leaf extract (chloroformic) against foot and mouth virus (Younus et al., 2016); aqueous extract of neem against Dengue virus; extract from bark against HSV-1 and 2 (Tiwari et al., 2009).

## 2.13 GILOY (*TINOSPORA CORDIFOLIA*)

*Tinospora cordifolia*, which is referred to as "giloy," is a climbing vine of the *Menispermaceae* family and an eminent resource of several bioactive compounds like alkaloids, polysaccharides and terpenes. These bioactive compounds are capable of eliciting immune-modulatory, antioxidant, anti-inflammatory as well as antidiabetic activities. Giloy is blessed with various compounds of efficient antiviral potential, e.g., tinosporin makes it effective against HIV, HTLV (Singh and Chaudhuri, 2018); silver nanoparticles make it effective against chikungunya virus (Sharma et al., 2019) and tinosponone enables it with the potential of being viricidal against SARS Coronavirus (Krupanidhi et al., 2020).

## 2.14 GOTU KOLA/JALBHRAMI (*CENTELLA ASIATICA*)

*Centella asiatica*, also referred to as "gotu kola," belongs to the *Apiaceae* family and is also known by different names such as 'Jal Bhrami,' 'Indian Pennywort' or 'Mandookaparni' in India. The plant is a perennial herb and is reported to have antibacterial and antifungal activities. Triterpenoids, a bioactive compound from gotu kola is reported to inhibit SARCoV-2, while crude water extracts of mixtures of gotu kola and mango are known to be anti-HSV. Gotu kola possesses several important properties against leprosy, feedant, stress, tuberculosis, wound healing, antibacterial for atherosclerosis and has fungicidal activity also. Crude water extracts of mixtures of *Centella* and mango (*Mangifera indica*) confirmed to be effective against HSV (Yoosook et al., 2000). Triterpenoids from *C. asiatica* successfully inhibits SARS-CoV-2 also (Table 2.1).

**TABLE 2.1**
**Summary of Magnificent Bioactive Compounds from Plants with Antiviral Potential Against Major Viral Diseases**

| Plant | Bioactive Compounds | Potential Against | Mechanisms of Action |
|---|---|---|---|
| Turmeric (*Curcuma longa* L.) | Curcuminoids, Sesquiterpenes, Steroids and Polyphenol | SARS Coronavirus | Disruption of viral envelope, inhibition of protease and spike glycoprotein-RBD |
| | | Herpes Simplex Virus | Inhibition of viral replication |
| | | Human Immunodeficiency Virus | Suppression of gene expression immediately by HIV-1LTR, whereas the cell viability remains intact |
| | | Influenza Virus | Inhibition of protease and integrase |
| | | Zika Virus | Inhibition of viral entry |
| Ginger (*Zingiber officinale*) | Zingerones, Gingerols, Paradols, 6-Gingerol | SARS Coronavirus | Blocking the spike protein with receptor ACE2 or inhibiting main protease |
| | | Herpes Simplex Virus | Disruption of viral envelope |
| | | Human Respiratory Syncytial Virus | Secretion of IFN-β from mucosal cells which counteracts viral infection |
| | | Influenza A Virus | Activation of macrophage for production of TNF-α |

*(Continued)*

**TABLE 2.1 *(Continued)***
**Summary of Magnificent Bioactive Compounds from Plants with Antiviral Potential Against Mmajor Viral Diseases**

| Plant | Bioactive Compounds | Potential Against | Mechanisms of Action |
|---|---|---|---|
| Black pepper (*Piper nigrum*) | Piperine, Piperidine, Piperanine, Isoscutellarein | SARS Coronavirus | Inhibition of viral entry |
| | | Dengue Virus | Inhibition of methyltransferase |
| | | Ebola Virus | Inhibition of VP35 interferon domain |
| | | Zika Virus | Inhibition of RNA polymerase activity |
| Black Cumin (*Nigella sativa*) | Thymoquinone, Nigellidine and A-Hederin | SARS Coronavirus | A5 shock protein's substrate-binding domain β |
| | | Hepatitis C Virus | Inhibition of viral entry |
| | | Human Immunodeficiency Virus | Inhibition of viral replication |
| Capsicum Species | Violaxanthin, Capsaicinoids, Lutein, β-Cryptoxanthin and β-Carotene | SARS Coronavirus | Initiates immune response against virus |
| | | Herpes Simplex Virus 1 and 2 | Inhibition of viral growth |
| | | Lassa Virus | Inhibition of viral entry |
| Black Turmeric (*Curcuma caesia*) | Curcumin | Influenza A | Disruption of viral envelope |
| Clove (*Syzygium aromaticum*) | Eugenol, Kaempferol, Bicornin, Euginiin and Biflorin | SARS Coronavirus | Inhibition of protease |
| | | Herpes Simplex Virus 1 | Inhibition of DNA polymerase |
| | | Influenza A | Inhibition of viral replication |
| | | Ebola Viruses | Inhibition of viral replication |
| Cinnamon (*Cinnamomum cassia*) | Procyanidin Type A | SARS-CoV-2 | Inhibition of viral replication |
| | | Human Respiratory Syncytial Virus | Inhibition of viral entry |
| | | Avian influenza Virus Subtype H7N3 | Inhibition of viral replication |
| Garlic (*Allium sativum* L.) | Organosulfur Compounds, Diallyl Disulphide, Allicin | SARS-CoV-2 | Inhibition of viral replication |
| | | Herpes Simplex Virus 1 and 2 | Disruption of viral envelope |
| | | Human Immunodeficiency Virus | Inhibition of viral proliferation, inhibition of integrin dependent processes |
| Basil (*Ocimum sanctum* Linn.) | Ursolic Acid, Terpenoid, Rosmarinic Acid, Ursolic Acid | SARS-CoV-2 | Inhibition of protease activity |
| | | Human Immunodeficiency Virus | Inhibition of viral replication |
| | | Coxsackie Virus, Bovine Viral Diarrhea Virus, H9N2 Virus | Disruption of viral envelope |
| Neem (*Azardirachata indica)* | Azadirachtin, Terpenoids, Rutin, Chloroformic | SARS-CoV-2 | Regulation of immune response |
| | | Foot and Mouth Virus | Disruption of viral envelope |
| Giloy (*Tinospora cordifolia*) | Alkaloids, Polysaccharides, Terpenes, Tinosponone | SARS-CoV-2 | Disruption of viral envelope |
| | | Chikungunya Virus | Inhibition of protease |
| Gotu Kola/Jalbhrami (*Centella asiatica*) | Triterpenoids | SARS-CoV-2 | Inhibition of viral replication |
| | | Herpes Simplex Virus | Disruption of viral envelope |

## 2.15 CONCLUSION

The major challenges of combating the viral diseases involve narrow therapeutic window, undesirable side effects and the emergence of resistance to existing drugs. In this scenario, plant-derived phytocompounds could serve as a library for screening antiviral properties since the plant-derived drugs have the major advantages of low side effects, easy access and easy adaptability with the human system. Hence, plant-derived compounds could be explored more extensively to tap into their potential in combating viral diseases in the coming times. The utilization of the latest techniques for the exclusive purification and characterization of these essential compounds will ensure a better procurement of these natural gems for human health.

## REFERENCES

Aboubakr, H.A.; Nauertz, A.; Luong, N.T.; Agarwal, S.; El-Sohaimy, S.; Youssef, M.M.; Goyal, S.M. In-vitro antiviral activity of clove and ginger aqueous extracts against feline calicivirus, a surrogate for human norovirus. *Journal of Food Protection*. 2016, 79(6), 1001–1012. DOI:10.4315/0362-028X.JFP-15-593.

Aggarwal, B.B.; Sung, B. Pharmacological basis for the role of curcumin in chronic diseases: an age-old spice with modern targets. *Trends Pharmacological Science*. 2009, 30, 85–94.

Ahkam, A.H.; Hermanto, F.E.; Alamsyahz, A.; Aliyyah, I.H.; Fatchiyah, F. Virtual prediction of antiviral potential of ginger (*Zingiber officinale*) bioactive compounds against spike and MPro of SARS-CoV2 protein. *Journal of Biological Researches*. 2020, 25(2), 52–57.

Ahmad, A.; Javed, M.R.; Rao, A.Q.; Husnain, T. Designing and screening of universal drug from neem (*Azadirachta indica*) and standard drug chemicals against influenza virus nucleoprotein. *BMC Complementary and Alternative Medicine*. 2016, 16, 519. DOI:10.1186/s12906-016-1,469-2.

Alejandria, M.M. Dengue hemorrhagic fever or dengue shock syndrome in children. *BMJ Clinical Evidence*. 2015, 0917, PMID:25860404.

Alijabre, S.H.M.; Randhawa, M.A.; Akhtar, A.; Alkalify, O.M.; Alqurashi, A.M.; Aldossary, A. Anti-dermatophyte activity of ether extract of *Nigella sativa* and its active principle, thymoquinone. *Journal of Ethnopharmacology*. 2005, 101(1–3), 116–119.

Ammayappan, L.; Moses, J.J. Study of antimicrobial activity of aloevera, chitosan, and curcumin on cotton, wool, and rabbit hair. *Fibers and Polymers*. 2009, 10(2), 161–166.

Anand, P.; Kunnumakkara, A.B.; Newman, R.A.; Aggarwal, B.B. Bioavailability of curcumin: problems and promises. *Molecular Pharmaceutics*. 2007, 4(6), 807–818.

Balasubramanyam, K.; Varier, R.A.; Altaf, M. Curcumin, a novel p300/CREB-binding protein-specific inhibitor of acetyl transferase, represses the acetylation of histone/nonhistone proteins and histone acetyltransferase-dependent chromatin transcription. *The Journal of Biological Chemistry*. 2004, 279(49), 51163–51171.

Bashir, F.; Afrin, Z. Zanjabeel (*Zingiber offcinale*) transformation of culinary spice to a multi-functional medicine. *Journal of Drug Delivery and Therapeutics*. 2019, 9, 721–725.

Borkotoky, S.; Banerjee, M. A computational prediction of SARS-CoV-2 structural protein inhibitors from *Azadirachta indica* (Neem). *Journal of Biomolecular Structure and Dynamics*. 2020, 1–17. DOI:10.10 80/07391102.2020.1774419.

Carole, K.A.; Valery, A.E.; Kouadio, Y.; Clement, K.K.; Guillaume, Y.Y.; Rose, K.; Guillaume, L.Y. Cytotoxic and antiviral activity of methanolic, ethanolic and acetate extracts of six varieties of Capsicum. *International Journal of Current Microbiology and Applied Sciences*. 2015, 4, 76–88.

Chang, J.S.; Wang, K.C.; Yeh, C.F.; Shieh, D.E.; Chiang, L.C. Fresh ginger (*Zingiber officinale*) has antiviral activity against human respiratory syncytial virus in human respiratory tract cell lines. *Journal of Ethnopharmacology*. 2013, 145(1), 146–51. DOI:10.1016/j.jep.2012.10.043.

Chen, D.Y.; Shien, J.H.; Tiley, L. Curcumin inhibits influenza virus infection and haemagglutination activity. *Food Chemistry*. 2010, 119(4), 1346–1351.

Dai, J.P.; Zhao, X.F.; Zeng, J.; Wan, Q.Y.; Yang, J.C.; Li, W.Z.; Chen, X.X.; Wang, G.F.; Li, K.S. Drug screening for autophagy inhibitors based on the dissociation of Beclin1-Bcl2 complex using BiFC technique and mechanism of eugenol on anti-influenza A virus activity. *PLoS ONE*. 2013, 8, e61026.

Damanhouri, Z.A.; Ahmad, A. A review on therapeutic potential of *Piper nigrum* L. (black pepper): The king of spices. *Medicial Aromatic* Plants. 2014, 3, 161. DOI:10.4172/2167-0412.1000161.

Devi, A.; Umasankar, M.E.; Babu, S. A comparative study of anti-oxidant properties in common Indian spices. *International Research Journal of Pharmacy (IRJP)*. 2012, 3(5), 465-468.

Elfiky, A.A. Natural products may interfere with SARS-CoV-2 attachment to the host cell. *Journal of Biomolecular Structure and Dynamics*. 2020, 39, 3194–3203.

Gharaibeh, A.; Gharaibeh, M.M.; Gharaibeh, S. Relation between Covid-19 and high capsaicin diets. *Magna Scientia Advanced Research and Reviews*. 2020, 1(1), 30–32.

Ghoke, S.S.; Sood, R.; Kumar, N.; Pateriya, A.K.; Bhatia, S.; Mishra, A.; Singh, V.P. Evaluation of antiviral activity of *Ocimum sanctum* and *Acacia arabica* leaves extracts against H9N2 virus using embryonated chicken egg model. *BMC Complementary and Alternative Medicine*. 2018, 18, 174. DOI:10.1186/s12906-018-2238-1.

Ghosh, D. A Cinnamon derived procyanidin type -A compound: A potential candidate molecule against coronaviruses including COVID-19. *Journal of Ayurveda Case Report*. 2020, 3(4), 122–126.

Goldstein, G.; Shumaker, A.G. Cinnamon extract and cinnamaldehyde inhibit the replication of T2 bacteriophage in *E. coli*: Potential for use in antiviral and anticancer therapy. *Research & Reviews: A Journal of Biotechnology*. 2019, 9(3), 8–17.

Gupta, S.C.; Patchva, S.; Aggarwal, B.B. Therapeutic roles of curcumin: Lessons learned from clinical trials. *The American Association of Pharmaceutical Scientists Journal*. 2013, 15(1), 195–218. DOI:10.1208/s12248-012-9,432-8).

Hafiz, T.; Mubaraki, M.; Dkhil, M.; Quraishy, S. Antiviral activities of *Capsicum annuum* methanolic extract against herpes simplex virus 1 and 2. *Pakistan Journal of Zoology*. 2017, 49(1), 267–272. DOI:10.17582/journal.pjz/2017.49.1.267.272.

Han, S.; Yang, Y. Antimicrobial activity of wool fabric treated with curcumin. *Dyes and Pigments*. 2005, 64(2), 157–161.

Hayakawa, H.; Minanyia, Y.; Ito, K.; Yamamoto, Y.; Fukuda, T. Difference of curcumin content in *Curcuma longa* L., (*Zingiberaceae*) caused by hybridization with other Curcuma species. *American Journal of Plant Sciences*. 2011, 2(2), 111–119.

Hilgenfeld, R. From SARS to MERS: Crystallographic studies on coronaviral proteases enable antiviral drug design. *The FEBS Journal*. 2014, 281(18), 4085–4096. DOI:10.1111/febs.12936

Hosny, I.M.; El Kholy, W.I.; Murad, H.A.; El Dairout R.K. Antimicrobial activity of Curcumin upon pathogenic microorganisms during manufacture and storage of a novel style cheese 'Karishcum'. *Journal of American Science*. 2011, 7, 611–618.

Kaur, S.; Modi, N.H.; Panda, D.; Roy, N. Probing the binding site of curcumin in *Escherichia coli* and *Bacillus subtilis* Fts Z-a structural insight to unveil antibacterial activity of curcumin. *European Journal of Medicinal Chemistry*. 2010, 45(9), 4209–4214.

Khanal, S.; Ghimire, P.; Dhamoon, A.S. The repertoire of adenovirus in human disease: The innocuous to the deadly. *Biomedicines*. 2018, 6(1), 30.

Khazdair, M.R.; Ghafari, S.; Sadeghi, M. Possible therapeutic effects of Nigella sativa and its thymoquinone on COVID-19. *Pharm Biol*. 2021, 59(1), 696–703.

Koch, C.; Reichling, J.; Schneele, J.; Schnitzlera, P. Inhibitory effect of essential oils against herpes simplex virus type 2. *Phytomedicine*. 2008, 15(1–2), 71–78.

Krupanidhi, S.; Peele, K.A.; Venkateswarulu, T.C.; Ayyagari, V.S.; Bobby, M.N.; Babu, D.J.; Aishwarya, G. Screening of phytochemical compounds of Tinospora cordifolia for their inhibitory activity on SARS-CoV-2: An in-silico study. *Journal of Biomolecular Structure and Dynamics*. 2020, 1–5. DOI:10.1080/07391102.2020.1787226.

Kumar, A. Molecular docking of natural compounds from Tulsi (*Ocimum sanctum*) and neem (*Azadirachta indica*) against SARS-CoV-2 protein targets. *Research Square*. 2020. DOI:10.21203/rs.3.rs-27151/v1.

Kurokawa, M.; Hozumi, T.; Tsurita, M.; Kadota, S.; Namba, T.; Shiraki, K. Biological characterization of eugeniin as an anti-herpes simplex virus type 1 compound in vitro and in vivo. *Journal of Pharmacology and Experimental Therapeutics*. 2001, 297(1), 372–379.

Kutluay, S.B.; Doroghazi, J.B.; Roemer, M.E.; Triezenberg, S.J. Curcumin inhibits Herpes simplex virus immediate-early gene expression by a mechanism independent of p300/CBP histone acetyltransferase activity. *Virology*. 2008, 373(2), 239–247.

Lemoine, M.; Nayagam, S.; Thursz, M. Viral hepatitis in resource-limited countries and access to antiviral therapies: Current and future challenges. *Future Virology*. 2013, 8(4), 371–380.

Maideen, N. Miracle Herb to Cure HIV-Black Seeds (*Nigella Sativa*): A Review. *International Journal of Medical Reviews*. 2021, 8(3), 116–12.

Mehrbod, P.; Ideris, A.; Amini, E.; Eslami, T.; Bande, F.; Kheiri, M. Assessment of direct immunofluorescence assay in detection of antiviral effect of garlic extract on influenza virus. *African Journal of Microbiology Research*. 2013, 7, 2608–2612.

Moghadamtousi, S.Z.; Kadir, H.A.; Hassandarvish, P.; Tajik, H.; Abubakar, S.; Zandi, K. A review on antibacterial, antiviral, and antifungal activity of curcumin. *BioMed Research International*. 2014, 186864. DOI:10.1155/2014/186864.

Mohideen, A.K.S. Molecular docking analysis of phytochemical thymoquinone as a therapeutic agent on SARS-Cov-2 envelope protein. *Biointerface Research in Applied Chemistry*. 2021, 11, 8389–8401.

Mosbauer, K.; Fritsch, V.N.; Adrian, L.; Bernhardt, J.; Gruhlke, M.; Slusarenko, A.; Niemeyer, D.; Antelmann, H. Allicin inhibits SARS-CoV-2 replication and abrogates the antiviral host response in the Calu-3 proteome. *bioRxiv preprint*. 2021. DOI:10.1101/2021.05.15.444275.

Nag, A.; Chowdhury, R.R. Piperine, an alkaloid of black pepper seeds can effectively inhibit the antiviral enzymes of Dengue and Ebola viruses, an in silico. *Virus Disease*. 2020, 31(3), 308–315.

Omosa, L.K.; Midiwo, J.O.; Kuete, V. "*Curcuma longa*," In *Therapeutic Potential Against Metabolic, Inflammatory, Infectious and Systemic Diseases*, ed. Kuete V. (Cambridge, MA: Academic Press), pp. 425–435, 2017.

Panpatil, V.V.; Tattari, S.; Kota, N.; Nimgulkar, C.; Polasa, K. In-vitro evaluation on antioxidant and antimicrobial activity of spice extracts of ginger, turmeric and garlic. *Journal of Pharmacognosy and Phytochemistry*. 2013, 2, 143–148.

Parvez, M.K.; Rehman, M.T.; Alam, P.; Al-Dosari, M.S.; Alqasoumi, S.I.; Alajmi, M.F. Plant-derived antiviral drugs as novel hepatitis B virus inhibitors: Cell culture and molecular docking study. *Saudi Pharmaceutical Journal*. 2019, 27(3), 389–400.

Rajagopal, K.; Byran, G.; Jupudi, S.; Vadivelan, R. Activity of phytochemical constituents of black pepper, ginger, and garlic against coronavirus (COVID-19): An in-silico approach. *International Journal of Health and Allied Sciences*. 2020, 9, 43–50.

Rathinavel, T.; Palanisamy, M.; Palanisamy, S.; Subramanian, A.; Thangaswamy, S. Phytochemical 6-Gingerol - A promising drug of choice for COVID-19. *International Journal of Advanced Science and Engineering*. 2020, 6(4), 1482–1489.

Rauf, A.; Abu-Izneid, T.; Thiruvengadam, M.; Imran, M.; Olatunde, A.; Shariati, M.A.; Bawazeer, S.; Naz, S.; Shirooie, S.; Sanches-Silva, A.; Farooq, U.; Kazhybayeva, G. Garlic (Allium sativum L.): Its Chemistry, Nutritional Composition, Toxicity, and Anticancer Properties. *Curr Top Med Chem*. 2022, 22(11), 957–972. DOI: 10.2174/1568026621666211105094939.

Rehman, M.; Al-Ajmi, M.F.; Hussain, A. Natural compounds as inhibitors of SARS-CoV-2 main protease (3CLpro): A molecular docking and simulation approach to combat COVID-19. *Current Pharmaceutical Design*. 2020, 27(33), 3577–3589.

Rojas, D.F.; Souza, C.R.; Oliveira, W.P. Clove (*Syzygium aromaticum*): A precious spice. *Asian Pacific Journal of Tropical Biomedicine*. 2014, 4(2), 90–96.

Salim, B.; Noureddine, M. Identification of compounds from *Nigella Sativa* as new potential inhibitors of 2019 Novel Coronasvirus (Covid-19): Molecular Docking study. *ChemRxiv*. 2020. DOI:10.26434/chemrxiv.12055716.v1.

Sharma, V.; Kaushik, S.; Pandit, P.; Dhull, D.; Yadav, J.P.; Kaushik, S. Green synthesis of silver nanoparticles from medicinal plants and evaluation of their antiviral potential against chikungunya virus. *Applied Microbiology and Biotechnology*. 2019, 103(2), 881–891.

Singh, D.; Chaudhuri, P.K. A review on phytochemical and pharmacological properties of Holy basil (*Ocimum sanctum* L.). *Industrial Crops and Products*. 2018, 118, 367–382.

Singh, N.A.; Kumar, P., Jyoti, Kumar, N. Spices and herbs: Potential antiviral preventives and immunity boosters during COVID-19. *Phytotherapy Research*. 2021, 1–13. DOI:10.1002/ptr.7019.

Srinivasan, D.; Venkatesan, M. Activity of phytochemical constituents of black pepper and rosemary against Zika Virus: An In silico Approach. *Letters in Applied Nano Bio Science*. 2022, 11(2), 3393–3404. DOI:10.33263/LIANBS112.33933404.

Straface, G.; Selmin, A.; Zanardo, V.; De Santis, M.; Ercoli, A.; Scambia, G. Herpes simplex virus infection in pregnancy. *Infectious Diseases in Obstetrics and Gynecology*. 2012, 2012, 385697. DOI:10.1155/2012/385697.

Szolcsanyi, J. Forty years in capsaicin research for sensory pharmacology and physiology. *Neuropeptides*. 2004, 38, 377–384.

Tang, K.; Zhanga, X.; Guo, Y. Identification of the dietary supplement capsaicin as an inhibitor of Lassa virus entry. *Acta Pharmaceutica Sinica B*. 2020, 10(5), 789–798.

Thimmulappa, R.K.; Mudnakudu-Nagaraju, K.K.; Shivamallu, C.; Subramaniam, K.J.T.; Radhakrishnan, A.; Bhojraj, S.; Kuppusamy, G. Antiviral and immunomodulatory activity of curcumin: A case for prophylactic therapy for COVID-19. *Heliyon. Science Direct.* 2021, 7(2), e06350.

Tiwari, V.; Darmani, N.A.; Yue, B.Y.J.T.; Shukla, D. In-vitro antiviral activity of neem (*Azardirachta indica* L.) bark extract against herpes simplex virus type-1 infection. *Phytotherapy Research.* 2009, 24(8), 1132–1140.

Ungphaiboon, S.; Supavita, T.; Singchangchai, P.; Sungkarak, S.; Rattanasuwan, P.; Itharat, A. Study on antioxidant and antimicrobial activities of turmeric clear liquid soap for wound treatment of HIV patients. *Songklanakarin Journal of Science and Technology.* 2005, 27(2), 269–578.

Varaprasad, K.; Vimala, K.; Ravindra, S.; Reddy, N.N.; Reddy, V.S.G.; Raju, M.K. Fabrication of silver nanocomposite films impregnated with curcumin for superior antibacterial applications. *Journal of Materials Science: Materials in Medicine.* 2011, 22(8), 1863–1872.

Vicidomini, C.; Roviello, V.; Roviello, G.N. Molecular basis of the therapeutical potential of *Clove Syzygiumaromaticum* L. and clues to its Anti-Covid 19 *Utility. Molecules.* 2021, 26(7), 1880.

Walls, A.C.; Park, Y.J.; Tortorici, M.A.; Wall, A.; McGuire, A.T.; Veesler, D. Structure, function and antigenicity of the SARS-CoV-2 spike glycoprotein. *Cell.* 2020, 181(2), 281–292.

Wang, L.; Jiao, H.; Zhao, J.; Wang, X.; Sun, S.; Lin, H. Allicin alleviates reticuloendotheliosis virus-induced immunosuppression via ERK/mitogen-activated protein kinase pathway in specific pathogen-free chickens. *Frontiers in Immunology.* 2017, 8, 1856, 1856.

Weber, N.D.; Andersen, D.O.; North, J.A.; Murray, B.K.; Lawson, L.D.; Hughes, B.G. In Vitro virucidal effects of *Allium sativum* (Garlic) extract and compounds. *Planta Medicine.* 1992, 58(5), 417–423.

Wen, C.C.; Kuo, Y.H.; Jan, J.T.; Liang, P.; et al. Specific plant terpenoids and lignoids possess potent antiviral activities against severe acute respiratory syndrome coronavirus. *Journal of Medicinal Chemistry.* 2007, 50, 4087–4095. DOI:10.1021/jm070295s.

Yeh, C.F.; Chang, J.S.; Wang, K.C.; Shieh, D.E.; Chiang, L.C. Water extract of *Cinnamomum cassia* Blume. inhibited human respiratory syncytial virus by preventing viral attachment, internalization, and syncytium formation. *Journal of Ethnopharmacology.* 2013, 147(2), 321–326.

Yoosook, C.; Bunyapraphatsara, N.; Boonyakiat, Y.; Kantasuk, C. Anti-herpes simplex virus activities of crude water extract of Thai Medicinal Plants. *Phytomedicine: International Journal of Phytotherapy and Phytopharmacology.* 2000, 6(6), 411–419.

Younus, I.; Siddiq, A.; Ishaq, H.; Anwer, L.; Badar, S.; Ashraf, M. Evaluation of antiviral activity of plant extracts against foot and mouth disease virus in vitro. *Pakistan Journal of Pharmaceutical Sciences.* 2016, 29(4), 1263–1268.

Zhang, J.; Wang, H.; Xiang, Z.; Shu, S.; Fang, F. Allitridin inhibits human cytomegalovirus replication in vitro. *Molecular Medicine Report.* 2013, 7(4), 1343–1349. DOI:10.3892/mmr.2013.1328.

# 3 Endophytic Secondary Metabolites, a Novel Source for Antiviral

*Monika Thakur and Swati Sharma*
Shoolini Institute of Life Science and Business Management

## 3.1 INTRODUCTION

Microorganisms known as endophytes inhabit the interior tissues of plants and do not produce any obvious symptoms (Stone et al., 2000). The word "endophyte" was coined by De Bary in 1866 and was first used to describe any organism found inside a plant that produces asymptomatic infections entirely inside plant tissues without any signs of disease (De Bary, 1866; Wilson, 1995). According to Wilson (1995), endophytes are fungi that live inside and are asymptomatic for at least a portion of their lifespan. Wilson also discussed the symptomless occupancy of plant tissue by endophytes and their symbiotic and mutualistic interactions with their hosts. However, they can also act as highly infectious or invasive saprophytes. Endophytes are plant-associated prokaryotes that invade their host plants' interior tissues and form associations with them, which has made them useful for agribusiness as a technique to enhance crop performance (Azevedo et al., 2000).

Later, this term was extended to include bacteria and fungi, especially actinomycetes, which spend the entire or a portion of their life cycle growing inside the healthy, living tissues of the host, sometimes without displaying any external signs of disease. The Greek words "endon" (meaning within) and "phyton" (meaning plant) are the origins of the term "endophyte." By definition, an endophytic fungus spends at least significant time in mycelial form in biological interaction with a living plant. As a result, the presence of a fungus' hyphae in living tissue should serve as the prerequisite for classifying it as an endophyte (Kaul et al., 2012). Complex microbial communities inhabit crops, some of which are harmful and spread diseases, while others support plant growth, facilitate nutrient uptake, and improve plant tolerance to biotic and abiotic stress factors through a variety of mechanisms (De Bary, 1866). Endophytes are fungus or bacteria that develop inside the tissue of plants without harming the host (Hirch and Broun, 1992). The majority of plant species present in uncontrolled and controlled environments form associations with them. Compared to plants without this symbiosis, endophytes are regarded as significant plant partners that play a significant role in strengthening the host's ability to withstand stress (Wilson and Roberts, 2012).

### 3.1.1 Role of Endophytes

Bary reported the existence of endophytes in the 19th century (Azevedo, 1998), and since then, they have been described in various ways. Due to the interactions between endophytes and plants, these microbes, known as endophytes, develop within host tissue and exhibit a wide range of variations. The diversity of endophytes may be attributed to factors such as endophytic behavior, endophyte associations with plants, endophytic colonization, invasion and endophytic associations with other microorganisms including bacteria, fungus, actinomycetes, yeast, and algae (Kennedy et al., 2020; Kaushik and Coloma, 2020).

DOI: 10.1201/9781003329169-4

Currently, endophytes are employed to produce critical agricultural, industrial, pharmaceutical, chemical and medicinal bioactive compounds that sustain human life. High-value pharmaceutical substances found in endophytes include taxol, podophyllotoxin, camptothecin, vinblastine, and vincristine, among many others. Endophytes serve as cost-effective sources of bioactive constituents compared to plants, which are the usual method but uncommon and uneconomical (Kumaran et al., 2010; Kumar et al., 2013). It is well-known that plants harboring endophytes contain compounds that contribute to disease resistance. Additionally, endophytes produce secondary metabolites such as antibiotics and lytic enzymes that confer resistance against many pathogens. Endophytes play a crucial role in agricultural pest and insect management (Ahmed et al., 2012; Dos Santos et al., 2003; Arnold et al., 2003). Plants containing endophytes have been found to adapt and tolerate abiotic conditions, including heat, salt, and drought more readily than non-symbiotic plants (Jalgaonwala et al., 2011). The abiotic stress tolerance (drought, salinity, high temperature, high $CO_2$, and metal toxicity) and biotic stress tolerance (against pathogens and pests) conferred by fungal endophytic colonization on plants are increased through higher photosynthesis rates, promoted shoot and root growth, improved uptake and nutrient use efficiency, and increased tolerance to these stresses (Harman and Uphoff, 2019; Grabka et al., 2022; Xu et al., 2022). Another area that needs attention is the lack of comprehensive data regarding the performance of the inserted endophyte in the field and the alterations in functioning brought on by climate changes. Lack of knowledge of both general and specific plant-endophyte interactions, physio-chemical barriers that prevent the stimulation of the plant immune response, the pathogen-plant immune defense system, gene action, and signaling pathways in agriculture are some potential obstacles to large-scale commercial incorporation of endophytes (Ma et al., 2016; Ma et al., 2019). Due to their long association with the hosts, the fungal endophytes typically exhibit a close interaction with them and thus tend to create secondary metabolites resembling those of the hosts (Jayaram et al., 2021).

Over 200 secondary metabolites from endophytic fungi have been identified thus far. Secondary metabolites called terpenoids are released as essential protective mechanisms under abiotic and biotic stress conditions. They are primarily known to interfere with the respiratory route to produce antibacterial action (Alibi et al., 2021). Studies have suggested the absence of flavonoids and saponins, whereas the presence of terpenoids, coumarins, and cardiac glycosides has been detected (Maliehe et al., 2022). In other words, new bioactive chemicals with application potential in various commercial sectors, including the food, agricultural, and pharmaceutical industries, are stored in endophytic bacteria. Several fungal endophytes produce bioactive substances, some of which are antibiotics with antifungal, antibacterial, and insecticidal activities strictly limiting the development of other microbes, including plant diseases (Gunatilaka, 2006). Endophytic fungi with many forms of bioactivity, such as antagonistic, antioxidant, and anti-mutagenic properties, were abundant in Thai mangrove trees. Studies have suggested that ethyl acetate extract showed the presence of several extracellular enzymes, such as lipase, cellulase, protease, and pectinase (Sopalun, et al., 2021; Khiralla et al., 2015). Microorganisms living in relationship with plant hosts contribute significantly to medicinal potential, surpassing that of actual plants (Subbulakshmi et al., 2012; Pimentel et al., 2011; Singh and Dubey, 2015; Nair and Padmavathy, 2014). Enzymes and pharmaceuticals are some of the bioactive substances produced by endophytes, in addition to phytohormones (Joseph and Priya, 2011; Parthasarathi et al., 2012). Each plant that has been researched so far has shown to include one or more forms of endophytes, according to researchers (Strobel and Daisy, 2003).

## 3.2 PLANT-ENDOPHYTE INTERACTIONS

### 3.2.1 Endophytic Fungi from Medicinal Plants

A study conducted in the Iran revealed the presence of L-asparaginase, which is an important anticancer enzyme. The findings of the study suggested that endophytic fungus *Fusarium proliferatum*, obtained from medicinal plants of the Asteraceae family, is found to be potent for the production of anticancer enzymes (Hatamzadeh et al., 2020). In Brazil, a medicinal plant called *Indigofera suffruticosa* Miller, traditionally used in medicine to cure various illnesses, was found to host the endophytic fungus *N. sphaerica* in healthy leaves. This fungus exhibited the ability to produce bioactive agents with pharmaceutical potential (Santos et al., 2015). The endophytic fungus *N. sphaerica*, isolated from *Vinca rosea* leaves, generated the steroid ergosta-7, 9 (14), 22-triene-3-ol, which demonstrated antifungal activity against *Cryptococcus* (Metwaly et al., 2014). Endophytic fungi present in the leaves of the medicinal plants such as *Acalypha ornata*, *Albizia zygia*, *Alchornea cordifolia*, *Chrysophyllum albidum*, *Ficus exasperata*, *Gomphrena celosioides*, *Millettia thonningii*, and *Newbouldia laevis* showed the presence of phenols, tocopherols, sterols, terpenoids, and fatty acids, which exhibited antioxidant activity (Ibrahim et al., 2021). Fungal endophytes, including *Phoma* sp., *Alternaria* sp., *Bipolaris* sp., and *Cladosporium* sp., were isolated and identified from the medicinal plants like *Caralluma acutangula*, *Rhazya stricta*, and *Moringa peregrine*. These endophytes exhibited significant free radical scavenging and anti-lipid peroxidation activity. They also contributed to the production of Indole acetic acid, which is helpful in improving seed germination and mitigating oxidative stress (Khan et al., 2017). Another study suggested that endophytic fungi *Talaromyces* sp. isolated from medicinal plants contributed to the antioxidant and anti-inflammatory activity (Santos et al., 2020). Egyptian medicinal plants associated endophytic fungi *Alternaria alternate* and *Humicola grisea*, isolated from *Salvia aegyptiaca*, were found to be effective sources of bioactive constituents and contributed to antibacterial, antioxidant, and anticancer potential (El-Bondkly et al., 2020).

## 3.3 BIO APPLICATIONS OF ENDOPHYTIC MEDICINAL PLANT COMPOUNDS FOR ANTIVIRAL THERAPIES

Viruses are responsible for causing serious outbreaks in all continents, leading to high rate of mortality and economic burden on society. Due to their high mutation rate and emergence of new serotypes, combating these viruses is a significant challenge. Viruses can be categorized as enveloped and non-enveloped. Enveloped viruses is made up of lipids which is less stable after the treatment with lipids solvents and in case of non-enveloped their genome is covered within the protein coat, making them more stable. Thus, epidemics caused by non-enveloped viruses like enteroviruses and noroviruses are challenging to control. They also don't exhibit much susceptibility to chemical disinfectants (Wutzler and Sauerbrei, 2004; Chan and Bakar, 2005). It is evident that viruses, especially RNA viruses, play a dominant role on our planet. RNA viruses are one of over a billion viruses existing globally. The World Health Organization (WHO) has listed few viruses as global health threats, including human immunodeficiency virus (HIV), Ebola virus (EBOV), and other highly dangerous viruses causing illness and hemorrhagic fevers. These threats also include Middle Eastern Respiratory Syndrome (MERS) caused by Zika virus (ZIKV), Nipah virus, and Severe Acute Respiratory Syndrome Coronavirus (SARS-CoV), Coronavirus (COVID-19), which caused pandemic in 2019 (Adamson et al., 2021). As a result, the development of both vaccines and antivirals is necessary to treat viral infections. Although just a few viruses are now treated by vaccines, it is challenging to produce immunizations against a wide range of newly emerging viral serotypes. Additionally, vaccines are ineffective if the virus has already infected the body.

## 3.4 ROLE OF ENDOPHYTIC METABOLITES/COMPOUNDS AGAINST ALL TYPES OF VIRAL DISORDERS

So far, only a few plants have been examined for their endophytic diversity and their ability to produce bioactive compounds (Table 3.1) (Chanda et al., 2013). When investigated on virus-infected animal models, many natural and synthetic medicines with antiviral activity were significantly less effective (Martinez et al., 2015; Takizawa and Yamasaki, 2018; Mukherjee, 2019). Additionally,

**TABLE 3.1**
**Antiviral Activity of Bioactive Compounds Produced by Fungal Endophytes and Their Hosts**

| S. No. | Fungal Endophyte | Host | Compound | Antiviral Activity of Compound | References |
|---|---|---|---|---|---|
| | *Microthyriaceae* | Paspalum conjugatum | Sterigmatocystin | Moderate antiviral activity by the inhibition of HSV-2 | Han et al. (2020) |
| | | | | Significant anti-HIV activities | Chen et al. (2015) |
| | *Dothiorella* sp. | *Avicennia marina* | Sterigmatocystin | IAV infection and to protect from excessive inflammatory lung injuries following infection | Egarnes et al. (2017) |
| | *Entrophospora infrequens* | *Nothapodytes foetida* | | Camptothecin had a protective effect against FAdV-4, indicating its potential as an antiviral drug against FAdV-4 (Fowl adenovirus serotype 4 infection). | Yin et al. (2022) |
| | | | Cytosporone B | Against SARS-CoV-2, by potentially blocking the interaction of the spike glycoprotein with the angiotensin-converting enzyme 2 (ACE2) receptor found on host cells | Devasia et al. (2021) |
| | *Cephalosporium* sp. | *Paris polyphylla var. yunnanensis* | Camptothecin | Antiviral activity *in vitro* against hepatitis C virus | Wang et al. (2011) |
| | *Alternaria* sp. | *Sinopodophyllum hexandrum* | | Podophyllotoxin and its derivatives were examined for inhibitory effects on the replication of herpes simplex virus types 1 and 2 (HSV-1 and HSV-2) | Sudo et al. (1998) |
| | *Eupenicillium parvum* | *Azadirachta indica, Azadirachta juss* | Diosgenin | In silico molecular docking study suggested that Azadirachtin A could be a potential inhibitor of SARS-CoV-2 main protease | Muhammed et al. (2021) |
| | *Aspergillus niger* | Wine grape Carbernet Sauvignon | Podophyllotoxin | • Influenza Virus<br>• Epstein-Barr Virus<br>• Herpes Simplex Virus<br>• Respiratory Syncytial Virus<br>• Human Immunodeficiency Virus<br>• Varicella Zoster and Enterovirus 71 | Palamara et al. (2005); Leo et al. (2012); Espinoza et al. (2012); Docherty et al. (1999); Docherty et al. (2004); Docherty et al. (2005); Xie et al. (2012); Clouser et al. (2012); Zhang et al. (2015); and Docherty et al. (2006) |
| | *Alternaria* | *Catharanthus roseus* | Azadirachtin A | Standard chemotherapy regimen for Hodgkin's lymphoma (HL), information on the results of this therapy in human immunodeficiency (HIV)-related HL is scarce | Xicoy et al. (2007) |

*(Continued)*

**TABLE 3.1** (*Continued*)
**Antiviral Activity of Bioactive Compounds Produced by Fungal Endophytes and Their Hosts**

| S. No. | Fungal Endophyte | Host | Compound | Antiviral Activity of Compound | References |
|---|---|---|---|---|---|
| | *Fusarium solani* | *Camptotheca accuminata, Apodytes dimidiata* | Resveratrol | *In vitro* antiviral efficacies of series I–IV of representative members from camptothecin analogs on herpes simplex virus type 2 (HSV-2) | |
| | *Entrophospora infrequens* | *Nothapodytes foetida* | Vinblastine | | Liu et al. (2010) |
| | *Neurospora crassa* | *Camptothecaacuminata* | Camptothecin | | |
| | *Fusarium solani* | *Camptothecaacuminata* | Camptothecin 2 | | |
| | *Fusarium oxysporum* | *Ephedra fasciculata* | Camptothecin 2 | Potent inhibitors of HIV-1 integrase. Antiviral activity | Shin et al. (2009) |
| | *Chaetomium chiversii* | *Ephedra fasciculata* | Camptothecin 2 | Radicicol, a heat shock protein 90 (Hsp90) inhibitor that effectively suppressed CHIKV replication by blocking the synthesis of both positive- and negative-strand viral RNA as well as expression of viral proteins | Nam et al. (2021) |
| | *Phialocephala fortinii* | *Podophyllum peltatum* | Beauvericin 29 | Inhibitory effects on the replication of herpes simplex virus types 1 and 2 (HSV-1 and HSV-2) | Sudo et al. (1998) |
| | *Alternaria* sp. | *Polygonum senegalense* | Radicicol 38 | Inhibition of HCV hepatitis C | Hawas et al. (2015) |
| | *Alternaria* sp. | *Polygonum senegalense* | Podophyllotoxin | | |
| | *Aspergillus parasiticus* | *Sequoia sempervirens* | Alternariol 51 | • Antiviral activity against influenza A virus<br>• (H1N1) | Zhng et al. (2011) |
| | *Thielavia subthermophila* | *Hypericum perforatum* | Altenusin 54 | • Toxicity toward certain viruses notably the class of enveloped viruses that includes human immunodeficiency virus (HIV) | Miskovsky (2002) |
| | *Thielavia subthermophila* | *Hypericum perforatum* | Sequoiamonascin B 65 | • Emodin significant activity against human coronavirus NL63 (HCoV-NL63)<br>• Coxsackievirus group B (CVB) and respiratory syncytial virus (RSV)<br>• Emodin blocks the SARS coronavirus spike protein and angiotensin-converting enzyme 2 interaction | Horvat et al. (2021); Liu et al. (2015); and Ho et al. (2007) |

there are challenges with safety and cost associated with the extraction of natural compounds from plants and the chemical production of synthetic drugs. Moreover, conventional medications fail to treat viral infections, and the emergence of certain viral resistances to these medications is a common occurrence (Linnakoski et al., 2018; Mulwa et al., 2018; Ma et al., 2020). Therefore, researchers need to explore alternate sources of natural antiviral medicines that are both effective and affordable. Viral diseases pose a major threat worldwide, with an increasing number of deaths annually. The rising mortality rate presents a global challenge to explore newer sources of antiviral drugs. There is a growing interest in discovering natural sources for the development of antiviral agents. These natural sources offer the potential of finding new structural classes with different bioactivities for antiviral therapy. Endophytic fungi can be considered as a rich source of bioactive metabolites that can be further processed to produce promising analogs for antiviral therapy. To achieve this target of developing new antiviral drugs, strategies for the isolation, characterization, and improvement of fungal strains must be developed to understand their mechanisms of action. Literature studies have suggested that endophytic fungal metabolites have the potential to serve as novel sources of antiviral agents when tested against various viruses.

Recent epidemics of a novel coronavirus (COVID-19) has significantly impacted public health, causing respiratory infection. Due to the lack of a suitable treatment, the emergence of drug resistance, and the recurrence of viral infections, medicinal herbs employed in practices provide significant potential to develop effective alternatives against viral diseases. Concerns regarding the use of plants and associated phytochemicals, especially in the context of COVID-19, as treatments for viral illnesses has increased. One well-known Ascomycetes filamentous fungus is *Aspergillus*. These fungi are extremely aerobic and can flourish in oxygen-rich conditions, and many of them can also grow in conditions deprived of essential nutrients (Lee et al., 2013). Endophytic *Aspergillus* species have been recognized in several publications as renewable, endless sources of bioactive secondary metabolites, including sterols, alkaloids, anthraquinone glycosides, and cytochalasins, all of which hold significance for the pharmaceutical and commercial industries (Wang et al., 2018). The endophytic fungus *Aspergillus terreus* is reported to have distinctive metabolites, such as itaconic acid, lovastatin, and the antibiotics sulochrin and terrain (Elkhayat et al., 2016). Additionally, *Aspergillus terreus* has been found to produce novel metabolites butenolides, anthraquinones, and quinones, which possess potential antibacterial and antiviral properties (Gao et al., 2013; Lu et al., 2017; Olesen et al., 2014).

## 3.5 CURRENT CHALLENGES OF EXPLOITING ENDOPHYTIC FUNGI

Fungi play a significant role in natural and agricultural ecosystems by developing interactions with plants, which can range from harmful to beneficial (mutualistic) associations. These relationships can have profound impacts on the environment, agriculture, and ultimately the economy. Fungi may, in fact, influence the composition of plant communities by suppressing certain species and favoring others (Bennett and Cahill, 2016). Endophytic fungi are present universally in close relationship with plant living tissues. While humans have been exploiting various plant sources for the development of herbal medicines since ancient times, only a limited number of microbial populations associated with plants have been explored (Jackson et al., 2015). Researchers from various fields are increasingly focused on understanding the roles and mechanisms of the fungal endophytes. They suggest that in early phases of infestation, plant parts are symptomless without showing any disease symptoms (Kaul et al., 2016; Marsberg et al., 2017; White et al., 2019). However, histological studies can later be conducted to visualize the internal microbial associations (Tadych et al., 2019). Molecular techniques have provided insights into the characterization and identification of endophytes (Emerson et al., 2008). Endophytes have, in fact, been widely thought of as a likely source of biocontrol agents (White et al., 2019), but there is a lack of evidence on the processes behind these interactions between plants and microbes, and the potential of these endophytic "plant pathogens" as biocontrol agents remains largely unexplored. The understanding of many molecular, physiological, and ecological aspects of the interactions between plants and beneficial microbes has been studied so far, and a substantial body of research has been produced; this suggests a crucial function to increase tolerance of plants and resistance to the various environmental stresses (Kaul et al., 2016; Balestrini et al., 2021). Although many aspects of these interactions, such as the processes behind the control of plant defense in the presence of beneficial microbes, still require clarification, overall functional potential of plant-associated microbes remains still largely unclear.

## 3.6 CONCLUSIONS: FUTURE CONSIDERATIONS: RESOLVING THE PRESENT CHALLENGES

Although there has been much research on the relationship between endophytes and their hosts for more than a century, many aspects remain unknown and unexplored. Modern molecular biology approaches might provide value-added insights into biodiversity of endophytes metabolism, along with the regulation of fungal metabolism. Additionally, investigating the impact and presence of bacterial or viral endosymbionts on endophyte-host interaction pattern is an exciting and very important field of study. Greater knowledge of the interactions between hosts and endophytes at the molecular and genetic levels, as well as environmental changes and the effects of culture conditions on gene expression, will be further helpful in better understanding the optimization of secondary metabolites production under lab conditions. The optimization of secondary metabolites production conditions can be a turning point in achieving a high yield of bioactive constituents, which are of medicinal and agricultural importance.

## REFERENCES

Adamson CS, Chibale K, Goss RJ, Jaspars M, Newman DJ, Dorrington RA. Antiviral drug discovery: Preparing for the next pandemic. *Chemical Society Reviews*. 2021;50(6):3647–55.

Ahmed M, Hussain M, Dhar MK, Kaul S. Isolation of microbial endophytes from some ethnomedicinal plants of Jammu and Kashmir. *Journal of Natural Product and Plant Resources*. 2012;2(2):215–20.

Alibi S, Ramos-Vivas J, Selma WB, Mansour HB, Boukadida J, Navas J. Virulence of clinically relevant multidrug resistant Corynebacterium striatum strains and their ability to adhere to human epithelial cells and inert surfaces. *Microbial Pathogenesis*. 2021 Jun 1;155:104887.

Arnold AE, Herre EA. Canopy cover and leaf age affect colonization by tropical fungal endophytes: ecological pattern and process in Theobroma cacao (Malvaceae). *Mycologia*. 2003 May 1;95(3):388–98.

Azevedo JL. *Microorganismos as endofiticos*. In: Melo IS and Azavedo JL, editors. *Ecologia Microbiana* 1998 (pp. 117–137). Editora Embrapa, Jaguariuna, Sao Paulo.

Azevedo JL, Maccheroni JW, Pereira JO, Luiz de Araujo W. Endophytic microorganisms: A review on insect control and recent advances in tropical plants. *Electronic Journal of Biotechnology*. 2000; 3(1):40–65.

Balestrini R, Brunetti C, Cammareri M, Caretto S, Cavallaro V, Cominelli E, De Palma M, Docimo T, Giovinazzo G, Grandillo S, Locatelli F. Strategies to modulate specialized metabolism in Mediterranean crops: From molecular aspects to field. *International Journal of Molecular Sciences*. 2021 Mar 12;22(6):2887.

Bennett JA, Cahill Jr JF. Fungal effects on plant-plant interactions contribute to grassland plant abundances: Evidence from the field. *Journal of Ecology*. 2016 May;104(3):755–64.

Chan YF, Bakar A. Virucidal activity of Virkon S on human enterovirus. *The Medical Journal of Malaysia*. 2005 Jun 1;60(2):246–8.

Chanda S, Menpara D, Desai D. Antimicrobial activity of *Terminalia bellerica* leaf and stem collected from two different sites. *American Journal of Phytomedicine and Clinical Therapeutics*. 2013;1(9):721–33.

Chen C, Zhu H, Wang J, Yang J, Li XN, Wang J, Chen K, Wang Y, Luo Z, Yao G, Xue Y. Armochaetoglobins K-R, Anti-HIV pyrrole-based cytochalasans from Chaetomium globosum TW1-1. *European Journal of Organic Chemistry*. 2015 May;2015(14):3086–94.

Clouser CL, Chauhan J, Bess MA, Van Oploo JL, Zhou D, Dimick-Gray S, Mansky LM, Patterson SE. Anti-HIV-1 activity of resveratrol derivatives and synergistic inhibition of HIV-1 by the combination of resveratrol and decitabine. *Bioorganic & Medicinal Chemistry Letters*. 2012 Nov 1;22(21):6642–6.

De Bary A. *Morphologie und physiologie der pilze, flechten und myxomyceten*. Engelmann; 1866.

Devasia RM, Altaf M, Alrefaei AF, Manoharadas S. Enhanced production of camptothecin by immobilized callus of Ophiorrhiza mungos and a bioinformatic insight into its potential antiviral effect against SARS-CoV-2. *Journal of King Saud University-Science*. 2021 Mar 1;33(2):101344.

Docherty JJ, Fu MM, Hah JM, Sweet TJ, Faith SA, Booth T. Effect of resveratrol on herpes simplex virus vaginal infection in the mouse. *Antiviral Research*. 2005 Sep 1;67(3):155–62.

Docherty JJ, Fu MM, Stiffler BS, Limperos RJ, Pokabla CM, DeLucia AL. Resveratrol inhibition of herpes simplex virus replication. *Antiviral Research*. 1999 Oct 1;43(3):145–55.

Docherty JJ, Smith JS, Fu MM, Stoner T, Booth T. Effect of topically applied resveratrol on cutaneous herpes simplex virus infections in hairless mice. *Antiviral Research*. 2004 Jan 1;61(1):19–26.

Docherty JJ, Sweet TJ, Bailey E, Faith SA, Booth T. Resveratrol inhibition of varicella-zoster virus replication in vitro. *Antiviral Research*. 2006 Dec 1;72(3):171–7.

Dos Santos, R. M. G., Rodrigues-Fo, E., Rocha, W. C., & Teixeira, M. F. S. Endophytic fungi from Melia azedarach. *World journal of Microbiology and Biotechnology*. 2003;19(8):767–770.

Egarnes B, Blanchet MR, Gosselin J. Treatment with the NR4A1 agonist cytosporone B controls influenza virus infection and improves pulmonary function in infected mice. *PLoS One*. 2017 Oct 20;12(10):e0186639.

El-Bondkly AA, El-Gendy MM, El-Bondkly EA, Ahmed AM. Biodiversity and biological activity of the fungal microbiota derived from the medicinal plants Salvia aegyptiaca L. and Balanties aegyptiaca L. *Biocatalysis and Agricultural Biotechnology*. 2020 Sep 1;28:101720.

Elkhayat ES, Ibrahim SR, Mohamed GA, Ross SA. Terrenolide S, a new antileishmanial butenolide from the endophytic fungus Aspergillus terreus. *Natural Product Research*. 2016 Apr 2;30(7):814–20.

Emerson D, Agulto L, Liu H, Liu L. Identifying and characterizing bacteria in an era of genomics and proteomics. *BioScience*. 2008 Nov 1;58(10):925–36.

Espinoza JL, Takami A, Trung LQ, Kato S, Nakao S. Resveratrol prevents EBV transformation and inhibits the outgrowth of EBV-immortalized human B cells. *PLoS One*. 2012 Dec 10;7(12):e51306.

Gao H, Guo W, Wang Q, Zhang L, Zhu M, Zhu T, Gu Q, Wang W, Li D. Aspulvinones from a mangrove rhizosphere soil-derived fungus Aspergillus terreus Gwq-48 with anti-influenza A viral (H1N1) activity. *Bioorganic & Medicinal Chemistry Letters*. 2013 Mar 15;23(6):1776–8.

Grabka R, d'Entremont TW, Adams SJ, Walker AK, Tanney JB, Abbasi PA, Ali S. Fungal endophytes and their role in agricultural plant protection against pests and pathogens. *Plants*. 2022 Jan 30;11(3):384.

Gunatilaka AL. Natural products from plant-associated microorganisms: Distribution, structural diversity, bioactivity, and implications of their occurrence. *Journal of Natural Products*. 2006 Mar 24;69(3):509–26.

Han X, Tang X, Luo X, Sun C, Liu K, Zhang Y, Li P, Li G. Isolation and identification of three new sterigmatocystin derivatives from the fungus Aspergillus versicolor guided by molecular networking approach. *Chemistry & Biodiversity*. 2020;17(6):e2000208.

Harman GE, Uphoff N. Advantages and methods of using symbiotic microbes to enhance plant agriculture and the environment. *Scientifica*. 2019 Oct;2019(9106395):9106395.

Hatamzadeh S, Rahnama K, Nasrollahnejad S, Fotouhifar KB, Hemmati K, White JF, Taliei F. Isolation and identification of L-asparaginase-producing endophytic fungi from the Asteraceae family plant species of Iran. *PeerJ*. 2020 Jan 14;8:e8309.

Hawas UW, El-Desouky S, El-Kassem A, Elkhateeb W. Alternariol derivatives from Alternaria alternata, an endophytic fungus residing in red sea soft coral, inhibit HCV NS3/4A protease. *Applied Biochemistry and Microbiology*. 2015 Sep;51(5):579–84.

Hirsch G, Braun U. Communities of parasitic microfungi. In: Winterhoff W, editor. *Handbook of vegetation science* 1992 (pp. 225–250). Kluwer Academic Publisher.

Ho TY, Wu SL, Chen JC, Li CC, Hsiang CY. Emodin blocks the SARS coronavirus spike protein and angiotensin-converting enzyme 2 interaction. *Antiviral Research*. 2007 May 1;74(2):92–101.

Horvat M, Avbelj M, Durán-Alonso MB, Banjanac M, Petković H, Iskra J. Antiviral activities of halogenated emodin derivatives against human coronavirus NL63. *Molecules*. 2021 Nov 11;26(22):6825.

Ibrahim M, Oyebanji E, Fowora M, Aiyeolemi A, Orabuchi C, Akinnawo B, Adekunle AA. Extracts of endophytic fungi from leaves of selected Nigerian ethnomedicinal plants exhibited antioxidant activity. *BMC Complementary Medicine and Therapies*. 2021 Dec;21(1):1–3.

Jackson CR, Stone BW, Tyler HL. Emerging perspectives on the natural microbiome of fresh produce vegetables. *Agriculture*. 2015 Jun;5(2):170–87.

Jalgaonwala RE, Mohite BV, Mahajan RT. A review: Natural products from plant associated endophytic fungi. *Journal of Microbiology and Biotechnology Research*. 2011;1(2):21–32.

Jayaram A, Deer E, Amaral LM, Campbell N, Vaka VR, Cunningham M, Ibrahim T, Cornelius DC, LaMarca BB. The role of tumor necrosis factor in triggering activation of natural killer cell, multi-organ mitochondrial dysfunction and hypertension during pregnancy. *Pregnancy Hypertension*. 2021 Jun 1;24:65–72.

Joseph B, Priya RM. Bioactive compounds from endophytes and their potential in. *American Journal of Biochemistry and Molecular Biology*. 2011;1(3):291–309.

Kaul S, Gupta S, Ahmed M, Dhar MK. Endophytic fungi from medicinal plants: A treasure hunt for bioactive metabolites. *Phytochemistry Reviews*. 2012;11:487e505.

Kaul S, Sharma T, K. Dhar M. "Omics" tools for better understanding the plant-endophyte interactions. *Frontiers in Plant Science*. 2016 Jun 29;7:955.

Kaushik N, Coloma AG. Therapeutic agents from endophytes. *Current Medicinal Chemistry*. 2020 Apr 1;27(11):1813–4.

Kennedy RK, Nagarajaprakash R, Pathma J. Therapeutic potency of bioactive compounds from fungal endophytes. In Joginder Singh and Praveen Gehlot, editors. *New and Future Developments in Microbial Biotechnology and Bioengineering* 2020 Jan 1 (pp. 51–70). Elsevier.

Khan AL, Gilani SA, Waqas M, Al-Hosni K, Al-Khiziri S, Kim YH, Ali L, Kang SM, Asaf S, Shahzad R, Hussain J. Endophytes from medicinal plants and their potential for producing indole acetic acid, improving seed germination and mitigating oxidative stress. *Journal of Zhejiang University-Science B*. 2017 Feb;18(2):125–37.

Khiralla A, Mohamed I, Thomas J, Mignard B, Spina R, Yagi S, Laurain-Mattar D. A pilot study of antioxidant potential of endophytic fungi from some Sudanese medicinal plants. *Asian Pacific Journal of Tropical Medicine*. 2015 Sep 1;8(9):701–4.

Kumar A, Patil D, Rajamohanan PR, Ahmad A. Isolation, purification and characterization of vinblastine and vincristine from endophytic fungus Fusarium oxysporum isolated from Catharanthus roseus. *PLoS One*. 2013 Sep 16;8(9):e71805.

Kumaran RS, Kim HJ, Hur BK. Taxol promising fungal endophyte, Pestalotiopsis species isolated from Taxus cuspidata. *Journal of bioscience and bioengineering*. 2010 Nov 1;110(5):541–6.

Lee YM, Kim MJ, Li H, Zhang P, Bao B, Lee KJ, Jung JH. Marine-derived Aspergillus species as a source of bioactive secondary metabolites. *Marine Biotechnology*. 2013 Oct;15(5):499–519.

Linnakoski R, Reshamwala D, Veteli P, Cortina-Escribano M, Vanhanen H, Marjomäki V. Antiviral agents from fungi: Diversity, mechanisms and potential applications. *Frontiers in Microbiology*. 2018 Oct 2;9:2325.

Liu YQ, Liu ZL, Tian X, Yang L. Anti-HSV activity of camptothecin analogues. *Natural Product Research*. 2010 Apr 10;24(6):509–14.

Liu Z, Ma N, Zhong Y, Yang ZQ. Antiviral effect of emodin from Rheum palmatum against coxsakievirus B5 and human respiratory syncytial virus in vitro. *Journal of Huazhong University of Science and Technology [Medical Sciences]*. 2015 Dec;35(6):916–22.

Lu YH, Jin LP, Kong LC, Zhang YL. Phytotoxic, antifungal and immunosuppressive metabolites from Aspergillus terreus QT122 isolated from the gut of dragonfly. *Current Microbiology*. 2017 Jan;74(1):84–9.

Ma J, Gu Y, Xu P. A roadmap to engineering antiviral natural products synthesis in microbes. *Current Opinion in Biotechnology*. 2020 Dec 1;66:140–9.

Ma Y, Rajkumar M, Zhang C, Freitas H. Beneficial role of bacterial endophytes in heavy metal phytoremediation. *Journal of Environmental Management*. 2016 Jun 1;174:14–25.

Ma Y, Vosátka M, Freitas H. Beneficial microbes alleviate climatic stresses in plants. *Frontiers in Plant Science*. 2019 May 16;10:595.

Maliehe TS, Mbambo M, Nqotheni MI, Senzo NS, Shandu JS. Antibacterial effect and mode of action of secondary metabolites from fungal endophyte associated with Aloe ferox Mill. *Microbiology Research*. 2022 Feb 3;13(1):90–101.

Marsberg A, Kemler M, Jami F, Nagel JH, Postma-Smidt A, Naidoo S, Wingfield MJ, Crous PW, Spatafora JW, Hesse CN, Robbertse B. Botryosphaeria dothidea: A latent pathogen of global importance to woody plant health. *Molecular Plant Pathology*. 2017 May;18(4):477–88.

Martinez JP, Sasse F, Brönstrup M, Diez J, Meyerhans A. Antiviral drug discovery: Broad-spectrum drugs from nature. *Natural Product Reports*. 2015;32(1):29–48.

Metwaly AM, Kadry HA, Atef A, Mohammad AE, Ma G, Cutler SJ, Ross SA. Nigrosphaerin A a new isochromene derivative from the endophytic fungus Nigrospora sphaerica. *Phytochemistry Letters*. 2014 Feb 1;7:1–5.

Miskovsky P. Hypericin-a new antiviral and antitumor photosensitizer: Mechanism of action and interaction with biological macromolecules. *Current Drug Targets*. 2002 Feb 1;3(1):55–84.

Muhammed D, Odey BO, Alozieuwa BU, Alawode RA, Okunlola BM, Ibrahim J, Lawal A, Berinyuy EB. Azadirachtin-A a bioactive compound from Azadirachta indica is a potential inhibitor of SARS-CoV-2 main protease. *AROC in Pharmaceutical and Biotechnology*. 2021;1(1):1–8.

Mukherjee PK. *Antiviral Evaluation of Herbal Drugs. Quality Control and Evaluation of Herbal Drugs.* Elsevier, 599–628, 2019. DOI:10.1016/B978-0-12-813374-3.00016-8

Mulwa LS, Jansen R, Praditya DF, Mohr KI, Wink J, Steinmann E, Stadler M. Six heterocyclic metabolites from the myxobacterium Labilithrix luteola. *Molecules.* 2018 Feb 28;23(3):542.

Nair DN, Padmavathy S. Impact of endophytic microorganisms on plants, environment and humans. *The Scientific World Journal.* 2014.

Nam S, Ga YJ, Lee JY, Hwang WY, Jung E, Shin JS, Chen W, Choi G, Zhou B, Yeh JY, Go YY. Radicicol inhibits chikungunya virus replication by targeting nonstructural protein 2. *Antimicrobial Agents and Chemotherapy.* 2021 Apr 26;65(7):e00135–21.

Olesen SH, Ingles DJ, Yang Y, Schönbrunn E. Differential antibacterial properties of the MurA inhibitors terreic acid and fosfomycin. *Journal of Basic Microbiology.* 2014 *Apr*;54(4):322–6.

Palamara AT, Nencioni L, Aquilano K, De Chiara G, Hernandez L, Cozzolino F, Ciriolo MR, Garaci E. Inhibition of influenza A virus replication by resveratrol. *Journal of Infectious Diseases.* 2005 May 15;191(10):1719–29.

Parthasarathi S, Sathya S, Bupesh G, Samy RD, Mohan MR, Kumar GS, Manikandan M, Kim CJ, Balakrishnan K. Isolation and characterization of antimicrobial compound from marine Streptomyces hygroscopicus BDUS 49. *World Journal of Fish and Marine Sciences.* 2012;4(3):268–77.

Pimentel MR, Molina G, Dionísio AP, Maróstica Junior MR, Pastore GM. The use of endophytes to obtain bioactive compounds and their application in biotransformation process. *Biotechnology Research International.* 2011;2011:576286.

Santos IP, Silva LC, Silva MV, Araújo JM, Cavalcanti MD, Lima VL. Antibacterial activity of endophytic fungi from leaves of Indigofera suffruticosa Miller (Fabaceae). *Frontiers in Microbiology.* 2015 May 7;6:350.

Santos TF, dos Santos Carvalho C, de Almeida MA, Delforno TP, Duarte IC. Endophytic fungi isolated from Brazilian medicinal plants as potential producers of antioxidants and their relations with anti-inflammatory activity. *3 Biotech.* 2020 May;10(5):1–8.

Shin CG, An DG, Song HH, Lee C. Beauvericin and enniatins H, I and MK1688 are new potent inhibitors of human immunodeficiency virus type-1 integrase. *The Journal of Antibiotics.* 2009 Dec;62(12):687–90.

Singh R, Dubey AK. Endophytic actinomycetes as emerging source for therapeutic compounds. *Indo Global Journal of Pharmaceutical Sciences.* 2015;5(2):106–16.

Sopalun K, Laosripaiboon W, Wachirachaikarn A, Iamtham S. Biological potential and chemical composition of bioactive compounds from endophytic fungi associated with Thai mangrove plants. *South African Journal of Botany.* 2021 Sep 1;141:66–76.

Stone JK, Bacon CW, White Jr JF. An overview of endophytic microbes: Endophytism defined. In Bacon CW and White Jr. JF, editors. *Microbial Endophytes* 2000 (pp. 17–44). Marcel-Dekker.

Strobel G, Daisy B. Bioprospecting for microbial endophytes and their natural products. *Microbiology and Molecular Biology Reviews.* 2003 Dec;67(4):491–502.

Subbulakshmi GK, Thalavaipandian A, Ramesh V, Rajendran A. Bioactive endophytic fungal isolates of Biota orientalis (L) Endl., Pinus excelsa Wall. and Thuja occidentalis L. *International Journal of Advanced Life Sciences (IJALS).* 2012;4:9–15.

Sudo K, Konno K, Shigeta S, Yokota T. Inhibitory effects of podophyllotoxin derivatives on herpes simplex virus replication. *Antiviral Chemistry and Chemotherapy.* 1998 Jun;9(3):263–7.

Tadych M, White JF, Moselio S. *Endophytic microbes.* In M. Schaechter, editor. *Encyclopedia of Microbiology* 2009 (pp. 431–42). Elsevier.

Takizawa N, Yamasaki M. Current landscape and future prospects of antiviral drugs derived from microbial products. *The Journal of Antibiotics.* 2018 Jan;71(1):45–52.

Wang P, Yu JH, Zhu K, Wang Y, Cheng ZQ, Jiang CS, Dai JG, Wu J, Zhang H. Phenolic bisabolane sesquiterpenoids from a Thai mangrove endophytic fungus, Aspergillus sp. xy02. *Fitoterapia.* 2018 Jun 1;127:322–7.

Wang YJ, Pan KL, Hsieh TC, Chang TY, Lin WH, Hsu JT. Diosgenin, a plant-derived sapogenin, exhibits antiviral activity in vitro against hepatitis C virus. *Journal of Natural Products.* 2011 Apr 25;74(4):580–4.

White JF, Kingsley KL, Zhang Q, Verma R, Obi N, Dvinskikh S, Elmore MT, Verma SK, Gond SK, Kowalski KP. Endophytic microbes and their potential applications in crop management. *Pest Management Science.* 2019 Oct;75(10):2558–65.

Wilson D. Fungal endophytes which invade insect galls: Insect pathogens, benign saprophytes, or fungal inquilines? *Oecologia.* 1995 Aug;103(2):255–60.

Wilson SA, Roberts SC. Recent advances towards development and commercialization of plant cell culture processes for the synthesis of biomolecules. *Plant Biotechnology Journal.* 2012 *Apr*;10(3):249–68.

Wutzler P, Sauerbrei A. Virucidal activity of the new disinfectant monopercitric acid. *Letters in Applied Microbiology.* 2004 Aug;39(2):194–8.

Xicoy B, Ribera JM, Miralles P, Berenguer J, Rubio R, Mahillo B, Valencia ME, Abella E, López-Guillermo A, Sureda A, Morgades M. Results of treatment with doxorubicin, bleomycin, vinblastine and dacarbazine and highly active antiretroviral therapy in advanced stage, human immunodeficiency virus-related Hodgkin's lymphoma. *Haematologica*. 2007 Feb 1;92(2):191–8.

Xie XH, Zang N, Li SM, Wang LJ, Deng Y, He Y, Yang XQ, Liu EM. Resveratrol inhibits respiratory syncytial virus-induced IL-6 production, decreases viral replication, and downregulates TRIF expression in airway epithelial cells. *Inflammation*. 2012 Aug;35(4):1392–401.

Xu H, Gao J, Portieles R, Du L, Gao X, Borras-Hidalgo O. Endophytic bacterium Bacillus aryabhattai induces novel transcriptomic changes to stimulate plant growth. *PloS One*. 2022 Aug 3;17(8):e0272500.

Yin D, Yin L, Wang J, Shen X, Dai Y, Zhao R, Hu X, Hou H, Zhang D, Wang G, Qi K. Antiviral and virucidal activities of camptothecin on fowl adenovirus serotype 4 by blocking virus replication. *Frontiers in Cellular and Infection Microbiology*. 2022:442.

Zhang L, Li Y, Gu Z, Wang Y, Shi M, Ji Y, Sun J, Xu X, Zhang L, Jiang J, Shi W. Resveratrol inhibits enterovirus 71 replication and pro-inflammatory cytokine secretion in rhabdosarcoma cells through blocking IKKs/NF-κB signaling pathway. *PLoS One*. 2015 Feb 18;10(2):e0116879.

# 4 Phytochemicals and Phyto-Essential Oils as Potent Antiviral Agents

*Sukhminderjit Kaur, Babita Thakur, Nitu, and Gurleen Kaur*
University Institute of Biotechnology, Chandigarh University

*Rajinder Kaur*
University of Idaho

## 4.1 INTRODUCTION

Viral infections continue to be the major cause of pandemics and large-scale morbidities. Influenza, severe acute respiratory syndrome, acquired immuno deficiency syndrome and Ebola have been the most dreadful viral infections ever seen by mankind. Although these diseases cause significant mortalities in human beings, still there is no specific vaccine available to cure the viral infections. Ebola virus disease has led to more than 7,500 mortalities in tropical countries. Influenza virus has caused an alarmingly increasing number of more than 500,000 deaths annually (Gasparini et al., 2012; Šafratová et al., 2018). The human immunodeficiency virus is reported to cause 38% increase in infections each year, hence continuing to pose a major global public health challenge. Recently, widespread destruction caused by Severe Acute Respiratory Syndrome Coronavirus-2 has put whole world on halt, causing more than 218 million infections and over 5.2 million deaths (WHO, 2020). The goal of developing effective vaccines and therapeutic drugs against these diseases remains a challenging task.

Approximately 90 different antiviral drugs are available, but these viral infections have under-passed the approved drugs (Brijesh et al., 2009). These antiviral drugs are costly, have side effects and often lead to viral resistance. Considering the fact that traditional antiviral drugs that show effective *in vitro* efficiency are usually unsuccessful in patients, there is a need to explore safe and effective antiviral strategies to cure these diseases. The use of plants and their naturally derived extracts have shown the presence of many phytochemicals that exhibit remarkable bioactive potentials. Various plants have been used since ancient times for therapeutic purposes. Literature studies have shown the isolation, purification and identification of various phytochemicals such as alkaloids, flavonoids, glycosides and terpenes that are effective against parainfluenza virus, avian influenza virus and HSV1 (Orhan et al., 2013). Rutin and quercetin, flavonoid glycosides commonly found in many plants, are effective against the replication of many deadly viruses such as rhinovirus, dengue virus, Adino virus, polio virus, mayaro virus, respiratory syncytial virus, Japanese encephalitis virus and Epstein-Barr virus (Ben-Shabat et al., 2020). These molecules inhibit the activity of HSPs (Heat Shock Proteins) and viral genome transcription. Quercetin is known to inhibit the dengue virus infection by reducing endocytosis and protein synthesis. Another flavonoid, myricetin, commonly found in vegetables, fruits,

 DOI: 10.1201/9781003329169-5

nuts, wild plants and berries, is known to inactive influenza viruses by inhibiting the action of DNA polymerase enzymes (Park et al., 2013). Apigenin, another flavonoid, has exhibited broad-spectrum antiviral activity against foot and mouth disease viruses, swine flu virus and influenza virus. Other flavonoids such as biochanin A, baicalein, kaempferol, naringenin and luteolin are active against influenza H5N1 virus, Enterovirus, Newcastle disease virus, human immunodeficiency virus and hepatitis B virus.

Essential oils from plants are also currently gaining importance in research due to their remarkable pharmacological activities. They are composed of terpenoids, phenolic compounds, sulfured derivatives and aromatic compounds that have antioxidants, anti-cancerous, antimicrobial and antiviral activities. The essential oils derived from plants have shown strong inhibition against herpes simplex virus, polio virus, dengue virus, influenza virus and junin virus (Allahverdiyev et al., 2004). In this chapter, the authors have tried to comprehensively review all the literature to find out biological molecules from plant sources that inhibit various viruses. The detailed discussion about various phytochemicals and phyto-essential oils has been provided for a better understanding of the mechanism of viral replication.

## 4.2 PHYTOCHEMICALS AS AN ANTIVIRAL AGENTS

Phytochemicals have traditionally been known to inhibit viral replication and viral transcription. Different classes of phytochemicals have been known to exert antiviral effects. These phytochemicals are popularly referred to as non-enzymatic antioxidants and are capable of attenuating ROS-induced oxidative damage. Different classes of phytochemicals that have established antiviral properties are discussed in the following sections.

### 4.2.1 Anthocyanins

Anthocyanin belongs to a structural class of glycoxylated polyphenols that are capable of producing anthocyanidins under certain favourable conditions. Anthocyanins are basically present in the petals of flower and in seeds. When present in petals, they are associated with the role of attracting pollinators, and when they are present in the seeds, they are responsible for dispersal. They are also used for the protection purposes against various factors, including UV radiation, which can be considered harmful to the plant and its growth (Castañeda-Ovando et al., 2009). The antiviral effects of anthocyanins have proven their importance in the pharmaceutical industry. There are various plants that are rich in producing anthocyanins and are capable of exhibiting antiviral properties.

Functional anthocyanins include malvidin, pelargonidin, peonidin, petunidin, delphinidin, cyanidin and cyanidin, which have 3-O-rutinosides and 3-O-glucosides. The compound 3-O-arabinoside has been discovered in a blackcurrant berry extract (Ribes nigrum L.). Studying the effects of *R. nigrum* extracts on InfV A and B led to the discovery of four anthocyanin-containing fractions with substantial antiviral activity. According to the results, one of the fractions included 3-O—L—D-glucopyranosyl-cyanidin and 3-O—D-glucopyranosyl-cyanidin anthocyanins, while the other contained 3-O—L—D-glucopyranosyl-delphinidin and 3-O—D-glucopyranosyl-delphinidin anthocyanins. Significant antiviral effectiveness against InfV A and B is present in both fractions (Rytel et al., 2019).

Anthocyanins inhibit viral adhesion and entry into cells, as well as viral discharge from infected cells. *In vitro* trails have shown that *R. nigrum* extracts interfered with viral internalisation, decrease overall InfV A growth and prevent InfV from adhering to the cell surface. Blackcurrant reduces replication while preventing InfV A and B from adhering to cell surfaces in high concentrations. Overall, it was discovered that blackcurrant extracts exhibit a variety of anti-InfV characteristics. The creation of innovative antiviral drugs to treat infections caused by the InfV virus may benefit from this extract.

### 4.2.2 Alkaloids

Alkaloids are secondary metabolites that feature a nitrogen atom in their ring structure. Higher plants contain alkaloids, which have a variety of biological effects, such as anticancer, antibacterial, antiviral and anti-asthmatic qualities. A total of 43 alkaloids have been reported to have antiviral properties against the influenza virus. The antiviral effect of alkaloids against the influenza virus may be attributed to the activation of immune system interferons. Some alkaloids encourage the activity of macrophages, enabling the phagocytosis and eradication of the virus. Alkaloids inhibit infection at several phases of influenza virus development. Some of them prevent the creation of viral proteins or interfere with other replication mechanisms (Ghidoli et al., 2021).

Homonojirimycin (HNJ), produced from *Commelina communis*, has substantial inhibitory effect against influenza virus infection. Additionally, *in vivo* experiments have been conducted in which mice infected with the H1N1 virus had a high survival rate when given HNJ, demonstrating that the virus induced a potent immune response against influenza. Swamy (2020) investigated the antiviral properties of dendrobine, an alkaloid found in the plant root of *Dendrobium nobile*. The researchers found that dendrobine had antiviral action against influenza A viruses. It is claimed that dendrobium failed to suppress the virus during the viral replication stage. Dendrobine inhibits viral replication by attaching to the virus' nucleoprotein, hence limiting viral export. It has also been demonstrated that dendrobine and zanamivir, a medication used to treat and prevent influenza, work in concert to reduce infections of influenza A virus. Recently, numerous synthetic substances, including alkaloid analogues, have been created and studied for their potential to inhibit the influenza virus.

These substances demonstrated antiviral action against various influenza A virus subtypes and efficiently reduced viral infection, as found by the researchers (H1N1, H3N2 and H5N2). The antiviral activity of alkaloids from the marine-derived fungus *Dichotomomyces cejpii* F31-1 was examined against the dengue virus. Additionally, alkaloids generated by plants such as *Castanospermum australe*, *Houttuynia cordata*, *Myrtopsis corymbosa* and *Uncaria tomentosa* were found to have antiviral action. Alkaloids were identified in the *Phaleria macrocarpa* plant, along with several other phytocompounds, during a preliminary screening of its phytochemicals (Naithani et al., 2010). These plant extracts have been shown to be efficient against herpes simplex virus (HSV) and safe for Vero cells. The methanol extract from *Ziziphus jujube* plant contained eight cyclopeptide alkaloids. Three alkaloids inhibited the swine epidemic diarrhoea virus in cytotoxicity and cytopathic effect inhibition tests (PEDV). *Enantia chlorantha* (Oliver), a traditional medicine used in Nigeria to treat malaria caused by *Plasmodium* spp, has antiviral properties that are ineffective against hepatitis A, B, C and D. Iftikhar et al. (2013) evaluated the antiviral potential of alkaloids found in the stem bark of *E. chlorantha* against the Newcastle disease virus, and the results showed that the alkaloids were harmful to the viruses.

### 4.2.3 Terpenes

The most typical type of secondary metabolite is terpenes, composed of five carbon isoprene units joined by simple hydrocarbons. On the other hand, terpenoids are essentially modified terpenes with different functional groups at various locations. Terpenoids, sometimes referred to as isoprenoids, are naturally occurring plant substances that play a crucial role in the creation of pharmaceuticals, therapies and herbal remedies. Terpenoids including diterpenoids, sesquiterpenoids and triterpenoids have a variety of therapeutic applications.

Lima et al. (2021) investigated the antiviral efficacy of betulin and dihydroquinopimaric acid derivatives against the H7N1 influenza A virus, the HSV-1 herpes simplex virus, the ECHO-6 non-enveloped RNA virus and human immunodeficiency virus (HIV) type 1. Betulic acid has been modified at C3 and C28 to exhibit antiviral action against HSV-1 and HIV-1. Furthermore, altering the C4 position increased the anti-H7N1 action of dihydroquinopimaric acid. Given the absence of vaccination for hepatitis C virus (HCV) and the lack of cost-effective treatment, researchers are exploring phytochemical-based therapy, which may prove to be both affordable and efficient. Saikoponins such as SSa, SSb2, SSc and SSd, isolated from the *Bupleurum kaoi* root, have been shown to reduce HCV infection. Betulin and its dihydroquinopimaric acid derivatives possess antiviral properties. SSb2 was found to be the most effective alkaloid for suppressing HCV infection, and the effectiveness of *Phyllanthus urinaria* plant extract against early HCV entry was discovered.

They discovered that loliolide, a monoterpene, prevented viral entry by inhibiting viral adhesion to the host membrane. The diterpenes 12-O-decanoylphorbol-13-acetate and 12-O-decanoyl-7-hydroperoxy-phorbol-5-ene-13-acetate, as along with loliolide, vomifoliol, dehydrovomifoliol, annuionone D and bluemol C, were isolated from the leaves of *Croton mauritianus.* Two phytoconstituents with demonstrated anti-chikungunya efficacy are 12-O-decanoylphorbol-13-acetate and 12-O-decanoyl-7-hydroperoxy-phorbol-5-ene-13-acetate. In research, the antiviral capacity of *Aglaia* sp. was examined. Terpenes have been found to block HIV-1 replication by generating a cytotoxic environment (N. Powers and Setzer, 2016). A wide range of viruses, including retroviruses like the simian immunodeficiency viruses (SIVs) and the murine leukaemia virus, were also suppressed by terpenes. By interfering with the cell cycle between the S and G2/M phases, the triterpenoid 3,4-secodammarane has been demonstrated to inhibit viral growth and spread. Terpenes found in the methanolic extract of *Marrubium vulgare* were tested for their ability to combat HSV-1 using a plaque reduction assay. By inhibiting HSV-1 from attaching to the cell membrane, terpenes have been shown to suppress HSV-1 viral replication. Terpenes (oleanolic terpenoids) were also isolated from the flowers of *Camellia japonica* (Ninfali et al., 2020).

### 4.2.4 Polysaccharide

Polysaccharides are the most prevalent kind of carbohydrate in the diet. They consist of lengthy, polymeric polysaccharides comprised of monosaccharide building blocks connected by glycosidic bonds. Amylase enzymes can aid in the digestion of this carbohydrates, leading to the formation of component sugars, such as monosaccharides or oligosaccharides. Their structure can be either simple or intricate. Polysaccharides may suppress viruses including hepatitis B, Herpes simplex, CMV, papillomaviruses and others.

AIDS (acquired immunodeficiency syndrome) is a pandemic immunosuppressive sickness that can lead to opportunistic infections and malignancies. Research has shown that carbohydrates are organic substances that can combat HIV infection. Several antivirally active plant extracts, including different polysaccharides, such as sulphated polysaccharides, have been demonstrated to possess these properties. Many of these medicinal plants are also employed as anti-infectives in traditional medicine. For instance, a (polysaccharides) PS(MAR-10) isolated from the aqueous extract of *Hyssop officinalis* L. was discovered to inhibit the production of syncytia and the HIV-1 p24 antigen, therefore preventing HIV-1 replication. A PS with good anti-HIV activity was produced by the common medicinal plant *Aspalathos linearis* Burm., which is found in South Africa, Europe and Japan. In terms of blocking HIV-1 binding to MT-4 cells, this medication exhibited essentially perfect (Ghidoli et al., 2021).

In a screening test of Thai plants for anti-HIV activity, 114 extracts from 81 species belonging to 32 groups were found to exhibit activity. *Merremia peltata* (L.) Merrill extracts were the most active among the active samples. The majority of anti-HIV drugs discovered were anionic polysaccharides. Sulphated derivatives of paramylon, a (1–3)-P-D-glucan isolated from the tropical climbing shrub *Ipomoea cairica* (L.) were found to have significant cytopathic effect on both HIV-1 and HIV-2. Additionally, two naturally occurring polysaccharides isolated from the tropical climbing shrub *Ipomoea cairica* (L.) were found to strongly inhibit HIV-1 replication *in vitro.*

It has long been known that natural sulphated polysaccharide (PS), including heparin, pentosan, chondroitin and dextran, possess powerful anti-HIV medications. These drugs block not only viral adsorption to cells but also virus-induced syncytium (giant cell) development by specifically targeting the interaction between the CD4 receptor and the viral envelope glycoprotein gpl20. Through a particular interaction with the V3 region of gpl20, dextran sulphate and its congeners seem to prevent viral binding, replication and syncytium formation. Additionally, sulphated polysaccharides may specifically prevent HIV particles from adhering to the heparin sulphate proteoglycan on the cell surface, regardless of whether this impact happens independently or in conjunction with contact between the viral envelope and the CD4 receptor (Gopal et al., 2021).

Other antiviral therapies lack several of the distinctive qualities found in this class of antiviral medications, including exceptional broad-spectrum antiviral efficacy against HIV-1, HIV-2 and a variety of other enveloped viruses. They also can prevent the formation of syncytium between HIV-infected and healthy CD4 T lymphocytes, which significantly increases HIV infectiousness while producing minimal viral treatment resistance. It's important to note that the sulphate groups in PS, which stop virus-cell contact, are responsible for conferring the substance with its anti-HIV effects. High molecular weight polysaccharide dextran is mostly composed of linked D-glucose units.

HSV infections can affect anybody, ranging from minor cold sores and fever blisters to life-threatening conditions of the central nervous system. Due to the development of effective antiviral medications, rapid identification has become an essential practice in general healthcare. Appropriate therapy can considerably lessen the psychological and physical effects of herpes infections, improving many people's quality of life. Novel anti-HSV medications, especially those with mechanisms of action different from acyclovir, are therefore urgently required. For many years, it has been known that several phytochemicals and synthetic antiviral medications with various molecular targets can prevent infections caused by members of the *Herpesviridae* family of viruses.

Interest in phyto-polysaccharides, which function as strong inhibitors of various enveloped viruses, including members of the *Herpesviridae* family, has been rekindled by recent research. For instance, the neutral polysaccharide scleroglucan derived from the *Sclerotium glucanicum* Halleck plant may be antiviral in nature. The third component is a pl,3-linked glucopyranose backbone with a single p-l,6-linked glucopyranosyl branch. Its highest inhibitory impact was observed when administered during the attachment phase, blocking HSV infection occurred in the very early stages of the viral multiplication cycle (Singh and Sinha, 2012).

A sulphated galactan isolated from *Pterocladia capillacea* (Gmelim), and mannans and xylogalactans isolated from the red seaweed *Nothogenia fastigiata* Turner (Lam.) were found to possess antiviral potential against HSV-1 and HSV-2 (Carlucci et al., 1997). Furthermore, it has been investigated that the antiviral activity of sulphated polysaccharides is known to rise with their molecular weight or degree of sulphation (Wang et al., 2016). Anticoagulant properties of sulphated, on the other hand, may limit their use as antiviral medications. Duarte et al. (2001) discovered that sulphated galactans from the marine alga *Bostrychia montagriei* have antiviral and anticoagulant effects. The results of the activated partial thromboplastin time test and the thrombin time assay revealed that three natural sulphated PS had very little anticoagulant activity, showing that antiviral and anticoagulant properties are unrelated. Rhamnan sulphate from *Monostroma latissimum* Wittrock and calcium spirulina from the blue-green alga calcium spirulina are two more seaweeds that contain anti-HSV sulphated polysaccharides.

### 4.2.5 Polyphenols

Polyphenols are secondary metabolites generally involved in defence mechanism. Virus families such as retrovirus hepadnaviridae, hespervirides, HIV, influenza, herpes simplex, dengue, polio, diarrhoea and others are all susceptible to the antiviral effects of polyphenols and bioflavonoids. The bioflavonoid family of polyphenols, includes quercetin, kaempferol and rutin, which have antiviral activities. It has been discovered that flavonoids have more antiviral activity than flavones against the HIV virus and type 1 HSV. There is evidence that flavonoids and certain antiviral medications interact. For instance, quercetin enhances the anti-HSV and anti-pseudorabies actions of acyclovir and 5-ethyl-2'-dioxyuridine. The antiviral effectiveness of acyclovir against some viruses is also increased by apigenin (Umesalma and Sudhandiran, 2010). A few examples of antiviral polyphenols are gallic acid, catechin, epicatechin, epigallocatechin, ellagic acid, rutin, paraascorbic acid, quercetin, kaempferol, morin, galagenin, apigenin, naringenin, chrysin, luteolin, analgin and rosmarinic acid.

In addition to stimulating enzymatic systems involved in xenobiotic detoxification, polyphenols have the ability to block the inactivation of carcinogens, mutant gene expression and enzyme activity involved in procarcinogen activation. All these processes can contribute to the development of malignant tumours. When associated with other carcinogens, co-carcinogens or promoters of skin carcinogenesis, many polyphenols could also be mutagenic. Enols, particularly flavonoids, have been shown in several studies to prevent the genesis, growth and spread of tumours while also functioning as antioxidants for DNA and gene expression (Lin et al., 2014).

Antiviral activity requires the presence of hydroxyl and ester groups. Anti-herpetic activity of phenolic compounds, notably various alkyl-esters of gallic acid, appears to be constant and unrelated to the length of the lateral carbonic chain. Some gallates, such as methyl gallate and its derivatives, show anti-HSV types 1 and 2, as well as anti-cytomegalovirus activity, indicating that hydroxyl groups, in addition to the alkalic chain, may impact viral adsorption and penetration and be responsible for the antiviral action. Additionally, gallic acid-derived galloyl moieties have been postulated. Galloyl glucose and flavanol glycoside are HIV-1 integrase 3 processor inhibitors. Catechin, epicatechin and quercetin have been shown to have anti-rabies action due to the presence of five hydroxyl groups.

The presence of hydroxyl groups (epigallocatechin), for example, might be a contributing component to anti-rabies potency. A subsequent study employing a different approach indicated that ascorbic acid may be employed to inactivate rabies, and the researchers hypothesised that this molecule might be auto-oxidized, resulting in the form that was responsible for the observed activity (Soldati, 2018). The anti-HSV activities of flavonoids have also been established, demonstrating that free hydroxyl groups are necessary for this action. Recent study suggests that the form responsible for the reported activity may be produced by replacing a 5'-tert butyl dimethylsilyl group with an acyl, aromatic or cyclic moiety. The anti-HSV activities of flavonoids have also been established, demonstrating that free hydroxyl groups are necessary for this action. The presence of 3,4,5-trihydroxy and 3,4,5-trimethoxy derivatives of gallic acid is responsible for its anti-rabies activity.

### 4.2.6 Lectin

A protein called lectin, which binds carbohydrates and is present in mushrooms, has garnered significant interest. Eighty-two percent of fungal lectins have been found in mushrooms, 15% in moulds and 3% in yeasts. Animals, plants, marine algae, bacteria, viruses, insects, vertebrate body fluid, lower vertebrate body fluid, mammalian cell membranes and fungi have all been shown to contain large amounts of carbohydrates-binding proteins. They are used as diagnostic tools in clinical practice as well as for a range of therapeutic purposes. Molecular and physiological characteristics of lectins, which are oligomeric proteins, include structural variety, molecular size, metal requirements, glycosylation and carbohydrate selectivity (Mani et al., 2020).

The HIV-inhibiting potential of mushroom crude extracts and secondary metabolites, such as lectin, has not been extensively researched. The HIV reverse transcriptase enzyme is inhibited by trichogin protein from *Tricholoma giganteum* mushrooms, which has the lowest half-maximal inhibitory concentration (IC50) value of 83 nM. Next in line are the lectin from *Russula delica* and the laccase from *Tricholoma mongolicum*, which have IC50 values of 0.26 and 0.62 M, respectively. Although method of action of lectins remains unknown, researchers have recently become interested in the antiviral activity of lectins, particularly those from macrofungi. By adhering to virus surface glycoproteins, blocking the host receptor or inhibiting the viral polymerase enzyme by interacting with its active area, this protein may reduce viral activity. By causing virus conformational rearrangements, lectins inhibit virus fusion and entry by attaching to the glycoproteins of the viral envelope. When using lectins for medicinal purposes, one challenge is the difficulty of introducing a foreign protein into the human body. However, the lectins griffithsin and cyanovirin-N were found in red algae and cyanobacterium, respectively.

These proteins are physiologically active, and subcutaneous injection does not affect their bioavailability. Well-studied lectins such as griffithsin and cyanovirin-N have demonstrated strong antiviral activity against a number of viruses, including the Zaire ebolavirus, HIV, HIV-1, HIV-2 and HCV, as well as the coronaviruses responsible for the initial SARS-CoV pandemic. The purification process, N-terminal, sugar and metal specificities, along with its physicochemical characteristics of the studied mushroom lectin, were compiled (Venkidasamy et al., 2019). The antiviral activity of mushroom-derived lectin against HIV has drawn the most interest. These lectins act as anti-HIV agents by attaching to group 120, preventing viral entry, or by inhibiting the reverse transcriptase enzyme, which is crucial for the HIV life cycle. An effective lectin against HIV-1 is the MBL. Its antiretroviral effectiveness against HIV is demonstrated by its interaction with the surface glycoproteins gp120 (rich in mannose) and gp41, making the virus susceptible to host antibodies and other immune cells. The MBL lectin *Hygrophorus russula* contains three different mannose binding site residues. Different *Boletus ediols* and *Agaricus bisporus* lectins have different ligand affinities for galactose and T antigen. These lectins significantly inhibit cancer growth (Jassim and Naji, 2003).

It is interesting to note that the alpha and beta N-acetyl-D-galactosamine binding site residues and the *Boletus ediols* lectin establish hydrogen bonds. MBL-resistant HIV strains emerged due to mutations in the N-glycosylation region of the gp120-gp41 complex. It was recommended to employ two carbohydrate-binding medications that target different N-glycan structures on gp120 in a synergistic combinational therapy. It is noteworthy that differing specificities of the same lectin for various carbohydrate moieties might result in a range of therapeutic benefits. The most potent anti-HIV-RT lectin was generated by *Russula delica*, followed by *Pleurotus citrinopileatus*, with IC50 values of 0.26 and 0.93 M, respectively. Nucleoside RT inhibitors (NRTIs) and non-nucleoside RT inhibitors (NNRTIs) are two different categories of HIV-RT inhibitor medications (NNRTIs). The lectin from *Boletus speciosus* suppresses HIV-RT with an IC50 value of 7.1 M. The previous lectin was classified as an NNRTI because it inhibits HIV-RT by interacting with proteins.

*Agrocybe aegerita* lectin (AAL) has been found to be a safe and effective adjuvant for the H9N2 influenza vaccine in mice. As it binds to the envelope glycosylated proteins neuraminidase and hemagglutinin, galectin AAL specifically binds to galactose and can boost humoral immunity against influenza virus. By promoting cellular and humoral immunity, *Pleurotus ostreatus* lectin (POL), which displays immunogenic activity, can be used as a secure and efficient adjuvant in hepatitis B virus (HBV) DNA vaccines. POL may drastically reduce the detectable amount of HBV surface antigen (HBs Ag) and increase the generation of HBV surface protein antibody in HBV surface antigen transgenic (HBVsAg-Tg) mice (HBVsAb). According to another study, POL may be utilised to treat chronic HBV by triggering the innate immune response and encouraging the development of the Toll-like receptor-6 (TLR6) signal pathway in dendritic cells. Additionally, POL was also able to reduce HBV infection by increasing the number of released HBV-specific antibodies and the responsiveness of follicular helper T.

An efficient combination of existing treatment options with another mode of action might address the problem and restrict the spread of viral illnesses. Entry inhibitory agents are potential therapeutic targets for viral prevention and treatment that might be derived from natural sources such as mushroom lectin. These agents target the conserved section of the viral spikes or compete with the virus on host receptors, making this therapeutic technique applicable to all virus genotypes. Mushrooms are valuable and easy to cultivate, making them a prospective supply for large-scale production of new antiviral medicines. However, much of the research on mushroom lectins is limited to basic screening.

### 4.2.7 Proteins

Until recently, it was believed that dietary proteins were substances that gave the body the essential amino acids needed for life support. However, it has been revealed that they can convey a variety of bioactive peptides to the organism, including those with antiviral activity, as demonstrated recently and conclusively. Human cytomegalovirus, HSV, poliovirus and rotavirus have all been shown to be suppressed *in vitro* by the dietary protein lactoferrin, demonstrating the value of naturally occurring food proteins as possible antiviral medicines. Some dietary proteins, such as a-lactalbumin and b-lactoglobulin, are particularly intriguing since they lack antiviral activity but may be changed to acquire antiviral properties through straightforward chemical interactions, a highly successful approach. The antiviral activity of methylation -lactalbumin was investigated using fixed or suspended Vero cell lines (Met-ALA) (Jassim and Naji, 2003).

## 4.3 PHYTO-ESSENTIAL OILS AS ANTIVIRALS

The rise in disease occurrence, the development of pathogenic resistance to the available treatments and chemotherapeutics, opportunistic infections in immune compromised individuals and economical constraints in developing countries have created a global demand for safe, effective and cost-effective alternate options for the prevention and treatment of diseases (Sun, 2018). As a result, the search for competing strategies continues, with natural phytochemicals derived from plants used in traditional medicines considered excellent substitutes for synthetic chemicals. Herbal remedies are a blessing of "Mother Nature" to mankind. Secondary metabolites and essential oils of therapeutic use abound in medicinal plants (Cragg and Newman, 2013).

Essential oils (EOs) are volatile secondary metabolites obtained from raw plant extracts through steam distillation, dry distillation or a mechanical process without the use of heat. These aren't oils in the traditional sense, but they share the same low water solubility characteristics as oils. EOs are utilized in food flavouring and perfumery because they have an odour (Bhardwaj et al., 2020). Aroma oils, vegetable oils, absolutes and concretes are all different types of oils. EOs are typically dense combinations containing hundreds of different fragrance components. Among the approximately 17,000 aromatic plant species, families such as *Lamiaceae*, *Verbenaceae*, *Myrtaceae*, *Zingiberaceae*, *Poaceae*, *Rutaceae*, *Piperaceae*, *Asteraceae*, *Cupressaceae*, *Lauraceae* and *Umbelliferae* are the most common EO producers(Reddy, 2019). Table 4.1 provides a list of some medicinal and aromatic plant families that produce EOs. Due to their beneficial properties in industries such as medicine, pharmaceutics and cosmetics, EOs are gaining popularity. By the year 2022, the EO market is predicted to reach USD 11.67 billion with Eco-SMART, the world's top EO-based industry, producing numerous bioactive formulations such as Eugenol-ethoxylate, Eugenol-Tween, ActiVin, Herbalox and Pycnogenol (Mihai et al., 2020).

### 4.3.1 Types of Phyto-Essential Oil

EOs are synthesized and stored majorly in secretary parts and epidermal cells of plants. About 85% of these major stored compounds consist of terpenes, terpenoids, aromatic and aliphatic compounds, with examples such as limonene and phellandrene making up to 35% in the case of Anethum leaf oil, and 59% menthol and 19% menthone present in *Mentha piperita* oil (Shaaban et al., 2012). According to a study, the main chemicals obtained from the *Aloysia tryphila* EO were geranial, neral and limonene accounting for 21.83%, 17.45% and 11.03%, respectively. *Thymus vulgaris* EO contained thymol 79.15%, caravrol 4.63% and *p*-cimene 3.27%. The *Ocimum* species showed different composition of oils. Thymol was major component of *O. applii* (64.5%) and *O. vulgare* (38.0%), while eugenol was from *O. gratissimum* (93.9%) and *O. basilicum* (28%) (Chouhan et al., 2017). Several factors contribute to the variation in the chemical compositions of EOs. Endogenous factors include the plant's anatomical, physiological behaviour and biosynthetic process of secondary

**TABLE 4.1**
**Antiviral Activity of Different Phenolic Compounds and Their Plants Source**

| Source | Property | Antiviral Compound | Chemical Characteristics | Reference |
|---|---|---|---|---|
| *Citrus sinensis* | Antiviral | Rotenone | Isoflavanone | Prakash and Gupta (2011) |
| *Prunus domestica*, *Quercus* spp. | Antiviral | Quercetin | Flavanol | Kopustinskiene et al. (2020) |
| *Zizia aptera* | Antiviral | Angelicin | Coumarin | Kapoor et al. (2017) |
| *Acaci* spp. | Antiviral | Hesperidin | Flavanol | Patra (2012) |
| *Citrus* spp. | Antiviral | Motin | Flavanol | Patel et al. (2021) |
| *Sorbus ancuporiaral* | Antiviral | Parascorbic acid | Phenolic acid | Girija (2017) |
| *Melissa officinalis* | Antiviral | Rosmarinic acid | Phenolic acid | Lima et al. (2021) |
| *Citrus* spp. | Antiviral | Fistin | Flavanone | Ghani et al. (2019) |
| *Olive* spp. | Antiviral | Benzoic acid | Pyran derivative | Gopal et al. (2021) |
| *Spondias* spp. | Antiviral | Rutin | Flavanone | Ninfali et al. (2020) |
| *Croton lechleri* | Antiviral | Proanthocyanin | Anthocyanins | Venkidasamy et al. (2019) |
| *Millettia erythrocalyx* | Antiviral | Ovilifolin, pongal methyl ether | Flavones | Jassim and Naji (2003) |
| *Artocarpus lakoocha* | Antiviral | Oxyresveratrol | Stilbene | Castañeda-Ovando et al. (2009) |
| *Populus* spp. | Antiviral | Salioperoside, Salicin | Phenolic glycoside | Jaberian et al. (2013) |
| *Acacia* spp. | Antiviral | Narinogenin | flavones | Soldatkina and Novotna (2017) |

metabolites. which may vary in different plant tissues or under different geographical conditions. Over a long period, extrinsic elements might alter some genes accountable for the volatile synthesis, resulting in chemotype or ecotype within the same species of plant (Barra, 2009). Some of the important chemical components of EOs have been widely studied, as discussed here.

### 4.3.2 Sources of Antiviral Phyto Essential Oils

Viral diseases are prevalent globally and hence, are a major concern. Due to the narrow spectrum of antiviral drugs, drug resistance has been observed in drugs such as Acyclovir, Amantadine and Zidovudine, leading to the need to discover novel therapeutic alternatives capable of overcoming resistance. Therefore, EOs derived from plants are considered potential therapeutic agents. EOs derived from several aromatic and herbal plants have been shown to have virucidal effects on a variety of viruses mentioned in Table 4.2. EOs exhibit both the extracellular and intracellular inhibitory properties as demonstrated in Figure 4.1. In the case of extracellular inhibitory mechanism, the envelope structure of the free viruses is modified by the EOs, thereby preventing the entry into the host cell. On the other hand, EOs also can interfere with viral uncoating, recombination, assembly or release. For this purpose, infected cells are treated with EO at different time intervals of the viral infection lifecycle. According to the study by Garozzo et al. (2011), tea tree EO and its main component, terpinen-4-ol, suppress the viral replication by interfering with the acidity of the intra-lysosomal compartment, which is required for virus uncoating.

Furthermore, the viral infection causes intracellular glutathione depletion as well as a redox change. Targeting the redox signalling system, EO-derived components, such as piperitenone oxide, could block the late stage of the HSV-1 lifecycle.

The key components of almost all EOs are terpenes, which are converted to terpenoids by enzymatic addition of oxygen and removal of methyl group (Armaka et al., 1999). Some of the important terpenes are listed in Table 4.3 (Chouhan et al., 2017). Isoborneol, a monoterpene, has demonstrated

**TABLE 4.2**
**Some Plant Families Producing Essential Oils**

| Family | Plant Name |
|---|---|
| Asteraceae | *Anthem nobilis* (roman chamomile), *Achillea millefolium* (yarrow) |
| Myrtaceae | *Eucalyptus globulus, E. alba, E. citriodora, E. Propinqua, E. Salgina, E. Robusta, E. Urophylla, Melaleuca alternifolia* (Tea tree), *Syzygium aromaticum* (clove) |
| Lamiaceae | *Lavandula officinalis, Mentha piperita, Rosmarinus officinalis, Thymus vulgaris, Melissa officinalis* (lemon balm), *Salvia officinalis* (sage) |
| Rutaceae | *Citrus limonum, Citrus bergamia, Citrus aurantifolia* |
| Labiatae | *Ocimum gratissimum* (Ram Tulsi), *Ocimum canum* (DulalTulsi), *Ocimum basilicum* (Ban Tulsi), *Ocimum kilimandscharicum, Ocimum americanum, Ocimum camphora* and *Ocimum sanctum* |
| Umbilliferae | *Trachyspermum copticum (Ajowan), Anethum sowa, Anethum graveolens, Cuminum cyminum, Petroselinum sativum, Allium sativum* |

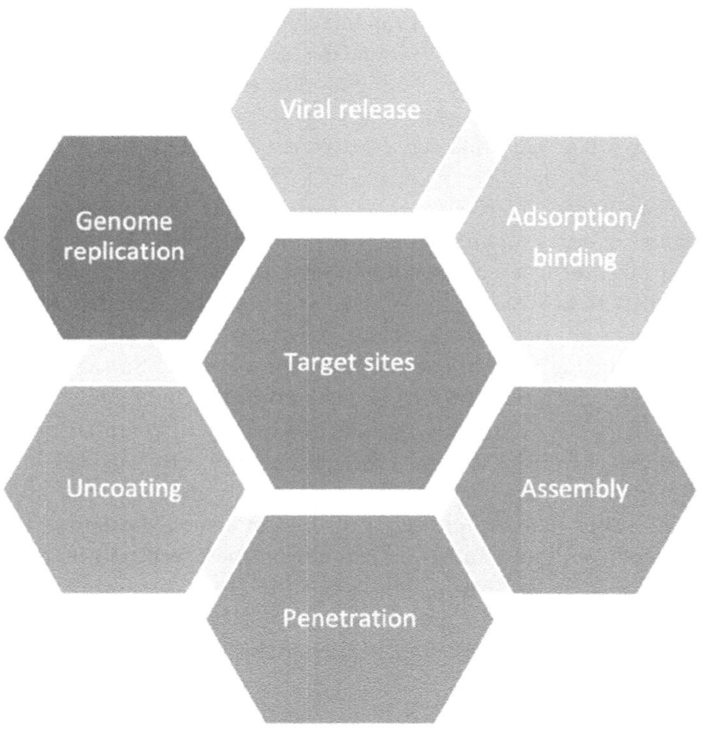


**FIGURE 4.1** Possible targeting sites of EOs in viral lifecycle.

potent anti-herpes simplex virus-1 (anti-HSV-1) activity while exhibiting low cytotoxicity. Its activity is based on interactions between the hydroxyl groups in isoborneol and the lipids in the virus envelope. Furthermore, there is evidence that isoborneol can inhibit viral replication and the glycosylation of viral proteins, making HSV-1 less infectious (Armaka et al., 1999). Cineol and borneol, stereoisomer of isoborneol, are two examples of monoterpenes that have shown potent anti-HSV-1 effects. Diterpene analysis also suggests that the compound putranjivain A from Euphorbia jolkini (*Euphorbiaceae*) may alter late stages of HSV-2 reproduction by blocking viral attachment and cell penetration (Paduch et al., 2007). Additionally, the triterpenes moronic acid (from *Rhus javanica*) and meliacine (from *Melia azeolarch*) have exhibited anti-HSV-1 actions. They supressed the

**TABLE 4.3**
**List of Various Terpenoids**

| Class | Essential Oils Component |
|---|---|
| Monoterpenes | limonene, $\alpha$-pinene, $p$-cymene and $\alpha$-terpinene |
| Oxygenated mono-terpenes | Camphor, eugenol, carvacrol and thymol, beta-fenchol |
| Diterpenes | Cembrene C, camphorene and kaurene |
| Sesqui-terpene | $\beta$-caryophyllene, humulene and germacrene D |
| Oxygenated sesqui-terpenes | Spathulenol, caryophyllene oxide, alpha-eudesmol |
| Monoterpene alcohols | Geraniol, linalool and nerol |
| Sesqui-terpene alcohol | Patchoulol |

polypeptides produced by infected cells late in the infection process and reduced viral DNA synthesis. Both lupenone and its saponin were effective in inhibiting the replication of herpes simplex viruses 1 and 2. Inhibiting viral plaque development, they affected HSV-1 DNA synthesis. Anti-HSV-1 action was found in the triterpenoid saponins glycyrrhizin, glycyrrhetic acid and oleanan-type triterpenoid saponins. Three terpenoids found in *Ganoderma pfeifferi* oil, namely, lucialdehyde D, ganoderon A and ganoderon C, were also highly effective HSV inhibitors (Niedermeyer et al., 2005). In conclusion, terpenoids represent a promising class of naturally occurring compounds with broad and targeted antiviral activity that could enhance the therapeutic benefits of current antiviral treatment.

Carvacrol is the most extensively researched hydrophobic monoterpene found in the essential oils of *Origanum vulgare*, thyme, pepperwort and bergamot. It is a monoterpenic phenol known for its broad range efficacy against pathogenic fungi, yeast and bacteria. Carvacrol has demonstrated the ability to effectively inhibit the infectiousness of murine norovirus (MNV), a human norovirus mimic and rotavirus (Sánchez and Aznar, 2015). In a 2012 study conducted in Iran by Mardani et al., the antiviral impact of the essential oils of Shirazi thyme was evaluated. Carvacrol was one of the chemicals studied for its antiviral activity against HSV-1. In this pilot investigation, vero cells were used to assess the cytotoxicity of essential oils at various doses. The results revealed that the dose that killed 50% of the vero cells was 0.0676, while the dosage that blocked viral plaque detection was 0.0059. Additionally, essential oils studied at concentrations of 0.01 and 0.02 were found to entirely prevent the virus (Mardani et al., 2012). Another investigation conducted by Pilao and his colleagues on the essential oils of *Mentha pulegium* revealed that carvacrol is the primary component; therefore, its antiviral effects on humans and animals were explored. *In vitro*, they determined that these essential oils can inhibit many human and animal viruses, including rotavirus, bovine diarrhoea virus and respiratory syncytial virus (Pilau et al., 2011).

Thymol, a naturally occurring monoterpenoid generated from cymene, is a key component of thyme (*Thymus vulgaris*). Thymol attacks membrane proteins and degrades both the inner and outer membranes. Most acute respiratory infections are caused by rhinoviruses and influenza viruses. No preventative or curative methods are available for rhinovirus infections, and despite the availability of influenza medications, upper and lower respiratory tract infections are often treated with pharmaceuticals that reduce symptoms. Thyme extract is one of the herbal remedies used to alleviate symptoms. Walther and Schmidtke (2020) conducted an *in vitro* investigation to determine the efficacy of fluid thyme extract against two rhinovirus and influenza A virus strains. The result showed the antiviral efficacy against the influenza virus, but not against rhinovirus. Depending on the dose, thyme extract at non-cytotoxic concentrations of 0.03%–0.33% v/v reduced the cytopathic effect of influenza.

HSV-1 and HSV-2 herpes viruses were successfully treated with *T. capitatus* by Toujani et al. (2018) (Feriotto et al., 2018). The most effective antiviral agent against both viruses tested was found in the ethanol extract which included sitosterol, cinnamaldehyde and carvacrol. Sitosterol exhibited the highest antiviral activity. The active chemicals recovered from this extract could directly inactivate HSV-2 virions, reducing the virus's capacity to infect new cells and its ability to spread. According to Feriotto et al. (2018), three essential oils from *T. vulgaris*, *Cymbopogon citrate* (DC.) and *Rosmarinus officinalis* L. showed antiviral activity against HIV-1. The Eos disrupted the Tat protein, a critical target for antiviral drugs, thus disrupting the HIV replication cycle. This trans-activator, which increases the P-TEFb complex's affinity for the transactivating portion of the viral RNA (trans-activating region), can lead to excessively long viral transcripts and emergence from the latent phase. *T. vulgaris* (at concentrations ranging from 3 to 6 g/mL) and *C. citratus* (at concentrations ranging from 6 to 12 g/mL) EOs demonstrated a clear inhibitory capacity for binding Tat to TAR RNA. There was also evidence of inhibition of viral transcription. At concentrations of 0.30 and 0.65 ng/mL, the EOs of *T. vulgaris* and *C. citratus* inhibited HIV-1 transcription by 52% and 60%, respectively. *Mentha* sp. essential oil also contains menthol, a terpenoid. Menthol's antiviral potency has been studied extensively. Coxsackievirus B infections can be greatly reduced with this treatment. It inhibits TRPV1-mediated proviral mitochondrial fission by activating TRPM8 activity (Taylor et al., 2020).

## 4.4 VIRAL DISEASES AND PLANT-DERIVED ESSENTIAL OILS

Some of the viral diseases and plant-derived EOs are listed in this section.

### 4.4.1 Human Herpes Virus

In the past decade, a significant focus has been placed on enveloped DNA viruses known as human herpes viruses (e.g., HSV-1 and HSV-2). *In vitro* studies have demonstrated that EOs isolated from *Eucalyptus caesia* possess high anti-HSV-1 activity. Furthermore, *Mentha suaveolens* and Australian tea tree exhibit antiviral properties. HSV-1 was found to be susceptible to a wide range of aromatic plants in antiviral screening trials employing EOs from Umbelliferae, Labiatae, Myrtaceae and Lamiaceae plants. Caryophyllene-containing plants, such as black pepper, cannabis, cinnamon, oregano and cloves, were found to be potential sources for selecting HSV antiviral agents (Ma and Yao, 2020; Kowalczyk et al., 2020). Apart from L-bornyl acetate and D-limonene, a wide variety of common components in EOs have been evaluated for their anti-herpes properties and all have exhibited substantial antiviral activity at concentrations of 0.025–0.8 g/mL.

### 4.4.2 Influenza Virus

Influenza viruses are the RNA viruses mainly transmitted either through airborne droplets or through contaminated surface. According to a study by Choi (2018), the EOs derived from tested marjoram, clary sage and anise oils were effective against the influenza A/WS/33 virus. The active component identified was linalool. Furthermore, EOs from *Thymus vulgaris*, *Cinnamomum zeylanicum* and *Citrus bergamia*, at concentration of 3.1 μL/mL demonstrated 100% inhibitory effects and thus can be employed for influenza therapy (Vimalanathan and Hudson, 2014). Similarly, with no cytotoxic effect, the vapour of *E. globulus* EO reduced influenza viral infection by about 78% (Madia et al., 2022). Another *in vitro* study showed that the Maqia EOs which is used as flavoring agent also demonstrated that inhibitory activity against both influenza A and B viruses (Raza et al., 2022). Additionally, germacone has also been proven to suppress numerous strains of feline caliciviruses, which are non-enveloped RNA viruses (Ma and Yao, 2020; Kowalczyk et al., 2020). These findings suggest that germacrone, a significant component of the traditional Chinese herb *rhizoma curcuma*, could be harnessed as a promising broad-spectrum medicinal agent or a reference for the development of innovative antiviral drugs.

### 4.4.3 Non-Enveloped Viruses

Non-enveloped viruses have been relatively neglected in recent years, possibly because most EOs demonstrated the extracellular inhibitory property and are highly effective against enveloped viruses. Non-enveloped viruses have fewer active sites that EOs can target. With an IC50 of 2.24 g/mL and a SI of 789.84, the EO from *Osmunda regalis*, a Tunisian fern, displayed outstanding anti-Coxsackie activity, significantly beyond the threshold. Furthermore, the EOs from *Eucalyptus bicostata* and *Dysphania ambrosioides* exhibited excellent anti-Coxsackie activity. Oregano EO and its major component carvacrol significantly inhibited human rotavirus and murine norovirus. These researches revealed that EOs may possess antiviral properties against non-enveloped viruses. Unfortunately, information on precise mechanisms of action and the bioactive components in the EO mixture responsible for antiviral activity remains insufficient (Kowalczyk et al., 2020).

### 4.4.4 SARs-CoV-2 Virus

The Coronavirus Infection of 2019 (COVID-19) has emerged as a major public health concern worldwide. Efforts are underway to find a vaccination or treatment for the illness it causes. FDA-approved drugs that have been used in the past are being re-evaluated for their effectiveness against SARS-CoV-2. Meanwhile, investigations into potential replacement therapies continue. This is facilitated by employing naturally occurring compounds of EOs, which have multiple biological effects, including antiviral activity and molecular docking strategies. Molecular docking is a method for predicting how two or more molecules with known structures will connect to one another and how strongly they will attach to each other in a complex. Protein-ligand docking has become an important part of modern structure-based drug design, allowing for informed predictions about the bonding and affinity of a complex between a protein receptor and a ligand, serving as an alternative to costly and time-consuming experiments. Several compounds found in EOs, including thymol, have been studied using this technique. Researchers have relied on the recently determined crystal structure of the SARS-CoV-2 spike glycoprotein (S) for these investigations (Walls et al., 2020).

S1 and S2 are the two subunits of the S protein. S1, which contains the receptor-binding domain (RBD), interacts with the peptidase domain of angiotensin-converting enzyme 2 (ACE2), establishing a connection between the virus and the host cell surface. The second subunit, S2, oversees the membrane that separates the virus from its host. Clinically, the S1 subunit is essential because blocking RBD leads to structural alterations in the S protein, which is the first line of defence against viral infections. The ligands carvacrol, anethole, cinnamaldehyde and L-4-terpineol have been shown to have higher binding affinities and docking scores than the other ligands tested. These ligands exhibited increased binding affinities with the RBD of the S protein due to their H-bond and hydrophobic interactions with the amino acid residues in the binding site. Strong binding capabilities were also seen for thymol, camphene, pulegone, ocimene and menthol. However, they did not engage in hydrogen-binding interactions with the protein. The stability of the protein-ligand complex may be attributed in part to hydrophobic interactions between the many hydrophobic residues surrounding the ligands in the binding pocket (Ma and Yao, 2020). Thymol is an active component in COVID-19 formulations and has been included on the lists of disinfectants with evidence for use compiled by the United States Environmental Protection Agency (EPA) and the Government of Canada.

## 4.5 SCOPE OF PHYTOCHEMICALS IN FUTURE MEDICINES

Despite remarkable high technology improvement in traditional approaches and therapies, cancer remains a leading cause of mortality. While significant advancement has been made in treating cancer, numerous restrictions still remains, therefore necessitating future development. Chemotherapies, which interfere with cancer's healing mechanisms, are associated with many negative effects and complications (Khan et al., 2022). Since primordial era, medicinal compounds derived from plants were extensively applied in clinical practice due to their effectiveness and lower adverse effects. The contemporary research activities and superior determination procedures of biologically active components in plants have facilitated invention of phytochemicals for the cure and prevention of disease, including cancer. Different phytochemicals including curcumin, vinblastine, colchicine, paclitaxel, lycopene and vincristine have demonstrated anticancerous potential (Khan et al., 2022).

Apart from their application in the treatment of cancer, phytochemicals also possess the ability to deal with severe acute respiratory syndromes such as coronavirus. Researchers worldwide are actively working on developing vaccines and drugs to combat Covid-19 (Renjith and Sankar, 2020). Phytonutrients and phytochemicals are widely acknowledged for their importance in preventing several microbial infections, cancer and coronary heart diseases. Through drug repurposing techniques such as molecular docking, various phytochemicals have illustrated considerable antiviral potential against coronavirus. These include coumarins, lignans, tannins, proteins, alkaloids, polyphenols, anthocyanidins, flavonoids and lecithins, which are known to show antiviral properties against severe acute respiratory syndrome coronaviruses (Kapoor et al., 2017).

## 4.6 SCOPE OF PHYTO-ESSENTIAL OILS IN FUTURE MEDICINE

Phyto-EOs demonstrate significant proficiency as curative candidates against numerous diseases and can be employed as effective medicines. These oils possess broader characteristics, such as antimicrobial potential and anti-inflammatory aspects, which contribute to their use in curing dental diseases. The biopharmaceutical importance of EOs relies on the development of effective delivery systems, ensuring that effective fraction of essential oil administered to the body reaches the circulatory system. EOs are encapsulated in different carrier materials to enhance the biomedical applications of EOs and protect them from undesirable environmental conditions. EOs are also complexed with cyclodextrins for improving stability and solubility.

Identifying active substances in the EOs to elucidate the underline mechanism is crucial for studying the therapeutic potential of EOs. Further research should be conducted to explore the possibility of combining antiviral properties of EOs with prescribed classical drugs. Another importance of EOs lies in the fact that EOs could be administered both in form of aerosols and oral route. Inhalation of EOs by COVID-19 patients has proved to be an adjuvant therapy due to the presence of active volatile phytochemicals that facilitate easier breathing and improve lung function.

## 4.7 CONCLUSION

Literature surveys depicting the role of phytochemicals and phtyo EOs suggest that these compounds could be effectively used in drug development against various pathogenic viruses. Phytochemicals and phtyo EOs can disrupt cell membrane of virus, inhibit DNA polymerase activity and stop protein synthesis. Further, EOs have shown a remarkable potential in inhibiting virus attachment to the host cells. Due to these proven benefits of phytochemicals and EOs, they could prove beneficial for viral diseases and suppress various symptomatic infections. Food and Drug Administration (FDA) closely monitors the use of EOs for pharmaceutical use. The FDA has provided guidelines for determination of different parameters such as chromatography profiles, acid value, phenolic content and moisture content of EOs in order to maintain quality standards for individual EOs. Although, having a wide pharmaceutical importance, it is crucial to note that phytochemical and EOs can be toxic at high concentrations. Additionally, a nontoxic phytochemical could be toxic to another person due to sensitization and genetic reasons. Therefore, a thorough understanding of bioactive compounds is essential before their legitimate use can be established.

## REFERENCES


Armaka, M., Papanikolaou, E., Sivropoulou, A., & Arsenakis, M. (1999). Antiviral properties of isoborneol, a potent inhibitor of herpes simplex virus type 1. *Antiviral Research*, *43*(2), 79–92.

Allahverdiyev, A., Duran, N., Ozguven, M., & Koltas, S. (2004). Antiviral activity of the volatile oils of *Melissa officinalis L.* against Herpes simplex virus type-2. *Phytomedicine : international journal of phytotherapy and phytopharmacology*, *11*(7-8), 657–661. https://doi.org/10.1016/j.phymed.2003.07.014

Barra, A. (2009). Factors affecting chemical variability of essential oils: a review of recent developments. *Natural Product Communications*, *4*(8), 1934578X0900400827.

Ben-Shabat, S., Yarmolinsky, L., Porat, D., & Dahan, A. (2020). Antiviral effect of phytochemicals from medicinal plants: applications and drug delivery strategies. *Drug Delivery and Translational Research*, *10*(2), 354–367.

Bhardwaj, K., Islam, M. T., Jayasena, V., Sharma, B., Sharma, S., Sharma, P., ... & Bhardwaj, P. (2020). Review on essential oils, chemical composition, extraction, and utilization of some conifers in Northwestern Himalayas. *Phytotherapy Research*, *34*(11), 2889–2910.

Brijesh, S., Daswani, P., Tetali, P., Antia, N., & Birdi, T. (2009). Studies on the antidiarrhoeal activity of Aegle marmelos unripe fruit: Validating its traditional usage. *BMC Complementary and Alternative Medicine*, *9*(1), 1–12.

Castañeda-Ovando, A., de Lourdes Pacheco-Hernández, M., Páez-Hernández, M. E., Rodríguez, J. A., & Galán-Vidal, C. A. (2009). Chemical studies of anthocyanins: A review. *Food chemistry*, *113*(4), 859–871.

Carlucci, M. J., Scolaro, L. A., Errea, M. L., Matulewicz, M. C., & Damonte, E. B. (1997). Antiviral activity of natural sulphated galactans on herpes virus multiplication in cell culture *PlantaMed*, *63*, 429–432.

Chouhan, S., Sharma, K., & Guleria, S. (2017). Antimicrobial activity of some essential oils-present status and future perspectives. *Medicines*, *4*(3), 58.

Choi, H.-J. (2018). Chemical constituents of essential oils possessing anti-influenza A/WS/33 virus activity. *Osong Public Health and Research Perspectives*, *9*, 348–353.

Cragg, G. M., & Newman, D. J. (2013). Natural products: A continuing source of novel drug leads. *Biochimica et Biophysica Acta (BBA)-General Subjects*, *1830*(6), 3670–3695.

Duarte, M. E. R., Noseda, D. G., Noseda, M. D., Tulio, S., Pujol, C. A., & Damonte, E. B. (2001). Inhibitory effect of sulfated galactans from the marine alga Bostrychia montagnei on herpes simplex virus replication in vitro. *Phytomedicine*, *8*(1), 53–58.

Feriotto, G., Marchetti, N., Costa, V., Beninati, S., Tagliati, F., & Mischiati, C. (2018). Chemical composition of essential oils from Thymus vulgaris, Cymbopogon citratus, and Rosmarinus officinalis, and their effects on the HIV-1 tat protein function. *Chemistry & Biodiversity*, *15*(2), e1700436.

Garozzo, A., Timpanaro, R., Stivala, A., Bisignano, G., & Castro, A. (2011). Activity of Melaleuca alternifolia (tea tree) oil on influenza virus A/PR/8: Study on the mechanism of action. *Antiviral Research*, *89*(1), 83–88.

Gasparini, R., Amicizia, D., Lai, P. L., & Panatto, D. (2012). Clinical and socioeconomic impact of seasonal and pandemic influenza in adults and the elderly. *Human Vaccines & Immunotherapeutics*, *8*(1), 21–28.

Ghani, U., Naeem, M., Rafeeq, H., Imtiaz, U., Amjad, A., Ullah, S., ... & Qasim, F. (2019). A novel approach towards nutraceuticals and biomedical applications. *Scholars International Journal of Biochemistry*, *2*, 245–252.

Ghidoli, M., Colombo, F., Sangiorgio, S., Landoni, M., Giupponi, L., Nielsen, E., & Pilu, R. (2021). Food containing bioactive flavonoids and other phenolic or sulfur phytochemicals with antiviral effect: Can we design a promising diet against COVID-19? *Frontiers in Nutrition*, *8*, 303.

Girija, A. R. (2017). Future Industries Institute. In: Sotirios Koutsotoulos (Ed.) *Peptide Applications in Biomedicine, Biotechnology and Bioengineering* (p. 157). University of South Australia, Mawson Lakes, SA, Australia.

Gopal, S. K., Rangasamy, M., & Balasubramaniam, N. (2021). In-silico docking analysis of phytochemicals from mimosa pudica l. Leaves as an antiviral agent against herpes simplex virus type I. *International Journal*, *1*, 1–9.

Iftikhar, S., Shahid, A. A., Javed, S., Nasir, I. A., Tabassum, B., & Haider, M. S. (2013). Essential oils and latices as novel antiviral agent against Potato leaf roll virus and analysis of their phytochemical constituents responsible for antiviral activity. *Journal of Agricultural Science*, *5*(7), 167.

Jaberian, H., Piri, K., & Nazari, J. (2013). Phytochemical composition and in vitro antimicrobial and antioxidant activities of some medicinal plants. *Food Chemistry*, *136*(1), 237–244.

Jassim, S. A., & Naji, M. A. (2003). Novel antiviral agents: A medicinal plant perspective. *Journal of Applied Microbiology*, *95*(3), 412–427.

Kapoor, R., Sharma, B., & Kanwar, S. S. (2017). Antiviral phytochemicals: An overview. *Biochemistry & Physiology*, *6*(2), 7.

Khan, A. W., Farooq, M., Haseeb, M., & Choi, S. (2022). Role of plant-derived active constituents in cancer treatment and their mechanisms of action. *Cells*, *11*(8), 1326.

Kopustinskiene, D. M., Jakstas, V., Savickas, A., & Bernatoniene, J. (2020). Flavonoids as anticancer agents. *Nutrients*, *12*(2), 457.

Kowalczyk, A., Przychodna, M., Sopata, S., Bodalska, A., & Fecka, I. (2020). Thymol and thyme essential oil-new insights into selected therapeutic applications. *Molecules*, *25*(18), 4125.

Lima, N. M., Andrade, T. D. J. A., Acquah, K. S., de Oliveira, M. A., Gois, K. F., & Medeiros, L. C. M. (2021). Phytochemicals as potential antiviral agents in SARS-CoV-2 therapy: An update. *Química Nova*, *44*, 667–672.

Lin, Y., Jain, R., & Yan, Y. (2014). Microbial production of antioxidant food ingredients via metabolic engineering. *Current Opinion in Biotechnology*, *26*, 71–78.

Mani, J. S., Johnson, J. B., Steel, J. C., Broszczak, D. A., Neilsen, P. M., Walsh, K. B., & Naiker, M. (2020). Natural product-derived phytochemicals as potential agents against coronaviruses: A review. *Virus Research*, *284*, 197989.

Madia, V. N., Toscanelli, W., De Vita, D., De Angelis, M., Messore, A., Ialongo, D., & Costi, R. (2022). Ultrastructural damages to H1N1 influenza virus caused by vapor essential oils. *Molecules*, *27*(12), 3718.

Ma, L., & Yao, L. (2020). Antiviral effects of plant-derived essential oils and their components: An updated review. *Molecules*, *25*(11), 2627.

Mardani, M., Motamedifar, M., & Hoseinipour, R. (2012). A study of the antiviral effect of the essential oil of zataria multiflora boiss on herpes simplex Type 1 in vero cell culture. *Journal of Dentistry*, *13*(Supplement-September-2012), 414–420.

Mihai, A. D., Chircov, C., Grumezescu, A. M., & Holban, A. M. (2020). Magnetite nanoparticles and essential oils systems for advanced antibacterial therapies. *International Journal of Molecular Sciences*, *21*(19), 7355.

N Powers, C., & N Setzer, W. (2016). An in-silico investigation of phytochemicals as antiviral agents against dengue fever. *Combinatorial Chemistry & High Throughput Screening*, *19*(7), 516–536.

Naithani, R., Mehta, R. G., Shukla, D., Chandersekera, S. N., & Moriarty, R. M. (2010). Antiviral Activity of Phytochemicals: A Current Perspective. In Ronald Ross Watson, Sherma Zibadi, & Victor R. Preedy (Eds.) *Dietary Components and Immune Function* (pp. 421–468). Humana Press, Totowa, NJ.

Niedermeyer, T. H., Lindequist, U., Mentel, R., Gördes, D., Schmidt, E., Thurow, K., & Lalk, M. (2005). Antiviral Terpenoid constituents of Ganoderma p feifferi. *Journal of Natural Products*, *68*(12), 1728–1731.

Ninfali, P., Antonelli, A., Magnani, M., & Scarpa, E. S. (2020). Antiviral properties of flavonoids and delivery strategies. *Nutrients*, *12*(9), 2534.

Orhan, I. E., Senol, F. S., Ozturk, N., Celik, S. A., Pulur, A., & Kan, Y. (2013). Phytochemical contents and enzyme inhibitory and antioxidant properties of *Anethum graveolens L.*(dill) samples cultivated under organic and conventional agricultural conditions. *Food and Chemical Toxicology*, *59*, 96–103.

Paduch, R., Kandefer-Szerszen, M., Trytek, M., & Fiedurek, J. (2007). Terpenes: substances useful in human healthcare. *Archivum Immunologiae et Therapiae Experimentalis*, *55*, 315–327.

Park, S., Kim, J. I., Lee, I., Lee, S., Hwang, M. W., Bae, J. Y., ... & Park, M. S. (2013). Aronia melanocarpa and its components demonstrate antiviral activity against influenza viruses. *Biochemical and Biophysical Research Communications*, *440*(1), 14–19.

Patel, C. N., Goswami, D., Jaiswal, D. G., Parmar, R. M., Solanki, H. A., & Pandya, H. A. (2021). Pinpointing the potential hits for hindering interaction of SARS-CoV-2 S-protein with ACE2 from the pool of antiviral phytochemicals utilizing molecular docking and molecular dynamics (MD) simulations. *Journal of Molecular Graphics and Modelling*, *105*, 107874.

Patra, A. K. (Ed.). (2012). *Dietary Phytochemicals and Microbes*. Springer Science & Business Media, Dordrecht.

Pilau, M. R., Alves, S. H., Weiblen, R., Arenhart, S., Cueto, A. P., & Lovato, L. T. (2011). Antiviral activity of the Lippia graveolens (Mexican oregano) essential oil and its main compound carvacrol against human and animal viruses. *Brazilian Journal of Microbiology*, *42*, 1616–1624.

Prakash, D., & Gupta, C. (2011). Role of phytoestrogens as nutraceuticals in human health. *Pharmacologyonline*, *1*, 510–523.

Raza, Q. S., Saleemi, M. K., Gul, S., Irshad, H., Fayyaz, A., Zaheer, I., & Khan, A. (2022). Role of essential oils/volatile oils in poultry production—A review on present, past and future contemplations. *Agrobiological Records*, *7*, 40–56.

Reddy, D. N. (2019). Essential Oils Extracted from Medicinal Plants and Their Applications. In: Mallappa Kumara Swamy & Mohd Sayeed Akhtar (Eds.) *Natural Bio-Active Compounds* (pp. 237–283). Springer, Singapore.

Renjith, M. R. D., & Sankar, M. (2020). Scope of phytochemicals in the management of COVID-19. *Pharmaceutical Research*, *3*(1), 26–29.

Rytel, E., Tajner-Czopek, A., Kita, A., Sokół-Łętowska, A., Kucharska, A. Z., & Hamouz, K. (2019). Effect of the production process on the content of anthocyanins in dried red-fleshed potato cubes. *Italian Journal of Food Science*, *31*(1) 150–160.

Šafratová, M., Hošťálková, A., Hulcová, D., Breiterová, K., Hrabcová, V., Machado, M., & Cahlíková, L. (2018). Alkaloids from *Narcissus poeticus* cv. Pink Parasol of various structural types and their biological activity. *Archives of Pharmacal Research*, *41*, 208–218.

Sánchez, G., & Aznar, R. (2015). Evaluation of natural compounds of plant origin for inactivation of enteric viruses. *Food and Environmental Virology*, *7*, 183–187.

Shaaban, H. A., El-Ghorab, A. H., & Shibamoto, T. (2012). Bioactivity of essential oils and their volatile aroma components. *Journal of Essential Oil Research*, *24*(2), 203–212.

Singh, J., & Sinha, S. (2012). Classification, regulatory acts and applications of nutraceuticals for health. *International Journal of Pharma and Bio Sciences*, *2*(1), 177–187.

Soldatkina, L., & Novotna, V. (2017). Removal of anthocyanins from aqueous berry extracts by adsorption on bentonite: Factorial design analysis. *Adsorption Science & Technology*, *35*(9–10), 866–877.

Soldati, L. (2018). The influence of diet on anti-cancer immune responsiveness. *Journal of Translational Medicine*, 16(1), 1–18. https://doi.org/10.1186/s12967-018-1448-0.

Sun, J. (2018). *The Repurposing of Pharmaceutical Agents to Improve Treatment Outcomes in Antibiotic-Resistant Bacterial Infections.* University of California, San Diego.

Swamy, M. K. (Ed.). (2020). *Plant-Derived Bioactives: Production, Properties and Therapeutic Applications.* Springer Nature, Singapore.

Taylor, D. J., Hamid, S. M., Andres, A. M., Saadaeijahromi, H., Piplani, H., Germano, J. F., ... & Sin, J. (2020). Antiviral effects of menthol on coxsackievirus B. *Viruses*, *12*(4), 373.

Toujani, M. M., Rittà, M., Civra, A., Genovese, S., Epifano, F., Ghram, A., ... & Donalisio, M. (2018). Inhibition of HSV-2 infection by pure compounds from Thymus capitatus extract in vitro. *Phytotherapy Research*, *32*(8), 1555–1563.

Umesalma, S., & Sudhandiran, G. (2010). Differential inhibitory effects of the polyphenol ellagic acid on inflammatory mediators NF-κB, iNOS, COX-2, TNF-α, and IL-6 in 1, 2-dimethylhydrazine-induced rat colon carcinogenesis. *Basic & Clinical Pharmacology & Toxicology*, *107*(2), 650–655.

Vimalanathan, S., & Hudson, J. (2014). Anti-influenza virus activity of essential oils and vapors. *American Journal of Essential Oils and Natural Products*, *2*(1), 47–53.

Venkidasamy, B., Selvaraj, D., Nile, A. S., Ramalingam, S., Kai, G., & Nile, S. H. (2019). Indian pulses: A review on nutritional, functional and biochemical properties with future perspectives. *Trends in Food Science & Technology*, *88*, 228–242.

Walls, A. C., Park, Y. J., Tortorici, M. A., Wall, A., McGuire, A. T., & Veesler, D. (2020). Structure, function, and antigenicity of the SARS-CoV-2 spike glycoprotein. *Cell*, *181*(2), 281–292.

Walther, C., Döring, K., & Schmidtke, M. (2020). Comparative in vitro analysis of inhibition of rhinovirus and influenza virus replication by mucoactive secretolytic agents and plant extracts. *BMC complementary medicine and therapies*, *20*(1), 380. https://doi.org/10.1186/s12906-020-03173-2

Wang, J., Guleria, S., Koffas, M. A., & Yan, Y. (2016). Microbial production of value-added nutraceuticals. *Current Opinion in Biotechnology*, *37*, 97–104.

# 5 Herbal or Poly-Herbal Formulations in Various Viral Diseases

*Gursharan Kaur and Palak Rana*
University Institute of Biotechnology, Chandigarh University

*Vagish Dwibedi*
University Institute of Biotechnology, Chandigarh University
Volcani Research Center, Agricultural Research Organization

*Santosh Kumar Rath*
DIT University

## 5.1 INTRODUCTION

Viral diseases represent a significant global health burden, causing widespread morbidity and mortality. Conventional antiviral therapies often face challenges such as limited efficacy, the emergence of drug-resistant strains and adverse side effects. Consequently, there is a growing interest in exploring alternative therapeutic approaches that complement or provide alternatives to conventional treatments (Dhama et al., 2020; Sharma et al., 2020). One such method is using herbal or poly-herbal formulations, which have been employed for centuries in traditional medicine systems worldwide. Herbal medicine, also known as botanical medicine or phytomedicine, utilizes various plant parts, including leaves, stems, roots and flowers, for their therapeutic properties. These plants contain a rich array of phytochemicals, such as alkaloids, flavonoids, terpenoids, phenolic compounds and polysaccharides, which possess diverse biological activities, including antiviral effects. The use of herbal formulations in the treatment of viral diseases is rooted in traditional knowledge and ethnobotanical practices, where specific plants or combinations of plants have been traditionally used for their antiviral properties (Othman et al., 2019; Parham et al., 2020; Prasathkumar et al., 2021).

In recent years, there has been a resurgence of interest in herbal medicine research, driven by the need for novel antiviral therapies. Modern scientific investigations have sought to validate the traditional use of herbal formulations and elucidate their mechanisms of action against viral pathogens. Studies have demonstrated the antiviral activities of various plant extracts, decoctions and standardized herbal products against a wide range of viruses, including influenza, herpesviruses, human immunodeficiency virus (HIV), hepatitis viruses and respiratory syncytial virus (RSV), among others (Li et al., 2021; Anjum et al., 2023; Wang et al., 2022). The potential advantages of herbal or poly-herbal formulations in viral diseases are manifold. First, these formulations often exhibit a broad-spectrum antiviral activity, targeting multiple steps in the viral replication cycle or exerting immunomodulatory effects to enhance the host's defense mechanisms. This multi-targeted approach may reduce the risk of viral resistance development, a common concern with single-target antiviral drugs. Second, herbal formulations often contain synergistic combinations of multiple plant constituents, which can work together to enhance their antiviral effects through additive or synergistic interactions (Wang et al., 2022). Additionally, herbal formulations may offer a more

DOI: 10.1201/9781003329169-6

holistic and patient-centered approach, considering the individual's overall well-being and addressing other aspects of health beyond viral suppression (Borchers et al., 2013; Newman et al., 2016).

Despite the promising potential of herbal formulations, challenges exist in their utilization. Standardization and quality control of herbal products are crucial to ensure consistent potency, efficacy and safety. Regulatory frameworks also need to be established to govern the production, distribution and usage of herbal formulations in a regulated and evidence-based manner. Furthermore, collaborative efforts among traditional healers, herbal medicine researchers and modern healthcare providers are vital for bridging the knowledge gap and integrating herbal medicine into mainstream healthcare systems (Ekor, 2014; Ampomah et al., 2020; Balekundri & Mannur, 2020).

The aim of this review is to provide a comprehensive overview of the current understanding of herbal or poly-herbal formulations in various viral diseases. It will explore the mechanisms of action of these formulations, discuss their efficacy in specific viral infections, summarizes the evidence from clinical studies, highlight synergistic combinations and adjuvant therapies, and address the challenges and future perspectives of integrating herbal medicine into the management of viral disease.

## 5.2 OVERVIEW OF HERBAL MEDICINE IN VIRAL DISEASES

Numerous herbal extracts or their derivatives have demonstrated promising antiviral effectiveness, indicating the potential of natural products and their compounds in treating viral infections. Traditional medicine, originating in approximately 2600 BCE, has long emphasized using plants as remedies. For example, in Mesopotamia, there were documented instances of using oils extracted from plants. Medicinal plants, such as *Cupressus arizonica* Greene, *Cedrus libani* (A. Rich), *Commiphora acuminata* Mattick, *Papaver somniferum* L. and *Glycyrrhiza glabra* L., continue to be utilized in treating various ailments, ranging from common colds, inflammation to parasitic diseases (Jana et al., 2021). The Egyptian pharmaceutical tradition, dating back approximately to 2900 BCE, holds a prominent place in the history of conventional medicine. One of the most well-known records of this ancient practice is the "Ebers Papyrus," dating from 1500 BCE, which catalogs over 700 drugs, majorly derived from plant sources (Zebroski et al., 2016).

The field of Indian medicinal plants holds great promise for treating various diseases. Ayurveda and Siddha practices, originating in India, continue to be extensively utilized by the Indian population (Adhikari et al., 2018). Since ancient times, Indians have referred to their medicine as Ayurveda, derived from the combination of two Sanskrit words: Ayur, meaning "life," and Veda meaning "knowledge." Ayurvedic medicine evolved from the wisdom found in the Vedas, India's four renowned books of ancient wisdom (Shrivastava et al., 2020). The Rig Veda, dating back 4,500 years, is the oldest among the four Vedas and comprises 1,028 Hindu hymns. These hymns were recited orally from 2500 BCE and later transcribed approximately 1,000 years later. Remarkably, the Rig Veda includes elaborate depictions of practices such as eye surgery, limb amputation and the formulation of medicinal remedies utilizing 67 herbs, including well-known ones like ginger, cinnamon and senna. Similarly, the Atharva Veda, which is approximately 3,200 years old, also provides insights into numerous medicinal herbs and their healing properties (Castleman, 2017).

Furthermore, the identification of phytochemical constituents within medicinal plants has the potential to aid in alleviating infections. Indian traditional medicine, as the bedrock of ancient medical practices worldwide, has been pivotal in providing vital healthcare services and promoting well-being since its inception. Some of the most commonly used Indian herbal formulations that are in daily use today include Triphala, Chyawanprash, RoohAfza and more. Therefore, Indian medicinal plants present a novel avenue for exploring their potential in combating viral transmission (Ahmad et al., 2020).

Chinese traditional medicine has gained widespread recognition over the centuries. Its documented history dates back approximately 2,000 years, with notable records including the Wu Shi Er Bing Fang (1100 BCE; containing 52 medicines), the herbal Shennong (100 BCE; featuring 365 medicines) and the herbal Tang (659 CE; comprising 850 medicines). In comparison, the records of Indian traditional medicine trace back more than 5,000 years, with the Charaka and Sushruta Samhitas containing 341 and 395 herbal medications, respectively, dating back to around 1000 BCE (Adhikari et al., 2018). The Romans and Greeks also made significant contributions to the advancement of traditional herbal medications in the ancient Western world. Notably, Dioscorides, a Greek physician from around 100 CE, documented the gathering, preservation and appropriate utilization of traditional medicine across the "known world" at that time. In modern times, natural products (NPs) sold as herbal or food supplements account for approximately half the top 50 drugs available in European pharmacies (Jana et al., 2021). Traditional systems of herbal medicine are widely practiced across the globe, encompassing a variety of opinions and social influences. Among these archaic practices, conventional Indian medicine holds the distinction of being one of the oldest.

Even in modern times, these herbal formulations are practiced as plant extracts containing numerous secondary metabolites with effective therapeutic potential. However, in the present day, the focus has shifted toward harnessing the medicinal properties of herbal medicines for drug development processes. Extracts, tinctures, essential oils and expressed juices are all examples of herbal compositions. Chemically defined components or sets of components in an herbal drug, preparation, or medicinal product are known as "markers," which can be used for research purposes even if they have no therapeutic value. As long as the markers in the herbal substance or preparations have been quantifiably determined, their use in herbal formulations can be calculated (Kumari and Kotecha, 2016). In the quest to combat COVID-19, scientific efforts have been directed toward two different approaches. However, it is crucial to emphasize the importance of exploring the medical potential of plant extracts in preventing and managing such viral diseases. Focusing on the study of herbal extracts can offer valuable insights and potential solutions in the fight against these viral diseases (Singh & Agarwal, 2021). More than 450 herbal formulations and compounds have shown the capability to hinder the coronavirus-host protein pathways and disrupt various stages of the virus life cycle. This includes impeding viral entry into host cells, inhibiting membrane fusion, interfering with processes like transcription and translation, impeding viral replication, hindering viral assembly and preventing the release of the virus (Fuzimoto & Isidoro, 2020). *Polygoni multiflora radix*, *Rhei Radix et Rhizoma*, *Polygoni multiflora Caulis* and specifically the phytochemical *emodin* derived from them have demonstrated potent antiviral activity against ACE-2 (angiotensin converting-enzyme-2), the host target protein of coronavirus. Their inhibitory concentration ($IC_{50}$) against the virus was measured to be $1^{10}$mg/mL, while emodin exhibited an $IC_{50}$ of 200 mM (Wu et al., 2020). In another research, scientists have confirmed that *theaflavin3,30-di-O-gallate*, a compound derived from *Camellia sinensis* (green tea), exhibits significant inhibition against the Covid-19 RNA-dependent RNA polymerase and chymotrypsin-like protease. Additionally, *epigallocatechin gallate*, another biochemical compound found in *C. sinensis*, effectively blocks the papain-like protease. These findings highlight the potential of these compounds in combating Covid-19 and inhibiting key viral enzymes involved in replication and proteolytic processing (Fuzimoto & Isidoro, 2020).

A survey revealed that approximately 65%–80% of the worldwide population of developing countries prefer traditional herbal medications as their first choice for treatment (Patel et al., 2021). Currently, no FDA-approved vaccine exists for HSV; therefore, it becomes important to identify novel anti-herpes simplex virus (HSV) drugs. Numerous herbal medications have been reported for their potential in this regard. Several FDA-approved plant-derived drugs exemplify the therapeutic potential of herbs. For instance, the alkaloid drug deserpidine, developed from the medicinal plants *Rauwolfia canescens*, acts as an anti-hypertensive drug. Veregen, another FDA-approved drug, is utilized in treating genital warts caused by human papillomavirus (HPV), synthesized from the dried leaf of *Camellia sinensis* (Luo et al., 2020). Furthermore, in the case of HSV, the infusion of *J. australis* and *Er. japonica* has demonstrated effective results against HSV-1 and HSV-2 (Bhagat & Mhaiskar, 2023).

Creating effective and combined herbal formulations that can effectively inhibit multiple pathways of the virus life cycle would undoubtedly be advantageous in combating various viral infections. Exploring the potential synergies between conventional drugs and natural substances would also be highly valuable. Additionally, many herbs and modified herbal formulations traditionally employed by experienced herbalists for treating viral infections may target specific proteins of the host. The herbal formulas, consisting of multiple herbs with numerous compounds, can potentially target various pathways, including multiple protein pathways, while mediating antiviral, anti-inflammatory, anti-complement, antipyretic, immune-modulatory, anti-toxicity, anti-pulmonary fibrosis and analgesic responses, including those that are yet to be fully understood. Such multifaceted approaches hold promise in the search for effective treatments against viral infections (Fuzimoto & Isidoro, 2020).

However, additional research beyond the initial drug discovery phase is necessary to gain a comprehensive understanding of the effectiveness and safety of these herbs and isolated substances when used alone or in combination with other herbs or drugs. It is important to note that the results obtained from *in vitro* studies may not necessarily translate to the same clinical effects when applied to human patients (Atanasov et al., 2021). The screening of herbal products usually entails testing a collection of extracts derived from natural sources. Nevertheless, this approach may not always be complemented with conventional target-based assays. One of the challenges lies in the identification of the specific bioactive compounds of interest. Furthermore, obtaining an adequate amount of biological material for isolating and characterizing a bioactive NP can also pose a challenge (Nasim et al., 2022).

## 5.3 MECHANISMS OF ACTION OF HERBAL AND POLY-HERBAL FORMULATIONS AGAINST VIRUSES

### 5.3.1 Immunomodulatory Effects

Herbal formulations have been found to exhibit immunomodulatory effects. These effects refer to the ability of certain herbs or herbal combinations to modulate or regulate the immune system's response. Herbal formulations can enhance or downregulate immune activity as needed, depending on the individual's condition. These immunomodulatory properties can contribute to various health benefits, including boosting immune function, reducing inflammation and supporting overall wellness (Shi et al., 2021). The specific mechanisms through which herbal formulations exert their immunomodulatory effects can vary and may involve interactions with the immune cells of the immune system (De Wolf et al., 2017).

Ashwagandha (*Withania somnifera*), an herb widely used in Ayurvedic medicine, has been shown to possess immunomodulatory effects by enhancing the activity of immune cells, such as natural killer cells and macrophages, and regulating immune responses (Tamoli et al., 2022). In a computational study, the immunomodulatory and anti-SARS-CoV-2 potential of key phytoconstituents from Ayurveda-based formulation, namely Ashwagandha, Shatavari and Guduchi, were investigated using system pharmacology and molecular docking approaches. Among these three herbs, Ashwagandha exhibited the highest potency as an immunomodulator. It was found to have the ability to regulate T cell proliferation, enhance NK (natural killer) cell cytotoxicity and modulate signaling pathways involved in B cell, T cell and NOD-like receptor functions (Varma et al., 2020).

Similarly, Ginseng (*Panax ginseng*) is a popular herb known for its adaptogenic properties (Ratan et al., 2021). It contains numerous bioactive components, including phytosterols, amino acids, saponins, nitrogenous substances, polyphenolic compounds, acidic polysaccharides and polyacetylenes. These constituents have been shown to possess therapeutic properties, making them beneficial in the treatment, alleviation and protection against various diseases (Nguyen & Nguyen, 2019). In a study focused on human rhinovirus (HRV) infection, Korean Red Ginseng

(KRG) was investigated. The study measured the levels of inflammatory cytokines interleukin IL-8 and IL-6 at the mRNA and protein level using real-time polymerase chain reaction and enzyme-linked immunosorbent assay. The results showed that KRG significantly reduced the HRV-induced increase in IL-8 and IL-6 mRNA and protein levels in primary human nasal epithelial (HNE) cells. Furthermore, KRG treatment inhibited the activation of NF-κB and MAP kinase, which are involved in the rhinovirus-induced inflammatory response (Alsayari et al., 2021). Another example is Echinacea (*Echinacea* spp.), an herb commonly used to support the immune system. It has been shown to stimulate immune cell activity, increase the production of cytokines and enhance the function of immune cells involved in the body's defense against infections (Brendler et al., 2021).

### 5.3.2 Antiviral Activities and Viral Replication Inhibition

Herbal formulations have been investigated for their antiviral activities and potential to inhibit viral replication. Certain herbal compounds can directly inhibit viral replication by targeting specific stages of the viral life cycle. For example, they may interfere with viral attachment, entry, fusion, or replication within host cells. By strengthening the immune response, these formulations can indirectly inhibit viral replication by enhancing the body's ability to fight off infections. Some herbs contain antioxidants that can scavenge free radicals and reduce oxidative stress, helping protect cells from viral-induced damage and support the body's defense mechanisms against viral infections (Rouf et al., 2020).

For example, Elderberry (*Sambucus* spp.) have been found to inhibit viral replication by preventing viral attachment and entry into host cells. The bioactive compound caffeic acid of elderberry has been shown to inhibit the replication of coronavirus effectively. It specifically targets the initial attachment of the virus, demonstrating significant inhibitory activity (IC50 = 8.1 μM). These findings highlight the potent anti-Human Coronavirus V-NL63 potential of the ethanol extract derived from the stem of *Sambucus formosana Nakai* (Weng et al., 2019). Numerous studies have demonstrated the antiviral properties of garlic (*Allium sativum* L.) and its constituents, such as allicin, diallyl sulfide and ajoene, which can effectively inhibit viral replication against various pathogenic viruses. In the context of the influenza A virus, it has been observed that fresh garlic extract (GE) can inhibit viral replication by interfering with the formation of viral antigens. Remarkably, this inhibitory effect occurs even at an early stage of infection in MDCK cells, indicating the potential of garlic as an early intervention against viral infections (Rouf et al., 2020).

Similarly, Licorice (*Glycyrrhiza glabra*) contains compounds such as glycyrrhizin and glycyrrhetinic acid, which have demonstrated antiviral activities against various viruses, including HSV, HIV and RSV. Liquorice extracts can inhibit viral replication and have immune-stimulating effects (Pastorino et al., 2018).

Andrographis (*Andrographis paniculata*) extracts have shown antiviral effects against several viruses, including influenza, dengue and herpes viruses. A study has shown that the maximum non-toxic dose of isolated andrographolide from *A. paniculata* extract was determined to be 15.62 μg/mL on the C6/36 cell line using the MTT assay. It exhibited significant anti-dengue activity, with a potency of 97.23% against the dengue-2 virus. Molecular docking studies revealed that andrographolide interacted with the NS5 protein of the dengue virus, demonstrating a strong binding affinity with a maximum binding energy of –7.35 kcal/mol (Kaushik et al., 2021). In laboratory experiments, it was observed that theaflavins (TFs), formed as oxidation products of *Camellia sinensis* (black tea), exhibited potent antiviral activity against HSV-1. Specifically, treatment with 50 μM of theaflavin-3,3'-gallate (TF-3) for 1 hour resulted in an impressive inhibition rate of up to 99% against viral replication of HSV-1 (Wang et al., 2021).

These herbs have shown potential antiviral effects in studies; further research is needed to fully understand their mechanisms of action and ensure their safety and effectiveness (Figure 5.1).

**FIGURE 5.1** Medicinal Herbs responsible for viral replication inhibition.

### 5.3.3 Viral Attachment and Entry Inhibition

Medicinal herbal formulations have the potential to inhibit viral attachment and entry, which refers to their ability to prevent viruses from attaching to and entering host cells. This pivotal stage in the viral life cycle determines the initiation of infection. Within herbal formulations, specific compounds can disrupt the attachment of viruses to host cells, thereby impeding the entry process (Tomas et al., 2022).

For instance, the bioactive compound capsaicin, found in chili peppers, has been found to inhibit the entry of Lassa virus (LASV) that causes Lassa hemorrhagic fever by interfering with the pH-dependent fusion process. It specifically targets the stable signal peptide (SSP)-GP2 transmembrane ($GP2_{TM}$) region on the surface glycoprotein of LASV. Through further investigations involving mutational studies, specific residues (Ala25, Phe434, Val431 and Val435) within the SSP-$GP2_{TM}$ region were identified as the reason for capsaicin's antiviral effect (Tang et al., 2020). Another study by Alvarenga et al. has reported the extract of *Stachytarpheta cayennensis* demonstrates inhibition of SARS-CoV-2 entry and attachment due to its favorable selectivity index among various natural products tested. The extract of *Stachytarpheta cayennensis* exhibited a 50% cytotoxic concentration ($CC_{50}$) of 693.5 μg/mL, indicating the concentration at which it caused toxicity to 50% of the cells tested. Additionally, it demonstrated a half-maximal inhibitory concentration ($IC_{50}$) of 91.65 μg/mL, which represents the concentration required to inhibit 50% of the viral entry process of SARS-CoV-2. Hence, the selectivity index (SI) calculated as CC50 divided by IC50 was found to be 7.57, which is a measure of the compound's safety and potential as an inhibitor of SARS-CoV-2 entry. A higher SI value suggests that the extract is more selective in inhibiting the virus without causing significant toxicity to host cells (Alvarenga et al., 2022).

The herbs *Picrorhiza kurroa*, *Cedrus deodara*, *Terminalia chebula*, *Commiphora wightii* and *Ocimum tenuiflorum* have exhibited inhibitory effects on Chikungunya virus in the plaque reduction assay. This suggests that these plants have the potential to act as viral attachment inhibitors, thereby preventing or reducing viral infection. These findings highlight the antiviral properties of these plants and their potential use in the development of treatments against Chikungunya virus infections (Raghavendhar et al., 2019). Curcumin, derived from the rhizome of the *Curcuma longa* plant, exhibits a diverse range of properties, including its ability to act as an antiviral compound. Notably, it has been found to possess antiviral activity by inhibiting viral attachment and entry in different types of viruses such as Zika Virus (ZIKV), Human Respiratory Syncytial Virus (HRSV) and Vesicular Stomatitis Virus (VSV) (Praditya et al., 2019). Du et al. (2017) documented in their study on Porcine reproductive and respiratory syndrome virus (PRRSV) that when 15 μM of curcumin was incubated with different strains of PRRSV and Marc-145 cells (which are susceptible to PRRSV infection), it effectively prevented the infection. Interestingly, curcumin did not affect the expression of surface receptors. However, it was observed that curcumin inhibited the fusion of PRRSV with the cell membrane and its subsequent internalization. This indicates that curcumin acts by interfering with the fusion process, preventing the virus from entering the cells (Alsayari et al., 2021).

### 5.3.4 Interference with Viral Proteins and Enzymes

During the life cycle of a viral particle, the genetic information is often organized efficiently, utilizing features like overlapping genes and maximizing the encoding of both homo- and hetero-oligomers as biologically active entities. Essentially, for a structure to be classified as a virus, it should consist of a small, unencapsulated DNA or RNA molecule with a few thousand bases. This molecule should have the ability to enter a living cell, undergo replication and translation using the cell's enzymes, and ultimately exit the cell to infect other cells. Thus, a virus can be seen as an obligate parasite that relies on the host's energy production and protein synthesis functions (Louten, 2016). Phytochemicals from various herbs can be employed as antiviral agents against various viruses due to their ability to inhibit viral replication and entry through different mechanisms. They can directly target the viral entry point or disrupt replication processes, or interfere with the synthesis of viral proteins and enzymes (Ben-Shabat et al., 2020).

Coumarin is a naturally occurring compound found in various parts of plants, such as seeds, leaves and roots, primarily belonging to the Apiaceae and Rutaceae families. Due to its desirable characteristics of stability, solubility and low toxicity, coumarin has gained attention as a potential drug candidate (Stefanachi et al., 2018). It has shown inhibitory effects against several viral infections, including HIV, influenza, enterovirus 71 (EV71) and coxsackievirus A16 (CVA16). The mechanisms by which coumarin exerts its antiviral activity involve multiple pathways. It can inhibit proteins that are crucial for viral entry, replication and infection. Additionally, coumarin can modulate cellular pathways such as Akt-Mtor (mammalian target of rapamycin) and NF-κB (nuclear factor kappa-light-chain-enhancer of activated B cells). It also exhibits an antioxidant effect by regulating the NrF-2 (nuclear factor erythroid 2 (NFE2)-related factor 2) pathway. These mechanisms collectively contribute to the antiviral activity of coumarin against various viruses (Figure 5.2). Its

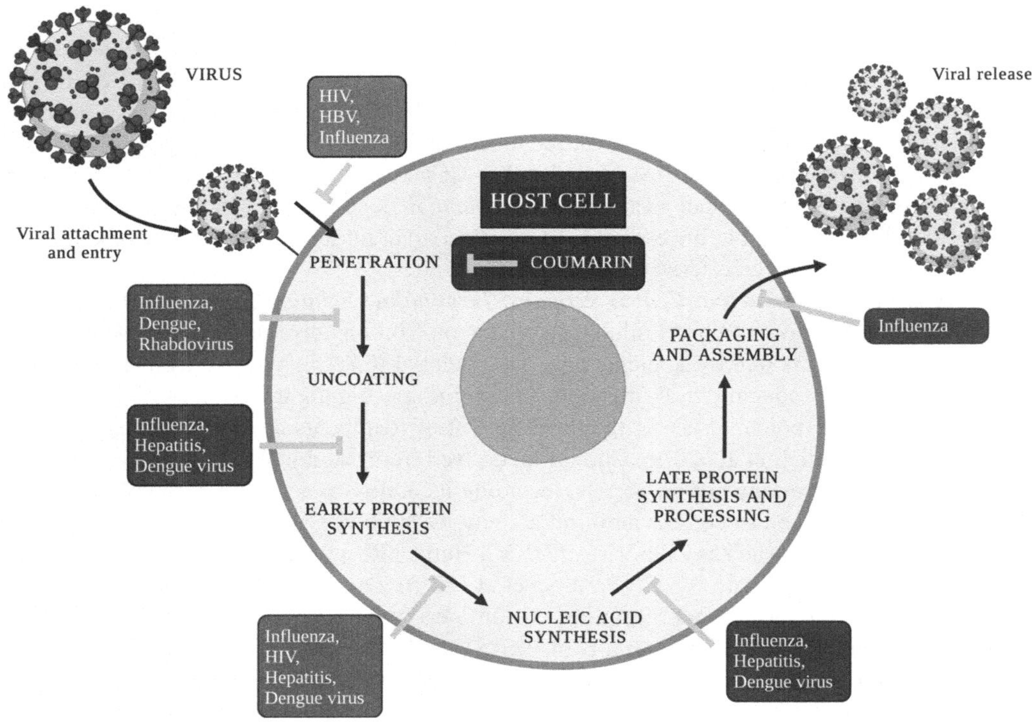


**FIGURE 5.2** Various inhibitor mechanisms of coumarin against different viruses.

ability to target essential viral proteins and regulate cellular pathways involved in viral infection makes it a promising candidate for further research and development of antiviral therapies (Mishra et al., 2020).

A study has shown that coumarin has also been identified as an inhibitor of HIV proteases. They synthesized a series of novel N-benzylated coumarin-AZT conjugates that demonstrated inhibitory activity against both HIV protease and reverse transcriptase enzymes (Xu et al., 2021). Similarly, herbs *Plantago asiatica* and *Clerodendrum trichotomum* have demonstrated anti-RSV activity. In a study conducted on these herbs, it was found that lower concentrations of the extracts derived from *P. asiatica* and *C. trichotomum* significantly inhibited the replication and transcription of RSV genes. This inhibition led to a decrease in the synthesis of RSV proteins, ultimately resulting in reduced RSV-induced cell death. These findings suggest that the extracts from these herbs have the potential to be used as antiviral agents against RSV infections (Chathuranga et al. 2019).

### 5.3.5 Induction of Apoptosis in Infected Cells

Inducing apoptosis and downregulating viral oncogenes can be a viable approach for effectively managing viral infections. By promoting apoptosis, the programmed cell death, infected cells can be eliminated, preventing the spread of the virus and reducing the overall viral load. For instance, bioactive compounds like Epigallocatechin-3-gallate (EGCG), a polyphenol derived from green tea, have shown an impact on CaSki, a cervical cancer cell line associated with HPV-16. It was observed that the anticarcinogenic properties of EGCG were attributed to the cell cycle arrest in the G1phase. This cell cycle arrest subsequently triggered programmed cell death through apoptosis (Pennisi et al., 2021). Further investigations have demonstrated that the stem extract of *Cudrania tricuspidata* induces apoptosis and exhibits cytotoxic effects, specifically on HPV-16 positive SiHa cancer cells. Importantly, this cytotoxic effect was not observed in HaCaT human normal keratinocytes when exposed to concentrations ranging from 0.125 to 0.5 mg/mL (Kwon et al., 2016). Similar studies have investigated the antiviral activity of specific types of Ginseng, namely protopanaxadiol (PD)-type ginsenosides and protopanaxatriol (PT)-type ginsenosides against rhinovirus infection. It was found that PT-type ginsenosides provided protection to HeLa cells against cell death induced by human rhinovirus 3 (HRV3), as demonstrated by the viability of cells assessed using sulforhodamine B staining and morphological evaluation. In contrast, PD-type ginsenosides did not exhibit protective effects and instead promoted HRV3-induced cell death. This suggests that the antiviral activity of ginsenosides against HRV3 is structure-dependent, with PT-type ginsenosides showing a favorable effect (Alsayari et al., 2021).

## 5.4 HERBAL AND POLY-HERBAL FORMULATIONS IN SPECIFIC VIRAL DISEASES

### 5.4.1 Influenza Virus Infections

Infectious disease caused by the influenza virus is commonly known as "the flu." Influenza causes affliction and fatality worldwide, particularly in developing countries with tropical climates like India, where the incidence of influenza-like illness (ILI) among young and older individuals is highest (Rangarajan et al., 2019).

Viruses A, B and C are responsible for the three primary strains of flu. Once you've been infected with type C, you're protected. Our bodies can't develop resistance to types A and B since they're always evolving (Petrova & Russell, 2018). For the first 1–2 days after exposure, the virus is highly contagious and very hard to contain. Uncomplicated ILI is treated with antipyretics, antihistamines and cough syrup, which have negative effects. During illness, patients are urged to rest and stay hydrated and gradually return to full activity. Viruses modify antigenicity to avoid immune-mediated eradication. This method leads to new strains of influenza virus, and past immunity fails to protect against new infections. This makes developing vaccinations for these infections difficult. The unavailability of ideal medication to limit illness development, ease recovery and empower the immune system has led to the research of traditional and alternative therapies to support herbal medication in ILI (Beressa et al., 2021).

In a study, extracts of five different medicinal plants were evaluated for their antiviral activity against the influenza A virus. Out of the five plants, the results showed that the methanol leaf extract of *Rapanea melanophloeos* exhibited the most potent antiviral effect, with an EC50 value of 113.3 µg/mL. Additionally, the methanol extract of *Pittosporum viridiflorum*, along with its 100% and 30% ethanol and acetone leaf extracts, demonstrated significant antiviral activity, with EC50 values of 3.6, 3.4, 19.2 and 82.3 µg/mL, respectively. These findings indicate that the extracts derived from *Rapanea melanophloeos* and *Pittosporum viridiflorum* possess strong antiviral properties against the influenza A virus. The results highlight the potential of these plant extracts as a source for developing natural antiviral agents. Further research and exploration of these plant species and their bioactive compounds could lead to the development of effective treatments or preventive measures against influenza infections (Mehrbod et al., 2018).

The clinical presentation of ILI can be compared to the Ayurvedic concept of Vata-KaphaJvara, which is characterized by fever with a predominance of Vata and Kapha doshas (Thakar et al., 2023). In Ayurvedic therapy for fever (Jvara), the focus is on eliminating ama, which signifies partially digested metabolic waste, through langhana (fasting or restricted calorie intake) and the use of formulations containing herbs with tikta rasa (bitter taste), ushnavirya (hot potency), katuvipaka (pungent post-digestive taste) and amapachaka (anti-inflammatory) properties (Gundeti et al., 2022).

In traditional Chinese medicine, the antiviral and anti-inflammatory activity of Lianhua-Qingwen, a Chinese herbal formulation comprising 11 medicinal plants (*Armeniacae Semen*

*Amarum, Glycyrrhizae Radix et Rhizoma, Ephedrae Herba, Lonicerae Japonicae Flos, Isatidis Radix, Houttuyniae Herba, Forsythia Fructus, Pogostemonis Herba, Rhodiolae Crenulatae Radix et Rhizoma, Dryopteris Crassirhizomatis Rhizoma* and *Rhei Radix & Rhizoma*), along with Gypsum Fibrosum (a mineral medicine) and menthol, was analyzed with regards to its inhibitory effects against influenza virus (Ding et al., 2017). This formulation was traditionally used for cough, sneezing, muscle ache, fever, bronchitis and fatigue. Clinical trials of the Lianhua-Qingwen drug are currently underway in the United States for checking its inhibitory activity against COVID-19 (Gao et al., 2020). The Chinese government has officially endorsed the use of the Lianhua-Qingwen drug for both the prevention and treatment of COVID-19 (Yang et al., 2020).

### 5.4.2 Herpes Virus Infections

A wide range of human disorders, ranging from minor to life-threatening, are caused by herpes viruses, especially in immunocompromised people and newborns. Because new drug-resistant strains of viruses are emerging at an alarming rate, viral infections are regarded to be one of the world's leading causes of mortality. Numerous traditional herbal medications are found to be effective against the herpes virus (Bandara & Samaranayake, 2019).

Genital ulcers are most commonly caused by HSV-2 infection, a type of herpes simplex virus. Ulcers have also been linked to HIV infection, which is not surprising given the widespread nature of the virus (Johnston & Corey, 2016). The treatment of HSV-2 infection is complicated by complications, medication resistance and adverse effects of antiviral medicines. Recent research suggests that HSV-2 and HIV infections may be treatable with herbal treatments. A poly-herbal gel formulation, which includes extracts from *Terminalia chebula*, *Lagerstroemia speciosa*, *Phllanthusemblica* and *Acacia catechu*, has been produced in this area. It contains 50% ethanolic extracts from each. There was considerable virucidal efficacy and HSV-2 infections in the gel formula, which had IC50s of 55.93 and 27.26 μg/μL. For HSV-2 attachment and penetration, it had an IC50 of 46.55 + 1.25 mL, which was much lower than acyclovir's 46.55 + 1.25 mL. When tested post-infection, acyclovir had an IC50 value of $0.065 \pm 0.01$ g/mL, whereas gel formulation had a value of $469.05 \pm 16.65$ g/mL under the same circumstances. *Lactobacilli*, on the integrity of the Caco-2 cell monolayer and human vaginal keratinocyte cells (Vk2/E6E7), were not inhibited by gel formulation. The mutagenic index and pro-inflammatory cytokines were not significantly increased by gel composition (Mishra et al., 2018).

Studies have shown that the free form of the HSV virus is highly susceptible to the effects of PEO, and it also inhibits the HSV either prior to or during its entry into host cells. It is believed that PEO reduces the infectivity of the virus by directly interacting with the viral envelope and glycoprotein. Additionally, it also disrupts the later stages of the HSV-1 life cycle, leading to the suppression of viral replication (Zhao et al., 2022).

### 5.4.3 Human Immunodeficiency Virus (HIV) Infections

Complimentary treatments, such as herbal medications, are frequently sought out by patients with HIV/AIDS for reasons such as disappointing results, excessive costs, lack of availability or bad effects of traditional drugs. HIV remains a worldwide public health problem, and Highly Active Antiretroviral Therapy (HAART), currently used for treating HIV patients, has reduced AIDS-related deaths. However, in undeveloped and growing nations, where the supply of medications is short and frequently not well maintained, HAART can be a costly regime. Anti-HIV chemicals from herbs might be a viable alternative to the current treatment, which is limited by the development of multi-drug resistance (Eggleton & Nagalli, 2020).

As HIV entrance and replication inhibitors, methanolic extracts from aerial portions of *A. aspera* and leaves of *R. centifolia* show potential anti-HIV1 activity. These experimental molecules hold promise for HIV/AIDS prevention and management. Methanolic extracts of *R. centifolia* leaves showed efficient safety and preserved epithelial reliability on HeC-1A cells. Various components present in plant extracts and the cytotoxicity of several chemicals may obscure plant extract's anti-HIV1 potency (Palshetkar et al., 2020).

SPV30, a drug that may slow the progression of HIV illness in asymptomatic HIV-infected patients, has shown effectiveness. People with HIV symptoms may benefit from using the Chinese herbal drug IGM-1 (Mukherjee et al., 2021). A newly developed herbal formulation for modulating the immune system of HIV-infected patients has shown effective results. The formulation comprises a hydromethanolic extract obtained from five different plants, prepared by maintaining a temperature of 80°C–90°C and controlling the pH of the solution between 6 and 7. The active compound is then separated. This Ayurvedic formulation uses aqueous and methanolic extracts of *Hippophaerhamnoides* (Badriphal, Fruits), *Withaniasomnifera* (Ashwagandha; Root), *Convolvulus pluricaulis* (Shankhapushpi, whole plant), *Cynodondactylon* (Durva, Whole plant) and *Ocimum sanctum* (Tulsi, Leaves). The extracts are obtained using a mixture of aqueous and methanol in a ratio of 50:50 (Rajan et al., 2011). *Tinospora cordifolia* (Guduchi) is one of Ayurveda's greatest immunomodulators. It improves the immune system by sustaining white blood corpuscle numbers in HIV-infected people (Sharma et al., 2019). Similarly, *Andrographis paniculata*, commonly known as kalmegh, is an immune-boosting plant. Clinical research has explored the herb's role in HIV treatment. Regular consumption of kalmegh protects the liver against HIV medications (Jadhav & Karuppayil, 2021).

### 5.4.4 Hepatitis Virus Infections

Hepatitis virus is classified into types A, B, C, D and E, all of which generally cause liver inflammation. Among them, hepatitis types B, C and D can lead to persistent infection, while exposure to any of these viruses can result in a short-lived illness. Food and water contaminated by human waste are the most prevalent means of transmission for acute hepatitis A, which was previously known as infectious hepatitis A. Hepatitis C is the most frequent viral hepatitis. Blood and blood products (which are now tested for the virus) carry type C, which can remain in the body for years before damaging the liver (Mehta & Reddivari, 2022).

Herbal remedies for viral hepatitis traditionally relied on plants from the *Euphorbiaceae* family, particularly those in the genus Phyllanthus. *Phyllanthus amarus*, *P. urinaria* and *P. niruri* were also compared in clinical trials for their anti-hepatitis B properties. Combination treatments for involving multiple medicinal herbs have also been explored for specific conditions (Upadhyay & Tiwari, 2023). Combination therapy for viral hepatitis has also been tested. For example, *Ganoderma lucidum* liquid fermentation broth demonstrated substantial anti-hepatitis B virus activity *in vitro* and *in vivo* when supplemented with an aqueous extract of *Radix Sophorae Flavescentis* (Chinese herbal medicine). The antiviral impact of co-fermenting these two medicinal plants to produce antiviral broth was superior to that of simple mixing, as well (Wu et al., 2023). Similarly, the *Boehmeria nivea L.* plant has been shown to suppress the hepatitis B virus. In both *in vitro* and *in vivo* paradigms, a root extract of Boehmeria nivea suppressed HBV production (Indrasetiawan et al., 2019).

The antiviral activity of different extracts derived from *Polygonum cuspidatum* has been observed against a range of viruses. This effect is attributed to the presence of anthraquinones, a chemical group known for its antiviral and virucidal properties against various types of viruses. Specifically, the ethanolic extract of *P. cuspidatum* exhibited a dose-dependent inhibition of HBV replication in the stable hepatoblastoma cell line (Jasemi et al., 2021).

Many different poly-herbal formulations are available in the form of solid and liquid products in the market. Liverin, Livotone, Stimu liv, Liv-52, Jigrine and many other hepatoprotective poly-herbal formulations are even commercially available. However, only a few formulations have experimental and clinical evidence (Girish & Pradhan, 2017).

### 5.4.5 Respiratory Syncytial Virus (RSV) Infections

RSV remains a significant global healthcare concern, particularly during seasonal endemic outbreaks commonly occurring in winter across various regions. Although RSV can cause infections in individuals of all ages, it is more prevalent among infants (0–1 years). Nearly every child contracts this virus before reaching the age of two, making it a primary cause of acute respiratory tract infections. In severe cases, RSV can lead to bronchiolitis, pneumonia and, in some instances, may contribute to the development of asthma with symptoms such as wheezy cough (Efstathiou et al., 2020).

Initial explorations into the advantages of medicinal plants have unveiled the antiviral properties present in several plant species, including aloe vera, olive, turmeric and java grass. These plants encompass specific compounds, such as 7-O-Methylaloresin, oleuropein, curcumin and cyperen (3,14), with significant therapeutic value (Sharma et al., 2023). Isatidis Radix, from the *Cruciferae* family, is a traditional Chinese herbal medicine that has exhibited antiviral activity due to the presence of glucosinolate in its structure (Chen et al., 2022). It contains three main active ingredients: alkaloid compounds, including epigoitrin (a degradation product of glucosinolate isomers); organic acids, such as salicylic acid; and total lignans, with clemastanin being a representative compound (Tomas et al., 2022). Isatidis Radix has been reported to exhibit significant efficacy against RSV-induced pneumonia, although the exact mechanism of action is not yet fully understood. Xu et al. (2019) reported that the active ingredients of Isatidis Radix can display antiviral activity individually or synergistically when combined. Fresh ginger has demonstrated strong antiviral effects against HRSV and rhinovirus, making it a promising treatment option for respiratory viral infections. The aqueous extract of fresh ginger specifically hinders the attachment and entry of HRSV into human larynx epidermoid carcinoma cells and human lung carcinoma cell lines when administered 1–2 hours prior to virus inoculation. In contrast, dried ginger does not exhibit the same inhibitory effects. The extract derived from fresh ginger also stimulates the production of IFN-$\alpha$ and IFN-$\beta$ from infected epithelial cells, further contributing to its antiviral properties against HRSV (Jafarzadeh et al., 2021).

Similarly, bioactive compounds tangeretin and nobiletin, which are polyethoxylated flavones found in the pericarps of mandarin orange (Citrus reticulata), have been found to impact the intracellular replication of the RSV. Tangeretin has shown the ability to downregulate the expression of RSV phosphoprotein, an essential protein involved in the replication process of the virus. By inhibiting the expression of this protein, tangeretin may interfere with the viral replication cycle, potentially reducing the spread and severity of RSV infection (Choudhary et al., 2021).

### 5.4.6 Emerging and Re-emerging Viral Diseases

Viral infections continue to spread extensively across the globe, posing a significant threat to human health. The presence of existing and emerging viral infections, such as COVID-19, Ebola virus and ZIKV, presents a challenge, as the number of effective antiviral drugs remains limited. While new antiviral drugs are approved by regulatory authorities such as the FDA each year, controlling emerging viral infections remains a pressing global concern. Throughout history, herbs have played a crucial role in the treatment of viral diseases by enhancing the body's immune response to viral infections. Additionally, certain herbs possess antipyretic and analgesic properties, aiding in the management of fever and body aches associated with viral infections (Sivagami & Sailaja, 2021).

The recent global spread of the Severe Acute Respiratory Syndrome Coronavirus-2 (SARS-CoV-2), leading to the coronavirus disease (COVID-19) pandemic, has presented a significant challenge to humanity. Its rapid transmission across 210 countries has affected over 25 million individuals and caused approximately 1 million deaths worldwide. Currently, there is no definitive cure for the disease, and the available antiviral treatments have yielded unsatisfactory clinical outcomes. However, throughout history, traditional plant-based ethnomedicine has served as a valuable source of leads for treating infectious diseases (Patel et al., 2021).

A traditional poly-herbal formulation called Kabasura Kudineer (KSK), which has been used in India for a long time, has shown promising results in reducing the SARS-CoV-2 virus load in asymptomatic COVID-19 cases. In comparison to individuals receiving standard of care, those who received KSK exhibited a significant reduction in the viral load (Natarajan et al., 2021). In India, AYUSH systems of medicine emphasize the importance of lifestyle modifications, dietary management and prophylactic interventions to enhance immunity and prevent COVID-19 (Patel et al., 2021). AYUSH-64 is a poly-herbal formulation that has been extensively studied by CCRAS, Ministry of AYUSH, Government of India. Extensive clinical, pharmacological and toxicological studies have conclusively established the safety and efficacy of this treatment in addressing infectious febrile conditions like malaria and influenza. The components of AYUSH-64, namely Saptaparna (*Alstonia scholaris* R. Br.), Katuki (*Picrorhiza kurroa* Royle ex. Benth), Kiratatikta (*Swertia chirata* Pexbex. Karst) and Kuberaksha (*Caesalpinia crista* L.), have been found to exhibit anti-inflammatory, immunomodulatory, antioxidant and antiviral activities. These properties make them valuable in the context of enhancing the immune system, alleviating fever, combating oxidative stress, exerting antiviral effects and reducing inflammation (Panda et al., 2022). Chopra et al. (2023) have reported the findings of a Molecular Docking study demonstrating that AYUSH-64 contains Akuammicine N-Oxide, which exhibited the highest binding energy with the $M^{pro}$ protein associated with SARS-CoV-2. Additionally, there were 34 other phytoconstituents present in AYUSH-64 that showed similar antiviral activity against SARS-CoV-2. These results indicate that AYUSH-64 has the potential to be repurposed as a suitable drug or medicine for managing COVID-19.

Medicinal plants employ various mechanisms for targeting the life cycle of the viral particle. For example, 13 components of *Garcinia kola*, a medicinal herb, have demonstrated activity against the Ebola virus by potentially binding to membrane proteins, metalloproteases and Ser/Thr kinase. These components primarily target three essential proteins: cyclin-dependent kinases, matrix metalloproteinase and cannabinoid receptors, which are crucial for the virus's survival and replication. Additionally, the components may also interact with cathepsin, collagenase and other matrix metalloproteinases (David et al., 2017). Baicalin, derived from *Scutellariae Radix*, a bioactive compound, exerts its effects by acting on chemokine receptors, thereby inhibiting the entry of HIV into cells. The N-butanol fraction of *Bredelia micrantha* has demonstrated activity in inhibiting the reverse transcriptase enzyme. Furthermore, terpenes have shown inhibitory effects against the protease enzyme of HIV (Beressa et al., 2021).

Once a relatively unknown flavivirus, ZIKV is an epidemic pathogen primarily transmitted by mosquitoes. However, it can be transmitted through placental and sexual means as well (Bajpai et al., 2018). The use of natural compounds for the treatment of ZIKV is discussed. An extract from *Aphloia theiformis* (Vahl) Benn. (*Aphloiaceae*), an evergreen shrub, has been found to inhibit ZIKV entry into host cells at C50 = 100 μg and CC50 = 3,000 μg/mL; SI = 30 (Clain et al., 2018). In Vero E6 cells, the bioactive compound berberine demonstrated a significant reduction of nearly 80% in virus infectivity. This compound is naturally found in two commonly used herbs in natural therapies, *Coptis* and *Berberis* species. In addition to its antiviral properties, berberine is effective in regulating glycemia and lipid metabolism (Batista et al., 2019). A study reported, Cephalotaxine (CET), an alkaloid derived from Cephalotaxus drupacea, demonstrated efficacy in inhibiting the activity of the ZIKV in laboratory settings, specifically in Vero and A549 cell lines. CET was observed to reduce the levels of ZIKV RNA and viral protein expression, hinder ZIKV replication and inhibit the production of ZIKV mRNA and proteins (Lai et al., 2020).

## 5.5 SYNERGISTIC COMBINATIONS AND ADJUVANT THERAPIES

The funding support for substances derived from natural sources is generally limited due to challenges related to patent protection. This issue is not related to the potential or effectiveness of these substances but rather stems from the difficulty in safeguarding intellectual property rights. Pharmaceutical companies are reluctant to invest significant amounts of money in the development and clinical trials of natural substances if they cannot be adequately protected by patents. To address this challenge, alternative strategies are sometimes explored. These include trademarking the formulations, exploring synergistic combinations of natural substances, or chemically modifying the molecules to enhance their specificity, activity, or bioavailability. These approaches aim to enable patenting and provide some intellectual property protection. However, these strategies often result in increased costs for the eventual treatment without necessarily improving its efficacy (Van Brummelen & Van Brummelen, 2022). Combining substances in a synergistic manner can result in reduced dosages of the individual components, thereby minimizing potential reactivity or adverse effects (Schein, 2020).

Resveratrol, a naturally occurring polyphenol (trans-3,4′,5-trihydroxystibene) found in grape skin and the medicinal plant Japanese knotweed, has demonstrated intermediate inhibition of the N1L protein, which plays a role in the virulence of certain viral infections, including smallpox (de Souza Andrade et al., 2022). Additionally, resveratrol has been shown to have a synergistic effect with decitabine, an analog of cytosine, in inhibiting the infectivity of human immunodeficiency virus type 1 (HIV-1) without causing increased toxicity to the cells (Xiao et al., 2022).

Shufeng Jiedu Capsule (SFJDC), a patented herbal drug consisting of eight medicinal plants, is utilized for treating various viral diseases. In the case of acute exacerbation of chronic obstructive pulmonary disease (AECOPD) caused by influenza A virus (IAV), SFJDC is combined with oseltamivir for treatment. Through *in vivo* studies, the combination of SFJDC and oseltamivir has demonstrated improved survival rates, alleviation of clinical symptoms, weight gain, reduced lung damage and a significant reduction in IL-1β and IL-18 levels in serum and BALF. Additionally, it has shown a decrease in the expression of NLRP3-associated components and viral titers in lung homogenates (Ji et al., 2020).

## 5.6 CHALLENGES AND FUTURE PERSPECTIVES

Herbal medicines have a rich composition of chemical constituents (phytoconstituents) that have been recognized for their pharmacological effects on the body and utilized for centuries. The widespread global acceptance and utilization of herbal medicines serve as evidence of their perceived safety and efficacy. However, concerns regarding the lack of sufficient data on the safety and effectiveness of herbal medicines arise due to inadequate pharmacokinetic, pharmacological and clinical studies on the majority of herbal products. Meeting the legislative requirements for conducting research on herbal drugs presents a significant challenge, further complicating the regulation of these products. Criticisms from various corners of the world regarding herbal medicines have been eye-opening for herbal practitioners, prompting a call for a more rigorous scientific approach and closer collaboration between researchers and regulatory bodies. By addressing these concerns and conducting robust scientific research, the safety, efficacy and overall potential of herbal medicines can be better understood and harnessed for the benefit of global healthcare (Ekor, 2014; Balekundri & Mannur, 2020; Ampomah et al., 2020).

Research on herbal drugs presents various challenges that must be effectively addressed. These challenges encompass multiple aspects, including financial considerations, ethical considerations, product standardization (quality control), study design and regulatory requirements. Economic challenges often arise when conducting research on herbal drugs, as such studies require funding for expenses related to personnel, laboratory equipment, data analysis and other research-related costs. Securing adequate funding is crucial to ensure the successful completion of research projects. Ethical considerations play a vital role in herbal drug research, particularly concerning patient safety and informed consent. Researchers must adhere to ethical guidelines and obtain approval from institutional review boards or ethics committees before conducting studies involving human participants. Additionally, protecting intellectual property rights and the equitable sharing of benefits derived from herbal drug research are critical ethical considerations that require attention (Zhang et al., 2012; Parveen et al., 2015; Noviana et al., 2022).

Product standardization, or quality control, is a critical aspect of herbal drug research. Ensuring consistent and reliable quality of herbal products is essential for accurate and reproducible research outcomes. Standardization involves establishing methods to authenticate and identify herbal materials, determining appropriate extraction and formulation techniques, and implementing quality control measures throughout manufacturing. These steps help guarantee that the herbal products used in research are of known composition and meet appropriate quality standards (Zhang et al., 2012; Parveen et al., 2015; Noviana et al., 2022).

The design of herbal drug studies is another challenge that researchers face. Well-designed studies are necessary to generate reliable and meaningful data. Considerations include selecting appropriate study designs (e.g., randomized controlled trials), determining sample sizes, defining suitable endpoints and implementing robust statistical analyses. Careful planning and collaboration with experts in clinical research methodology are crucial for the successful execution of herbal drug studies. Regulatory requirements are an essential aspect of herbal drug research. Researchers must adhere to specific regulations and guidelines set forth by regulatory authorities in the respective countries or regions where the studies are conducted. These requirements include obtaining appropriate permits, licenses, or approvals, adhering to good clinical practice (GCP) guidelines, and fulfilling specific reporting and documentation obligations. Compliance with regulatory requirements ensures the ethical and legal conduct of herbal drug research and provides a framework for assessing safety and efficacy. In 2019, the World Health Organization (WHO) recognized the importance of addressing regulatory requirements for clinical trials of herbal products. The WHO issued operational guidelines that outline the necessary regulatory framework to support the conduct of clinical trials involving herbal products. These guidelines promote harmonization and consistency in evaluating and regulating herbal drug research globally (World Health Organization, 2019).

Addressing the challenges associated with herbal drug research requires collaboration among researchers, regulatory bodies, funding agencies and other stakeholders. By developing standardized protocols, providing adequate funding, ensuring ethical conduct and complying with regulatory requirements, the research community can overcome these challenges and advance the understanding of herbal drugs. This, in turn, will contribute to evidence-based practice and the integration of herbal medicine into mainstream healthcare.

## 5.7 CONCLUSION

While herbal medicines have a long history of use and widespread acceptance, there is a need for more scientific research and regulatory oversight to ensure their safety and efficacy. By conducting comprehensive pharmacokinetic, pharmacological and clinical studies, we can better understand the mechanisms of action, optimal dosages and potential risks associated with herbal medicines. Collaboration between stakeholders is essential to bridge the gap between traditional knowledge and modern regulatory requirements, ultimately ensuring the responsible and evidence-based use of herbal medicines in healthcare.

## REFERENCES

Adhikari, P.P.; Paul, S.B. History of Indian traditional medicine: a medical inheritance. *History* 2018, 11(1), 421.

Ahmad, S.; Zahiruddin; S.; Parveen, B.; Basist, P.; Parveen, A.; Parveen, R.; Ahmad, M. Indian medicinal plants and formulations and their potential against COVID-19-preclinical and clinical research. Front. Pharmacol. 2020, 11, 2470.

Alsayari, A.; Muhsinah, A.B.; Almaghaslah, D.; Annadurai, S.; Wahab, S. Pharmacological efficacy of Ginseng against respiratory tract infections. *Molecules* 2021, 26(13), 4095.

Alvarenga, N.; Eugenia, M.; Barúa, J.E.; Cristina, M.; Langjahr, P.; Sotelo, P.H. Screening of natural products inhibitors of SARS-CoV-2 entry. *Molecules* 2022, 27(5), 1743.

Ampomah, I.G.; Malau-Aduli, B.S.; Malau-Aduli, A.E.; Emeto, T.I. Effectiveness of integrated health systems in Africa: a systematic review. *Medicina* 2020, 56(6), 271.

Anjum, N.; Ridwan, Q.; Akhter, F.; Hanief, M. Phytochemistry and therapeutic potential of Berberislycium Royle, an endangered species of Himalayan region. *Acta Ecol. Sin.* 2023, 43(4), 577–584.

Atanasov, A.G.; Zotchev, S.B.; Dirsch, V.M.; Supuran, C.T. Natural products in drug discovery: advances and opportunities. *Nat. Rev. Drug Discov.* 2021, 20(3), 200–216.

Bajpai, V.K.; Chandra, V.; Kim, N.H.; Rai, R.; Kumar, P.; Kim, K.; Aeron, A.; Kang, S.C.; Maheshwari, D.K.; Na, M.; Rather, I.A. Ghost probiotics with a combined regimen: a novel therapeutic approach against the Zika virus; an emerging world threat. *Crit. Rev. Biotechn.* 2018, 38(3), 438–454.

Balekundri, A.; Mannur, V. Quality control of the traditional herbs and herbal products: a review. *Future J. Pharm. Sci.* 2020, 6, 1–9.

Bandara, H.M.H.N.; Samaranayake, L.P. Viral; bacterial; and fungal infections of the oral mucosa: types; incidence; predisposing factors; diagnostic algorithms; and management. *Periodontol.* 2000. 2019, 80(1), 148–176.

Batista, M.N.; Braga, A.C.; Campos, G.R.; Souza, M.M.; Matos, R.P.; Lopes, T.Z.; Candido, N.M.; Lima, M.L.; Machado, F.C.; Andrade, S.T.; Bittar, C.; Nogueira, M.L.; Carneiro, B.M.; Mariutti, R.B.; Arni, R.K.; Calmon, M.F.; Rahal, P. Natural products isolated from oriental medicinal herbs inactivate Zika virus. *Viruses* 2019, 11(1), 49.

Ben-Shabat, S.; Yarmolinsky, L.; Porat, D.; and Dahan, A. Antiviral effect of phytochemicals from medicinal plants: applications and drug delivery strategies. *Drug Deliv. Transl. Res.* 2020, 10(2), 354–367.

Beressa, T.B.; Deyno, S.; Mtewa, A.G.; Aidah, N.; Tuyiringire, N.; Lukubye, B.; Weisheit, A.; Tolo, C.U.; Ogwang, P.E. Potential benefits of antiviral African medicinal plants in the management of viral infections: systematic review. *Front. Pharmacol.* 2021, 12, 682794.

Bhagat, R.; Mhaiskar, B. *Potential Antiviral Herbal Therapeutics for Viral Infections.* Medtext Publications LLC, 2023.

Borchers, A.T.; Chang, C.; Gershwin, M.E.; Gershwin, L.J. Respiratory syncytial virus-a comprehensive review. *Clin. Rev. Allergy Immunol.* 2013, 45; 331–379.

Brendler, T.; Al-Harrasi, A.; Bauer, R.; Gafner, S.; Hardy, M.L.; Heinrich, M.; Hosseinzadeh, H.; Izzo, A.A.; Michaelis, M.; Nassiri-Asl, M.; Panossian, A.; Wasser, S.P.; Williamson, E.M. Botanical drugs and supplements affecting the immune response in the time of COVID-19: implications for research and clinical practice. Phytother. Res. 2021, 35(6), 3013–3031.

Castleman, M. The New Healing Herbs: The Essential Guide to More Than 130 of Nature's Most Potent Herbal Remedies. Rodale, 2017.

Chathuranga, K.; Kim, M.S.; Lee, H.C.; Kim, T.H.; Kim, J.H.; Gayan Chathuranga, W.A.; Ekanayaka, P.; Wijerathne, H.M.S.M.; Cho, W.K.; Kim, H.I.; Ma, J.Y. Anti-respiratory syncytial virus activity of Plantago asiatica and Clerodendrum trichotomum extracts in vitro and in vivo. *Viruses* 2019, 11(7), 604.

Chen, J.; Zhu, Z.; Gao, T.; Chen, Y.; Yang, Q.; Fu, C.; Zhu, Y.; Wang, F.; Liao, W. Isatidis Radix and Isatidis Folium: a systematic review on ethnopharmacology; phytochemistry and pharmacology. *J. Ethnopharmacol.* 2022, 283, 114648.

Chopra, A.; Tillu, G.; Chuadhary, K.; Reddy, G.; Srivastava, A.; Lakdawala, M.; Gode, D.; Reddy, H.; Tamboli, S.; Saluja, M.; Sarmukaddam; S. Co-administration of AYUSH 64 as an adjunct to standard of care in mild and moderate COVID-19: a randomized; controlled; multicentric clinical trial. PLoS One 2023, 18(3), 0282688.

Choudhary, R.; Borse, L.; Jadhav, A. Ayurvedic antiviral agents: overview of medicinal plants perspective. *Int. J. Pharm. Sci. Res.* 2021, 6214–6220.

Clain, E.; Sinigaglia, L.; Koishi, A.C.; Gorgette, O.; Gadea, G.; Viranaicken, W.; Krejbich-Trotot, P.; Mavingui, P.; Desprès, P.; Nunes Duarte dos Santos, C.; Guiraud, P. Extract from Aphloia Theiformis; an edible indigenous plant from Reunion Island; Impairs Zika virus attachment to the host cell surface. Sci. Rep. 2018, 8, 10856.

David, S.; Toluwase, F.; Oluwasegun, O. Bioinformatics analysis of Garcinia Kola active components and glycoproteins of Ebola virus (Zaire Ebolavirus). *J. Chem. Pharm.* 2017, 9(4), 364–370.

De Souza Andrade, M.M.; Leal, V.N.; Fernandes, I.G.; Gozzi-Silva, S.C.; Beserra, D.R.; Oliveira, E.A.; Teixeira, F.M.; Yendo, T.M.; Sousa, M.D.G.T.; Teodoro, W.R.; Oliveira, L.D.M. Resveratrol downmodulates neutrophil extracellular trap (NET) generation by neutrophils in patients with severe COVID-19. *Antioxidants* 2022 11(9), 1690.

De Wolf, C.; Van De Bovenkamp, M.; Hoefnagel; M. Regulatory perspective on in vitro potency assays for human mesenchymal stromal cells used in immunotherapy. *Cytotherapy* 2017, 19(7), 784–797.

Dhama, K.; Khan, S.; Tiwari, R.; Sircar, S.; Bhat, S.; Malik, Y.S.; Singh, K.P.; Chaicumpa, W.; Bonilla-Aldana, D.K.; Rodriguez-Morales, A.J. Coronavirus disease 2019-COVID-19. *Clin. Microbiol. Rev.* 2020, 33(4), 10–1128.

Ding, Y.; Zeng, L.; Li, R.; Chen, Q.; Zhou, B.; Chen, Q.; Cheng, P.L.; Yutao, W.; Zheng, J.; Yang, Z.; Zhang, F. The Chinese prescription lianhuaqingwen capsule exerts anti-influenza activity through the inhibition of viral propagation and impacts immune function. *BMC Complement. Altern. Med.* 2017, 17, 1–11.
Du, T.; Shi, Y.; Xiao, S.; Li, N.; Zhao, Q.; Zhang, A.; Nan, Y.; Mu, Y.; Sun, Y.; Wu, C.; Zhang, H. Curcumin is a promising inhibitor of genotype 2 porcine reproductive and respiratory syndrome virus infection. BMC Vet. Res. 2017, 13(1), 1–9.
Efstathiou, C.; Abidi, S.H.; Harker, J.; Stevenson, N.J. Revisiting respiratory syncytial virus's interaction with host immunity; towards novel therapeutics. *Cell. Mol. Life Sci.* 2020, 77, 5045–5058.
Eggleton, J.S.; Nagalli; S. *Highly Active Antiretroviral Therapy (HAART).* StatPearls Publishing, 2020.
Ekor, M. The growing use of herbal medicines: issues relating to adverse reactions and challenges in monitoring safety. *Front. Pharmacol.* 2014, 4, 177.
Fuzimoto, A.D.; Isidoro, C. The antiviral and coronavirus-host protein pathways inhibiting properties of herbs and natural compounds - additional weapons in the fight against the COVID-19 pandemic? *J. Tradit. Complement. Med.* 2020, 10(4), 405–419.
Gao, D.; Niu, M.; Wei, S.Z.; Zhang, C.E.; Zhou, Y.F.; Yang, Z.W.; Li, L.; Wang, J.B.; Zhang, H.Z.; Zhang, L.; Xiao, X.H. Identification of a pharmacological biomarker for the bioassay-based quality control of a thirteen-component TCM formula (Lianhua Qingwen) used in treating influenza A virus (H1N1) infection. *Front. Pharmacol.* 2020, 11, 746.
Girish, C.; Pradhan, S. Herbal drugs on the liver. *Liver Physio.* 2017, 605–620.
Gundeti, M.S.; Bhurke, L.W.; Mundada, P.S.; Murudkar, S.; Surve, A.; Sharma, R.; Mata, S.; Rana, R.; Singhal, R.; Vyas, N.; Khanduri, S.; Sharma, B.S.; Srikanth, N.; Dhiman, K.S. AYUSH 64; a polyherbal ayurvedic formulation in influenza-like illness - results of a pilot study. *J. Ayurveda Integr. Med.* 2022, 13(1), 100325.
Indrasetiawan, P.; Aoki-Utsubo, C.; Hanafi, M.; Hartati, S.R.I.; Wahyuni, T.S.; Kameoka, M.; Yano, Y.; Hotta, H.; Hayashi, Y. Antiviral activity of cananga odorata against hepatitis B virus. *Kobe J. Med. Sci.* 2019, 65(2), E71.
Jadhav, A.K.; Karuppayil, S.M. Andrographis paniculata (Burm. F) Wall ex Nees: antiviral properties. *Phytother. Res.* 2021, 35(10), 5365–5373.
Jafarzadeh, A.; Nemati, M.; Saha, B.; Bansode, Y.D.; Jafarzadeh, S. Protective potentials of type III interferons in COVID-19 patients: lessons from differential properties of type I- and III interferons. *Viral Immuno.* 2021, 34(5), 307–320.
Jana, D.; Guchhait, K.C.; Ballav, S.; Panda, A.K.; Ghosh, C. Traditional herbal medicine practiced in Plateau-Fringe and Rarh districts of West Bengal; India. In *Evidence Based Validation of Traditional Medicines: A Comprehensive Approach*, R. Chakraborty, S. Sen, S.C. Mandal, Eds.Springer: Singapore, 2021, pp. 1019–1089.
Jasemi, E.; Momtaz, S.; Ghaffarzadegan, R.; Abdolghaffari, A.H.; Abdollahi, M. A narrative review of herbal preparations against RNA viruses. *J. Contemp. Med. Sci.* 2021, 7(1).
Ji, S.; Bai, Q.; Wu, X.; Zhang, D.W.; Wang, S.; Shen, J.L.; Fei, G.H. Unique synergistic antiviral effects of Shufeng Jiedu Capsule and oseltamivir in influenza A viral-induced acute exacerbation of chronic obstructive pulmonary disease. *Biomed. Pharmacother.* 2020, 121, 109652.
Johnston, C.; Corey, L. Current concepts for genital herpes simplex virus infection: diagnostics and pathogenesis of genital tract shedding. *Clin. Microbiol. Rev.* 2016, 29(1), 149–161.
Kaushik, S.; Dar, L.; Kaushik, S.; Yadav, J.P. Identification and characterization of new potent inhibitors of dengue virus NS5 proteinase from Andrographis paniculata supercritical extracts on in animal cell culture and in silico approaches. *J. Ethnopharmacol.* 2021, 267, 113541.
Kumari, R.; Kotecha; M. A review on the Standardization of herbal medicines. *Int. J. Pharm. Sci. Res.* 2016, 7(2), 97–106.
Kwon, S.-B.; Kim, M.-J.; Yang, J.M.; Lee, H.-P.; Hong, J.T.; Jeong, H.-S.; Kim, E.S.; Yoon, D.Y. Cudrania tricuspidata stem extract induces apoptosis via the extrinsic pathway in SiHa cervical cancer cells. *PLoS. One.* 2016, 11, e0150235.
Lai, Z.; Ho, Y.; Lu; J. Cephalotaxine inhibits Zika infection by impeding viral replication and stability. *Biochem. Biophys. Res. Commun.* 2020, 522(4), 1052–1058.
Li, L.; Wu, Y.; Wang, J.; Yan, H.; Lu, J.; Wang, Y.; Zhang, B.; Zhang, J.; Yang, J.; Wang, X.; Zhang, M. Potential treatment of COVID-19 with traditional Chinese medicine: what herbs can help win the battle with SARS-CoV-2?. *Engineering,* Elsevier. 2022, 139–152.
Louten, J. Virus replication. *Essent. Human Virol.* 2016, 49–70.
Luo, H.; Tang, Q.L.; Shang, Y.X.; Liang, S.B.; Yang, M.; Robinson, N.; Liu, J.P. Can Chinese medicine be used for prevention of corona virus disease 2019 (COVID-19)? A review of historical classics; research evidence and current prevention programs. *Chin. J. Integr. Med.* 2020, 26, 243–250.

Mehrbod, P.; Abdalla, M.A.; Njoya, E.M.; Ahmed, A.S.; Fotouhi, F.; Farahmand, B.; Gado, D.A.; Tabatabaian, M.; Fasanmi, O.G.; Eloff, J.N.; McGaw, L.J. South African medicinal plant extracts active against Influenza A virus. *BMC Complement. Altern. Med.* 2018, 18(1), 112.

Mehta, P.; Reddivari, A.K.R. Hepatitis. In *StatPearls [Internet].* StatPearls Publishing, 2022.

Mishra, N.N.; Kesharwani, A.; Agarwal, A.; Polachira, S.K.; Nair, R.; Gupta; S.K. Herbal gel formulation developed for anti-human immunodeficiency virus (HIV)-1 activity also inhibits in vitro HSV-2 infection. *Viruses* 2018, 10(11).

Mishra, S.; Pandey, A.; Manvati, S. Coumarin: an emerging antiviral agent. *Heliyon* 2020 6(1), e03217.

Mukherjee, P.K.; Banerjee, S.; Katiyar, C.K.; Sharma, S.; Chattopadhyay, N.. Traditional medical system (TMS) for sustainable healthcare in India. In *Drug Discovery and Drug Development: The Indian Narrative*, Madhu Dikshit, Ed.Springer: Singapore, 2021, pp. 1–36.

Nasim, N.; Sandeep, I.S.; Mohanty; S. Plant-derived natural products for drug discovery: current approaches and prospects. *Nucleus* 2022 65(3), 399–411.

Natarajan, S.; Anbarasi, C.; Sathiyarajeswaran, P.; Manickam, P.; Geetha, S.; Kathiravan, R.; Prathiba, P.; Pitchiahkumar, M.; Parthiban, P.; Kanakavalli, K.; Balaji, P. Kabasura Kudineer (KSK); a poly-herbal Siddha medicine; reduced SARS-CoV-2 viral load in asymptomatic COVID-19 individuals as compared to vitamin C and zinc supplementation: findings from a prospective; exploratory; open-labeled; comparative; randomized controlled trial; Tamil Nadu; India. Trials 2021, 22(1), 1–11.

Newman, D.J.; Cragg, G.M. Natural products as sources of new drugs from 1981 to 2014. *J. Nat. Prod.* 2016, 79(3); 629–661.

Nguyen, N.H.; Nguyen, C.T. Pharmacological effects of ginseng on infectious diseases. *Inflammopharmacology* 2019, 27, 871–883.

Noviana, E.; Indrayanto, G.; Rohman, A. Advances in fingerprint analysis for standardization and quality control of herbal medicines. *Front. Pharmacol.* 2022, 13, 853023.

Othman, L.; Sleiman, A.; Abdel-Massih, R.M. Antimicrobial activity of polyphenols and alkaloids in middle eastern plants. *Front. Microbiol.* 2019, 10, 911.

Palshetkar, A.; Pathare, N.; Jadhav, N.; Pawar, M.; Wadhwani, A.; Kulkarni, S.; Singh, K.K. In vitro anti-HIV activity of some Indian medicinal plant extracts. *BMC Complement. Med. Ther* 2020, 20, 1–11.

Panda, A.K.; Kar, S.; Rai, A.K.; Rao, B.; Srikanth, N. AYUSH-64: a potential therapeutic agent in COVID-19. *J. Ayurveda Integr. Med.* 2022, 13(2), 100538.

Parham, S.; Kharazi, A.Z; Bakhsheshi-Rad, H.R.; Nur, H.; Ismail, A.F.; Sharif, S.; RamaKrishna, S.; Berto, F. Antioxidant; antimicrobial and antiviral properties of herbal materials. *Antioxidants* 2020, 9(12), 1309.

Parveen, A.; Parveen, B.; Parveen, R.; Ahmad, S. Challenges and guidelines for clinical trial of herbal drugs. *J. Pharm. Bioallied. Sci.* 2015, 7(4), 329.

Pastorino, G.; Cornara, L.; Soares, S.; Rodrigues, F.; Oliveira, M.B.P. Liquorice (Glycyrrhiza glabra): a phytochemical and pharmacological review. *Phytother. Res.* 2018, 32(12), 2323–2339.

Patel, B.; Sharma, S.; Nair, N.; Majeed, J.; Goyal, R.K.; Dhobi, M. Therapeutic opportunities of edible antiviral plants for COVID-19. *Mol. Cell Biochem.* 2021, 476, 2345–2364.

Pennisi, R.; Mandalari, G.; Sciortino, M.T. Antiviral activity exerted by natural products against human viruses. *Viruses* 2021, 13(5), 828.

Petrova, V.N.; Russell, C.A. The evolution of seasonal influenza viruses. *Nat. Rev. Microbiol.* 2018, 16(1), 47–60.

Praditya, D.; Kirchhoff, L.; Brüning, J.; Rachmawati, H.; Steinmann, J.; Steinmann, E. Anti-infective properties of the golden spice curcumin. *Front. Microbiol.* 2019, 10, 912.

Prasathkumar, M.; Anisha, S.; Dhrisya, C.; Becky, R.; Sadhasivam, S. Therapeutic and pharmacological efficacy of selective Indian medicinal plants-a review. *Phytomedicine* 2021, 1(2), 100029.

Raghavendhar, S.; Tripati, P.K.; Ray, P.; Patel, A.K. Evaluation of medicinal herbs for anti-CHIKV activity. *Virology* 2019, 533, 45–49.

Rajan, S.K.S; Dubey, G.P; Agarwal, A.; Dubey, N.; Dubey, S.; Arun, V.; Rajan, A.K.S. A novel herbal formulation for the modulation of immune system of HIV infected patients and a process of preparation thereof. CA Pat. App CA2864816A, 2011

Rangarajan, P.; Mody, S.K.; Marathe, M. Forecasting dengue and influenza incidences using a sparse representation of Google trends; electronic health records; and time series data. *PLoS Comput. Biol.* 2019, 15(11), e1007518.

Ratan, Z.A.; Youn, S.H.; Kwak, Y.S.; Han, C.K.; Haidere, M.F.; Kim, J.K.; Min, H.; Jung, Y.J.; Hosseinzadeh, H.; Hyun, S.H.; Cho, J.Y. Adaptogenic effects of Panax ginseng on modulation of immune functions. *J. Ginseng Res.* 2021, 45(1), 32–40.

Rouf, R.; Uddin, S.J.; Sarker, D.K.; Islam, M.T.; Ali, E.S.; Shilpi, J.A.; Nahar, L.; Tiralongo, E.; Sarker, S.D. Antiviral potential of garlic (Allium sativum) and its organosulfur compounds: a systematic update of pre-clinical and clinical data. *Trends Food Sci Technol.* 2020, 104, 219–234.

Schein, C.H. Repurposing approved drugs on the pathway to novel therapies. *Med. Res. Rev.* 2020, 40(2); 586–605.

Sharma, A.; Tiwari, S.; Deb, M.K.; Marty, J.L. Severe acute respiratory syndrome coronavirus-2 (SARS-CoV-2): a global pandemic and treatment strategies. *Int. J. Antimicrob. Agents* 2020, 56(2), 106054.

Sharma, P.; Dwivedee, B.P.; Bisht, D.; Dash, A.K.; Kumar, D. The chemical constituents and diverse pharmacological importance of Tinospora cordifolia. *Heliyon* 2019, 5(9).

Sharma, R.;Bhattu, M.; Tripathi, A.; Verma, M.; Acevedo, R.; Kumar, P.; Rajput, V.D.; Singh, J. Potential medicinal plants to combat viral infections: a way forward to environmental biotechnology. *Environ. Res.* 2023, 227, 115725.

Shi, J.; Weng, J.H.; Mitchison, T.J. Immunomodulatory drug discovery from herbal medicines: insights from organ-specific activity and xenobiotic defenses. *Elife* 2021, 10, e73673.

Shrivastava, S.; Soundararajan P.; Agrawal A.; Ayurvedic approach in chronic disease management. In Diana Noland, Jeanne A. Drisko, Leigh Wagner (Eds.) *Integrative and Functional Medical Nutrition Therapy: Principles and Practices*. Springer Link. 2020, pp. 783–98.

Singh, D.; Agarwal, V. A review on the COVID-19; its history; diagnostic approaches; role of herbs and current world scenario. *Biomedicine* 2021, 41(2), 328–332.

Sivagami, B.; Sailaja, B. A review on analytical methods for antiviral Phytoconstituents. *J. Young Pharm.* 2021, 13(1),7.

Stefanachi, A.; Leonetti, F.; Pisani, L.; Catto, M.; Carotti, A. Coumarin: a natural; privileged and versatile scaffold for bioactive compounds. *Molecules* 2018, 23(2), 250.

Tamoli, S.; Gokarn, V.; Ibrahim, M.; Ahmad, S. Comparative investigation of Ashwagandha FMB extract and standardized extract for their antioxidant; anti-inflammatory and immunomodulatory potential. *Ann. Phytomedicine Int. J.* 2022, 11(1), 405–412.

Tang, K.; Zhang, X.; Guo, Y. Identification of the dietary supplement capsaicin as an inhibitor of Lassa virus entry. *Acta Pharm. Sin. B.* 2020, 10(5), 789–798.

Thakar, A.; Panara, K.; Goyal, M.; Kumari, R.; Sungchol, K. AYUSH (Indian System of Medicines) therapeutics for COVID-19: a living systematic review and meta-analysis (first update). *J. Complement. Integr. Med.* 2023, 29(3), 139–155.

Tomas, M.; Capanoglu, E.; Bahrami, A.; Hosseini, H.; Akbari-Alavijeh, S.; Shaddel, R.; Rehman, A.; Rezaei, A.; Rashidinejad, A.; Garavand, F.; Goudarzi, M. The direct and indirect effects of bioactive compounds against coronavirus. *Food Front.* 2022, 3(1), 96–123.

Upadhyay, R.; Tiwari, K.N. The antiviral potential of Phyllanthus species: a systematic review. *Arch. Virol.* 2023, 168(7), 177.

van Brummelen, R.; van Brummelen, A.C. The potential role of resveratrol as supportive antiviral in treating conditions such as COVID-19 - a formulator's perspective. *Biomed. Pharmacother.* 2022, 148, 112767.

Varma, V.; Parwe, S.; Nisargandha, M. Systematic review of anti-viral & immunomodulatory potential of Ashwagandha (Withania Somnifera. Linn) in light of covid-19 pandemic. *Int. J Botany Stud.* 2020, 5(5), 494–497.

Wang, Y.X.; Yang, Z.; Wang, W.X.; Huang, Y.X.; Zhang, Q.; Li, J.J.; Tang, Y.P.; Yue; S.J. Methodology of network pharmacology for research on Chinese herbal medicine against COVID-19: a review. *J. Integr. Med.* 2022.

Wang, Y.; Li, Q.; Zheng, X.; Lu, J.; Liang, Y. Antiviral effects of green tea EGCG and its potential application against COVID-19. *Molecules* 2021, 26(13), 3962.

Weng, J.; Lin, C.; Lai, H.; Lin, Y.; Wang, C.; Tsai, Y.; Wu, K.; Huang, S.; Lin, C. Antiviral activity of Sambucus FormosanaNakai ethanol extract and related phenolic acid constituents against human coronavirus NL63. *Virus Res.* 2019, 273, 197767.

World Health Organization. *WHO Global Report on Traditional and Complementary Medicine*. World Health Organization, 2019.

Wu, C.; Liu, Y.; Yang, Y.; Zhang, P.; Zhong, W.; Wang, Y.; Wang, Q.; Xu, Y.; Li, M.; Li, X.; Zheng, M.; Chen, L.; Li; H. Analysis of therapeutic targets for SARS-CoV-2 and discovery of potential drugs by computational methods. *Acta Pharm. Sin. B.* 2020, 10(5), 766–788.

Wu, S.; Zhang, S.; Peng, B.; Tan, D.; Wu, M.; Luo, H.; Wei, J.; Wang, Y. Ganoderma lucidum: a comprehensive review of phytochemistry; efficacy; safety and clinical study. *Food Sci. Hum. Wellness* 2023.

Xiao, J.; Liu, P.; Wang, Y.; Zhu, Y.; Zeng, Q.; Hu, X.; Ren, Z.; Wang, Y. A novel cognition of decitabine: insights into immunomodulation and antiviral effects. *Molecules* 2022, 27(6), p.1973.

Xu, H.Q.; He, L.W.; Chen, J.; Hou, X.B.; Fan, F.T.; Wu, H.Y.; Zhu, H.P.; Guo, Y.Q. Different types of effective fractions from Radix Isatidis revealed a multiple-target synergy effect against respiratory syncytial virus through RIG-I and MDA5 signaling pathways; a pilot study to testify the theory of superposition of traditional Chinese medicine efficacy. *J. Ethnopharmacol.* 2019, 239, 111901.

Xu, Z.; Chen, Q.; Zhang, Y.; Liang, C. Coumarin-based derivatives with potential anti-HIV activity. *Fitoterapia* 2021, 150, 104863.

Yang, Y.; Islam, M.S.; Wang, J.; Li, Y.; Chen, X. Traditional Chinese medicine in the treatment of patients infected with 2019-New Coronavirus (SARS-CoV-2): a review and perspective. *Int. J. Biol. Sci.* 2020, 16, 1708–1717.

Zebroski, B. *A Brief History of Pharmacy: Humanity's Search for Wellness.* Routledge: New York, 2016.

Zhang, J.; Wider, B.; Shang, H.; Li, X.; Ernst, E. Quality of herbal medicines: challenges and solutions. *Complement Ther. Med.* 2012, 20(1–2), 100–106.

Zhao, H.; Ren, S.; Yang, H.; Tang, S.; Guo, C.; Liu, M.; Tao, Q.; Ming, T.; Xu, H. Peppermint essential oil: its Phytochem.; biological activity; pharmacological effect and application. *Biomed. Pharmacother.* 2022, 154, 113559.

# Section B

## RNA Viruses and Bioactive Compounds

# 6 Medicinal Plants and Herbs in the Treatment of Dengue Virus

*Mohini Yadav, Avijit Malakar, Pritam Mukherjee, Shaon Mukherjee, Tanisha Mahajan, and Nancy George*
University Institute of Biotechnology, Chandigarh University

## 6.1 INTRODUCTION

Dengue fever is a mosquito-borne viral disease, that is spreading around the globe and becoming more common. The fever is spread by female mosquito, primarily *Aedesaegypti* (Kularatne, 2015). Infection with one of the virus's four varieties usually results in lifetime immunity for that specific type, but other viruses provide short-term immunity. Following infection with a variable kind (secondary heterologous dengue infections), the chance of serious consequences increases (WHO, 2017). Children under the age of 15 account for the great majority of dengue cases. Dengue fever can range from a moderate fever to a severe infection with shock and leakage of plasma. The most vulnerable to severe dengue sickness is seen in toddlers with secondary heterologous dengue infections. Serum viral components can be directly detected and can be used to provide a diagnosis of dengue fever within 5 days after the disease onset. The serologic diagnosis gives indirect evidence of dengue after day 5. Dengue infection is currently treated with no effective antiviral medicines. As a result, treatment focuses on close hematological monitoring, early detection of severe illness warning signs, therapy using replacement of fluid, and/or blood transfusions as needed. Taking account of the fatality, dengue vaccine development is a top priority for the health of humans (Verhagen et al., 2014).

Currently, antiviral vaccines are accessible for the human papillomavirus, varicella, measles, mumps, rubella, and rabies, but not for other viral infections, such as dengue. This is due to the emergence of novel viral strains and growing resistance to conventional antiviral treatment (WHO, 2017). The necessity to develop immunity against all four dengue serotypes slowed down the development of dengue vaccines, which started in the 1920s. One variant, referred to as CYD-TDV, has been commercialized under the trade name Dengvaxia and is commercially accessible as of 2021 (Verhagen et al., 2014). Only those who have formerly had dengue fever or live in the areas where the majority of people have already contracted the disease are advised to get the vaccination. Thus, scientific committee is now focusing on the discovery of cost-effective and precise antiviral treatments that include herbal treatment.

The rising popularity of natural products among people, particularly in developing nations, has resulted in the development and advancement of herbal therapies and phytochemicals to combat various infections and illnesses. Many medicinal plants, including DENV infection, have been introduced for treatment of viral infections in both people and animals. At various phases of viral development, it has been shown that a number of medicinal herbs exhibit potent antiviral properties (Hidari et al., 2008; Talarico et al., 2005). With more information on the impact of medical plant varieties and their wide range of bioactive agents on generation and progress of anti-DENV medications, the goal of this chapter is to give a brief evaluation of the possible medicinal plants for the treatment of DENV infection and their potential role in the management of DENV infection cases.

DOI: 10.1201/9781003329169-8

## 6.2 DENGUE VIRUS: STRUCTURE AND SEROTYPES

A virus in the family *Flaviviridae* and genus *Flavivirus*, dengue virus is a (+) sense, single-stranded RNA virus. It has a diameter of 50 nm and an approximately spherical shape. The viral core is a structure called the nucleocapsid, which is composed of the virus's DNA and C proteins. A lipid bilayer leads to the formation of viral envelope, which is produced from the host and surrounds the nucleocapsid (Figure 6.1). The viral envelope contains 180 copies of the unique proteins E and M, embedded within lipid bilayer. These proteins create a defense mechanism that keeps the virus from invading human cells.

The single-stranded (+) sense RNA of DENV is expressed and translated directly into proteins. There are ten genes in the viral genome. A single long polypeptide created by the translation of the DNA is then split into ten proteins.

The structural proteins of DENV comprises capsid (C), envelope (E), and membrane (M), whereas NS1, NS2A, NS2B, NS3, NS4A, NS4B, and NS5 are the non-structural proteins found in virus. These non-structural proteins help viruses replicate and assemble. Non-structural (NS) proteins are expressed in the host cell but not incorporated into the viral particle. The membrane proteins NS2A, NS4A, NS2B, and NS4B play highly significant role during viral replication (Gubler et al., 1998).

**NS1:** As DENV infects the cell, NS1 is attached to the lumen of endoplasmic reticulum (ER) vesicles membrane and aids in anchoring of the replication of viral complex. Additionally, it can be secreted outside the cell, probably to aid the virus evade the immune system (ECDC, 2022).

**NS2A:** NS2A has been inadequately studied so far. Studies reveal that viral RNA replication can be affected by mis-cleavage occurring between NS1 and NS2A, and virion assembly can be affected by mutations in the protein (Dwivedi et al., 2017).

**NS2B:** To evade the mitochondrial DNA during infection, the NS2B protease cofactor directs the DNA sensor cyclic GMP-AMP synthase (cGAS) for destruction of lysosome. As a result of its breakdown, production of interferon in the infected cell is inhibited (Dwivedi et al., 2017).

**NS3:** NS3 is a 618-residue protein with mainly two domains. First is an N-terminal protease that employs the NS2B cytoplasmic loop as a cofactor, and another a C-terminal helicase that unwinds double-stranded RNA (dsRNA) and an ATPase that caps newly generated genomic RNA. The NS3 protein's helicase domain interacts with NS4B's cytoplasmic loop (Lin et al., 2006).

**NS4A:** NS4A is thought to cause membrane changes by acting as a scaffold for the replication complex. It has been discovered that it may form oligomers, and that changes in the protein have an impact on virus replication (Teramoto et al., 2017).

**NS5:** NS5 non-structural protein is the largest (102 kDa) and significantly preserved among all the serotypes of DENV, which is expressed during infection of dengue. It comprises two domains: an N-terminal domain, i.e., methyltransferase (MTase) and RNA-dependent RNA polymerase (RdRp), which is a C-terminal (Lin et al., 2019).

### 6.2.1 DENV Serotypes

The virus is divided into four serotypes, the first four of which are known as DENV-1, DENV-2, DENV-3, and DENV-4 (Datan et al., 2016). In 2007, another serotype (fifth) was revealed, but very limited information is known about it. The antigenicity of the serotypes is used to distinguish them (WHO, 2022).

**DENV-1:** There are five different genotypes of DENV-1 (genotypes I–V). Recently, it has been found that a basal group of DENV-1 with significantly divergent sequences might represent a novel genotype (genotype VI). Dengue, the most prevalent virus-transmitted

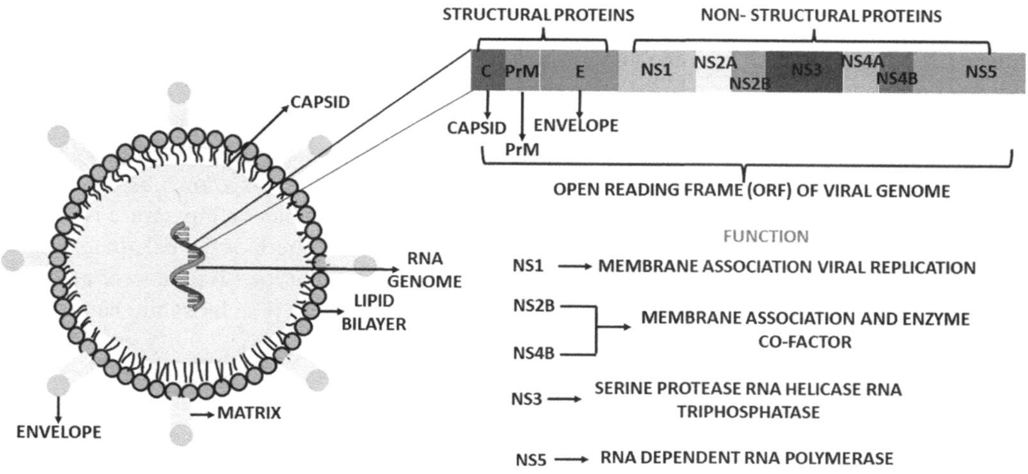


**FIGURE 6.1** Structure of DENV and structural and non-structural proteins coded by different viral genes.

disease in humans spread by mosquitoes, is caused by the most prevalent viral serotypes of the four, i.e., DENV-1. These viruses are currently widespread in more than 100 nations due to an increase in prevalence over the past few decades.

**DENV-2:** Native Dengue Virus Type 2 (DENV-2) is a concentration of virus particles derived from tissue culture supernatant. It is a viral infection spread by mosquitoes that can be asymptomatic or produce symptoms such as hemorrhagic fever, undifferentiated fever, shock syndrome, and fever (Rodenhuis-Zybert et al., 2010).

**DENV-3:** DENV-3 induced inoculum-dependent fatality, which was preceded by multiple hematological and biochemical alterations, as well as increased virus spread, all of which are symptoms of a severe disease in people (Rodenhuis-Zybert et al., 2010).

**DENV-4:** Showed clinical indications of sickness, had high virus loads in many tissues, and died 5 days later. The breakdown of tissue architecture activates cells, resulting in increased levels of circulating cytokines (Solomonides et al., 2010).

## 6.3 DENV: ETIOLOGY AND EPIDEMIOLOGY

Dengue fever has spread throughout the world since World War II and is now found in over 120 countries, primarily in Asia and South America (Ranjit et al., 2011). Each year, dengue causes about 400 million cases and 22,000 fatalities worldwide (Mustafa et al., 2015). In 2019, there was a considerable rise in frequency of cases. The first reports of an outburst date back to 1779 (Gubler et al., 1998). The Pan American Health Organization (PAHO) recorded 695,028 dengue cases in America as of 4 May 2022, including 285,432 confirmed cases and 209 related deaths. Dengue fever surveillance indicators are minimal or non-existent in the French Caribbean Islands (ECDC, 2022). The disease has now become common in Southeast Asia, South America and Pacific islands. During the early 20th century, the cause and dissemination of DENV had been identified.

### 6.3.1 DENV Transmission

The transmission of all four DENV serotypes is maintained in two cycles: sylvatic (transmission in wild animals) and human. The sylvatic cycle is ecologically and evolutionarily distinct from the human transmission cycle. This is maintained by non-human primates or by a monkey-Aedes-monkey cycle in the sylvatic environments of Southeast Asia, West Africa, peninsular Malaysia, and eastern Senegal. The primary and secondary vectors of DENV are mosquito *Aedes aegypti* and *Aedes albopictus*, respectively (Gubler et al., 1998). In humans, DENV transmission occurs via following ways.

#### 6.3.1.1 Transmission through Mosquito Bite

Female mosquitoes infected with the virus, usually the *Aedes aegypti* mosquito, transmit the virus to humans through bites. *Aedes* species other than *Aedes aegypti* can act as vectors, but their importance is secondary to that of *Aedes aegypti*. After feeding on a DENV-infected person, the virus replicates in the mosquito midgut before migrating to secondary tissues like the salivary glands. The time between ingesting the virus and transmitting it to a new host is known as the extrinsic incubation period (EIP). The EIP takes about 8–12 days when the ambient temperature is between 25°C and 28°C (Mustafa et al., 2015). A lot of factors influence the length of the EIP, including the amplitude of daily temperature variations (ECDC, 2022), virus genotype (Teramoto et al., 2017), and starting viral concentration (Simmons et al., 2012), among others. The mosquito can transmit the virus for the rest of its life after it has become infected.

#### 6.3.1.2 Human-to-Mosquito Transmission

Mosquitoes can be infected with DENV by people who are infected with the virus. This can include persons who have an asymptomatic dengue infection, those who have not yet developed an asymptomatic infection (pre-symptomatic), and those who exhibit no signs of sickness (they are asymptomatic) (Chen et al., 2010). Human-to-mosquito transmission can happen up to 2 days before symptoms appear and up to 2 days after the fever has subsided (Chen et al., 2010). High viraemia (increased presence of DENV in human body) and fever in the patient are linked to a higher chance of mosquito infection; on the other hand, high levels of DENV-specific antibodies are linked to a lower risk of mosquito infection. The majority of persons are viraemic for 4–5 days, but viremia can extend up to 12 days (Rodenhuis-Zybert et al., 2010).

#### 6.3.1.3 Maternal Transmission

Mosquito vectors are the principal method of transmission of DENV between people. However, there is evidence that maternal transmission is possible (from a pregnant mother to her baby). While the risk of vertical transmission appears to be modest, it appears to be linked to the time of the dengue infection during pregnancy (Datan et al., 2016). When a mother has a DENV infection while pregnant, her baby may be born prematurely, have a low birth weight, or have fetal distress.

#### 6.3.1.4 Other Transmission Modes

There have been a few reported cases of transmission by blood products, organ donation, and transfusions. The virus's transovarial transmission within mosquitoes has also been documented (Rodenhuis-Zybert et al., 2010).

## 6.4 SIGNS, SYMPTOMS AND TREATMENT STRATEGIES

Clinical symptoms of this viral infection might range from a slight fever to serious physiological problems. The symptoms of dengue fever include a rapid start of fever, a headache (often behind the eyes), pain in the muscles and joints, and a rash. Dengue fever is sometimes known as "breakbone fever" because of accompanying joint and muscle symptoms. The infection progresses through three stages: febrile, critical, and recovery (Rodenhuis-Zybert et al., 2010). The stages of dengue infection are outlined as follows:

a. **Febrile phase:** This phase is characterized by a high fever, up to 40°C (104°F), as well as general pain and a headache, and stays for 2–7 days (Datan et al., 2016). Sometimes the side effects might include nausea and vomiting (Simmons et al., 2012). A rash appears on initial days of infection characterized as flushed skin, or later from days 4 to 7 in the form of measles-like rash (ECDC, 2022) in 50–80% of those with symptoms (Chen et al., 2010). At this point, petechiae (small red spots which persist as the skin is squeezed and are produced by burst capillaries) may emerge (Datan et al., 2016), as well as some moderate bleeding from the mouth and nose mucous membranes (Ranjit et al., 2011). The fever is biphasic in character, meaning it breaks and then returns for 1 or 2 days (Chen et al., 2010).

b. **Critical phase:** As the fever fades, some people's condition progresses to a critical stage (Simmons et al., 2012). Plasma leaks from the blood arteries during this time, which usually lasts 1–2 days (Teramoto et al., 2017). Malfunction of organs and prominent bleeding, usually from the gastrointestinal tract, are also possible (Ranjit et al., 2011). Less than 5% of dengue infections result in shock (dengue shock syndrome) or bleeding (dengue hemorrhagic fever), although those patients who encountered previous infection with another dengue virus serotype ("secondary infection") are more vulnerable (Ranjit et al., 2011).
c. **Recovery phase:** The healing procedure starts when the spilled fluid is absorbed into the blood flow (Teramoto et al., 2017). This often lasts 2–3 days. Acute itching and a sluggish heart rate are typically present, and the improvement is frequently dramatic. It is possible to get another rash that looks maculopapular or vasculitic and is followed by skin peeling (Simmons et al., 2012). During this phase, a fluid overflow scenario could arise; if it impacts the brain, it can cause loss of awareness or convulsions. Adults may feel fatigued for weeks on end (Simmons et al., 2012).

Luke-warm sponging for rise in body temperature and use of antipyretics for managing body ache or fever are the mainstays of treatment for dengue virus infection. A specific antiviral medication for dengue is not available; however, research has been conducted on a number of sulfated polysaccharides isolated from natural resources, namely algae and seaweeds, which exhibit strong activity against DENV (Rodenhuis-Zybert et al., 2010). Ineffective and inadequate efforts for control of the vectors and increasing incidences of infection have necessitated the development of DENV vaccine. The key challenges for the creation of a vaccine are DENV's complicated pathophysiology, ADE impact, and genetic changes over time. The Sanofi Pasteur CYD-TDV (Dengvaxia candidate), DENVax, and TV005 vaccines are the three live-attenuated tetravalent DENV vaccine candidates now being examined in significant clinical trials (Diamond et al., 2015). DENVax, first created by the Centers for Disease Control and Prevention of the USA (CDC, USA) in collaboration with Inviragen and currently licensed to Takeda, is the other promising tetravalent recombinant live-attenuated dengue vaccine (Osaka, Japan). Up to 48 months following immunization, TAK-003 demonstrated antibody responses against all four serotypes without any risk of severity in either seropositive or seronegative patients (Tricou et al., 2020). However, it demonstrated various immunogenicities in relation to DENV serotype infection, and the cause of this is now unknown.

With the emergence of new viral strains and increasing resistance to standard antiviral therapy, as well as the fact that the majority of available antiviral therapies are non-specific for specific viruses, the primary focus of medical research is now on the development of specific and cost-effective antiviral regimens that include alternative treatment from herbal sources. The increasing tendency of humans toward natural products, particularly in developing countries, has resulted in the invention and advancement of herbal treatments and phytochemical to combat different alignments, in various infectious symptoms.

## 6.5 BIOACTIVE COMPONENTS FROM MEDICINAL PLANTS HAVING ANTI-DENGUE ACTIVITY

Biologically active components present in various plant species have been found to have varying inhibition degrees of activity against serotypes of DENV. The isolated components are classified as non-glycosylated polyphenols, quinones, sulfated polysaccharides, glycosylated, alkaloids, terpenoids, and flavonoids. In one of the studies, more than 50 biologically active molecules derived

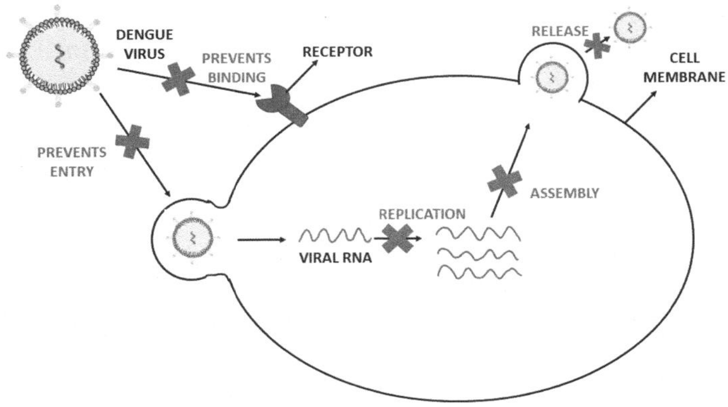


**FIGURE 6.2** Target biomolecules/pathways for inhibition of DENV by phytochemicals.

from different plant sources were shown to have varying degrees of anti-dengue effectiveness (Tricou et al., 2020). These bioactive phytochemicals effect viral life cycle in human in many ways (Figure 6.2). Plaque and focus formation tests employed in *in vitro* research are primary methods for testing the drugs for activity against dengue, and these methods are generally used to test the virus titer. Plaque test is employed to determine the amount of DENV (virions) via plaque formation, whereas the latter is a version of the former that uses immunostaining. This assay enumerates infected cells of host and infectious DENV virions prior to plaque formation during DENV-induced cell lysis (Abd Kadir et al., 2013). The important bioactive components from medicinal plants showing anti-dengue activity are outlined as follows.

### 6.5.1 Sulfated Polysaccharides

Polysaccharides, which belongs to carbohydrates, are long chains of monosaccharides joined with the help of glycosidic linkage, which results in the development of polymeric structures. These can be broadly divided into two groups: starch polysaccharides and non-starch polysaccharides (NSPs). Non-starch polysaccharides are further subdivided into cellulose, non-cellulosic polymers, and pectic polysaccharides (Lovegrove et al., 2017). Sulfated polysaccharides have been linked to a variety of biological functions, like antiviral activity. Sulfated polysaccharides exhibit antiviral effect against DENV. This is apparently due to polysaccharide restricting virus binding to the cell, and hence preventing virus penetration into the host cell's cytoplasm. Their antiviral action is mainly reliant on the virus serotype. This variance in susceptibility of DENV serotype to this class of carbohydrates could be attributed to changes in the ability of growth in Vero cells. Different RNA and DNA viruses (enveloped), including serious human infections like retroviruses and herpes viruses, have been shown to be inhibited by seaweed sulfated polysaccharides. For instance, the commercial substances heparins as well as two sulfated polysaccharides (DL-galactan hybrid C2S-3 and the kappa/iota/nu carrageenan G3d) derived from South American red seaweeds were particularly effective against DENV-2. The docking analysis disclosed that the sulfated group and carboxyl group interacts with DENV E protein and is attributed for antiviral effect (Sacchelli et al., 2019). Galactans extracted from *Cryptomenia crenulata* and *Gymnogongrus tolulosus* showed suppression of replication process of serotype DENV-2 in Vero cells. The drug was reported to be completely ineffective against DENV-1, but it showed strong inhibition against DENV-3 and DENV-2, with only a negligible effect against DENV-4 (Casciaro et al., 2020). The protective effects may have been caused by preventing the virus from adhering to surfaces or from internalizing into cells (Wang et al., 2014).

### 6.5.2 Terpenes and Terpenoids

Terpenes are a class of secondary metabolic compounds generated through enzymatic enucleation of primary metabolites. Terpenes are formed by assembling multiple isoprene units by various pathways. Terpenes are divided into different groups based on the structure and the quantity of isoprene units they hold. These are naturally occurring volatile, unsaturated 5-carbon cyclic substances that have a flavor or aroma to defend the plant from organisms that feed on particular plant species. Terpenoids are frequently used interchangeably with terpenes, but the two terms have a few subtle differences. Terpenes serve a variety of purposes in plants, including thermo-protection, signaling, taste, and solvents. They are in charge of giving plants their flavor, aroma, and color. They also have a number of therapeutic uses. Terpenoids are divided into monoterpenes, diterpenes, triterpenes, sesterpenes, and sesquiterpenes. Particularly, the antiviral ability of monoterpenes is the subject of extensive research. Triterpenes, on the other hand, make up the majority of the class and are biologically diverse because of their complex structure, offering numerous sites for linking and modifications as well. In the case of triterpenoid, antiviral action may occur from the creation of an H-bond and a hydrophobic bond with the DENV's NS2B-NS3pro non-structural protein, which inhibits the virus (Dwivedi et al., 2017). Additionally, the hydroxyl group at the ring's C3 position is responsible for triterpenoid's larvicidal ability, also known as mosquito-killing ability (Dhinakaran et al., 2019). Many studies have scientifically proven that those essential oils primarily containing mixtures of terpenes have larvicidal or mosquito-killing effects against DENV vectors (Islam et al., 2019). In Malpighian tubules, forskolin, a labdane diterpene generated by the Indian Coleus plant, increases intracellular cAMP levels during the peak of NaCl dieresis (Premaratna et al., 2007). Labdane diterpenes extracted from the oil-resin fractions of *Copaifera reticulata* have been shown by da Silva and his colleagues to have strong larvicidal activity (Valotto et al., 2011). Similarly, midgut cell death in *Aedes aegypti* larvae was induced by the labdane diterpenoid of the same species (Morrison et al., 2006). A combination of monoterpenes might function as a more effective antiviral agent than a single monoterpene (Casciaro et al., 2020).

### 6.5.3 Alkaloids

Alkaloids are organic acid salts found in plant tissues that are naturally occurring substances with the elements carbon, hydrogen, nitrogen, and oxygen (e.g., oxalic, malic, acetic, citric, lactic). Alkaloids derived from plants can be found as crystalline, amorphous, flavorless, and non-volatile substances (Wang et al., 2014). It is a group of secondary metabolites that spans a wide range in chemical structure and are exceedingly diverse. However, each compound's unique properties are due to the several functional groups that make up its structure. Experimental evidence supports the ability of numerous alkaloids to obstruct the synthesis of DNA and RNA in a variety of virus families through various pathways. For example, lycorine blocks the dengue virus's RNA-dependent polymerase (RdRp) (DENV) (Gómez-Calderón et al., 2017). They have been discovered to show understanding toward the NS proteins of DENV, supporting alkaloid's ability to suppress viral replication (Jia et al., 2010). The NS2B/NS3 protease, playing role in prevention of viral replication in host cells, was also suggested to have inhibitory effect by some molecular docking studies (Hishiki et al., 2017). Alkaloids are also reported to hinder multiplication DENV at both early and late stages (Su et al., 2009), perhaps by interfering with ribosome activity, which prevents eukaryotic cells from synthesizing proteins (Quintana et al., 2016). Additionally, they reduced viral maturation and impacted the release of viral particles for extracellular dispersion (Pan et al., 1993). For example, Castanospermine, a natural alkaloid, is generated from the Moreton Bay chestnut tree or the black bean (*Castanospermum australae*). It is water soluble and easily separated in large quantities using purification methods (Diamond et al., 2002). It was examined *in vivo* using A/J mice and *in vitro* using BHK-21 cells against all serotypes. Castanospermine functions as an ER-glycosidase I inhibitor *in vitro*, where it inhibits the secretion and infectivity of DNA viruses and a subset of enveloped RNA. However, *in vivo*, it could only prevent the death of the mouse by inhibiting DENV-2 (Whitby et al., 2005). For Castanospermine to be deemed a viable anti-DENV agent, it must be extremely active against all DEN serotypes. Treatment of BHK-21 cells with this phytochemical significantly reduced viral production and cell-to-cell transmission of virus particles, of all four DENV serotypes (Ferreres et al., 2010).

### 6.5.4 Flavonoids

As phyto-constituents, flavonoids are classified as having a C-15 skeleton made up of two phenyl rings joined together and a heterocyclic ring. In essence, flavonoids are phenolic chemicals with low molecular weight that are present in a variety of plants. Fruits, roots, nuts, seeds, bark, stems, and flowers of plants contain a variety of flavonoids. Different types of flavonoids have been discovered with antiviral properties like daidzein, naringin, hesperetin, and quercetin.

One of these compounds is quercetin, which is found in a variety of foods and fruits, including apples, onions, citrus, tomatoes, green and black tea, and several other plants (Zandi et al., 2011). Quercetin has the strongest antiviral efficacy against DENV-2. Daidzein exhibited very little anti-dengue action. When administered after DENV-2 had adhered to the Vero cells, naringin and hesperetin showed no anti-dengue action. Numerous methods of action, including virucidal activity, complex formation with viral RNA, and inhibition of RNA polymerase, which stops viral replication, have been shown to produce antiviral activity (Raekiansyah et al., 2018). Additionally, Qaddir et al. (2017) highlighted the activity of isoflavones like derrisin, mundulinol, against the DENV4-NS4B protein in docking studies in 2017. Srivarangkul et al. reported that the flavanone subclass of flavonoids inhibits the DENV virus by focusing on the role of the E protein. The NS2B and NS3 proteases' activity is targeted by the proprotein convertase furin enzyme. The anti-DENV-2 activities of quercetin and quercetirin from the fraction of ethyl acetate of *Houttuynia cordata* aerial leaves showed that quercetin has a stronger virucidal impact than its glycoside quercetirin. Hyperoside, chlorogenic acid, quercetin, and quercetirin, four components from the same plant's ethyl acetate fraction when combined demonstrated higher antiviral activity. *Epigallocatechin* (EGCG), a catechin found in green tea, has the ability to prevent virus entry by targeting the virion and ultimately preventing the virus from attaching the host cell (Ichsyani et al., 2017). Curcumin, an active ingredient derived from *Curcuma longa*, is also known to be an antiviral that is effective against dengue virus (Cheung et al., 2014).

### 6.5.5 Glycosides

Glycosides are made up of sugar molecules linked to a functional group via a glycosidic bond. Recently, it was discovered that numerous (–vestrand) RNA viruses, including the influenza virus, and Newcastle disease virus, are inhibited by lanatoside C, a glycoside having cardiac role, works by suppressing the $Na^+$ -$K^+$-ATPase pump (Whitby et al., 2005). It is also a powerful inhibitor of dengue virus infection. According to a recent report, it can avert a variety of negative-strand RNA viruses. The early phases of the replication cycle of DENV-2 were ruled out as potential targets since pre-treatment with lanatoside C did not suppress its replication. It was only when lanatoside C was added at or before 24 hours post-injection that DENV-2 become inactive, suggesting that the inhibition effect is completed before 48 hours post-injection (hpi). These findings led to the conclusion that lanatoside C specifically targets viral protein translation, viral replication complex assembly, and viral RNA production during the post-entry stages of the DENV replication cycle. Expressions of viral proteins as well as the synthesis of viral RNA by DENV-2 were both successfully inhibited by lanatoside C (Premaratna et al., 2007).

Most of the phytochemicals act on the NS proteins of DENV. There aren't any known plant compounds with anti-DENV characteristics that might operate by inhibiting helicase and protease activity. Most research projects are still in the activity screening stage. In order to identify prospective candidates for DENV treatment, more research must be conducted to examine the anti-DENV potential with their mechanisms, like suppression of viral genomes and the destruction of viral encoding enzymes.

### 6.5.6 Essential Oils

Oils extracted from *Curcuma longa* exhibits excellent insect repellent (Jia et al., 2010) and anti-feeding components (Hishiki et al., 2017), that is, turmerone and arturmerone (Hishiki et al., 2017) curcuminoids. Other essential oils attained from neem (*Azadirachtaindicia*), galingale (*Alpiniagalangal* L.), and clove (*Syzyaiumaromaticum* L.) have mosquito repellents properties (Quintana et al., 2016). Several essential oils have been shown to only be efficient when tested using olfactometry or on lab mice that were *Aedes aegypti* (L.) hosts (Pan et al., 1993). When tested against *A. aegypti*, *Culex quinquefasciatus*, and *Anopheles dirus*, other essential oils from *Citrus sinensis*, *Eucalyptus citriodora*, and *Syzygiumaromaticum* revealed that *C. citratus* oil was highly significant to control all tested mosquito species. When the effectiveness of *Cinnamomumosmophloeum* essential oils was evaluated against *A. aegypti*, it was discovered that mosquitoes larvicidal assay had a superb inhibitory impact. The findings showed that *C. osmophloeum*'s components can successfully suppress *A. aegypti* (Diamond et al., 2002). Azadirachtin, nimbin, nimbidin, and nimbolides are present in neem (*A. indica*) seeds and exhibit suppression of fertility, ovicidal effect, insect growth regulation, antifeedant, and pest-repellent activities (Whitby et al., 2005). In several breeding sites under natural and field settings, the results showed that neem oil formulation was particularly successful at controlling *Anopheles*, *Culex*, and *Aedes*. Consequently, *A. indica* can be used as a substitute for synthetic pesticides to reduce mosquitoes (Ferreres et al., 2010).

## 6.6 MEDICINAL PLANTS USED IN TREATMENT

Natural products with a plant-based origin have been the most significant source of bioactive ingredients needed for developing new drugs to treat a wide range of illnesses. Numerous herbal remedies are now understood to be extremely important in the management of impending pandemics. This conventional way of disease management is still known as a preferred way of treatment in many communities. Owing to easy availability, low cost and safety concerns, people depend on these natural resources in many parts of the world. Important medicinal plants possessing anti-dengue activity and DENV serotype they effect are listed in Table 6.1. Various parts of these plants (leaves, roots, stems, and flowers) are used in a wide range of medical formulations. Various plant varieties having anti-dengue effect are discussed in the following.

### 6.6.1 *Alternanthera philoxeroides*

Family: Amaranthaceae

Location: South African, Chinese, and South American perennial plant

Common name: Alligator weed

In 2005 scientist Jiang and his fellow researchers tested *A. philoxeroides* extracts in ethyl acetate, petroleum ether, ethyl ether, and coumarin for antiviral activity. Based on their research, all extracts contain anti-dengue activity, and with toxic dosage ($TD_{50}$) ratios of 47.43, 86.67, 97.63, and 106.63, respectively, it suppressed DENV-2. *A. philoxeroides* coumarin extract exhibited the lowest toxicity on cells, however, *A. philoxeroides* petroleum ether extract had a high inhibitory effect on the dengue virus (Ferreres et al., 2010). They also analyzed anti-DENV effect of extracts using different solvents and showed that extract in petroleum ether was most effective.

**TABLE 6.1**
**Important Medicinal Plants Having Anti-dengue Activity and DENV Serotype They Effect**

| Name of the Medicinal Plant | Antiviral Compound | Dengue Type |
|---|---|---|
| *Andrographis paniculata* | Andrographolide | DENV-1 |
| *Azidarachta indica* | Azadirachtin | DENV-2 |
| *Boesenbergia rotunda* | 4 hydroxypanduratin A, panduratin A | DENV-2 |
| *Carica papaya* | Carpaine, Kaempferol, Quercetin | DENV-2 |
| *Cladosiphon okamuranus* | Fucoidan | DENV-2 |
| *Cryptonemia crenulate* | Galactan | DENV-2 |
| *Euphorbia hirta* | Quercetin | DENV |
| *Gymnogongrus griffithsiae* | Kappa carrageenan | DENV-2 |
| *Gymnogongrus torulosus* | Galactan | DENV-2 |
| *Hippophae rhamnoides* | Fucoidan | DENV-2 |
| *Houttuynia cordata* | Hyperoside | DENV-2 |
| *Leucaena leucocephala* | Galactomannan | DENV-1 |
| *Meristiellagelidium* | Kappa carrageenan | DENV-2 |
| *Mimosa scabrella* | Galactomannan | DENV-1 |
| *Momordica charantia* | Fucoidan | DENV-1 |
| *Psidium guajava* | Galactan | DENV-2 |
| *Zostera marina* | P-sulfoxycinnamic acid, zosteric acid | DENV-2 |

### 6.6.2 *Andrographis paniculate*

Family: Acanthaceae

Location: China, India, and Southeast Asia

Common name: King of bitters

*Andrographis paniculata* is an herbaceous plant having a bitter taste. It mainly treats flu, respiratory infections, and sore throats. Secondary metabolites found in this plant are diterpenoids, diterpenes, flavonoids, flavonoid glycosides, and lactones. The primary secondary molecule derived from *A. paniculata* with antiviral activity is andrographolide. Andrographolide (a diterpenoid) and its derivative have shown to have antiinflammatory, antibacterial, antidiabetic, anticancer, and antiviral properties (Lim et al., 2020). The maximum nontoxic dose (MNTD) of methanol extract was 0.050 mg/mL and inhibits the virus in Vero E6 cell lines by approximately 75%. It is also effective against DENV-1 serotype (Ranjit et al., 2011). According to the investigation, DENV-2 contaminated HepG2 and HeLa cells and DENV-4 contaminated HepG2 cells are treated through the antiviral activity of andrographolide. It mainly reduces viral production and cellular infection with 50% Effective Concentrations ($EC_{50}$). For HepG2 and HeLa, the $EC_{50}$ of andrographolide is 21.304 and 22.739 M, respectively. It also decreases the copy number of DENV-4 genome and prevents the expression of DENV-4 E protein from day 1 of post infection in HepG2 cells (Ranjit et al., 2011).

There is a study depending on the addition of andrographolide to: (i) pre-treat cells for 1 hour, 30 minutes, or 15 minutes, before the infection, and (ii) post treat cells for 0, 1, 2, 3, 4, 6, 9, and 12 hours after the infection. This study revealed that in post treat cells, andrographolide showed significant effects as compared to the pre-treat cells. In another study, andrographolide showed disruptive effects on actin and several cellular pathways by binding with DENV-2 and DENV-4 E protein. The Glucose Regulated Protein 78 (GRP78) of HepG2 cells, which are treated with andrographolide, show downregulation of DENV-2 infection, when compared with the control. GRP78 is a chaperone that helps protein folding in the endoplasmic reticulum and a mediator of unfolded protein response manipulated by DENV, which helps in viral production (Lim et al., 2020).

### 6.6.3 *Azidarachta indica*

Family: Mahogany

Location: India, Pakistan, Sri Lanka, and Thailand

Common name: Neem

It is a very common plant in India and Pakistan, called neem, and has many medicinal benefits (antibacterial, antifungal, antitumor, antiinflammatory, antipyretic, antiarthritic, etc.). *A. indica* is a fast-growing tree (reaches a maximum of 15–20 m) and highly valued medicinal plant. It has an inhibitory effect on DENV-2 replication. Cytotoxicity studies were performed to determine the MNTD of extract, and it is found that a concentration of 1.897 mg/mL completely inhibited 100–10,000 tissue culture infective dose ($TCDI_{50}$) of virus (Raekiansyah et al., 2018). The triterpenoids found in neem extract is supposedly responsible for anti-dengue effect by binding with NS2B and NS3. Intracerebral inoculation of neem leaf (NL) extract was used to test the antiviral effects of neem leaf aqueous extract on day-old suckling mice. In RT-PCR experiments, the aqueous extract inhibited the virus at nontoxic doses of 120–130 mg/mL, as evidenced by the absence of 511 bp dengue group-specific amplicons (Kularatne, 2015).

### 6.6.4 *Boesenbergia rotunda*

Family: Zingiberaceae

Location: Southeast Asia, India, China, and Sri Lanka

Common name: Chinese ginger

This plant is traditionally used in the treatment of fever, stomach ache, flatulence, gout, and inflammation (Lim et al., 2020). Pinostrobin, pinocembrin, alpinetin, cardamonin, panduratin A, and 4-hydroxypanduratin A, were among the six compounds isolated from *Boesenbergia rotunda* that were tested for their inhibitory effects against DENV-2 NS3 protease. Panduratin A and 4-hydroxypanduratin A inhibited DENV-2 NS2B/3 protease cleavage most effectively, blocking 65% activity at a concentration of 80 ppm. Pinocembrin and cardamonin had lesser inhibitory action at all doses examined, but when combined, they had a strong inhibitory impact on DENV-2 protease, indicating a synergistic effect (Lim et al., 2020).

### 6.6.5 *Carica papaya*

Family: Caricaceae

Location: Central America, Mexico, and other tropical countries

Common name: Papaya, also called paw paw

Dengue can be controlled by *Carica papaya* at both the transmission and host levels (Lim et al., 2020). The active compounds in *C. papaya* are chymopapain and papain. Alkaloids, phenolics, flavonoids, and amino acids were among the secondary compounds screened out of this plant (Ranjit et al., 2011). If a dengue patient takes two tablespoons of papaya leaf juice three times per day at 6 hours of intervals, then it is shown to increase the platelet count within 24 hours (Raekiansyah et al., 2018). *C. papaya* capsules with thrombocytopenia for 7 days shows an increase in platelet counts in dengue patients (Yadav et al., 2018). Leaf extract treat DENV by lowering the pro-inflammatory cytokine level of interferon gamma (INF-$\gamma$), TNF-$\alpha$, interleukine-6 (IL-6), IL-4. In spite of anti-DENV properties of *C. papaya* leaf extract (CPLE), there are some studies that reported that, administration of CPLE or capsule, not increases a patient's platelet count (Nuri et al., 2016). The actual mechanism of CPLE to dengue fever is unknown and all patients showed an increment in platelet counts after consuming the CPLE (Yadav et al., 2018). The majority of the aforementioned studies claimed that *C. papaya* had anti-dengue effects based on the reduction of patient's symptoms.

A new Guinea C strain which is a mouse model, nourished with 1,000 mg/kg FCPLJ (Freeze-dried *Carica papaya* Leaf Juice) from post infection day 1 to day 3, showed a change in its expression of gene. Those altered genes are (i) upregulation of chemokine ligand 2 (CCL-2), (ii) downregulation of Integrin neta-3, Intercellular Adhesion Molecule 1, and Fibronectin 1 (FN1) (Hopkins & Groom, 2002). After treating the dengue-infected AG128 mice with FCPLJ shows downregulation of inflammatory cytokine genes like chemokine ligand 12, chemokine ligand 6, chemokine ligand 17, chemokine ligand 8, Interleukin 1 receptor type I (IL1R1), nicotinamide phosphoribosyltransferase (NAMPT), and platelet factor 4 (PF4). These data recommend the function of *Carica papaya* leaf juice in dengue infection (Norahmad et al., 2019). Sprague Dawley (SD) rats that are infected with DENV-2, after administration of leaf extract show an increase in platelet counts IL6 and thrombopoietin levels. Similarly, a study found that administering leaf extract directly to dengue platelet rich plasma and regulate platelet poor plasma leads to a decrease in platelet aggregation. In DENV-infected THP-1 cells, which are human leukemia monocytic cell lines, the extract drastically lowers the production of NS1 proteins and viral load (Lim et al., 2020).

### 6.6.6 *Cladosiphon okamuranus*

Family: Chordariaceae

Location: Okinawa, Japan

Fucoidan a sulfated polysaccharide extracted from *Cladosiphon okamuranus* can potentially bind with DENV-2 envelope glycoprotein (EGP). Fucoidan's inhibitory impact on viral infection is dependent on the difference of EGP structures based on amino acid residues, which may alter the virus's interaction with fucoidan. Four dengue virus serotypes were combined with fucoidan at different doses before being injected into BHK-21 cells. Fucoidan's effects on DENV-2 infection were far stronger than those on the other dengue serotypes studied (Hidari et al., 2008).

### 6.6.7 *Cryptonemia crenulate*

Family: Halymeniaceae

Location: Atlantic Islands, North America, Caribbean Islands, Western Atlantic, South America

It is mainly found in the ocean or saline water (marine environment). Sulfated polysaccharides, i.e., Galactan collected from the *C. crenulata* are the main inhibitors of DENV-2. They mainly restrict the multiplication of DENV-2 with $IC_{50}$ (Inhibitory Concentration) of 1.0 µg/mL (Kularatne, 2015). It shows minimum antiviral properties against DENV-3 ($IC_{50}$ of 13.9–14.2 µg/mL) and DENV-4 ($IC_{50}$ of 29.3–50 µg/mL), but is ineffective against DENV-1 (Lim et al., 2020).

### 6.6.8 *Cymbopogon citratus*

Family: Poaceae

Location: India, Sri Lanka, and many Asian countries

Common name: Lamon grass

*Cymbopogon citratus* is antifungal, antiinflammatory, and antiprotozoal medicinal plant (Brügger et al., 2019). Its root methanol extracts inhibited DENV-2 by 98.9% *in vitro* at a dosage of 20 µg/mL, with $CC_{50}$ and $EC_{50}$ values of 183.74 and 29.37 µg/mL, respectively (Rosmalena et al., 2019). The extract was found to be nontoxic against human cell lines.

### 6.6.9 *Euphorbia hirta*

Family: Euphorbiaceae

Location: Philippines, Australia, Java, Sumatra, Peninsular Malaysia

Common name: Dudhi

The Philippines uses a decoction of *Euphorbia hirta* leaves for treating dengue fever, known locally as gatas-gatas (Verhagen et al., 2014). It is presently utilized in mosquito repellents, with an effective concentration of 272.36 ppm of petroleum ether extract made from plant (Ranjit et al., 2011). It contains immunomodulatory alkenes, triterpenes, phytosterols, tannins, polyphenols, and flavonoids (Lim et al., 2020). The plant extract contains numerous flavonoids, including euphorbianin, leucocyanidol, camphol, quercitrin, and quercitol (Ranjit et al., 2011). After 24 hours, dengue patients who drank its herbal water experienced an increase in platelet counts, as well as a marked improvement in fever and flu-like symptoms. Patients recovered more quickly after taking *Euphorbia hirta*, which helped stimulate platelet production and prevent further bleeding. Locals in the Philippines also used capsules made from this plant to treat dengue (Lim et al., 2020).

### 6.6.10 *FLAGELLARIA INDICA*

Family: Flagellariaceae

Location: India, Southeast Asia, and Australia

Common name: False rattan

Its leaves are mainly used to treat coughs, vomiting, wounds, and as a health tonic (Lim et al., 2020). *Flagellaria indica* was studied in Vero cells for its anti-dengue properties. 12.5 μg/mL of ethanol extract of this plant shows inhibition of DENV-2 by 45.52% in an *in vitro* antiviral assay. The MTT assays are showed the cytotoxic nature of this plant. *F. indica* shows 50% cytotoxic concentration ($CC_{50}$) of its ethanol extract was 312 μg/mL. As a result of this research, *F. indica* appears to have a crucial effect on DENV and has the potential to be used in DENV treatment (Verhagen et al., 2014).

### 6.6.11 GYMNOGONGRUS GRIFFITHSIAE

Family: Phyllophoraceae

Location: Europe, the Atlantic Islands, North America, South America, Africa, South Asia, Australia, and New Zealand

Common name: Red-seaweed

Kappa carrageenan, a sulfated polysaccharide from *Gymnogongrus griffithsiae*, shows inhibitory property in the case of DENV-2, which was tested in the Vero cells. This kappa carrageenan inhibits the DENV-2 multiplication at half-maximal inhibitory concentration ($IC_{50}$) value of 0.9 μg/mL, the compound, effectively inhibiting DENV-2 multiplication. A commercial polysaccharide, DS8000 has the same value of $IC_{50}$ (Verhagen et al., 2014). *G. griffithsiae* shows less potency in the case of DENV-3 and DENV-4. In the case of DENV-3, the $IC_{50}$ is between 13.9 and 14.2 μg/mL, and in the case of DENV-4, $IC_{50}$ is more than 29.3 to less than 50 μg/mL (Lim et al., 2020). It was not potent against the DENV-1, but in the case of DENV-2, it was potent in the early infection stage when combined with the virus (Lim et al., 2020).

### 6.6.12 *GYMNOGONGRUS TORULOSUS*

Family: Phyllophoraceae

Location: Australia and New Zealand

Common name: Red-seaweed

Carrageenans are sulfated galactans extracted from *G. torulosus*. In *in vitro* situation, *G. torulosus* secretes hybrids of the secondary compound DL-galactan that interfere with viral surface envelope glycoprotein by binding to the cell receptor. This DL-galactan has $IC_{50}$ value between 0.19 and 1.7 μg/mL, which is effective for viral plaque reduction assay of Vero cells in the case of DENV-2 (Pujol et al., 2002). The DL-galactans were also anticoagulant and had no cytotoxic effects on stationary cells as well as actively dividing cells (Lim et al., 2020). Researchers also reported that extracts showed complete inhibition of DENV when exposed during the absorption phase, indicating their role in interfering binding to host cell.

### 6.6.13 *Hippophae rhamnoides*

Family: Elaeagnaceae

Location: Europe, United Kingdom, Asia including Japan, and the Himalayas

Common name: Sea Buckthorn

It is beneficial nutritionally and medicinally, serving as a liver tonic, antioxidant anticancer, antiulcer, and antiinflammatory (Lim et al., 2020). The leaf extract of *H. rhamnoides* exhibited anti-dengue activity against the DENV-2, which was investigated in infected human macrophages derived from blood. With a decrease and an increase in TNF-α and IFN-γ, respectively, *H. rhamnoides* leaf extract-treated cells were able to maintain dengue-infected cells' viability on par with Ribavirin, a commercial antiviral drug (George & Debroy, 2020). When compared to control cells, the viral titer was significantly reduced in infected macrophages after receiving a 50 mg/mL treatment of an alcoholic extract from the leaves (Lim et al., 2020).

### 6.6.14 *Houttuynia cordata*

Family: Saururaceae

Location: Japan, Korea, Southern China, and Southeast Asia

Common Name: Heart leaf or Fish leaf

*H. cordata*is has been consumed and applied as medicine by indigenous community for 1,000 years. In present times, it is also cultivated for food and medicine in the Yarlung Zangbo Valley in Assam, India (Pujol et al., 2002). *Houttuynia cordata* is a perennial plant that is indigenous to mountain areas of eastern Asia and spreads endlessly as a creeping rhizome in damp areas. This plant exhibits antiviral effect against clinically significant enveloped viruses such as herpes simplex virus type I (HSV-1), influenza virus, and human immunodeficiency virus type I *in vitro.* A study demonstrated that the ethyl acetate fraction of *H. cordata* is capable of inhibiting the infection by DENV (Teramoto et al., 2017). The research suggests that flavonoids found in *H. cordata* extracts contribute to its effective antiviral properties with no cytotoxicity *in vitro* and *in vivo* as well.

### 6.6.15 *Leucaena leucocephala*

Family: Fabaceae

Locations: Northern Central America, Southern Mexico, West Indies

Common Name: Mimosoid tree, White lead tree

Galactomannans isolated from the seed of *Leucaena leucocephala* has shown action against the DENV-1 and YFV (Yellow Fever Virus) in both *in vitro* and *in vivo* tests. Galactomannans are polysaccharides made of a mannose backbone and side groups connected to galactose; more precisely, they have α (1–4)-linked β-D-mannopyranose backbone and branchpoint from their 6-positions linked to α-D-galactose (i.e., 1–6-linked α-D-galactopyranose). A study demonstrated that the dosage of DENV-1 required to cause a 100-fold drop in viral titer in C6/36 cell culture tests was 37 mg/L, according to *in vitro* investigations (Hidari et al., 2008). In addition, trypsin inhibitors present in plant extract also exhibits antilarvicidal effect.

### 6.6.16 *Lippia alba* and *Lippia citriodora*

Family: Verbenaceae

Locations: Central and South America, The Caribbean, Southern Texas, Mexico

Common Name: Bushy matgrass

In Vero cells, the dengue virus serotype replication was significantly inhibited by the essential oils from *Lippia alba* and *Lippia citriodora* (Solomonides et al., 2010). With *L. alba* oil, viral plaque numbers were reduced by 50% between 0.4 and 32.6 μg/mL, whereas the $IC_{50}$ values for

*L. citriodora* oil ranged from 1.9 to 33.7 μg/mL. While *L. citriodora* essential oil's virucidal efficacy against DENV-1, 2, and 3 was comparable to but less potent than that against DENV-4, *L. alba* essential oil was more effective against DENV-2 than other serotypes. At 100 μg/mL, it was found that *L. alba* essential oil eliminated the yield (ECDC, 2022).

### 6.6.17 *Meristiella gelidium*

Family: Solieriaceae family

Locations: North America, South America, Atlantic Island, The Caribbean

*Meristiella gelidium*'s kappa carragenan's antiviral capacity against DENV-2 was assessed. Carrageenans that were isolated from *M. gelidium* had an $IC_{50}$ that ranged from 0.14 to 1.6 μg/mL. The findings demonstrate that, in comparison to reference polysaccharides (heparin and DS 8000), the extract and fraction obtained from *M. gelidium* were more efficient DENV-2 inhibitors (Kularatne, 2015).

### 6.6.18 *Mimosa scabrella*

Family: Fabaceae family

Locations: Southeastern Brazil's cool, subtropical plateaus

Common Name: Bracatinga

A secondary compound called galactomannan isolated from *M. scabrella*'s seed plays a useful role as DENV-1 and YFV inhibitors (Simmons et al., 2012). It also has a lot of industrial applications such as foods, paints, cosmetics, and pharmaceuticals. In 87.7% of the mice with YFV infection, *M. scabrella* demonstrated survival protection. DENV-1 virus titers were shown to drop 100-fold at a concentration of 347 mg/L using C6/36 cell culture assays (Lim et al., 2020).

### 6.6.19 *Momordica charantia*

Family: Cucurbitaceae family

Location: South America, Middle East, Africa, Amazon, Caribbean, and New Zealand

Common Name: Balsam pear, bitter gourd, karela & kugua

*Momordica charantia* is used to treat and prevent conditions such as diabetes, jaundice, pneumonia, psoriasis, dysmenorrhea, eczema, and scabies (George & Debroy, 2021). Luteolin, kaempferol, and quercetin were some of the flavonoids present in *M. charantia*. In Vero cell lines, the MNTD concentration in methanolic extract was 0.20 mg/mL. It demonstrated about 50% inhibition of DENV-1 against dengue activity (Rodenhuis-Zybert et al., 2010).

### 6.6.20 *Ocimum sanctum*

Family: Labiatae family

Location: Tropical Asia and America

Common Name: Tulsi or 'holy basil'

It is a useful herb for treating and preventing cold, flu, and ulcers. It is also believed to have hepatoprotective, antiinflammatory, and antiviral properties. Many flavonoids were found in *Ocimum sanctum* extract, including orientin, vicenin, and luteolin (Ranjit et al., 2011). Traditional tea is made from boiling leaves and works as a preventative treatment for dengue fever (Verhagen et al., 2014). The extract had a Maximum Treatment Dose (MTD) of 0.10 mg/mL and cytotoxic values of 1.5 mg/mL. In cell lines, it inhibits the DENV-1 serotype (Abd Kadir et al., 2013). The MTD value obtained from HepG2 cells, on the other hand, was 0.023 mg/mL, which was much lower than the MTD value obtained from the Vero E6 cells. This demonstrates that the extracts are more cytotoxic to HepG2 cells than to Vero cells (Ranjit et al., 2011).

### 6.6.21 *Piper retrofractum*

Family: Piperaceae family

Location: Southeast Asia, India, Cambodia, Thailand, and Indonesia

Common Name: Dei phei

*Piper retrofractum* roots can be chewed or decocted to treat colic, dyspepsia, and gastralgia (Lim et al., 2020). Dichloromethane ethanol extract was tested for inhibitory efficacy against DENV-2-infected cells using the MTT technique. The roots of *Piper retrofractum* may be chewed or brewed into a decoction to treat colic, dyspepsia, and gastralgia. The dried fruits may be used to treat diarrhea and cough, while the leaf extract have antipyretic effect (Lim et al., 2020). The DENV-2 activity was inactivated by 100 μg/mL ethanol extracts of plant by 84.93%, whereas the $CC_{50}$ values of the ethanol and dichloromethane extracts were 625 and 156.25 μg/mL, respectively (Lovegrove et al., 2017).

### 6.6.22 *Psidium guajava*

Family: Myrtaceae family

Location: India, Central America, Asia, Africa, Mexico

Common Name: Guava

The main component of *Psidium guajava* are ascorbic acid and citric acid, exhibiting antioxidant, anticancer, antiviral, and antiinflammatory properties. In four out of five from *Psidium guajava* that displayed high selective inhibition on DENV-2 replication, catechin, and the most effective phytochemical, achieved viral suppression of more than 90% both *in vitro* and *in vivo* (George & Debroy, 2021). Post-treatment analysis revealed that all phytochemicals, with the exception of hesperidin, significantly reduced DENV-2 infection. The cultures treated with catechin, gallic acid, quercetin, and naringin showed the highest levels of viral inhibition, which were 91.8%, 67.3%, 100%, and 64.5%, respectively. Its bark ethanolic extracts had a $CC_{50}$ of 1,000.0 μg/mL and an $EC_{50}$ of 7.8 μg/mL when they interacted with DENV-2-infected Vero cells (Trujillo et al., 2019).

### 6.6.23 *Quercus lusitanica*

Family: Fagaceae

Location: Greece, Asia Minor, Syria, and Iran

Common name: Quercus infectoria

The main chemical compounds of this plant are gallic acid and ellagic acid (Ranjit et al., 2011). A dose of 180 mg/mL of methanol extracts of *Quercus lusitanica* completely suppress DENV infection at 1–1,000 Tissue Culture Infectious Dose ($TCID_{50}$), without cytopathic effect (CPE) (Lim et al., 2020). Another research found that a modest dosage of *Quercus lusitanica* methanol extract at 0.032 mg/mL demonstrated 100% inhibition with 10 $TCID_{50}$ of virus, 50% inhibition with 100 $TCID_{50}$, and 25% inhibition with 1000 $TCID_{50}$. A proteomics analysis showed that *Q. Lusitanica* extract reduced infected C6/36 cells' expression of NS1 protein (Verhagen et al., 2014).

### 6.6.24 *Rhizophora apiculate*

Family: Rhizophoraceae

Location: Australia, India, Southeast Asia, and Sri Lanka

Common name: Ture mangroves

The plant is rich phytochemicals exhibiting antimicrobial effect. Additionally, insecticidal effect of plant extract has also been demonstrated. The development of viral, bacterial, and fungal infections in humans was prevented by extracts of the roots, leaves, or stems of *Rhizophora apiculata*. Ethanolic extract of *R. apiculata* at concentrations of 12.5 and 100 μg/mL reduced activity of DENV-2, 56.14% and 41.5%, respectively, and inactivated viral particle activity. According to the $CC_{50}$ measurement on DENV-infected Vero cells, the $CC_{50}$ of ethanol extract of *R. apiculate* was 625 μg/mL (Lim et al., 2020).

### 6.6.25 *Uncaria tomentosa*

Family: Rubiaceae

Location: Colombia, South and Central America, Ecuador, including Peru, Venezuela, and Guyana

Common name: Cat's claw

*U. tomentosa* is mostly utilized as a traditional medicine in the Peruvian jungle (Kularatne, 2015). The highest potent activity was found in the alkaloidal fraction of *U. tomentosa*. It is mainly used in the treatment of abscesses, arthritis, side effects from chemotherapy, asthma, stomach, ulcer, fever, recovery after childbirth, inflammations, and wounds. At 1 μg/mL concentrations, the alkaloid-enriched extract of *U. tomentosa* was shown to be the most efficient at reducing DENV-Ag detection among monocytes (Kularatne, 2015). The *in vitro* anti-DENV-2 and immunomodulating properties of *U. tomentosa* alkaloids reduced the proportion of infected cells, suppressed cytokine activities, and reduced TNF-α, IL-10, and IFN-α production levels. Human monocytes infected with DENV-2, showed a drop in IL-10 levels from $572 \pm 219$ to $244 \pm 60$ pg/mL after treatment with *U. tomentosa*.

### 6.6.26 *Zostera marina*

Family: Zosteraceae

Location: North America and Eurasia

Common name: Eelgrass

*Z. marina* produces antiadhesive components like zosteric acid, p-sulfoxycinnamic acid, which had a low $IC_{50}$ against DENV-2 of about 2.3 mm. They also inhibited the other DENV serotypes. The most effective compound was CF 238; it has $IC_{50}$ values for DENV-1, DENV-2, DENV-3, and DENV-4 of 24, 46, 14, and 47 μM, respectively (Verhagen et al., 2014).

## 6.7 CONCLUSION

Dengue fever, an ailment caused by spread of dengue virus by mosquitoes, is growing widespread worldwide. The acute pyretic condition during DENV infection can be fatal in critical cases, resulting in hemorrhagic shock. Cases of dengue fever continue to rise in the community despite efforts to control the spread of the dengue virus. While there are currently antiviral vaccines available for many viral infections such as human papillomavirus, measles, mumps, and rubella, there is no vaccine yet developed for few other viral infections, including dengue fever. The development of novel viral strains and modification in viral protein structure has led to the development of resistance toward normal antiviral therapy. This, in turn, has led to a shift toward the usage of natural plant compounds as an antiviral agent to treat or prevent the spread of dengue. This chapter summarizes the knowledge that is now accessible for different plant parts and extract types that are used to treat dengue. Some plants, however, may have a wide range of therapeutic uses that have not yet been completely investigated. To discover more effective and less dangerous anti-dengue medications, new anti-dengue products made from plant-based chemicals are required. This approach may highlight potential molecules that can be improved and used to create brand-new anti-dengue compounds.

## REFERENCES

Abd Kadir, S. L.; Yaakob, H.; Mohamed Zulkifli, R. Potential Anti-Dengue Medicinal Plants: A Review. *Journal of Natural Medicines* 2013, 67 (4), 677–689.

Brügger, B. P.; Martínez, L. C.; Plata-Rueda, A.; Castro, B. M.; Soares, M. A.; Wilcken, C. F.; Carvalho, A. G.; Serrão, J. E.; Zanuncio, J. C. Bioactivity of the Cymbopogon Citratus (Poaceae) Essential Oil and Its Terpenoid Constituents on the Predatory Bug, PodisusNigrispinus (Heteroptera: Pentatomidae). *Scientific Reports* 2019, 9 (1), 8358.

Casciaro, B.; Mangiardi, L.; Cappiello, F.; Romeo, I.; Loffredo, M. R.; Iazzetti, A.; Calcaterra, A.; Goggiamani, A.; Ghirga, F.; Mangoni, M. L.; Botta, B.; Quaglio, D. Naturally-Occurring Alkaloids of Plant Origin as Potential Antimicrobials against Antibiotic-Resistant Infections. *Molecules* 2020, 25 (16), 3619.

Chen, L. H.; Wilson, M. E. Dengue and Chikungunya Infections in Travelers. *Current Opinion in Infectious Diseases* 2010, 23 (5), 438–444.

Cheung, Y. Y.; Chen, K. C.; Chen, H.; Seng, E. K.; Chu, J. J. Antiviral Activity of Lanatoside C against Dengue Virus Infection. *Antiviral Research* 2014, 111, 93–99.

Datan, E.; Roy, S. G.; Germain, G.; Zali, N.; McLean, J. E.; Golshan, G.; Harbajan, S.; Lockshin, R. A.; Zakeri, Z. Dengue-Induced Autophagy, Virus Replication and Protection from Cell Death Require Er Stress (PERK) Pathway Activation. *Cell Death & Disease* 2016, 7(3), e2127.

Dengue and severe dengue - fact sheet - updated April 2017- world. https://reliefweb.int/report/world/dengue-and-severe-dengue-fact-sheet-updated-april-2017?gclid=CjwKCAjw8JKbBhBYEiwAs3sxN5Tt7QXhM-3aLVbVG9h2RTMdyag5QivSl5XDL4y7ATGJmt62lZhC1BoCy2kQAvD_BwE (accessed Nov 4, 2022).

Dengue guidelines for diagnosis, treatment, prevention and Control: New Edition. https://apps.who.int/iris/handle/10665/44188 (accessed Nov 4, 2022).

Dengue worldwide overview. https://www.ecdc.europa.eu/en/dengue-monthly#:~:text=In%202022%20and%20as%20of,and%20Ecuador%20(4%20963) (accessed Nov 4, 2022).

Dhinakaran, S. R.; Mathew, N.; Munusamy, S. Synergistic Terpene Combinations as Larvicides against the Dengue Vector Aedes Aegypti Linn. *Drug Development Research* 2019, 80 (6), 791–799.

Diamond, M. S.; Pierson, T. C. Molecular Insight into Dengue Virus Pathogenesis and Its Implications for Disease Control. *Cell* 2015, 162 (3), 488–492.

Diamond, M. S.; Zachariah, M.; Harris, E. Mycophenolic Acid Inhibits Dengue Virus Infection by Preventing Replication of Viral RNA. *Virology* 2002, 304 (2), 211–221.

Dwivedi, V. D.; Tripathi, I. P.; Tripathi, R. C.; Bharadwaj, S.; Mishra, S. K. Genomics, Proteomics and Evolution of Dengue Virus. *Briefings in Functional Genomics* 2017, 16 (4), 217–227.

Ferreres, F.; Taveira, M.; Pereira, D. M.; Valentão, P.; B. Andrade, P. Tomato (Lycopersicon Esculentum) Seeds: New Flavonols and Cytotoxic Effect. *Journal of Agricultural and Food Chemistry* 2010, 58 (5), 2854–2861.

George, N.; Debroy, A. Artificial Intelligence for Covid-19 and Future Pandemics: A Mini-Review. *Journal of Acute Disease* 2021, 10 (4), 135.

George, N.; Debroy, A. Covid-19 Outbreak: An Analysis of Risk Factors and Potential Role of Medicinal Plants to Combat the Disease. *International Journal of Multidisciplinary Educational Research* 2020, 9(11), 143.

Gómez-Calderón, C.; Mesa-Castro, C.; Robledo, S.; Gómez, S.; Bolivar-Avila, S.; Diaz-Castillo, F.; Martínez-Gutierrez, M. Antiviral Effect of Compounds Derived from the Seeds of Mammea Americana and TabernaemontanaCymosa on Dengue and Chikungunya Virus Infections. *BMC Complementary and Alternative Medicine* 2017, 17 (1), 1–12.

Gubler, D. J. Dengue and Dengue Hemorrhagic Fever. *Clinical Microbiology Reviews* 1998, 11 (3), 480–496.

Hidari, K. I. P. J.; Takahashi, N.; Arihara, M.; Nagaoka, M.; Morita, K.; Suzuki, T. Structure and Anti-Dengue Virus Activity of Sulfated Polysaccharide from a Marine Alga. *Biochemical and Biophysical Research Communications* 2008, 376 (1), 91–95.

Hishiki, T.; Kato, F.; Tajima, S.; Toume, K.; Umezaki, M.; Takasaki, T.; Miura, T. Hirsutine, an Indole Alkaloid of UncariaRhynchophylla, Inhibits Late Step in Dengue Virus Lifecycle. *Frontiers in Microbiology* 2017, 8, 1674.

Hopkins, A. L.; Groom, C. R. The Druggable Genome. *Nature Reviews Drug Discovery* 2002, 1 (9), 727–730.

Ichsyani, M.; Ridhanya, A.; Risanti, M.; Desti, H.; Ceria, R.; Putri, D. H.; Sudiro, T. M.; Dewi, B. E. Antiviral Effects of Curcuma Longa L. against Dengue Virus *in Vitro* and in Vivo. *IOP Conference Series: Earth and Environmental Science* 2017, 101, 012005.

Islam, M. T.; Mubarak, M. S. Diterpenes and Their Derivatives as Promising Agents against Dengue Virus and Dengue Vectors: A Literature-Based Review. *Phytotherapy Research* 2019, 34 (4), 674–684.

Jia, F.; Zou, G.; Fan, J.; Yuan, Z. Identification of Palmatine as an Inhibitor of West Nile Virus. *Archives of Virology* 2010, 155 (8), 1325–1329.

Kularatne, S. A. Dengue Fever. *BMJ* 2015, 351, h4661.

Lim, S. Y.; Chieng, J. Y.; Pan, Y. Recent Insights on Anti-Dengue Virus (DENV) Medicinal Plants: Review on *in Vitro*, in Vivo and in Silico Discoveries. *All Life* 2020, 14 (1), 1–33.

Lin, C.-F.; Wan, S.-W.; Cheng, H.-J.; Lei, H.-Y.; Lin, Y.-S. Autoimmune Pathogenesis in Dengue Virus Infection. *Viral Immunology* 2006, 19 (2), 127–132.

Lovegrove, J. A.; Stainer, A.; Hobbs, D. A. Role of Flavonoids and Nitrates in Cardiovascular Health. *Proceedings of the Nutrition Society* 2017, 76 (2), 83–95.

Morrison, A. C.; Sihuincha, M.; Stancil, J. D.; Zamora, E.; Astete, H.; Olson, J. G.; Vidal-Ore, C.; Scott, T. W. Aedes Aegypti (Diptera: Culicidae) Production from Non-Residential Sites in the Amazonian City of Iquitos, Peru. *Annals of Tropical Medicine & Parasitology* 2006, 100 (supl), 73–86.

Mustafa, M. S.; Rasotgi, V.; Jain, S.; Gupta, V. Discovery of Fifth Serotype of Dengue Virus (DENV-5): A New Public Health Dilemma in Dengue Control. *Medical Journal Armed Forces India* 2015, 71 (1), 67–70.

Norahmad, N. A.; Mohd Abd Razak, M. R.; Mohmad Misnan, N.; Md Jelas, N. H.; Sastu, U. R.; Muhammad, A.; Ho, T. C.; Jusoh, B.; Zolkifli, N. A.; Thayan, R.; Mat Ripen, A.; Zainol, M.; Syed Mohamed, A. F. Effect of Freeze-Dried Carica Papaya Leaf Juice on Inflammatory Cytokines Production during Dengue Virus Infection in AG129 MICE. *BMC Complementary and Alternative Medicine* 2019, 19 (1), 1–10.

Nuri, T.; Ming, L. Papaya Leaves Juice as an Alternative Treatment for Dengue Fever. *Journal of Applied Pharmaceutical Science* 2016, 6 (3), 172–173.

Pan, Y. T.; Ghidoni, J.; Elbein, A. D. The Effects of Castanospermine and Swainsonine on the Activity and Synthesis of Intestinal Sucrase. *Archives of Biochemistry and Biophysics* 1993, 303 (1), 134–144.

Premaratna, R.; Bailey, M. S.; Ratnasena, B. G. N.; de Silva, H. J. Dengue Fever Mimicking Acute Appendicitis. *Transactions of the Royal Society of Tropical Medicine and Hygiene* 2007, 101 (7), 683–685.

Pujol, C. A.; Estevez, J. M.; Carlucci, M. J.; Ciancia, M.; Cerezo, A. S.; Damonte, E. B. Novel DL-Galactan Hybrids from the Red Seaweed Gymnogongrus Torulosus Are Potent Inhibitors of Herpes Simplex Virus and Dengue Virus. *Antiviral Chemistry and Chemotherapy* 2002, 13 (2), 83–89.

Quintana, V. M.; Piccini, L. E.; PanozzoZénere, J. D.; Damonte, E. B.; Ponce, M. A.; Castilla, V. Antiviral Activity of Natural and Synthetic β-Carbolines against Dengue Virus. *Antiviral Research* 2016, 134, 26–33.

Raekiansyah, M.; Buerano, C. C.; Luz, M. A.; Morita, K. Inhibitory Effect of the Green Tea Molecule EGCG against Dengue Virus Infection. *Archives of Virology* 2018, 163 (6), 1649–1655.

Ranjit, S.; Kissoon, N. Dengue Hemorrhagic Fever and Shock Syndromes*. *Pediatric Critical Care Medicine* 2011, 12 (1), 90–100.

Rodenhuis-Zybert, I. A.; Wilschut, J.; Smit, J. M. Dengue Virus Life Cycle: Viral and Host Factors Modulating Infectivity. *Cellular and Molecular Life Sciences* 2010, 67 (16), 2773–2786.

Rosmalena, R.; Elya, B.; Dewi, B. E.; Fithriyah, F.; Desti, H.; Angelina, M.; Hanafi, M.; Lotulung, P. D.; Prasasty, V. D.; Seto, D. The Antiviral Effect of Indonesian Medicinal Plant Extracts against Dengue Virus in Vitro and in Silico. *Pathogens* 2019, 8 (2), 85.

Sacchelli, B. A.; Faccin-Galhardi, L. C.; Ito, V. Y.; Lopes, J. L.; Dekker, R. F. H.; Barbosa-Dekker, A. M.; Orsato, A. Botryosphaeran and Sulfonated Derivatives as Novel Antiviral Agents for Herpes Simplex and Dengue Fever. *International Journal of Biological Macromolecules* 2019, 138, 334–339.

Simmons, C. P.; Farrar, J. J.; van Vinh Chau, N.; Wills, B. Dengue. *New England Journal of Medicine* 2012, 366 (15), 1423–1432.

Solomonides, T.; Blanquer, I.; Breton, V. *Healthgrid Applications and Core Technologies: Proceedings of Healthgrid 2010*. IOS Press: Amsterdam, 2010.

Su Yin Low, J.; Caiyun Chen, K.; Xing Wu, K.; Mah-Lee Ng, M.; Jang Hann Chu, J. Antiviral Activity of Emetine Dihydrochloride against Dengue Virus Infection. *Journal of Antivirals & Antiretrovirals* 2009, 01 (01), 062–071.

Talarico, L.; Pujol, C.; Zibetti, R.; Faria, P.; Noseda, M.; Duarte, M.; Damonte, E. The Antiviral Activity of Sulfated Polysaccharides against Dengue Virus Is Dependent on Virus Serotype and Host Cell. *Antiviral Research* 2005, 66 (2–3), 103–110.

Teramoto, T.; Balasubramanian, A.; Choi, K. H.; Padmanabhan, R. Serotype-Specific Interactions among Functional Domains of Dengue Virus 2 Nonstructural Proteins (NS) 5 and NS3 Are Crucial for Viral RNA Replication. *Journal of Biological Chemistry* 2017, 292 (23), 9465–9479.

Tricou, V.; Low, J. G.; Oh, H. M.; Leo, Y.-S.; Kalimuddin, S.; Wijaya, L.; Pang, J.; Ling, L. M.; Lee, T. H.; Brose, M.; Hutagalung, Y.; Rauscher, M.; Borkowski, A.; Wallace, D. Safety and Immunogenicity of a Single Dose of a Tetravalent Dengue Vaccine with Two Different Serotype-2 Potencies in Adults in Singapore: A Phase 2, Double-Blind, Randomised, Controlled Trial. *Vaccine* 2020, 38 (6), 1513–1519.

Trujillo-correa, A. I.; Quintero-gil, D. C.; Diaz-castillo, F.; Quiñones, W.; Robledo, S. M.; Martinez-gutierrez, M. In vitro and In Silico Anti-dengue Activity of Compounds Obtained from Psidium guajava through Bioprospecting. *BMC Complementary and Alternative Medicine* 2019 19 (298), 1–16.

Valotto, C. F.; Silva, H. H.; Cavasin, G.; Geris, R.; Rodrigues Filho, E.; Silva, I. G. AlteraçõesUltraestruturaisEmLarvas De Aedes Aegypti SubmetidasAoDiterpenoLabdano, Isolado De Copaifera Reticulata (Leguminosae), E à Uma Fração Rica EmTaninos De MagoniaPubescens (Sapindaceae). *Revista da SociedadeBrasileira de Medicina Tropical* 2011, 44 (2), 194–200.

Verhagen, L. M.; de Groot, R. Dengue in Children. *Journal of Infection* 2014, 69.

Wang, P.; Li, L.-F.; Wang, Q.-Y.; Shang, L.-Q.; Shi, P.-Y.; Yin, Z. Anti-Dengue-Virus Activity and Structure-Activity Relationship Studies of Lycorine Derivatives. *ChemMedChem* 2014, 9 (7), 1522–1533.

Whitby, K.; Pierson, T. C.; Geiss, B.; Lane, K.; Engle, M.; Zhou, Y.; Doms, R. W.; Diamond, M. S. Castanospermine, a Potent Inhibitor of Dengue Virus Infection in Vitro and in Vivo. *Journal of Virology* 2005, 79 (14), 8698–8706.

Yadav, J. P.; Kaushik, S.; Kaushik, S.; Sharma, V. Antiviral and Therapeutic Uses of Medicinal Plants and Their Derivatives against Dengue Viruses. *Pharmacognosy Reviews* 2018, 12 (24), 177.

Zandi, K.; Teoh, B.-T.; Sam, S.-S.; Wong, P.-F.; Mustafa, M. R.; AbuBakar, S. Antiviral Activity of Four Types of Bioflavonoid against Dengue Virus Type-2. *Virology Journal* 2011, 8 (1), 1–11.

# 7 Medicinal Plants and Herbs in Viral Hepatitis

*Shweta Kumari*
University Institute of Biotechnology, Chandigarh University

*Sahil Jain*
Tel-Aviv University

## 7.1 INTRODUCTION

Hepatitis is an inflammatory condition of the liver, primarily caused by five viruses belonging to distinct families, namely hepatitis A virus (HAV; *Picornaviridae*), hepatitis B virus (HBV; *Hepadnaviridae*), hepatitis C virus (HCV; *Flaviviridae*), hepatitis D virus (HDV; *Kolmioviridae*) and hepatitis E virus (HEV; *Hepeviridae*) (Malik et al., 2022; Rasche et al., 2019). These viruses emerged over a span of 25 years and are responsible for more than 435 million cases worldwide (Malik et al., 2022). HAV is responsible for 1.4 million new cases per annum globally, with 25,000 of these cases in the United States alone (Melnick, 1992; CDC, 2020). HBV is responsible for maximum number of global hepatitis cases (approximately 67% cases), with the Western Pacific and African regions bearing the highest burden, totaling around 175 million HBV cases (Castaneda et al., 2021; WHO, 2021). HCV infection results in around 1.75 million new cases per year, with maximum HCV prevalence being recorded in Eastern Mediterranean region. In the United States, HBV and HCV are responsible for nearly 22,000 and 50,000 new cases every year, respectively (CDC, 2020). HDV infects around 70 million people worldwide and records a high presence in various regions, such as Mediterranean basin, amazon basin, Indian subcontinent and parts of Africa (Castaneda et al., 2021; Chen et al., 2006; Lanini et al., 2018). HEV is responsible for nearly 20 million new cases each year, and is known to highly infect the American, Asian and African regions (Castaneda et al., 2021; mondiale de la Santé, 2015).

Hepatitis viruses may enter the body by various transmission modes. HAV may infect via fecal-oral route and same-sex intercourse between males (Guet et al., 2011; ECDC, 2016), while HBV is known to infect through iatrogenic transmission and contact with infected blood or body fluids (WHO, 2017; Busch et al., 2019; Jafari et al., 2012; Bixler et al., 2019; Dienstag, 2008; Liver, 2017). HCV may infect via immunoglobulin injections, hemodialysis, religious scarification and traditional circumcision (Luma et al., 2016; Layden et al., 2015; Murphy et al., 2000). Perinatal transmission, sexual intercourse with infected individual and organ transplantation from a positive donor are major transmission factors for both HBV and HCV, while drug abuse and usage of contaminated surgical/piercing needles are recognized as transmission factors for HAV, HBV as well as HCV (Jafari et al., 2012; Tovo et al., 2016, 2015; Benova et al., 2014; Foster et al., 2018). HBV infection is a pre-requisite for HDV infection, therefore, HDV either co-infects along with HBV or superinfects individuals already infected with HBV (Castaneda et al., 2021; Lanini et al., 2018). Parenteral transmission via infected blood or body fluids is the major cause of HDV infection (Ponzetto et al., 1987). Similar to HAV, HEV is known to spread by fecal-oral route (Xia et al., 2015; Andonov et al., 2019). Upon entry, these viruses may result in some common symptoms, such as fever, nausea, fatigue, malaise, vomiting and jaundice (Castaneda et al., 2021). Chronic

DOI: 10.1201/9781003329169-9

hepatitis state might be observed in case of HBV, HCV and some genotypes of HEV, which may ultimately lead to cirrhosis or hepatocellular carcinoma (HCC) (Castaneda et al., 2021; Chu and Liaw, 2007) (Figure 7.1). Chronic liver disease affects 18.5% of the world's population, while cirrhosis affects approximately 7% of the population, with 2 million people dying each year (Islam and Alam, 2019). Therefore, it is essential to find a multi-genotypic, global and safe therapy against various hepatitis viruses.

Hepatitis treatment depends on the type of virus causing infection, as well as the extent of liver damage (Dakio et al., 2020). No treatment for HAV infection is available yet, and the disease is mainly managed by symptomatic-relief therapy (Castaneda et al., 2021). In case of HBV and HBV/HDV infection, treatment involves the administration of PEG-interferon and nucleoside/nucleotide analogs, such as lamivudine, tenofovir, telvidudine and entecavir (Terrault et al., 2018; Yuen et al., 2018). Though PEG-interferon treatment has high potential, its application is limited due to the poor tolerance of the HBV patients (Tao et al., 2020). PEG-interferon has also been contemplated as a treatment for HCV infection (Dusheiko and Wedemeyer, 2012). Apart from PEG-interferon, various direct-acting antivirals (DAAs), such as boceprevir, grazoprevir, elbasvir, pibrentasvir, sofosbuvir and dasabuvir are administered for HCV treatment. The primary goal of all HCV treatments is to prevent progression to the cirrhosis or HCC state by achieving sustained viral remission (SVR) (Castaneda et al., 2021). Ribavarin, a guanosine analog, has shown only variably encouraging results in the treatment of HCV and immunosuppressed-HEV patients (Castaneda et al., 2021; Dalton and Kamar, 2016). Therefore, given the high number of individuals suffering from hepatitis and the limited, partially successful treatment options available, there is a need to explore natural resources for beneficial medicinal compounds. Such compounds may open a range of high efficacy treatment alternatives, which are devoid of side effects.

Natural herbal medicines have been used in the treatment of numerous ailments since ancient times (Ramsey et al., 2020; Barkat et al., 2021). Herbal drugs are widely popular owing to their low cost, less toxicity and high efficacy (Carmona and Pereira, 2013). Many currently available synthetic drugs preparations include plant-derived bioactive components as raw materials (Moghadamtousi et al., 2015). For example, Taxol (derived from *Taxus brevifolia*) is used for treatment of breast, ovarian and lung cancer (Gallego-Jara et al., 2020), morphine (derived from *Papaver somniferum*)

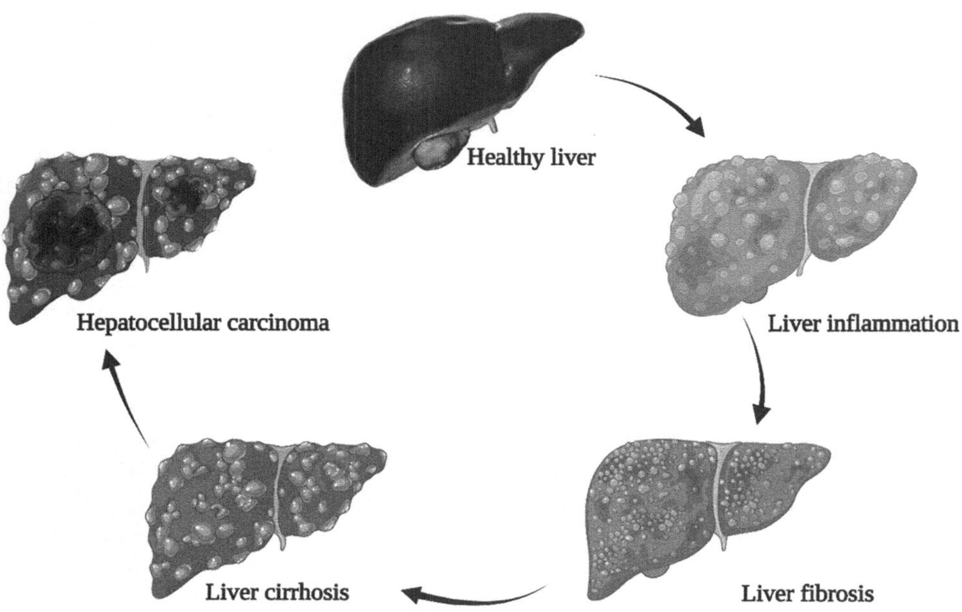


**FIGURE 7.1** Various stages of disease progression during hepatitis infection.

is used for relieving pain (Ofoegbu and Ettienne, 2021) and aspirin (derived from members of *Spiraea* genus) is administered to reduce fever and inflammation (Fiala and Pasic, 2020). The use of natural remedies for the treatment of liver diseases has a long history spanning different geographical regions and cultures, such as the Indian Ayurveda, the Chinese traditional medicines and the European medicinal plants (Thyagarajan et al., 2002). Numerous plants have been studied for successful therapy against hepatic illnesses, and for the recovery of damaged livers (Ali et al., 2019; Jain et al., 2020). In this chapter, we encapsulate the studies on different medicinal and ayurvedic plants reported to exhibit hepatoprotective effects (Table 7.1).

## 7.2 HEPATOPROTECTIVE MEDICINAL PLANTS

An extensive collection of various hepatoprotective medicinal plants, known to be beneficial against different hepatitis viruses, is presented in Table 7.2 and Figure 7.2. Additionally, the mechanism of viral inhibition is enlisted in Table 7.2. Here, we discuss some of the medicinal plants and their reported activity.

*Abelmoschus manihot* (commonly known as sunset muskmallow or sunset hibiscus) is an annual and perennial herbaceous flowering plant belonging to the Malvaceae family (Li et al., 2021). This edible form of hibiscus is widely distributed throughout the Asian continent, northern Australia and Europe (Luan et al., 2020). In 2007, Wu et al. (2007) assessed the anti-HBV activity of *Abelmoschus manihot*-derived hyperoside extract on Hep G2.2.15 (human hepatoma cell line) cells and duck hepatitis B virus (DHBV) model. In the *in vitro* validation conducted on Hep G2.2.15 cells, the authors reported a >80% inhibition rate of HBeAg (HBV viral protein) and HBsAg (HBV surface antigen) at a hyperoside extract dosage of 0.05 g/L. In the *in vivo* DHBV model, the extract attenuated the DHBV-DNA levels at a dose of 0.10 g/kg day, thus, supporting the *in vitro* analysis.

Sho-saiko-to (also known as *Xiǎocháihútāng*, XCHT) is a widely used traditional medicine in China and Japan (Fabbri et al., 2021). It comprises baicalin (a flavonoid found in *Scutellaria* genus plants), glycyrrhzin (a saponin found in *Glycyrrhiza glabra* roots), gingerol (a phenol found in *Zingiber officinale*), wogonin (a flavonoid found in *Scutellaria baicalensis*) and ginsenosides (a saponin found in *Panax* genus plants) among other components (Ohtake et al., 2004). This traditional supplement has multiple benefits such as anti-hepatotoxicity, immunomodulatory effect, anti-fibrotic effect and liver cell regeneration properties (Shimizu and Sho-saiko-to, 2000; Ohtake et al., 2005; Song et al., 2014). It was suggested that Sho-saiko-to could affect the expression of HBeAg, and could inhibit HBV production in Hep G2.2.15 cell line at a dose of 20 μg/mL (Chang et al., 2007). Another study suggested that Sho-saiko-to could inhibit hepatic fibrosis, and that baicalin and baicalein components were responsible for his anti-fibrogenic effect (Shimizu et al., 2000).

*Glycyrrhiza glabra* (commonly known as Liquorice) is a perennial herb belonging to the *Fabaceae* family (Pastorino et al., 2018). The root extracts of this plant have been traditionally employed to treat a diverse range of disorders, such as skin diseases, rheumatism, paralysis and sexual debility (Kaur et al., 2013; Batiha et al., 2020). Glycyrrhizin is the main component of the *Glycyrrhiza glabra* root extract (Batiha et al., 2020). A study in 2011 indicated that glycyrrhizin may reduce HCV titers and inhibit HCV core gene expression by 50% at 14–16 μg concentration. Encouragingly, glycyrrhizin was shown to exhibit a synergistic effect along with the standard treatment agent, IFN-α 2a (Ashfaq et al., 2011).

*Phyllanthus urinaria* L. (commonly known as gripeweed, chamber bitter or stonebreaker) is an annual perennial herb belonging to the *Phyllanthaceae* family. It is widely located in the Asian, American and Chinese regions (Geethangili and Ding, 2018). *Phyllanthus urinaria* L. extract was suggested to act as an inhibitor for HBV virus replication. A concentration of 0.2–0.8 g/L of the extract was indicated to inhibit HBV replication and expression with a favorable toxicology profile in an HBV-transfected HepG2 cell line. The treatment resulted in attenuation of HBV rcDNA, dsDNA and ssDNA in the transfected cells (Wu et al., 2015).

**TABLE 7.1**
**Some Basic Information about the Known Hepatitis Viruses and the Resultant Diseased Condition**

| Virus | HAV | HBV | HCV | HDV | HEV |
|---|---|---|---|---|---|
| **Discovery** | 1973 | 1965 | 1989 | 1977 | 1990 |
| **Family** | *Picornaviridae* | *Hepadnaviridae* | *Flaviviridae* | *Kolmioviridae* | *Hepeviridae* |
| **Genus** | *Hepatovirus* | *Orthohepadnavirus* | *Hepacivirus* | *Deltaviridae* | *Orthohepevirus* |
| **Genome** | ssRNA | dsDNA | ssRNA | ssRNA | ssRNA |
| **Envelope** | No | Yes | Yes | Yes | No |
| **Symptoms** | Fever, malaise, jaundice, weakness, fatigue, nausea, vomiting, dark urine, clay-colored stool, abdominal pain, arthralgias, myalgias, diarrhea and anorexia | Fatigue, nausea, vomiting, jaundice, right upper quadrant pain, liver necroinflammation, fibrosis, ALT and AST elevation, which might lead to cirrhosis or HCC. DNA elevation at later stages may indicate HDV superinfection. | Jaundice, nausea, dark urine, sleep disturbances, right upper quadrant pain, cell necrosis, hepatic steatosis, oxidative stress, diarrhea, ALT elevation, insulin resistance abdominal pain, anorexia, myalgia and arthralgia. May lead to cirrhosis and HCC. | Results in coinfection with HBV or superinfection after HBV infection. | Similar to HAV infection. Jaundice, fever, extreme fatigue and tiredness, extreme abdominal pain, nausea, vomiting, clay-colored stool, joint pain |
| **Chronic state** | Not reported | ~ 10% | ~ 80% | ~ 5% (coinfection) ~70% (superinfection) | Reported for certain genotypes |
| **Vaccine** | Yes | Yes | No | No[a] | No |
| **References** | WHO (2017, 2021); CDC (2020); Castaneda et al. (2021) | Melnick (1992); Rasche et al. (2019); Malik et al. (2022) | Melnick (1992); Rasche et al. (2019); Malik et al. (2022) | Rasche et al. (2019); Malik et al. (2022) | Melnick (1992); Rasche et al. (2019); Malik et al. (2022) |

[a] HBV vaccine is known to be protective against HDV.

*ALT, Alanine transaminase; AST, Aspartate aminotransferase; HCC, Hepatocellular carcinoma.*

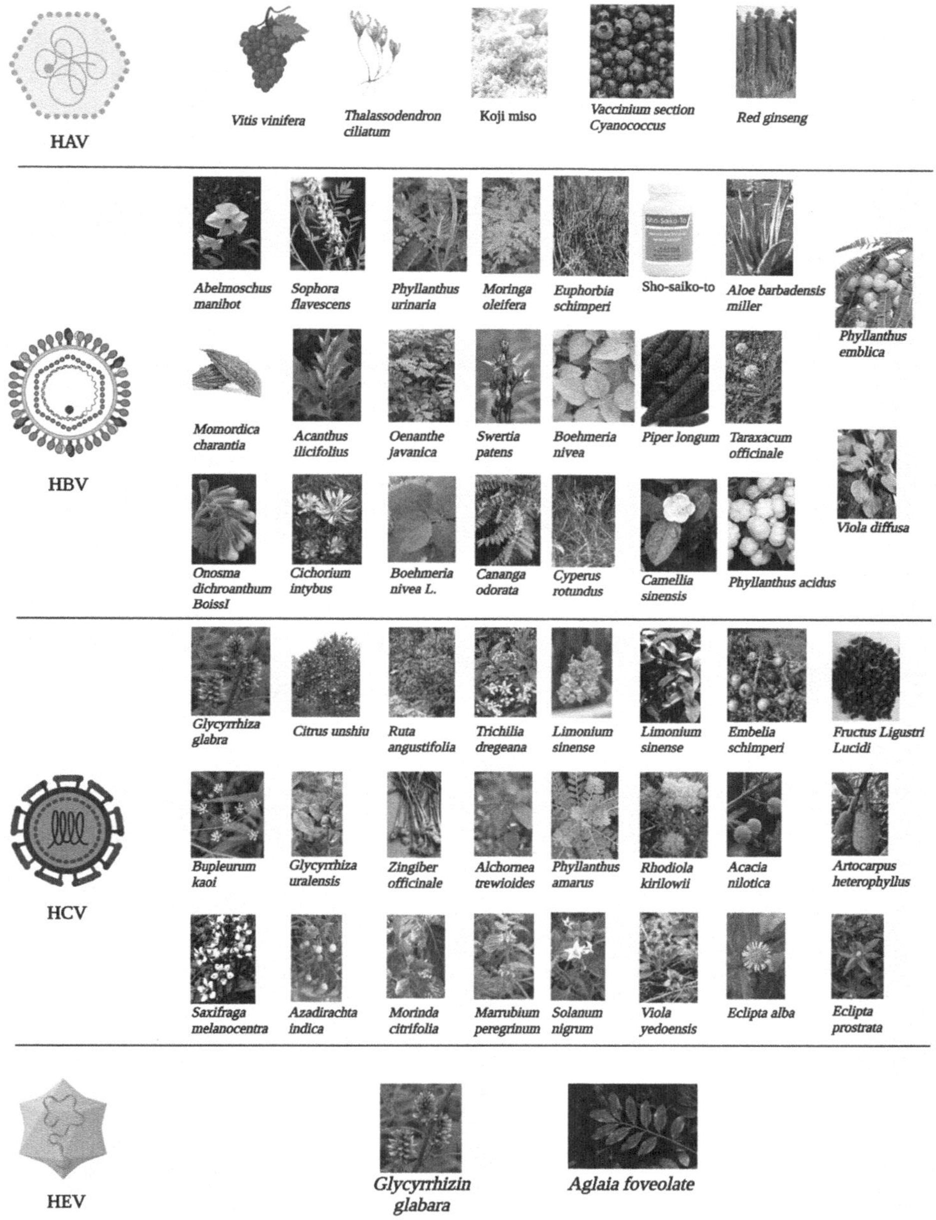


**FIGURE 7.2** A diagrammatic representation of various medicinal plants commonly used to treat different hepatitis viruses.

*Citrus unshiu* (commonly known as satsuma mandarin or miyagawa mandarin) is one of the most significant and widely cultivated plants in Japan (Akhlaghi Amiri, 2019). It has a juicy, seedless fruit, and is well-distributed in Australia, China, South Africa, Spain, etc. (Lado et al., 2016). In an *in vitro* study, it was shown that ethyl acetate extract of the *Citrus unshiu* peel effectively inhibited the absorption of HCV in MOLT-4 cells (a human lymphoblastoid leukemia cell line). This inhibition was observed at a dosage of 100 μg/mL. Further, the authors suggested that

**TABLE 7.2**
**Plants and Their Parts Used against Different Hepatitis Viruses**

| S. No. | Medicinal Plant | Common Name | Part of Plant Used | Bioactive Compound | Study Type | Plausible Inhibition Mechanism | References |
|---|---|---|---|---|---|---|---|
| | | | | **Hepatitis A Virus (HAV)** | | | |
| 1. | *Thalassodendron ciliatum* | Sickle-leaved cymodocea | Whole plant | Rutin, asebotin, 3-hydroxyasebotin, quercetin-3-O-β-dxylopyranoside | *In vitro* | Prooxidant activity | Hamdy et al. (2012) |
| 2. | Miso | – | – | – | *In vitro* | Elevation in GRP78 expression | Win et al. (2018) |
| 3. | *Vaccinium corymbosum* | Blue huckleberry | Fruit | Proanthocyanidins | *In vitro* | Prevention of viral adsorption via disruption of the virus capsid | Joshi et al. (2016) |
| 4. | *Panax ginseng* | Asian ginseng, Korean ginseng | Whole plant | Ginsenoside Rg1 and Rb1 | *In vitro* | Inhibition of viral replication | Lee et al. (2013) |
| | | | | **Hepatitis B Virus (HBV)** | | | |
| 5. | *Vitis vinifera* | European wine grape | Fruit | Triterpenes, flavonoids, and polysaccharides | *In vitro* | Protection of hepatocytes membranes, and cytokine regulation | Liu et al. (2010) |
| 6. | *Abelmoschus Manihot* | Sunset hibiscus, sunset muskmallow | Flower | Hyperoside | *In vitro, in vivo* | Inhibit HBsAg and HBeAg secretion | Wu et al. (2007) |
| 7. | *Sophora flavescens* | Shrubby sophora | Root | (+)-oxysophocarpine, (−)-sophocarpine (6), (+)-lehmannine (9), (−)-13,14-dehydrosophoridine | *In vitro* | Inhibit HBsAg and HBeAg secretion | Ding et al. (2006) |
| 8. | *Phyllanthus urinaria* L. | Gripeweed, chamber bitter, stonebreaker | Whole plant | – | *In vitro* | Inhibition of viral replication | Wu et al. (2015) |
| 9. | *Moringa oleifera* | Drumstick tree, Horseradish tree | Leaves | – | *In vitro* | Reduced HBsAg, CTGF, IFN-β1 and pgRNA levels | Feustel et al. (2017) |
| 10. | *Euphorbia schimperi* | Pencil cactus | Whole plant | Quercetin-3-O-glucuronide, quercetin-3-O-rhamnoside, kaempferol-3-O-glucuronide | *In vitro, in silico* | Interfering with viral Pol/RT and other core proteins | Parvez et al. (2019) |
| 11. | Sho-saiko-to | *Xiǎocháihútāng* | – | Baicalin and baicalein | *In vitro* | Inhibition of oxidative stress in hepatocytes | Shimizu (2000) |

*(Continued)*

**TABLE 7.2 (*Continued*)**
**Plants and Their Parts Used against Different Hepatitis Viruses**

| S. No. | Medicinal Plant | Common Name | Part of Plant Used | Bioactive Compound | Study Type | Plausible Inhibition Mechanism | References |
|---|---|---|---|---|---|---|---|
| 12. | *Aloe barbadensis miller* | *Aloe vera* | – | Emodin, chrysophanol, aloin B | *In vitro* | Anti-oxidant effects by elevating cytochrome P450 3A4 (CYP3A4) activity | Parvez et al. (2019) |
| 13. | *Momordica charantia* | Bitter melon, bitter gourd, *karela* | Leaves | – | *In vitro* | Inhibit HBV cccDNA | Waiyaput et al. (2012) |
| 14. | *Acanthus ilicifolius* | Sea holly, holy mangrove | Whole plant | – | *In vivo* | Reduce level of Alanine Aminotransferase (ALT) levels | Wei et al. (2015) |
| 15. | *Oenanthe javanica* | Water dropwort, Chinese celery, water celery | Whole plant | Flavone, especially hyperoside | *In vitro, in vivo* | Inhibition of the DNA-dependent DNA polymerase reaction, which effects DNA replication | Wang et al. (2005) |
| 16. | *Swertia patens* | – | Whole plant | (+)-dehydrodiconiferyl alcohol, dehydrozingerone | *In vitro* | Inhibit secretion of HBsAg | He et al. (2016) |
| 17. | *Boehmeria nivea* | China grass, rhea | Roots | – | *In vivo* | Inhibit the production of infectious virus | Chang et al. (2010) |
| 18. | *Piper longum* | Indian long pepper, *thippali* | Whole plant | Piperine | *In vitro* | Inhibit HBsAg and HBeAg secretion | Jiang et al. (2013) |
| 19. | *Taraxacum officinale* | Dandelion | – | Taraxasterol | *In vitro* | Decreased expression levels of polypyrimidine tract binding protein 1 (PTBP1) and sirtuin 1 (SIRT1) | Yang et al. (2020) |
| 20. | *Onosma dichroanthum Boiss I* | Havachoobe | Root | – | *In vitro* | Inhibit HBsAg secretion | Mohebbi et al. (2020) |
| 21. | *Cichorium intybus* | Common chicory | Leaves | Cichoric acid | *In- vitro* | Inhibit DNA replication, and HBsAg and HBeAg secretion | Zhang et al. (2014) |

(*Continued*)

**TABLE 7.2 (*Continued*)**
**Plants and Their Parts Used against Different Hepatitis Viruses**

| S. No. | Medicinal Plant | Common Name | Part of Plant Used | Bioactive Compound | Study Type | Plausible Inhibition Mechanism | References |
|---|---|---|---|---|---|---|---|
| 22. | *Cananga odorata* | Ylang-ylang, cananga tree | Leaves | – | *In vitro* | Inhibit HBsAg secretion | Indrasetiawan et al. (2019) |
| 23. | *Cyperus rotundus* | Coco grass, nut grass, purple nutsedge | Rhizomes | Cyperenoic acid, 10-epieudesm-11-ene-3β, 5α-diol, 3β, 4α-dihydroxy-7-epieudesm-11 (13), -ene, 7α (H), 10β-eudesm-4-en-3-one-11,12-diol, rhombitriol | *In vitro* | Inhibit DNA replication and HBsAg secretion | Xu et al. (2015) |
| 24. | *Camellia sinensis* | Tea plant, tea tree | – | Epigallocathechin-3-gallate | *In vitro* | Inhibit DNA replication by interaction with cellular proteins | Karamese et al. (2015) |
| 25. | *Phyllanthus acidus* | Otaheite gooseberry, Tahitian gooseberry, starberry | Roots, stems | Phyllanthacidoid A, phyllanthacidoid B, phyllanthacidoid C, phyllanthacidoid D, phyllanthacidoid F, phyllanthacidoid H, phyllanthacidoid I, phyllanthacidoid M | *In vitro* | Inhibit HBsAg and HBeAg secretion | Lv et al. (2014b) |
| 26. | *Phyllanthus emblica* | Indian gooseberry, myrobalan, amla | Roots | Phyllaemblicin G6, phyllaemblicin G7, phyllaemblicin G8, phyllaemblicin F | *In vitro* | Inhibit HBsAg and HBeAg secretion | Lv et al. (2014a) |
| 27. | *Viola diffusa* | Spreading white violet | Whole plant | 2β-hydroxy-3, 4-seco-friedelolactone-27-oic acid, 2β, 28β-dihydroxy-3,4-seco-friedelolactone-27-oic acid, 2β, 30β-dihydroxy-3,4-seco-friedelolactone-27-lactone, stigmast-25-ene-3β,5α,6β-triol | *In vitro* | Inhibit HBsAg and HBeAg secretion | Dai et al. (2015) |
| 28. | *Juzen-taiho-to* | Tj48 | – | – | *In vivo* | Inhibition of Kupffer cell-induced oxidative stress | Tsuchiya et al. (2008) |
| 29. | *Curcuma longa* | Turmeric, yellow ginger | Root | Curcumin | *In vitro* | Increase in p53 gene expression and p53 protein accumulation, ultimately, blocking the expression of HBx gene | Kim et al. (2009) |
| 30. | *Andrographis paniculata* | Green chiretta | – | 19-O-2'-thenoyl | *In vitro* | Inhibit DNA replication, and HBsAg and HBeAg secretion | Chen et al. (2014) |

(*Continued*)

**TABLE 7.2 (*Continued*)**
**Plants and Their Parts Used against Different Hepatitis Viruses**

| S. No. | Medicinal Plant | Common Name | Part of Plant Used | Bioactive Compound | Study Type | Plausible Inhibition Mechanism | References |
|---|---|---|---|---|---|---|---|
| | | | | **Hepatitis C Virus (HCV)** | | | |
| 31. | *Andrographis paniculata* | Green chiretta | | Andrographolide | *In vitro* | Upregulation of biliverdin levels, thus inhibiting DNA replication | Lee et al. (2014) |
| 32. | *Silybum marianum* | Milk thistle, Mary thistle, Scotch thistle | – | Silibinin | *In vitro* | By hindering clathrin-mediated endocytosis of HCV | Blaising et al. (2013) |
| 33. | *Glycyrrhiza glabra* | Liquorice | Root | Glycyrrhizin | *In vitro* | Reduce HCV titers, inhibit HCV core gene expression | Ashfaq et al. (2011) |
| 34. | *Citrus unshiu* | Satsuma mandarin, miyagawa mandarin | Peel | 3′,4′,5,6,7,8-hexamethoxyflavone | *In vitro* | Inhibition of HCV entry | Suzuki et al. (2005) |
| 35. | *Ruta angustifolia* | – | Leaves | Chalepin, pseudane IX | *In vitro* | Reducing viral replication and viral protein synthesis | Wahyuni et al. (2014) |
| 36. | *Trichilia dregeana* | Forest natal-mahogany | Root | – | *In vitro* | Inhibit HCV entry | Galani et al. (2015) |
| 37. | *Limonium sinense* | Sea lavender | Underground parts | Gallic acid | *In vitro* | Inhibit viral entry, attachment and fusion, inactivate cell-free virion | Hsu et al. (2015) |
| 38. | *Garcinia mangostana* | Mangosteen | Fruit peels | α-mangosteen, γ-mangosteen | *In vitro* | Inhibit DNA replication, downregulates protein expression | Choi et al. (2014) |
| 39. | *Embelia schimperi* | Vatke | Fruit | Embelin, 5-O-methylembelin | *In vitro* | Inhibit HCV protease | Hussein et al. (2000) |
| 40. | *Bupleurum chinense* | Chai Hu | Root | Saikosaponin b2 | *In vitro* | Inhibit HCV entry via interaction with HCV glycoproteins | Lin et al. (2015) |

*(Continued)*

**TABLE 7.2 (*Continued*)**
**Plants and Their Parts Used against Different Hepatitis Viruses**

| S. No. | Medicinal Plant | Common Name | Part of Plant Used | Bioactive Compound | Study Type | Plausible Inhibition Mechanism | References |
|---|---|---|---|---|---|---|---|
| 41. | *Glycyrrhiza uralensis* | Chinese liquorice | Root | Glycycoumarin, glycyrin, glycyrol, liquiritigenin | *In vitro* | Inhibit DNA replication, downregulates protein expression | Adianti et al. (2014) |
| 42. | *Phyllanthus amarus* | Gale of the wind, dukung anak, seed on the leaf | Root, leaves | – | *In vitro* | Inhibit HCV-NS3 (root extract) and HCV-NS5B (leaf extract) | Ravikumar et al. (2011) |
| 43. | *Rhodiola kirilowii* (Regel) Maxim | – | Rhizome | 3,3'-Digalloylproprodelphinidin B2, 3,3'-Digalloylprocyanidin B2, (–)-Epigallocatechin-3-O-gallate, (–)-Epicatechin-3-O-gallate | *In vitro* | Inhibit HCV-NS3 serine protease | Zuo et al. (2007) |
| 44. | *Acacia nilotica* | Babul, gum arabic tree, Egyptian acacia | Whole plant | – | *In vitro* | Inhibit HCV RNA and impede viral titer | Rehman et al. (2011) |
| 45. | *Saxifraga melanocentra* | Yellow spotted white saxifrage | Aerial parts | Gallated esters of D-glucose and rutin, flavonoids, gallic acid, Bergenin, 1,2,3,4,6-penta-*O*-galloyl-beta-D-glucoside | *In vitro* | Inhibit HCV-NS3 serine protease | Zuo et al. (2005) |
| 46. | *Azadirachta indica* | Indian lilac, neem | Leaves | 3-Deacetyl-3-cinnamoyl-azadirachtin | *In silico* | Interaction with HCV-NS3/4A protease | Ashfaq et al. (2016) |
| 47. | *Morinda citrifolia* | Noni, Indian mulberry | Fruit | – | *In vitro* | Inhibit DNA replication | Selvam et al. (2009) |
| 48. | *Marrubium peregrinum* | Horehound | – | Flavonoid, ladanein | *In vitro, In vivo* | Interferes with entry of HCV by preventing specific HCV-receptor interactions | Haid et al. (2012) |
| 49. | *Solanum nigrum* | European black nightshade | Seed | – | *In vitro* | Inhibit HCV-NS3 protease | Javed et al. (2011) |

(*Continued*)

**TABLE 7.2 (*Continued*)**
**Plants and Their Parts Used against Different Hepatitis Viruses**

| S. No. | Medicinal Plant | Common Name | Part of Plant Used | Bioactive Compound | Study Type | Plausible Inhibition Mechanism | References |
|---|---|---|---|---|---|---|---|
| 50. | *Viola yedoensis* | Chinese violet | Whole plant | 5,5'-bi (6,7dihydroxycoumarin) | *In vitro* | Inhibit HCV NS3/4A protease | Zhang et al. (2013) |
| 51. | *Eclipta alba* | Bhringaraja, Bhangara | Whole plant | Wedelolactone, luteolin, apigenin | *In vitro* | Inhibit DNA replication and RNA dependent RNA polymerase (RdRp) activity | Manvar et al. (2012) |
| 52. | *Artocarpus heterophyllus* | Jackfruit | Leaves | Terpenoids and steroids | *In vitro* | Inhibited viral entry via direct virucidal activity | Hafid et al. (2017) |
| 53. | *Ligustrum lucidum* | Chinese privet, wax-leaf privet | Fruit | Oleanolic acid, Ursolic acid | *In vitro* | Inhibit HCV replication, suppress HCV-NS5B RdRp activity | Kong et al. (2013) |
| 54. | *Alectryon serratus* | – | Leaves | Chlorophyll derivative | *In vitro* | Inhibit post HCV entry step | Tumewu et al. (2016) |
| | | | | **Hepatitis E Virus (HEV)** | | | |
| 55. | *Glycyrrhiza glabra* | Liquorice | Root | Glycyrrhizin | *In vivo* | Normalization in AST and ALT level | Tandon et al. (2002) |
| 56. | *Aglaia foveolata* | – | Bark | Silvestrol | *In vitro, in vivo* | Inhibit HEV replication | Todt et al. (2018) |

3′,4′,5,6,7,8-hexamethoxyflavone (nobiletin) is the main ingredient in the ethyl acetate extract, which plays a significant role in inhibition of HCV absorption (Suzuki et al., 2005).

*Silybum marianum* (commonly known as milk thistle, Mary thistle, or Scotch thistle) is a biennial plant belonging to the *Asteraceae* family (Porwal et al., 2019). It has been a part of traditional medicines since centuries (Tooiserkani et al., 2019). The extract of *Silybum marianum* is known as silymarin, and silibinin has been identified as the major constituent of this extract (Tajmohammadi et al., 2018). The effect of silibinin on HCV infection was estimated by experimenting on HepG2-NTCP-C4 cells, which are a HepG2 (human hepatoma) cell line, modified to over express the sodium taurocholate cotransporting polypeptide (NTCP; an HCV receptor). The study suggested that silibinin slowed down the endosomal trafficking of the virions, thus, hindering HCV entry. Mechanistically, silibinin was suggested to act on clathrin-coated pits and cellular vesicles, as well as affect transferrin trafficking (Blaising et al., 2013).

*Moringa oleifera* (commonly known as drumstick tree or horseradish tree) is a member of *Moringaceae* family and is commonly found in the Indian subcontinent (Foidl et al., 2001). Leaves of *Moringa oleifera* are suggested to be rich in vitamins, carotenoids, saponins, flavonoids, isothiocyanates, tannins and phytates, etc. (Abdull Razis et al., 2014). In traditional medicine, its leaves extract is applied for its high anti-oxidant activity (Leone et al., 2015). Feustel et al. (2017) evaluated the antiviral effect of the aqueous leaf extract in HBV-infected Huh7 (a human hepatoma cell line) cells. It was observed that HBsAg titers declined, independent of the extract dose. In addition, reduced expression of connective tissue growth factor (CTGF), transforming growth factor beta 1 (TGF-β1), IFN-β1 and pregenomic RNA (pgRNA) was also reported.

Juzen-taiho-to (JTT; Tj48) is one of the most famous Japanese traditional medicine formulae (Ikemoto et al., 2011). It consists of ten herbal ingredients, namely astragalus root (derived from plants of *Astragalus* genus), *Atractylodes lancea* rhizome, cinnamon bark (derived from *Cinnamomum cassia*), ginseng (derived from plants of *Panax* genus), poria sclerotium (derived from *Wolfiporia extensa*), rehmannia root (derived from plants of *Rehmannia* genus), peony root (derived from *Paeonia officinalis* L.), glycyrrhiza (derived from plants of *Glycyrrhiza* genus), cnidium rhizome (derived from plants of *Cnidium* genus) and Japanese angelica root (derived from *Aralia elata*) (Chang, 2006). In 2008, Tsuchiya et al. suggested that administration of Juzen-taiho-to could delay the recurrence of hepatocellular carcinoma in treated patients (Tsuchiya et al., 2008). Further, a reduction of Kupffer cell-induced oxidative stress and inflammatory cell infiltration was observed in Juzen-taiho-to-treated mice (Feustel et al., 2017).

*Ligustrum lucidum* (commonly known as Chinese privet or wax-leaf privet) is an evergreen tree belonging to the *Oleaceae* family (Deng et al., 2016). *Fructus Ligustri Lucidi* (FLL) is the dried fruit of this plant and is used against osteoporosis, rheumatic bones and various types of cancers (Pang et al., 2015; Chen et al., 2017). FLL comprises varied components, including flavonoids, phenolics, glucosides and triterpenes (Gao et al., 2015). In 2013, Lingbao et al. reported that FLL could suppress HCV nonstructural protein 5B (NS5B) RdRp, as well as inhibit HCV replication. Further, the authors suggested that two pentacyclic triterpenoids, oleanolic acid and ursolic acid, are the components of FLL responsible for its anti-HCV activity (Kong et al., 2013).

*Sophora flavescens* (commonly known as shrubby sophora) is an evergreen shrub belonging to the *Fabaceae* family (He et al., 2015). In an interesting study, four quinolizidine alkaloids, namely (+)-oxysophocarpine, (−)-sophocarpine (6), (+)-lehmannine (9) and (−)-13,14-dehydrosophoridine, isolated from *Sophora flavescens*, were suggested to exhibit anti-HBV activity. The authors observed that HBsAg and HBeAg secretion were reduced by nearly 48% and 24%–34%, respectively at 0.2 μmol/mL concentration of these alkaloids. Notably, these alkaloids exhibited higher anti-HBV activity than clinically proven anti-HBV compounds, (+)-oxymatrine and (+)-matrine (Ding et al., 2006). Thus, this study highlighted the potential of these four alkaloids in future anti-HBV drug development efforts.

*Vitis Vinifera* (commonly known as European wine grape or common grape) is a major source of wine production (Dwibedi et al., 2022) and belongs to the *Vitaceae* family (Felicio et al., 2001). Multiple studies have evaluated the fruit of *Vitis vinifera* for its hepatoprotective effect. In 2010, the fruit-derived triterpenes, flavonoids and polysaccharides were evaluated for their anti-HBV activity. It was suggested that a combination of triterpenes (20 μg/mL), flavonoids (50 μg/mL) and polysaccharides (50 μg/mL) can exhibit a synergistic affect and reduce secretion of HBsAg and HBeAg from HepG2.2.15 cells (Liu et al., 2010). In another study, the fruit-derived triterpenoids and flavonoids were observed to attenuate IFN-γ and IL-2 expression, while increasing the secretion of IL-4 and IL-10 in mice suffering from liver injury. In addition, the phytochemicals reduced the levels of hepatic nitric oxide and malondialdehyde, while alanine aminotransferase (ALT) and aspartate aminotransferase (AST) levels were increased (Liu et al., 2012).

*Thalassodendron ciliatum* (commonly known as sickle-leaved cymodocea) is a rhizomatous hydrogeophyte seagrass belonging to the *Cymodoceaceae* family (Abdelhameed et al., 2020). In an encouraging study, the crude extract of *Thalassodendron ciliatum*, containing rutin, asebotin, 3-hydroxyasebotin, quercetin-3-O-β-dxylopyranoside, a racemic mixture of (+)-catechins and *trans*-caffeic acid, was observed to effectively inhibit HAV infection. Also, quercetin-3-O-β-D-xylopyranoside was observed to display anti-HEPG (human hepatoma cell line) activity (7.25 ± 0.39 μM) (Hamdy et al., 2012).

Miso is a traditional Japanese seasoning, made by fermenting soybeans, barley, rice, wheat or cycad, with *Aspergillus oryzae* (Aidoo et al., 2006). It is typically rich in vitamins (such as vitamin A, riboflavin, folate and choline), minerals (such as calcium, phosphorus, potassium and sodium) and amino acids (such as leucine, arginine, aspartic acid and glutamic acid) (Nout, 2015). In a recent study, it was suggested that Japanese miso extracts can inhibit HAV replication in Huh7 and PXB (human hepatocyte cell line) cells. Mechanistically, elevation in GRP78 (a member of Hsp70 protein family) expression was contemplated as the reason for inhibition of HAV replication (Win et al., 2018).

*Vaccinium corymbosum* (commonly known as blue huckleberry, high blueberry or northern highbush blueberry) is a popular deciduous shrub belonging to the *Ericaceae* family (Valenzuela-Estrada et al., 2008). This plant is a rich source proanthocyanidins (flavonoids), consisting of 88–261 mg of these phytochemicals per 100 g (Joshi et al., 2016). In 2016, Joshi et al. evaluated the blueberry-derived proanthocyanidins against HAV infection. They observed attenuation of HAV titers within half hour of 2–5 mg/mL blueberry proanthocyanidins administration. Also, a lower dose (1 mg/mL) of blueberry proanthocyanidins was shown to be effective against HAV, albeit requiring a longer action time (Joshi et al., 2016).

*Panax ginseng* (commonly known as Asian ginseng or Korean ginseng) is a perennial herb belonging to *Araliaceae* family (Koren et al., 2003). The active compounds in *Panax ginseng*, known as ginsenosides or panaxosides, have been suggested to confer beneficial effects in case of Alzheimer's disease and erectile dysfunction (Jang et al., 2008; Lee et al., 2008). A study evaluated the effect of red ginseng on HAV infection, wherein Frhk-4 (fetal rhesus monkey kidney cell line) cells were pre-treated with red ginseng extract at a non-cytotoxic concentration of 0–10 μg/mL. Upon infection with HAV, the HAV titers appeared to be attenuated in the pre-treated cells, indicating the anti-HAV potential of red ginseng (Lee et al., 2013).

*Euphorbia schimperi* (commonly known as pencil cactus) is a perennial succulent shrub belonging to the *Euphorbiaceae* family (Alqahtani, 2018). *Euphorbia schimperi* extracts have been suggested to possess anti-oxidant and anti-microbial properties (Mothana et al., 2009; Ahmed et al., 2019). In a recent study, quercetin-3-O-glucuronide, quercetin-3-O-rhamnoside and kaempferol-3-O-glucuronide were isolated from extract of this plant, and tested for their anti-HBV potential. The compounds were able to attenuate viral HBsAg/HBeAg production in HepG2.2.15 cells by at least 60%. Interestingly, *in vitro* results were also supported by *in silico* analysis, wherein the authors suggested that the quercetin and kaempferol derivatives interfered with various HBV proteins to affect the anti-HBV activity (Parvez et al., 2021).

*Glycyrrhiza glabara* (commonly known as Liquorice) is an herbaceous perennial herb belonging to the *Fabaceae* family (El-Saber Batiha et al., 2020). It has been commonly suggested against sexual debility, skin disorders, epilepsy and hemorrhagic diseases (El-Saber Batiha et al., 2020; Damle, 2014). One of the major constituents of *Glycyrrhiza glabara* root is a saponin named glycyrrhizin (Nassiri and Hosseinzadeh, 2007). Tandon et al. (2002) estimated the benefits of glycyrrhizin therapy against HEV infection. A dose of 60 mL/day was administered to 22 patients with moderate or severe liver malfunction due to HEV infection. After 14 days, total bilirubin, AST and ALT levels appeared to be attenuated by 68.9%, 94% and 97%, respectively, while normal levels were observed after 30 days in majority of the patients. Additionally, the glycyrrhizin therapy was noted to have a favorable toxicological profile.

*Ruta angustifolia* is a member of the *Rutaceae* family and belongs to the genus of strongly scented evergreen subshrubs (Amtul et al., 2018). Members of this genus are known to possess anti-inflammatory, wound healing and analgesic properties (Loonat and Amabeoku, 2014; Velmurugan et al., 2021). Wahyuni et al. (2014) reported anti-HCV activity by two compounds, chalepin and pseudane IX, isolated from the leaves of *Ruta angustifolia*. Chalepin and pseudane IX were reported to exhibit an $IC_{50}$ of $1.7 \pm 0.5$ and $1.4 \pm 0.2$ µg/mL, respectively, in HCV-infected Huh7.5 (a subline of Huh7 cell line) cells. Interestingly, these observations were better than that for ribavirin ($2.8 \pm 0.4$ µg/mL), a popular anti-HCV agent. The authors suggested that mechanistically chalepin and pseudane IX were able to attenuate HCV infection by reducing viral replication and viral protein synthesis.

*Alectryon serratus* is a member of the *Sapindaceae* family (Edwards and Gadek, 2001). The fruits and seeds from this plant have been commonly used for their insecticidal property (Tumewu et al., 2016). Additionally, *Alectryon serratus* leaves extract has been suggested to exhibit anti-malarial and anti-oxidant activities (Khasanah et al., 2017; Hafid et al., 2014). In 2016, Tumewu et al. (2016) reported anti-HCV activity leaves extract, wherein an $IC_{50}$ value of 9.8–14.9 µg/mL was observed in Huh7 cells infected with different HCV genotypes. The inhibitory action was suggested to initiate at the post-entry step by a chlorophyll derivative present in the leaves extract.

*Aglaia foveolata* is an evergreen tree belonging to the *Meliaceae* family (Salim et al., 2007). It is popular for being the source of silvestrol, a member of the *flavagline* family, which is suggested to have anti-Zika, anti-Ebola and anti-cancer activity (Schulz et al., 2021). A recent *in vitro* and *in vivo* study was conducted to estimate the anti-HEV potential of silvestrol. The result suggested that silvestrol blocked the HEV replications of different sub-genomic replicons in dose-dependent manner in both HepG2 and Huh7.5 cell lines, as well as in mice. This indicated that silvestrol may be used for the treatment of HEV infection (Todt et al., 2018).

## 7.3 SOME MORE AYURVEDIC MEDICINAL PLANTS

### 7.3.1 *Curcuma longa*

*Curcuma longa* (commonly known as turmeric or yellow ginger) is a rhizomatous perennial herb belonging to the *Zingiberaceae* family (Labban, 2014). It is suggested to have beneficial effects against multiple conditions, such as helminthiasis, asthma and gonorrhea (Rakh et al., 2014; Manarin et al., 2019; Wessler et al., 2005). The primary bioactive curcuminoid extracted from turmeric is curcumin (Aggarwal et al., 2007), which is believed to possess anti-oxidant and anti-inflammatory properties (Calabrese et al., 2008; Kunnumakkara et al., 2008; Sarkar et al., 2010).

Various studies have reported that curcumin exposure results in the activation of heme oxygenase-1 (HO-1) expression, a crucial host cell anti-oxidant enzyme (Yang et al., 2017; Mhillaj et al., 2019). This property of curcumin might be beneficial as it may attenuate oxidative stress caused by HCV infection in hepatocytes (Zhu et al., 2008). In 2009, Kim et al. (2009) tested crude turmeric root extract against HBV-expressing HepG 2.2.15 cells. The study observed suppression of the DNA replication and HBsAg expression upon administration of the root extract. Mechanistically, the authors proposed that the anti-HBV activity of turmeric extract is due to an increase in p53 gene expression and p53 protein accumulation. Consequently, p53 was found to suppress the HBV enhancer I and X promoter, thus, blocking the expression of HBx gene (Kim et al., 2009). Another study highlighted the potential benefits of curcumin in the context of liver cirrhosis. In bile duct ligation (BDL) rat model, the administration was suggested to attenuate hydroxyproline and TGF-β levels (Reyes-Gordillo et al., 2008). Curcumin's efficacy in treating BDL was further demonstrated in another study, which showed a reduction in liver damage and oxidative stress (Erenoğlu et al., 2011). Another curcumin hepatoprotection mechanism study was conducted in C57BL/6 mice model, where the curcumin pretreatment protected the mice from T cell-mediated hepatitis, as demonstrated by lower ALT levels in blood (Wang et al., 2012). Curcumin pretreatment also drastically decreased hepatocytes oxidative stress and pro-inflammatory cytokines, such as TNF-α and IFN-α (Wang et al., 2012). Further, the release of high-mobility group box-1 (HMGB1) in liver tissues was also significantly reduced (Wang et al., 2012). These findings indicate that pretreatment with curcumin defends mice from Con A-induced liver injury by suppressing hepatocyte oxidative stress, inflammation and HMGB1 secretion (Wang et al., 2012).

### 7.3.2 *Zingiber officinale*

*Zingiber officinale* (commonly known as ginger) is a monocotyledonous perennial reed-like herb belonging to the *Zingiberaceae* family (Deme et al., 2021). Ginger consists of various bioactive components, such as 6-gingerol, 8-gingerol, 10-gingerol, α-curcumene, α-farnesene, gingerenone-A, 6-dehydrogingerdione, β-bisabolene, zingerone, zingiberene, quercetin and β-sesquiphellandrene. These components are suggested to be responsible for various properties, such as cardioprotective, neuroprotective, anti-inflammatory, anti-oxidant, anti-diabetes, anti-obesity, anti-cancer and antimicrobial properties (Mao et al., 2019).

In 2009, an *in vitro* analysis suggested that lyophilized ginger juice extract displays anti-HCV activity. In this study, the inhibition of viral replication was observed in HCV-infected HepG2 cell line upon administration of 100 μg/mL of the lyophilized extract (Abd El-Wahab et al., 2009). In an *in vivo* study, Ezeonu et al. (2011) suggested anti-hepatotoxic effect of ginger extract. They administered the extract in different hepatotoxic-induced mice models under three conditions: prior to, simultaneously with, and after hepatotoxicity induction. Under all conditions, they reported attenuation of serum glutamate oxaloacetate transaminase (SGOT), serum glutamate pyruvate transaminase (SGPT) and lipid peroxidase levels, albeit at different efficacies. This study confirmed the findings of a previous report, wherein Yusof et al. had suggested modulation of lipid peroxidation by ginger supplementation in Wister rats suffering from induced hepatocarcinogenesis (Yusof et al., 2009). Mechanistically, ginger extracts might result in attenuation of NF-κB and TNF-α levels, thus reducing the hepatocarcinogenic effects (Habib et al., 2008).

### 7.3.3 *Aloe barbadensis miller*

*Aloe barbadensis miller* (commonly known as *Aloe vera*) is a perennial, succulent, shrubby plant belonging to the *Asphodelaceae* family (Adom et al., 2022). A plethora of bioactive constituents have been identified in *Aloe vera*, including bradykinase (anti-inflammatory), alprogen (anti-allergic), C-glucosyl chromone (anti-inflammatory), emodin (analgesic), aloin (antiviral), lupeol (antiseptic), β-sisosterol (anti-inflammatory) and salicylic acid (anti-bacterial) (Surjushe et al., 2008).

In 2014, Bhatt et al. (2014) evaluated the effect of *Aloe vera* juice on patients suffering from acute hepatitis. The study reported a higher decrease in bilirubin, ALT, AST and serum alkaline phosphatase (ALP) levels in patients who supplemented with *Aloe vera* juice compared to those receiving regular treatment. In fact, the authors reported nearly 23% more decrease in ALP levels after daily supplementary intake of *Aloe vera* juice for 42 days (Bhatt et al., 2014). In a recent study, Parvez et al. tested the therapeutic efficacy of *Aloe vera* extract against HBV infection in HepG2.2.15 cells (Parvez, et al., 2019). Emodin was identified as the most potent inhibitor of viral antigen, followed by chrysophanol and aloin B. The authors observed that emodin performed at par with a known HBV inhibitor, lamivudine. Mechanistically, emodin could possibly exert anti-oxidant effects by elevating cytochrome P450 3A4 (CYP3A4) activity (Parvez, et al., 2019). A similar action mechanism was proposed for *Aloe vera* polysaccharides (APs). It was observed that treatment with APs led to attenuation of hepatotoxicity in rat models (Cui et al., 2017). This resulted in increased appetite, reduced impairment of organs and decreased elevated enzyme levels (Cui et al., 2017). Moreover, the authors suggested that APs could downregulate IL-1β and IL-6 expression, as well as block malondialdehyde production (Cui et al., 2017).

### 7.3.4 *Momordica charantia*

*Momordica charantia* (commonly known as bitter melon, bitter gourd, *karela*, or balsam-pear) is an annual tropical vine belonging to the *Cucurbitaceae* family (Daniel et al., 2014). It is suggested to contain anti-diabetic, anti-abortifacient, anthelmintic, anti-malarial and laxative components. It is recommended for addressing a range of health concerns such as dysmenorrhea, emmenagogue, eczema, gout, galactagogue, kidney stone, jaundice, leucorrhea, leprosy, pneumonia, piles, rheumatism and psoriasis-like ailments (Grover and Yadav, 2004). Recent studies have indicated even more beneficial properties, such as anti-oxidant (Wei et al., 2013), anti-cancer (Kwatra et al., 2013a, b), anti-inflammatory (Chao et al., 2014) and immuno-modulatory (Deng et al., 2014) properties.

In 2009, Chaudhari et al. (2014) estimated the effect of hydroalcoholic leaf extract obtained from *Momordica charantia* on albino Wistar rats suffering from induced hepatotoxicity. It was observed that the extract administration at 100–200 mg/kg lowered the elevated SGOT, SGPT, ALP and total bilirubin levels. In 2012, Waiyaput et al. (2012) conducted an *in vitro* study to analyze the efficacy of *Momordica charantia* treatment in HBV-infected HepG2 cells. Their findings supported the observations made by Chaudhari et al. (2014), as the hydroalcoholic leaf extract of the plant was reported to exhibit anti-HBV activity, with mild inhibition of HBV cccDNA was seen (Waiyaput et al., 2012). A recent study, similar to the work of Chaudhari et al. (2014), further ascertained the therapeutic potential of *Momordica charantia* in attenuating induced hepatotoxicity (Moharir et al., 2019). Moreover, in 2015, it was reported that *Momordica charantia* consumption by Long Evans rats helped in reducing induced hepatotoxicity (Sagor et al., 2015). Anti-oxidant activity of *Momordica charantia* was suggested, as well as a decrease in fibrosis, iron deposition and lipid peroxidation levels was observed (Sagor et al., 2015).

## 7.4 CONCLUSION

Hepatitis continues to annually present numerous cases around the world. Medicinal plants are a treasure trove for various natural bioactive compounds with hepatoprotective properties, in addition to their safer toxicity profile as compared to synthetic drugs. Thus, it is imperative to validate the easily accessible, potent, globally effective, anti-hepatotoxic natural compound/s for their suitability to treat hepatitis. In an effort to supplement such validatory studies, we herein enlist nearly 55 medicinal plants known to exhibit hepatoprotective effects. The enlisted plants belong to diverse geographical regions, and have presented encouraging results in preliminary *in vitro/ in vivo* analysis. We hope this chapter will reinforce the potential of natural compounds in the fight against hepatitis, and motivate further anti-hepatitis natural drug discovery studies.

## REFERENCES


Abd El-Wahab, A., H. El-Adawi, and M. El-Demellawy, In vitro study of the antiviral activity of Zingiber officinale. *Planta Medica*, 2009. 75(09): PF7.

Abdelhameed, R.F., et al., Thalassosterol, a new cytotoxic aromatase inhibitor ergosterol derivative from the Red Sea seagrass Thalassodendron ciliatum. *Marine Drugs*, 2020. 18(7): 354.

Abdull Razis, A.F., M.D. Ibrahim, and S.B. Kntayya, Health benefits of Moringa oleifera. *Asian Pacific Journal of Cancer Prevention*, 2014. 15(20): 8571–8576.

Adianti, M., et al., Anti-hepatitis C virus compounds obtained from Glycyrrhiza uralensis and other Glycyrrhiza species. *Microbiology and Immunology*, 2014. 58(3): 180–187.

Adom, D., P.A. Sekyere, and M.K. Krishnappa, The chemical constituents, anti-inflammatory, anti-oxidant, and ethnomedicinal properties of aloe barbadensis. In Mehdi Khosrow-Pour (Ed.) *Research Anthology on Recent Advancements in Ethnopharmacology and Nutraceuticals*. 2022, IGI Global, United States, pp. 454–471.

Aggarwal, B.B., et al., Curcumin: the Indian solid gold. *Adv Exp Med Biol*, 2007. 595: 1–75. doi: 10.1007/978-0-387-46401-5_1.

Ahmed, S., et al., Antidiabetic, antioxidant, molecular docking and HPTLC analysis of miquelianin isolated from Euphorbia schimperi C. Presl. *Saudi Pharmaceutical Journal*, 2019. 27(5): 655–663.

Aidoo, K.E., M. Rob Nout, and P.K. Sarkar, Occurrence and function of yeasts in Asian indigenous fermented foods. *FEMS Yeast Research*, 2006. 6(1): 30–39.

Akhlaghi Amiri, N., Vegetative characteristics of Miyagawa Satsuma mandarin on six different rootstocks in East of Mazandaran. *Iranian Journal of Horticultural Science*, 2019. 50(3): 683–695.

Ali, S.A., N.H. Sharief, and Y.S. Mohamed, Hepatoprotective activity of some medicinal plants in Sudan. *Evidence-Based Complementary and Alternative Medicine*, 2019. 2019.

Alqahtani, A.A.A., *Genomic Resources of Euphorbia schimperi: evolutionary and Medicinal*. 2018. Available from: https://repositories.lib.utexas.edu/handle/2152/85299

Amtul, Z., Nature's medicines to treat epileptic seizures. *Studies in Natural Products Chemistry*, 2018. 56: 129–150.

Andonov, A., et al., Rat hepatitis E virus linked to severe acute hepatitis in an immunocompetent patient. *The Journal of Infectious Diseases*, 2019. 220(6): 951–955.

Ashfaq, U.A., A. Jalil, and M.T. ul Qamar, Antiviral phytochemicals identification from Azadirachta indica leaves against HCV NS3 protease: an in silico approach. *Natural Product Research*, 2016. 30(16): 1866–1869.

Ashfaq, U.A., et al., Glycyrrhizin as antiviral agent against Hepatitis C virus. *Journal of Translational Medicine*, 2011. 9(1): 1–7.

Barkat, M.A., et al., Herbal medicine: clinical perspective and regulatory status. *Combinatorial Chemistry & High Throughput Screening*, 2021. 24(10): 1573–1582.

Batiha, G.E.-S., et al., Traditional uses, bioactive chemical constituents, and pharmacological and toxicological activities of Glycyrrhiza glabra L.(Fabaceae). *Biomolecules*, 2020. 10(3): 352.

Benova, L., et al., Estimation of hepatitis C virus infections resulting from vertical transmission in Egypt. *Hepatology*, 2015. 61(3): 834–842.

Benova, L., et al., Vertical transmission of hepatitis C virus: systematic review and meta-analysis. *Clinical Infectious Diseases*, 2014. 59(6): 765–773.

Bhatt, S., et al., Evaluation of hepatoprotective activity of Aloe Vera in acute viral hepatitis. *International Journal of Pharmaceutical Sciences and Research*, 2014. 5(6): 2479.

Bixler, D., et al., Hepatitis B and C virus infections transmitted through organ transplantation investigated by CDC, United States, 2014-2017. *American Journal of Transplantation*, 2019. 19(9): 2570–2582.

Blaising, J., et al., Silibinin inhibits hepatitis C virus entry into hepatocytes by hindering clathrin-dependent trafficking. *Cellular Microbiology*, 2013. 15(11): 1866–1882.

Busch, M.P., E.M. Bloch, and S. Kleinman, Prevention of transfusion-transmitted infections. *Blood*, 2019. 133(17): 1854–1864.

Calabrese, V., et al., Curcumin and the cellular stress response in free radical-related diseases. *Molecular Nutrition & Food Research*, 2008. 52(9): 1062–1073.

Carmona, F. and A.M.S. Pereira, Herbal medicines: old and new concepts, truths and misunderstandings. *Revista Brasileira de Farmacognosia*, 2013. 23(2): 379–385.

Castaneda, D., et al., From hepatitis A to E: a critical review of viral hepatitis. *World Journal of Gastroenterology*, 2021. 27(16): 1691.

CDC. *The ABCs of Hepatitis - for Health Professionals*. 2020 [cited 2022 19 June]; Available from: https://www.cdc.gov/hepatitis/resources/professionals/pdfs/abctable.pdf.

Chang, I.M., Juzen-taiho-to (Shi-Quan-Da-Bu-Tang): Scientific Evaluation and Clinical Application. *Evidence-Based Complementary and Alternative Medicine*, 2006. 3(3): 393–394.

Chang, J.-S., et al., Sho-saiko-to (Xiao-Chai-Hu-Tang) and crude saikosaponins inhibit hepatitis B virus in a stable HBV-producing cell line. *The American Journal of Chinese Medicine*, 2007. 35(02): 341–351.

Chang, J.M., et al., The anti-hepatitis B virus activity of Boehmeria nivea extract in HBV-viremia SCID mice. *Evidence-Based Complementary and Alternative Medicine*, 2010. 7(2): 189–195.

Chao, C.-Y., et al., Anti-inflammatory effect of Momordica charantia in sepsis mice. *Molecules*, 2014. 19(8): 12777–12788.

Chaudhari, B.P., et al., Hepatoprotective activity of Hydroalcoholic extract of Momordica charantia Linn. leaves against Carbon tetra chloride induced Hepatopathy in Rats. *International Journal of ChemTech Research*, 2009. 1(2): 355–358.

Chen, B., et al., Fructus ligustri lucidi in osteoporosis: a review of its pharmacology, phytochemistry, pharmacokinetics and safety. *Molecules*, 2017. 22(9): 1469.

Chen, C.J., et al., Risk of hepatocellular carcinoma across a biological gradient of serum hepatitis B virus DNA level. *JAMA*, 2006. 295(1): 65–73.

Chen, H., et al., Synthesis, structure-activity relationships and biological evaluation of dehydroandrographolide and andrographolide derivatives as novel anti-hepatitis B virus agents. *Bioorganic & Medicinal Chemistry Letters*, 2014. 24(10): 2353–2359.

Choi, M., et al., Mangosteen xanthones suppress hepatitis C virus genome replication. *Virus Genes*, 2014. 49(2): 208–222.

Chu, C.M. and Y.F. Liaw, Chronic hepatitis B virus infection acquired in childhood: special emphasis on prognostic and therapeutic implication of delayed HBeAg seroconversion. *Journal of Viral Hepatitis*, 2007. 14(3): 147–152.

Cui, Y., et al., Evaluating the hepatoprotective efficacy of Aloe vera polysaccharides against subchronic exposure of aflatoxins B1. *Journal of the Taiwan Institute of Chemical Engineers*, 2017. 76: 10–17.

Dai, J.J., et al., Anti-hepatitis B virus activities of friedelolactones from Viola diffusa Ging. *Phytomedicine*, 2015. 22(7–8): 724–729.

Dakio, B., et al., Medicinal Plants used in the Treatment of Hepatitis in Bobo-Dioulasso: studying the Availability and Analyzing the Phytochemical Properties of Combretum micranthum G. Don and Entada africana. *Academic Journal of Life Sciences*, 2020. 6(8): 101–107.

Dalton, H.R. and N. Kamar, Treatment of hepatitis E virus. *Current Opinion in Infectious Diseases*, 2016. 29(6): 639–644.

Damle, M., Glycyrrhiza glabra (Liquorice)-a potent medicinal herb. *International Journal of Herbal Medicine*, 2014. 2(2): 132–136.

Daniel, P., U. Supe, and M. Roymon, A review on phytochemical analysis of Momordica charantia. *International Journal of Advances in Pharmacy, Biology and Chemistry*, 2014. 3(1): 214–220.

Deme, K., et al., Importance, genetic diversity and prospects for varietal improvement of ginger (Zingiber officinale Roscoe) in Burkina Faso. *World Journal of Agricultural Research*, 2021. 9: 92–99.

Deng, Q. Investigation of Ligustrum lucidum Ait. Seeds extract as green corrosion inhibitors and bacteriostatic agent. In Mingliang Zhang and Zhiguo Zhu (Eds.) *2016 5th International Conference on Advanced Materials and Computer Science (ICAMCS 2016)*. 2016. Atlantis Press, China.

Deng, Y.-Y., et al., Immunomodulatory activity and partial characterisation of polysaccharides from Momordica charantia. *Molecules*, 2014. 19(9): 13432–13447.

Dienstag, J.L., Hepatitis B virus infection. *The New England Journal of Medicine*, 2008. 359(14): 1486–1500.

Ding, P.L., et al., (+)-12alpha-Hydroxysophocarpine, a new quinolizidine alkaloid and related anti-HBV alkaloids from Sophora flavescens. *Bioorganic & Medicinal Chemistry Letters*, 2006. 16(5): 1231–1235.

Dusheiko, G. and H. Wedemeyer, New protease inhibitors and direct-acting antivirals for hepatitis C: interferon's long goodbye. *Gut*, 2012. 61(12): 1647–1652.

Dwibedi, V., et al., Inhibitory activities of grape bioactive compounds against enzymes linked with human diseases. *Applied Microbiology and Biotechnology*, 2022. 106(4): 1399–1417.

ECDC. *Hepatitis A virus in the EU/EEA, 1975-2014*. 2016 [cited 2022 21 June]; Available from: https://www.ecdc.europa.eu/en/publications-data/hepatitis-virus-eueea-1975-2014.

Edwards, K. and Gadek, Evolution and biogeography of Alectryon (Sapindaceae). *Molecular Phylogenetics and Evolution*, 2001. 20(1): 14–26.

El-Saber Batiha, G., et al., Traditional uses, bioactive chemical constituents, and pharmacological and toxicological activities of Glycyrrhiza glabra L.(Fabaceae). *Biomolecules*, 2020. 10(3): 352.
Erenoğlu, C., et al., Protective effect of curcumin on liver damage induced by biliary obstruction in rats. *Balkan Medical Journal*, 2011. 2011(4): 352–357.
Ezeonu, C., et al., Antihepatotoxicity studies of crude extract of Zingiber officinale on CCl4 induced toxicity and comparison of the extract's fraction D hepatoprotective capacity. *Research Journal of Medical Sciences*, 2011. 5(2): 102–107.
Fabbri, C., et al., Drug repositioning for treatment-resistant depression: hypotheses from a pharmacogenomic study. *Progress in Neuro-Psychopharmacology and Biological Psychiatry*, 2021. 104: 110050.
Felicio, J., R.d.S. Santos, and E. Gonçalez, Chemical constituents from Vitis vinifera (Vitaceae). *Arquivos do Instituto Biológico*, 2001. 68(1): 47–50.
Feustel, S., et al., Protective effects of Moringa oleifera on HBV genotypes C and H transiently transfected Huh7 cells. *Journal of Immunology Research*, 2017. 2017: 6063850.
Fiala, C. and M.D. Pasic, Aspirin: bitter pill or miracle drug? *Clinical Biochemistry*, 2020. 85: 1–4.
Foidl, N., H. Makkar, and K. Becker, The potential of Moringa oleifera for agricultural and industrial uses. *What Development Potential for Moringa Products*, 2001. 20.
Foster, M., et al., Hepatitis A virus outbreaks associated with drug use and homelessness-California, Kentucky, Michigan, and Utah, 2017. *Morbidity and Mortality Weekly Report*, 2018. 67(43): 1208.
Galani, B.R., et al., Plant extracts from Cameroonian medicinal plants strongly inhibit hepatitis C virus infection in vitro. *Frontiers in Microbiology*, 2015. 6: 488.
Gallego-Jara, J., et al., A compressive review about taxol(r): history and future challenges. *Molecules*, 2020. 25(24): 5986.
Gao, L., et al., Ligustri Lucidi Fructus as a traditional Chinese medicine: a review of its phytochemistry and pharmacology. *Natural Product Research*, 2015. 29(6): 493–510.
Geethangili, M. and S.T. Ding, A review of the phytochemistry and pharmacology of Phyllanthus urinaria L. *Frontiers in Pharmacology*, 2018. 9: 1109.
Grover, J. and S. Yadav, Pharmacological actions and potential uses of Momordica charantia: a review. *Journal of Ethnopharmacology*, 2004. 93(1): 123–132.
Guet, L., et al., Investigation of a severe nosocomial outbreak of hepatitis A among healthcare workers and adult patients. *Journal of Hospital Infection*, 2011. 78(1): 70–71.
Habib, S.H.M., et al., Ginger extract (Zingiber officinale) has anti-cancer and anti-inflammatory effects on ethionine-induced hepatoma rats. *Clinics*, 2008. 63(6): 807–813.
Hafid, A.F., et al., Antiviral activity of the dichloromethane extracts from Artocarpus heterophyllus leaves against hepatitis C virus. *Asian Pacific Journal of Tropical Biomedicine*, 2017. 7(7): 633–639.
Hafid, A.F., et al., Free-radical scavenging activity screening of some Indonesian plants. *Free-Radical Scavenging Activity Screening of Some Indonesian Plants*, 2014. 6: 115–117.
Haid, S., et al., A plant-derived flavonoid inhibits entry of all HCV genotypes into human hepatocytes. *Gastroenterology*, 2012. 143(1): 213–222.e5.
Hamdy, A.H., et al., Bioactive phenolic compounds from the Egyptian Red Sea seagrass Thalassodendron ciliatum. *A Journal of Biosciences: Zeitschrift für Naturforschung C*, 2012. 67(5–6): 291–6.
He, K., et al., Two new secoiridoids and other anti-hepatitis B virus active constituents from Swertia patens. *Journal of Asian Natural Products Research*, 2016. 18(6): 528–534.
He, X., et al., Sophora flavescens Ait.: traditional usage, phytochemistry and pharmacology of an important traditional Chinese medicine. *Journal of Ethnopharmacology*, 2015. 172: 10–29.
Hsu, W.C., et al., Limonium sinense and gallic acid suppress hepatitis C virus infection by blocking early viral entry. *Antiviral Research*, 2015. 118: 139–147.
Hussein, G., et al., Inhibitory effects of sudanese medicinal plant extracts on hepatitis C virus (HCV) protease. *Phytotherapy Research*, 2000. 14(7): 510–516.
Ikemoto, T., et al., Japanese herbal medicine TJ-48 prevents autoimmune diabetes in NOD mice. *The American Journal of Chinese Medicine*, 2011. 39(04): 743–756.
Indrasetiawan, P., et al., Antiviral activity of Cananga odorata against Hepatitis B virus. *Kobe Journal of Medical Sciences*, 2019. 65(2): E71–E79.
Islam, R. and M.J. Alam, Evaluation of liver protective activity of Moringa oleifera bark extract in paracetamol induced hepatotoxicity in rats. *BioRxiv*, 2019: 513002.
Jafari, S., et al., Tattooing and risk of hepatitis B: a systematic review and meta-analysis. *Canadian Journal of Public Health*, 2012. 103(3): 207–212.

Jain, D., et al., Hepatoprotective activity of medicinal plants: a mini review. *Journal of Medicinal Plants*, 2020. 8(5): 183–188.

Jang, D.J., et al., Red ginseng for treating erectile dysfunction: a systematic review. *British Journal of Clinical Pharmacology*, 2008. 66(4): 444–450.

Javed, T., et al., In-vitro antiviral activity of Solanum nigrum against Hepatitis C Virus. *Virology Journal*, 2011. 8(1): 26.

Jiang, Z.Y., et al., Anti-HBV active constituents from Piper longum. *Bioorganic & Medicinal Chemistry Letters*, 2013. 23(7): 2123–2137.

Joshi, S.S., A.B. Howell, and D.H. D'Souza, Reduction of enteric viruses by blueberry juice and blueberry Proanthocyanidins. *Food and Environmental Virology*, 2016. 8(4): 235–243.

Karamese, M., et al., Preventive effects of a major component of green tea, epigallocathechin-3-gallate, on hepatitis-B virus DNA replication. *The Asian Pacific Journal of Cancer Prevention*, 2015. 16(10): 4199–4202.

Kaur, R., H. Kaur, and A.S. Dhindsa, Glycyrrhiza glabra: a phytopharmacological review. *International Journal of Pharmaceutical Sciences and Research*, 2013. 4(7): 2470.

Khasanah, U., et al., Antiplasmodial activity of isolated polyphenols from Alectryon serratus leaves against 3D7 Plasmodium falciparum. *Pharmacognosy Research*, 2017. 9(Suppl 1): S57.

Kim, H.J., et al., Antiviral effect of Curcuma longa Linn extract against hepatitis B virus replication. *Journal of Ethnopharmacology*, 2009. 124(2): 189–196.

Kong, L., et al., Oleanolic acid and ursolic acid: novel hepatitis C virus antivirals that inhibit NS5B activity. *Antiviral Research*, 2013. 98(1): 44–53.

Koren, O.G., V.V. Potenko, and Y.N. Zhuravlev, Inheritance and variation of allozymes in Panax ginseng CA Meyer (Araliaceae). *International Journal of Plant Sciences*, 2003. 164(1): 189–195.

Kunnumakkara, A.B., Anand, and B.B. Aggarwal, Curcumin inhibits proliferation, invasion, angiogenesis and metastasis of different cancers through interaction with multiple cell signaling proteins. *Cancer Letters*, 2008. 269(2): 199–225.

Kwatra, D., et al., Bitter melon extracts enhance the activity of chemotherapeutic agents through the modulation of multiple drug resistance. *Journal of Pharmaceutical Sciences*, 2013a. 102(12): 4444–4454.

Kwatra, D., et al., Methanolic extracts of bitter melon inhibit colon cancer stem cells by affecting energy homeostasis and autophagy. *Evidence-Based Complementary and Alternative Medicine*, 2013b. Available from: https://doi.org/10.1155/2013/702869

Labban, L., Medicinal and pharmacological properties of Turmeric (Curcuma longa): a review. *International Journal of Pharmaceutical and Bio-Medical Science*, 2014. 5(1): 17–23.

Lado, J., et al., Nutritional composition of mandarins. In *Nutritional Composition of Fruit Cultivars*. Monique S.J. Simmonds and Victor R. Preedy, Eds. 2016, Elsevier, United States. 419–443.

Lanini, S., et al., Global epidemiology of viral hepatitis and national needs for complete control. *Expert Review of Anti-infective Therapy*, 2018. 16(8): 625–639.

Layden, J.E., et al., High frequency of active HCV infection among seropositive cases in west Africa and evidence for multiple transmission pathways. *Clinical Infectious Diseases*, 2015. 60(7): 1033–1041.

Lee, J.C., et al., Andrographolide exerts anti-hepatitis C virus activity by up-regulating haeme oxygenase-1 via the p38 MAPK/Nrf2 pathway in human hepatoma cells. *British Journal of Pharmacology*, 2014. 171(1): 237–252.

Lee, M.H., et al., Reduction of hepatitis A virus on FRhK-4 cells treated with Korean red ginseng extract and ginsenosides. *Journal of Food Science* 2013. 78(9): M1412– M1415.

Lee, S.-T., et al., Panax ginseng enhances cognitive performance in Alzheimer disease. *Alzheimer Disease & Associated Disorders*, 2008. 22(3): 222–226.

Leone, A., et al., Cultivation, genetic, ethnopharmacology, phytochemistry and pharmacology of Moringa oleifera leaves: an overview. *International Journal of Molecular Sciences*, 2015. 16(6): 12791–12835.

Li, N., et al., Chemical constituents, clinical efficacy and molecular mechanisms of the ethanol extract of Abelmoschus manihot flowers in treatment of kidney diseases. *Phytotherapy Research*, 2021. 35(1): 198–206.

Lin, L.-T., et al., Saikosaponin b2 is a naturally occurring terpenoid that efficiently inhibits hepatitis C virus entry. *Journal of Hepatology*, 2015. 62(3): 541–548.

Liu, T., et al., Evaluation on anti-hepatitis viral activity of Vitis vinifer L. *Molecules*, 2010. 15(10): 7415–7422.

Liu, T., et al., Hepatoprotective effects of total triterpenoids and total flavonoids from vitis vinifera L against immunological liver injury in mice. *Evidence-Based Complementary and Alternative Medicine*, 2012. 2012: 969386.

Liver, E.A.F.T.S.O.T., *EASL 2017* Clinical practice guidelines on the management of hepatitis B virus infection. *Journal of Hepatology*, 2017. 67(2): 370–398.

Loonat, F. and G.J. Amabeoku, Antinociceptive, anti-inflammatory and antipyretic activities of the leaf methanol extract of Ruta graveolens L. (Rutaceae) in mice and rats. *African Journal of Traditional, Complementary and Alternative Medicines*, 2014. 11(3): 173–181.

Luan, F., et al., Traditional uses, chemical constituents, biological properties, clinical settings, and toxicities of Abelmoschus manihot L.: a comprehensive review. *Frontiers in Pharmacology*, 2020. 11: 1068.

Luma, H.N., et al., Characteristics of anti-hepatitis C virus antibody-positive patients in a hospital setting in Douala, Cameroon. *International Journal of Infectious Diseases*, 2016. 45: 53–58.

Lv, J.J., et al., Anti-hepatitis B virus activities and absolute configurations of sesquiterpenoid glycosides from Phyllanthus emblica. *Organic and Biomolecular Chemistry*, 2014a. 12(43): 8764–8774.

Lv, J.J., et al., Anti-hepatitis B virus norbisabolane sesquiterpenoids from Phyllanthus acidus and the establishment of their absolute configurations using theoretical calculations. *The Journal of Organic Chemistry*, 2014b. 79(12): 5432–47.

Malik, G.F., et al., Viral hepatitis-the road traveled and the journey remaining. *Hepatic Medicine: Evidence and Research*, 2022. 14: 13.

Manarin, G., et al., Curcuma longa L. ameliorates asthma control in children and adolescents: a randomized, double-blind, controlled trial. *Journal of Ethnopharmacology*, 2019. 238: 111882.

Manvar, D., et al., Identification and evaluation of anti hepatitis C virus phytochemicals from Eclipta alba. *Journal of Ethnopharmacology*, 2012. 144(3): 545–554.

Mao, Q.Q., et al., Bioactive Compounds and Bioactivities of Ginger (Zingiber officinale Roscoe). *Foods*, 2019. 8(6).

Melnick, J.L., Properties and classification of hepatitis A virus. *Vaccine*, 1992. 10: S24–S26.

Mhillaj, E., et al., Curcumin and heme oxygenase: neuroprotection and beyond. *International Journal of Molecular Sciences*, 2019. 20(10): 2419.

Moghadamtousi, S.Z., et al., Potential antiviral agents from marine fungi: an overview. *Marine Drugs*, 2015. 13(7): 4520–4538.

Moharir, G., et al., *Evaluation of Hepatoprotective Effect of Hydroalcoholic Extract of Momordica charantia Leaves in Carbon Tetrachloride-Induced Liver Toxicity in Wistar Rats*. 2019.

Mohebbi, A., et al., Havachoobe (Onosma dichroanthum Boiss) root extract decreases the Hepatitis B virus surface antigen secretion in the PLC/PRF/5 cell line. *Intervirology*, 2020: 1–5.

mondiale de la Santé, O. and W.H. Organization, Hepatitis E vaccine: WHO position paper, May 2015. *Weekly Epidemiological Record= Relevé épidémiologique hebdomadaire*, 2015. 90(18): 185–200.

Mothana, R.A.A., et al., Evaluation of the in vitro anticancer, antimicrobial and antioxidant activities of some Yemeni plants used in folk medicine. *Die Pharmazie-An International Journal of Pharmaceutical Sciences*, 2009. 64(4): 260–268.

Murphy, E.L., et al., Risk factors for hepatitis C virus infection in United States blood donors. *Hepatology*, 2000. 31(3): 756–762.

Nassiri Asl, M. and H. Hosseinzadeh, Review of antiviral effects of Glycyrrhiza glabra L. and its active component, glycyrrhizin. *Journal of Medicinal Plants*, 2007. 6(22): 1–12.

Nout, R., Quality, safety, biofunctionality and fermentation control in soya. In Wilhelm Holzapfel (Ed.) *Advances in Fermented Foods and Beverages*. 2015, Elsevier, Cambridge. 409–434.

Ofoegbu, A. and E. B. Ettienne, Pharmacogenomics and morphine. *The Journal of Clinical Pharmacology*, 2021. 61(9): 1149–1155.

Ohtake, N., et al., Separation and isolation methods for analysis of the active principles of Sho-saiko-to (SST) oriental medicine. *Journal of Chromatography B*, 2004. 812(1–2): 135–148.

Ohtake, N., et al., The herbal medicine Sho-saiko-to selectively inhibits CD8+ T-cell proliferation. *European Journal of Pharmacology*, 2005. 507(1–3): 301–310.

Pang, Z., et al., The advances in research on the pharmacological effects of Fructus Ligustri Lucidi. *BioMed Research International*, 2015. 2015.

Parvez, M.K., et al., Novel anti-hepatitis B virus activity of Euphorbia schimperi and its Quercetin and Kaempferol derivatives. *ACS Omega*, 2021. 6(43): 29100–29110.

Parvez, M.K., et al., The anti-hepatitis B virus therapeutic potential of anthraquinones derived from Aloe vera. *Phytotherapy Research*, 2019. 33(11): 2960–2970.

Pastorino, G., et al., Liquorice (Glycyrrhiza glabra): a phytochemical and pharmacological review. *Phytotherapy Research*, 2018. 32(12): 2323–2339.

Ponzetto, A., et al., Titration of the infectivity of hepatitis D virus in chimpanzees. *Journal of Infectious Diseases*, 1987. 155(1): 72–78.

Porwal, O., et al., Silybum marianum (Milk Thistle): review on Its chemistry, morphology, ethno medical uses, phytochemistry and pharmacological activities. *Journal of Drug Delivery and Therapeutics*, 2019. 9(5): 199–206.

Rakh, M.S., R.V. Pawar, and A.N. Khedkar, Anthelmintic potential of various extracts of the rhizomes of Curcuma amada Roxb. *Asian Pacific Journal of Tropical Disease*, 2014. 4: S276–S278.

Ramsey, J.T., et al., Focus: plant-based medicine and pharmacology: essential oils and health. *The Yale Journal of Biology and Medicine*, 2020. 93(2): 291.

Rasche, A., et al., Evolutionary biology of human hepatitis viruses. *Journal of Hepatology*, 2019. 70(3): 501–520.

Ravikumar, Y.S., et al., Inhibition of hepatitis C virus replication by herbal extract: phyllanthus amarus as potent natural source. *Virus Research*, 2011. 158(1–2): 89–97.

Rehman, S., et al., Antiviral activity of Acacia nilotica against Hepatitis C virus in liver infected cells. *Virology Journal*, 2011. 8(1): 220.

Reyes-Gordillo, K., et al., Curcumin prevents and reverses cirrhosis induced by bile duct obstruction or CCl4 in rats: role of TGF-β modulation and oxidative stress. *Fundamental & clinical pharmacology*, 2008. 22(4): 417–427.

Sagor, A.T., et al., Supplementation of fresh ucche (Momordica charantia L. var. muricata Willd) prevented oxidative stress, fibrosis and hepatic damage in CCl4 treated rats. *BMC Complementary and Alternative Medicine*, 2015. 15(1): 115.

Salim, A.A., et al., Constituents of the leaves and stem bark of Aglaia foveolata. *Tetrahedron*, 2007. 63(33): 7926–7934.

Sarkar, F., et al., Lesson learned from nature for the development of novel anti-cancer agents: implication of isoflavone, curcumin, and their synthetic analogs. *Current Pharmaceutical Design*, 2010. 16(16): 1801–1812.

Schulz, G., et al., Rocaglamide and silvestrol: a long story from anti-tumor to anti-coronavirus compounds. *Natural Product Reports*, 2021. 38(1): 18–23.

Selvam, P., et al., Studies of antiviral activity and cytotoxicity of Wrightia tinctoria and Morinda citrifolia. *Indian Journal of Pharmaceutical Sciences*, 2009. 71(6): 670–672.

Shimizu, I., Sho-saiko-to: Japanese herbal medicine for protection against hepatic fibrosis and carcinoma. *Journal of Gastroenterology and Hepatology*, 2000. 15(s1): 84–90.

Song, K.H., Y.H. Kim, and B.Y. Kim, Sho-saiko-to, a traditional herbal medicine, regulates gene expression and biological function by way of microRNAs in primary mouse hepatocytes. *BMC Complementary and Alternative Medicine*, 2014. 14: 14.

Surjushe, A., R. Vasani, and D.G. Saple, Aloe vera: a short review. *Indian Journal of Dermatology*, 2008. 53(4): 163–166.

Suzuki, M., et al., Anti-hepatitis C virus effect of citrus unshiu peel and its active ingredient nobiletin. *The American Journal of Chinese Medicine*, 2005. 33(01): 87–94.

Tajmohammadi, A., B.M. Razavi, and H. Hosseinzadeh, Silybum marianum (milk thistle) and its main constituent, silymarin, as a potential therapeutic plant in metabolic syndrome: a review. *Phytotherapy Research*, 2018. 32(10): 1933–1949.

Tandon, A., B.N. Tandon, and R.A. Bhujwala, Clinical spectrum of acute sporadic hepatitis E and possible benefit of glycyrrhizin therapy. *Hepatology Research*, 2002. 23(1): 55–61.

Tao, Y., et al., Present and future therapies for chronic hepatitis B. *Hepatitis B Virus Infection*, 2020: 137–186.

Terrault, N.A., et al., Update on prevention, diagnosis, and treatment of chronic hepatitis B: AASLD 2018 hepatitis B guidance. *Hepatology*, 2018. 67(4): 1560–1599.

Thyagarajan, S., et al., Herbal medicines for liver diseases in India. *Journal of Gastroenterology and Hepatology*, 2002. 17: S370–S376.

Todt, D., et al., The natural compound silvestrol inhibits hepatitis E virus (HEV) replication in vitro and in vivo. *Antiviral Research*, 2018. 157: 151–158.

Tooiserkani, F., A. Hormati, and H. Moradi, A Glimpse of Silybum marianum from the perspective of Iranian traditional medicine and modern studies. *Qom University of Medical Sciences Journal*, 2019. 13(1): 78–86.

Tovo, P.-A., et al., Vertically acquired hepatitis C virus infection: correlates of transmission and disease progression. *World Journal of Gastroenterology*, 2016. 22(4): 1382.
Tsuchiya, M., et al., Protective effect of Juzen-taiho-to on hepatocarcinogenesis is mediated through the inhibition of Kupffer cell-induced oxidative stress. *International Journal of Cancer*, 2008. 123(11): 2503–2511.
Tumewu, L., et al., Antihepatitis C virus activity of Alectryon serratus leaves extract. *Procedia Chemistry*, 2016. 18: 169–173.
Valenzuela-Estrada, L.R., et al., Root anatomy, morphology, and longevity among root orders in Vaccinium corymbosum (Ericaceae). *American Journal of Botany*, 2008. 95(12): 1506–1514.
Velmurugan, C., et al., Evaluation of anti-diabetic and wound healing potential of Ethiopia plant 'Ruta Graveolens' in diabetic induced rat. *NVEO-Natural Volatiles & Essential OILS Journal*, 2021: 5629–5640.
Wahyuni, T.S., et al., Inhibition of hepatitis C virus replication by chalepin and pseudane IX isolated from Ruta angustifolia leaves. *Fitoterapia*, 2014. 99: 276–283.
Waiyaput, W., et al., Inhibitory effects of crude extracts from some edible Thai plants against replication of hepatitis B virus and human liver cancer cells. *BMC Complementary and Alternative Medicine*, 2012. 12(1): 1–7.
Wang, C., et al., Curcumin inhibits HMGB1 releasing and attenuates concanavalin A-induced hepatitis in mice. *European Journal of Pharmacology*, 2012. 697(1–3): 152–157.
Wang, W.-n., et al., Effect of Oenanthe javanica flavone on human and duck hepatitis B virus infection. *Acta Pharmacologica Sinica*, 2005. 26(5): 587–592.
Wei, L., et al., Increase in the free radical scavenging capability of bitter gourd by a heat-drying process. *Food & Function*, 2013. 4(12): 1850–1855.
Wei, P.-H., et al., Effect of alcohol extract of Acanthus ilicifolius L. on anti-duck hepatitis B virus and protection of liver. *Journal of Ethnopharmacology*, 2015. 160: 1–5.
Wessler, S., et al., The anti-inflammatory compound curcumin inhibits Neisseria gonorrhoeae-induced NF-kappaB signaling, release of pro-inflammatory cytokines/chemokines and attenuates adhesion in late infection. *Biological Chemistry*, 2005. 386(5): 481–490.
WHO. *Global hepatitis report, 2017.* 2017 [cited 2022 22 June]; Available from: https://apps.who.int/iris/bitstream/handle/10665/255016/9789241565455-eng.pdf?sequence=1;.
WHO. *Global progress report on HIV, viral hepatitis and sexually transmitted infections, 2021.* 2021 [cited 2022 19 June]; Available from: https://www.who.int/publications/i/item/9789240027077.
WHO. *WHO estimates of the prevalence and incidence of HCV infection by WHO regions, 2015.* 2017 [cited 2022 21 June]; Available from: https://apps.who.int/iris/bitstream/handle/10665/277005/WHO-CDS-HIV-18.46-eng.pdf.
Win, N.N., et al., Inhibitory effect of Japanese rice-koji miso extracts on hepatitis A virus replication in association with the elevation of glucose-regulated protein 78 expression. *International Journal of Medical Sciences*, 2018. 15(11): 1153–1159.
Wu, L.-L., et al., In vivo and in vitro antiviral activity of hyperoside extracted from Abelmoschus manihot (L) medik. *Acta Pharmacologica Sinica*, 2007. 28(3): 404–409.
Wu, Y., et al., Extract from Phyllanthus urinaria L. inhibits hepatitis B virus replication and expression in hepatitis B virus transfection model in vitro. *Chinese Journal of Integrative Medicine*, 2015. 21(12): 938–943.
Xia, J., et al., Swine and rabbits are the main reservoirs of hepatitis E virus in China: detection of HEV RNA in feces of farmed and wild animals. *Archives of Virology*, 2015. 160(11): 2791–2798.
Xu, H.B., et al., Bioactivity-guided isolation of anti-hepatitis B virus active sesquiterpenoids from the traditional Chinese medicine: Rhizomes of Cyperus rotundus. *Journal of Ethnopharmacology*, 2015. 171: 131–140.
Yang, X., H. Jiang, and Y. Shi, Upregulation of heme oxygenase-1 expression by curcumin conferring protection from hydrogen peroxide-induced apoptosis in H9c2 cardiomyoblasts. *Cell & Bioscience*, 2017. 7: 20.
Yang, Y., et al., In vitro inhibition effects of hepatitis B virus by dandelion and taraxasterol. *Infectious Agents and Cancer*, 2020. 15(1): 44.
Yuen, M.-F., et al., Hepatitis B virus infection. *Nature Reviews Disease Primers*, 2018. 4(1): 1–20.
Yusof, Y., et al., Chemopreventive efficacy of ginger (Zingiber officinale) in ethionine induced rat hepatocarcinogenesis. *African Journal of Traditional, Complementary and Alternative Medicines*, 2009. 6(1).

Zhang, H.-L., et al., Evaluation of hepatocyteprotective and anti-hepatitis B virus properties of Cichoric acid from Cichorium intybus leaves in cell culture. *Biological and Pharmaceutical Bulletin*, 2014. 37(7): 1214–1220.

Zhang, L., et al., Isolation, identification, quantification and inhibitory activity on HCV protease of coumarins from Viola yedoensis. *Canadian Chemical Transactions*, 2013. 1: 157–164.

Zhu, Z., et al., Heme oxygenase-1 suppresses hepatitis C virus replication and increases resistance of hepatocytes to oxidant injury. *Hepatology*, 2008. 48(5): 1430–1439.

Zuo, G., et al., Activity of compounds from Chinese herbal medicine Rhodiola kirilowii (Regel) Maxim against HCV NS3 serine protease. *Antiviral Research*, 2007. 76(1): 86–92.

Zuo, G.Y., et al., In vitro anti-HCV activities of Saxifraga melanocentra and its related polyphenolic compounds. *Antiviral Chemistry & Chemotherapy*, 2005. 16(6): 393–398.

# 8 HIV/AIDS and Medicinal Plants/Herbs

*Srilekha Das Nag, Chaitali Mallick, and Nandan Sarkar*
School of Medical Sciences, Adamas University

## 8.1 INTRODUCTION

AIDS, commonly known as acquired immune deficiency syndrome, is primarily caused by human immunodeficiency virus (HIV). According to the World Health Organization (WHO), it is one of the most serious and life-threatening conditions affecting millions (Worku et al., 2020). It is estimated that around 75 million people across the globe is affected by HIV, with roughly around 37 million people living along with the infection. According to the data of 2016, it was depicted that approximately 2 million people were affected by HIV, resulting in the death of 1 million people. However, the number of cases has fortunately reduced since 2000. This is due to the decrease in the rate of mortality and infected patients by 50% and 35%, respectively. The deduction in the rate of HIV infection is mainly believed to be due to the factors like use of condoms to deduce the prevalence of STD (sexually transmitted disease), enhanced use of therapies such as three-drug antiretroviral therapy (ART). According to an estimation, the total number of patients receiving the ART was approximately found to be 685,000 and 20.9 million in the year 2000 and 2017, respectively.

Patients affected by AIDS may develop a remarkable risk of contracting a wide range of deadly infection and cancer. The developing cause of cancer is mainly because of the destruction of immune system. It is also found that the lifespan of those effected patients could reduce by 9–11 years, depending on the subtypes of HIV (Dhonde, 2022). Vital immune T helper cells, specifically, $CD4^+$ T cells, dendritic and MAC-like cells, and macrophages cells are the main targets of HIV infection. Following infection, these immune cells are destroyed through a variety of mechanisms, particularly in the case of $CD4^+$ T cells. These mechanisms included T helper cell pyroptosis, apoptosis (Lindegaard et al., 2004) in uninfected bystander cells, killing of cells with direct viral infection, and the destructive role played by $CD8^+$ cytotoxic lymphocytes on recognising the infected $CD4^+$ T cells. Lower levels of $CD4^+$ T cells result in immunosuppression due to a loss of cell-mediated immunity and make HIV patients more vulnerable to a variety of opportunistic infections, ultimately leading to the development of AIDS (Chinsembu et al., 2015). The mortality rate could be reduced in immunocompromised patient by early diagnosis; however, due to the lack of effective public health policies, therapeutic adhesion, and psychosocial support to patients, the disease remains underdiagnosed (Villela and Barbosa, 2017). The implementation of pharmacotherapy could reduce the number of morbidity and mortality in patients affected by AIDS. Discovery of zidovudine and protease inhibitor could effectively act against HIV, thereby overall reducing the infection level and aiding people to live a better life (Bonolo et al., 2007).

However, antiretroviral drugs have certain disadvantages due to several side effects. It includes lipodystrophy, thus leading to loss of peripheral fat and accumulation of central fat causing thin facial pads, thinner legs and arms, humps resembling buffaloes and potted bellies. This makes the patient stigmatised. The other disadvantages of ART include shortage in availability, resistance, toxicity, and low therapeutic effect. Considering the overall limitations, the potent alternative anti-HIV drugs are to be developed. According to various scientist, medicinal plants could be chosen as an alternative because of its effective antiviral potentiality, some of which are *Rheum palmatum* L.,

DOI: 10.1201/9781003329169-10

*Rheum officinal*, *Trigonostemagyphiophiloides*, *Vetica astrotricha*, *Vernonia amygdalina*, *Hypoxius pelargonium*, *Sidoides hemerocallidea* and *Sutherlandia frutescens*. Several phytoconstituents can be isolated from these medicinal plants such as alkaloids, flavonoids, tannins, saponins, phenolic compounds, and glycosides, which possess the potentiality to boost immunity and deduce the further replication of HIV. Different viruses may have different molecular mechanisms linked to plant extracts' antiviral effects. However, the ability of plant extract to strengthen the body's natural ability to fight off viruses may use common immune system pathways. The immunostimulatory effects of plant extracts with antiviral properties have been the subject of several studies in recent years (Webster et al., 2006).

In the macrophage activation assay, the root extracts of the medicinal plant *Heracleum maximum* Bartr. (Umbelliferae), which also has antifungal and antibacterial effects, stimulated the production of Interleukin 6 (IL-6), confirming the association between the antiviral effects and the immunostimulatory properties (Webster et al., 2006). Furthermore, *Plantago major* Linn. and *P. asiatica* Linn. (Plantaginaceae), two plants that are frequently used in Taiwanese folk medicine to treat infectious diseases, showed lymphocyte proliferation and low-level interferon-gamma (IFN-) secretion. Indicators of cell-mediated immune response modulation include interferon-gamma (IFN-) secretion and lymphocyte proliferation activity (Chiang et al., 2003). In a similar vein, Sambucol, a product isolated from *Sambucus nigra L.*, which is effective against a variety of influenza strains, has demonstrated to enhance immune responses by secreting inflammatory cytokines (IL-1 beta, TNF-alpha, IL-6, and IL-8) (Barak et al., 2012). Another intriguing finding is that plant extracts have broad-spectrum antiviral properties in addition to their immunomodulatory effects. This might relate to one phytochemical or a variety of different plant compounds. Among many of these observations, a few include an extract from the *Trifollium* species, which demonstrated antiviral activity against several contagious viruses, including the human papillomavirus, HIV, influenza, Marburg, HBV, and HCV (Ooi et al., 2004). Additionally, a lectin called pandanin was discovered in a saline extract of the *Pandanus amaryllifolius Roxb* leaves, exhibiting antiviral activity against HSV-1 and the H1N1 influenza virus (Gonzalez et al., 2019). Using the aforementioned premises as a foundation, this chapter aims to evaluate the use of medicinal plants, phytotherapy, and phytomedicines in HIV patients, including their use in conjunction with traditional antiretroviral medications (Klos et al., 2009).

## 8.2 HIV LIFE CYCLE

When the human immunodeficiency virus enters the body, it specifically targets important immune cells, such as T helper cells (more specifically, $CD4^+$ T cells), dendritic, and macrophage cells (Kanta et al., 2011). The viral envelope glycoprotein fuses the virus with the targeted cell membrane. The viral entry into the host cell, where $CD4^+$ and gp120 interaction occurs, is triggered by $CD4^+$ T cells, which also cause Ip120 changes (Miceli et al., 1993). Following the formation of the ternary CD4-coreceptor gp120 complex, the membrane fusion is confirmed by changes in the conformation of gp41. After the engagement of the receptors and $CD4^+$, fusion occurs at the low pH of the endosome compartment. The lipids that make up the targeted cells are crucial for the fusion of membranes. Lipid rafts, which are plasma membrane microdomains, are responsible for the fusion of the plasma membrane. Cholesterol and glycosphingolipid make up these microdomains (Alkhatib and Berger, 2007) (Figure 8.1).

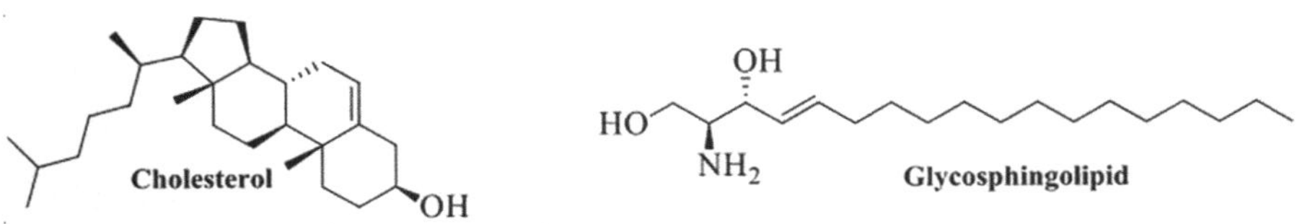


**FIGURE 8.1** Structures of plasma membrane microdomains responsible for the fusion of the plasma membrane.

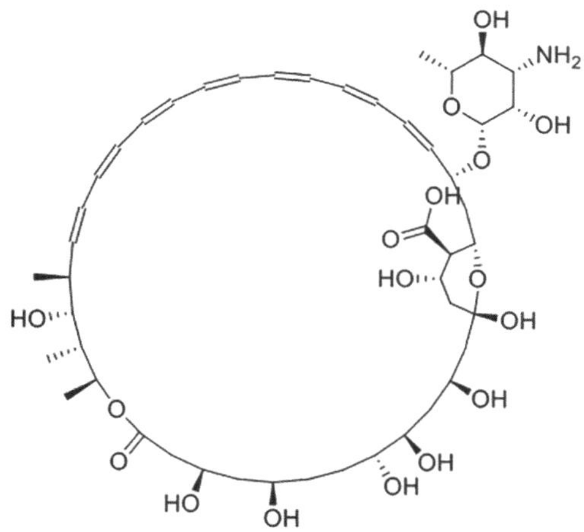


**FIGURE 8.2** Amphotericin B; a methyl ester derivative is crucial in preventing HIV fusion.

The fusion of the membranes may be interfered with by cholesterol particles in the target cell or by HIV. Amphotericin B methyl esters are crucial in preventing HIV fusion. Reverse transcriptase is responsible for transforming single-stranded HIV RNA into double-stranded HIV DNA(Sarafianos et al., 2009). The ribonuclear protein complex made up of NC, CA, RT, VPr, and IN, enters the cytosol and fuses there to form the reverse transcription complex (Figure 8.2).

RNA binds to the 5′ end of the DIS (dimerization initiation sequence). Retrovirus carries two strands of retroviral RNA, often referred to as pseudodiploid. Retroviruses use tRNA during the process of DNA synthesis. As soon as the virus enters the target cell, the retroviral ribonuclease H (retroviral RNase H), which is a catalytic domain of the retroviral reverse transcriptase (RT) enzyme, begins the process of converting RNA into DNA. The retroviral RNA genome is converted into complementary DNA (cDNA) using the RT enzyme (Garzino-Demo, 2007). DNA synthesis for a single strand takes place at the 3′-OH end of tRNA. DNA synthesis then continues in the direction of the 5′ end. The 3′ and 5′ ends of the minus strand contain a virtue of true repeated region, which is transferred for hybridization. As the RNA ScH mediator degrades, an RNA–DNA hybrid forms and moves toward the 5′ end of the genome. The residual, hybridised RNA fragments can be used for priming and strand synthesis in addition to being used as the positive strand (PPT). The plus strand produces the tRNA after RNase H is eliminated. As a result, the plus strand can hybridise with the 3′ end of its homologues (Ho, 1997). The RT synthesises additional templates from the two strands of RNA using one template. Every infection can cause this up to 30 times. Jumps between molecules cause deletion, duplication, and insertion mutations. Recombinants are produced by intermolecular jumps when two RNA packages are not identical. With the help of this recombinant, HIV can quickly develop drug resistance. A non-infectious virus-like particle is created from the precursors of the retroviral gag protein, which are synthesised by polyproteins like PR55. Enzymes encoded by pol, viral genomic RNA, and enveloped glycoproteins are not needed to form non-infectious particles (Gouws et al., 2006). Retroviral Gag proteins are in charge of locating the developing particles exported from the cell and forming the structural virion (Fanales-Belasio et al., 2010).

## 8.3 PATHOGENICITY OF HIV INFECTION; ENTRY OF HIV

Retroviruses lack deoxyribonucleic acid and are unable to replicate outside of living host cells (DNA). The host cellular environment, the viral life cycle, and the number of viruses present in the infected person all play a role in the pathogenesis of HIV infection. Once inside the body, the viral particle is drawn to cells that have the right CD4 receptor molecules, where it enters the cell by fusing or endocytosis with a cell membrane that is receptive to it. The likelihood of infection depends on both the quantity of infectious HIV virions present in the bodily fluid contacting the host and the availability of cells equipped with the necessary CD4 receptors (Sundquist and Kräusslich, 2012). Even in the absence of mucosal disruption, HIV infection can potentially spread through the oropharyngeal, cervical, vaginal, and gastrointestinal mucosal surfaces. HIV that is cell-free or associated with cells can enter the mucosal epithelium through microscopic tears brought on by sex, through paracellular passage following an epithelial disruption brought on by HIV, or by transcytosis (the movement of cells across epithelial cells). During contact between HIV-infected cells and the apical surface of an epithelial cell, HIV can transcytoses through a strong epithelial barrier. Mucus on epithelial surfaces prevents viruses from entering epithelial surfaces more effectively, especially in the endocervix, where there is only one layer of columnar epithelial cells. HIV entry may be influenced by immune cells expressing viral receptors and co-receptors, a local inflammatory environment, mucosal antiviral factors, HIV-cell interactions, hormonal levels, the make-up of the commensal microbiota, and pathogenic co-infections like STDs. Beyond the epithelial barrier, dendritic cells, Langerhans cells, gut microfold cells, and even fibroblasts in the submucosal space may transmit viral particles to $CD4^+$ T cells, primarily through trans-infection. Since dendritic cells can bind to gp120 via a C type lectin, it is possible that they will capture HIV and deliver it to underlying T cells, causing it to spread to lymphoid organs when they squeeze between "tight" epithelium (Keele and Estes, 2011; Gonzalez et al., 2019).

HIV and cells are oversized and not scaled in the above illustration. An epithelial dendritic cell, also known as a Langerhans cell, is the red cell with long processes in the illustration that can catch viruses that cross the epithelium. Viruses can also come into contact with and infect a CD4 lymphocyte. HIV-positive cells have the ability to enter lymphatics, where they are transported to local lymph nodes and systemic lymphoid tissues. In these locations, additional cells become infected with the virus, leading to productive HIV replication (Fanales-Belasio et al., 2010). Several host defence mechanisms contribute to the effectiveness of viral transmission and infection establishment. Lack of expression of the viral coreceptor CCR5, increased production of the chemokines like MIP-1/, RANTES, or SDF-1, and high expression of factors like SLPI, defensins, cathelicidin, TRIM5, APOBEC-3G, SAMHD-1, serpinal, and elafin are examples of chemical factors that can reduce transmission. Variations in the activity of natural killer cells, dendritic cells, $CD4^+$ cells, and $CD8^+$ cells are among the cellular defence factors. IgA antibodies that neutralise mucosal cells might be important. Despite numerous HIV exposures, some people do not contract the virus. These people, known as HIV-exposed seronegative individuals (HESNs), probably express more of these molecular barriers at their mucosal cellular interface, preventing productive HIV infection. A loss or disruption of tight junctions leads to the formation of gaps at the epithelial monolayers through which a virus may enter the submucosa, resulting in epithelial paracellular passage through mucosa. Occludin, claudin, and ZO-1 expression can be decreased by HIV gp120 binding to the CCR5 or CXCR4 co-receptors, or to galactosylceramide expressed on epithelial cells (Shaw et al., 2012). Transport through intracellular vesicles and endosomes occurs during epithelial transcytosis. Intestinal, vaginal, oral, and endometrial epithelial cells have all been found to engage in this process.

Viral particles are transported into the intracellular compartment during this process after binding to molecules on the surface of epithelial cells, such as heparan sulphate proteoglycans. Following translocation into the intracellular space and release into the external basal space, viral particles continue to be contagious and can either infect intraepithelial lymphocytes or be taken up and transmitted by antigen-presenting cells. However, transcytosis is extremely inefficient with an efficiency of only 0.01% of the initial viral inoculum. Thus, a healthy epithelium serves as a significant barrier against HIV infection, but the presence of inflammatory cells and antigen-processing cells facilitates HIV transmission (Shaw et al., 2012). Mucus on the skin's surface can trap and then release viruses. This efficient trapping is supported by the cervicovaginal mucus' acidic pH. Semen may raise the pH of the vagina. Due to the normal microbiota's dominant Lactobacillus species

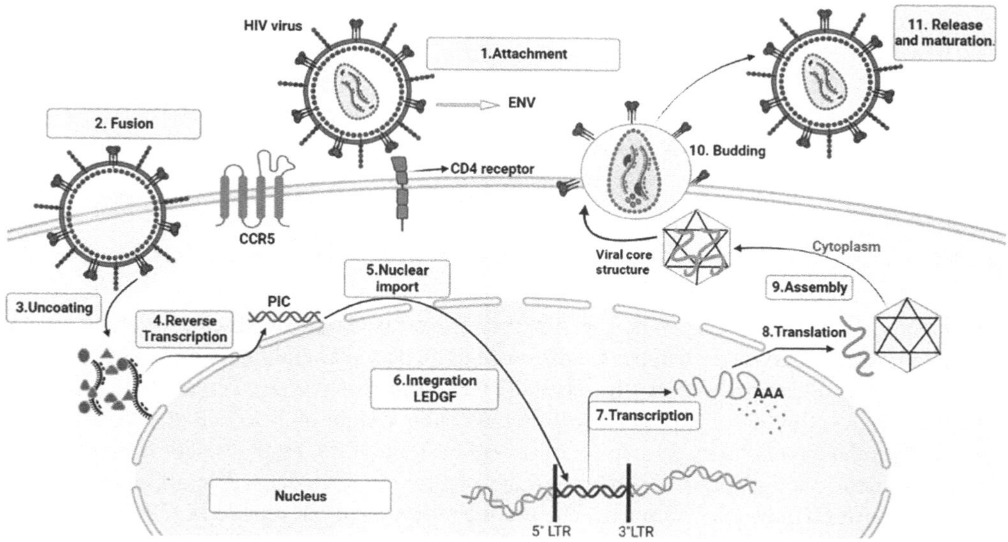


**FIGURE 8.3** Illustration of the HIV pathogenicity.

and tendency to promote low pH, the vaginal microbiome is involved. On the other hand, bacterial vaginosis caused by *Gardnerella*, *Prevotella*, and *Mobiluncus* spp. encourages higher pH and pro-inflammatory cytokines (Nazli et al., 2019). The production of pro-inflammatory cytokines, such as tumour necrosis factor (TNF), by genital epithelial cells can be increased in response to HIV-1 exposure. TNF weakens the cellular tight junction barrier, allowing HIV-1 and luminal bacteria to cross the epithelium (Lucas and Nelson, 2015). Following parenteral transmission of the virus as well as during HIV viremia after infection, the endothelium may also hold HIV virions. Virus gp120 binds to the syndecans' heparan sulphate chains to mediate the adsorption of HIV by endothelial cells. Syndecan increases infectivity and keeps virus infectivity for a week, even though it does not replace HIV entry receptors; unbound virus loses its infectivity in less than a day. Additionally, selectin binding can speed up the transfer of HIV to CD4 cells and improve virion attachment to endothelial cells (Gallay, 2004) (Figure 8.3).

### 8.3.1 Entry Mechanism for HIV

HIV is an enclosed virus that enters the host cell by fusing with the viral membrane. The encapsulating glycoprotein is the HIV fusion protein. The envelope (Env) gene codes for a polypeptide (gp160), which is cleaved into two fragments, a binding fragment (gp120) and a fusion fragment (gp41) during its transit through the Golgi apparatus of the host cell. A spike-like viral surface is created by three copies of each component. Gp120's interaction with the host T-lymphocyte CD4 receptor triggers a conformational shift in gp120 that exposes coreceptor binding sites, which is how entry is initiated. To enable the activation of these T-lymphocyte helper cells, CD4 interacts with MHC II molecules on the surface of antigen-presenting cells. HIV uses a CD4 receptor to enter the body (Chen, 2019). A coreceptor called a chemokine receptor, in addition to the CD4 receptor, facilitates HIV entrance into cells. Chemokine receptors are fusion-mediating molecules located on cell surfaces that are membrane-bound. These receptors belong to a family of roughly 50 tiny proteins known as chemokines, which are produced in tissues as part of the body's rapid reaction to tissue injury. About 20 G protein-coupled receptors (GPCRs), known as chemokine receptors, are a complimentary family found in the

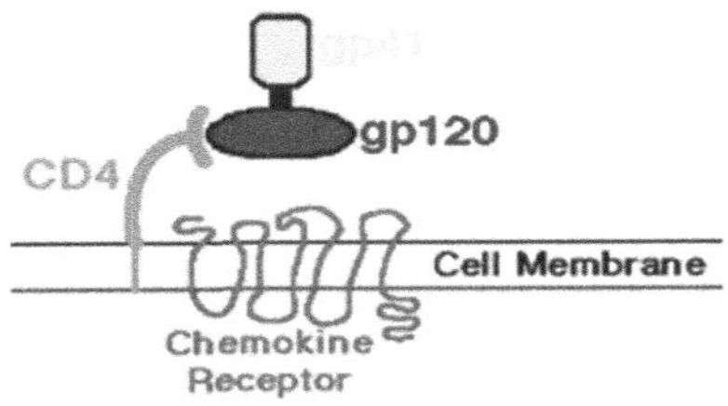


**FIGURE 8.4** Depiction of the connection between the CD4 receptor and the chemokine receptor.

plasma membranes of leukocytes. When chemokines are released into the vasculature, their receptors are activated, which encourages leukocyte adherence to the vascular endothelium, remodelling of the actin cytoskeleton, and migration into the tissues (Bhusal and Foster, 2020) (Figure 8.4).

Chemokines are tiny, soluble proteins with a molecular weight of 10 kDa that are characterised by two disulphides bonds formed by four conserved cysteine (Cys) residues. They are categorised according to their spacing and the existence of the first two Cys residues. According to the type of chemokines to which they react primarily, chemokine receptors are designated CCR1-10, CXCR1-8, CX3CR1, and XCR1. Their presence on cells may facilitate the binding of HIV envelope glycoproteins. A fully infectious virion has 10–14 spikes per particle, and three gp120–gp41 heterodimers can serve as a trimer spike. To enter target cells, HIV-1 must first attach to the host receptors CD4 and CCR5/CXCR4. This attachment sets off a series of conformational changes in the envelope glycoprotein trimer that fuse the viral and cell membranes by gp120, thus facilitating infection (Chen, 2019). The gp120 component undergoes conformational changes during the first binding of HIV to the CD4 receptor, but these modifications are insufficient to cause fusion. By causing structural changes and activating gp41's fusogenic ability, the chemokine receptors facilitate membrane fusion, enabling HIV to enter the body. By attaching to a cell surface receptor, chemokines cause the trafficking of immune cells and can cause inflammation. CXCR4 and CCR5 are two crucial G protein-coupled chemokine receptors for HIV transmission, with CCR5 serving as the main receptor for initial HIV entrance. Aside from the lymphoid organs' epithelium, endothelium, vascular smooth muscles, fibroblasts, and microglia, CCR5 is also expressed on lymphocytes, monocytes/macrophages, granulocytes, natural killer (NK) cells, and regulatory T (Treg) cells.

The fact that mononuclear cells have higher levels of CCR5 expression in chronically inflamed tissues implies that these cells are attracted to the areas of inflammation. The virus is anchored close to the cell surface by CCR5, which stabilises the shape of the Env protein caused by CD4s (Jasinska et al., 2022). The CCR5-Δ32 variation has a 32-bp deletion (Δ32), rendering the HIV entrance coreceptor inoperable and inhibiting fusion of HIV with the target-cell membrane. Most often seen in European populations, the CCR5–32 allele has an average frequency of 10% and homozygosity of 1%. Sub-Saharan Africans, Asians, and Native Americans seldom carry this gene. The level of protection against HIV-1 infection for those who are homozygous for this allele is almost 100% (Jasinska et al., 2022). Cytoskeletal actin protects cells like CD4 lymphocytes against pathogen invasion, but when HIV gp120 binds to CCR5 or CXCR4, signalling is activated, rearranging the actin to allow for viral entry. CXCR4 is produced more abundantly as the HIV infection develops. CXCR4 may be crucial for HIV entrance into dormant memory CD4 cells, which develop latency with persistent infection (Wang et al., 2017). Env interaction to cellular CD4 and chemokine receptors causes conformational changes in the envelope (Env) glycoprotein, which in turn mediates HIV entrance into host cells. Whether the host cell's CCR5 or CXCR4 chemokine co-receptors will be activated is determined by the V3 loop region of gp120. This variable area causes the host cell membrane to fuse to allow viral penetration. The V3 loop's crown closely resembles the beta2-beta3 loop in the CXC and CC chemokines, which are the actual ligands of CXCR4 and CCR5 (Cardozo et al., 2007).

The HR1 and HR2 helical sections and a fusion peptide make up the gp41 transmembrane coreceptor. Following activation of the chemokine coreceptor, the gp41 on the surface of HIV changes shape by opening the gp120 surface subunit, exposing the fusion peptide in the gp41 transmembrane subunit, and folding the gp41 heptad repeat regions HR1 and HR2. This conformational shift in gp41 caused by the interaction of HR1 and HR2 results in the creation of a stable structure that permits the fusing of the membranes of HIV and the host cell, with a fusion pore via which the viral core enters the host cell (Keller et al., 2018). Most HIV isolates discovered in samples taken during acute and chronic infections use the CCR5 coreceptor and are T cell-tropic (R5 T-tropic), adapting to and replicating in $CD4^+$ memory T cells. However, they require large densities of the CD4 receptor present on $CD4^+$ T cells for effective entrance and use the CCR5 coreceptor, which is most common on the memory fraction of $CD4^+$ T cells. A viral population evolves the capacity to utilise CXCR4 as a coreceptor in around 50% of late-stage HIV-1 infections. These CXCR4-using T cell-tropic (X4 T-tropic) viruses employ CXCR4 as their target because $CD4^+$ naïve T cells display greater concentrations of CXCR4 than $CD4^+$ memory T cells do and lower densities of CCR5 (Brenchley et al., 2004). In comparison to $CD4^+$ T cells, macrophages express less CD4. However, X4 M-tropic viruses have also been discovered.

The majority of macrophage-tropic (M-tropic) HIV variations employ the CCR5 coreceptor (R5 M-tropic). There are HIV strains known as dual tropic that may utilise several chemokine co-receptors (Arrildt et al., 2015). The dominance of CCR5-tropic virus strains, followed by CCR5/CXCR4 dual-tropic virus, and then the more cytopathic CXCR4-tropic strain predominance, may indicate that HIV mutations over time have increased the virus' capacity to infect host cells via these routes. Early in HIV infection, the CCR5 tropic virus predominates because it infects dendritic cells and macrophages more easily, replicates quickly, and is less visible to cytotoxic lymphocytes (Levy, 2009). As most CD4 cells express CCR5 in the gastrointestinal tract, so it is a preferred location for HIV infection (Lackner et al., 2009). Using these receptors to gain entry, HIV primarily infects cells that have CD4 cell surface receptor molecules. Although many different cell types share similar receptor epitopes, CD4 lymphocytes are crucial in this process.

The mononuclear phagocyte system, primarily blood monocytes and tissue macrophages, as well as T lymphocytes, NK lymphocytes, dendritic cells (epithelial Langerhans cells and follicular dendritic cells in lymph nodes), haematopoietic stromal cells, and microglial cells in the brain are examples of cells with CD4 receptors that are susceptible to HIV infection. As a mucosal epithelial receptor for gp41 on HIV, galactosyl ceramide is expressed by immature dendritic cells derived from human monocytes, as well as by dendritic cells isolated from blood and mucosal tissue and in situ on mucosal tissue (Sierra et al., 2005; Stebbing et al., 2004; Magérus-Chatinet et al., 2007). HIV can enter cells without interacting with CD4 receptors. This type of entry is less thorough and effective. Such entry has been described for renal tubules, gut enterocytes, vascular endothelium, cardiac myocytes, and astrocytes. HIV-related illnesses that develop in tissue sites where these cells are present may be influenced by the pathogenesis of their infection.

## 8.4 DIAGNOSIS OF HIV INFECTION

HIV infection is characterized by an acute symptomatic period that includes fever, general malaise, lymphadenopathy, rash, and myalgias, but serious side effects like meningitis have also been reported. HIV virus detection in the blood is typically measured as viral RNA load (CDC, 2019). The highest plasma levels of HIV RNA are present during the acute infection period, and the viral load level is inversely correlated with the severity of symptoms. It has been proposed that viral properties and viral load control viral replication and pathogenesis. Therefore, in addition to the host, the viral genotype also affects the clinical outcomes and the course of the disease (CDC, 2019). HIV establishes a dormant or latent infection within memory $CD4^+$ T cells, which have a stem-cell-like capacity for self-renewal, making it challenging to completely eradicate. If that cell is alive, the HIV virus can repeatedly start replication once the virus' DNA has been incorporated into the host chromatin. While ART can control the infection from spreading to new cells, it cannot eliminate the infection once the DNA has successfully entered the target cell. Due to the limited penetration of antiretroviral medications and the host's inability to eliminate the virus, the lymph nodes act as a reservoir for the virus when treatment is stopped or interrupted. Prior to the viral reservoir depleting, it has been suggested that ART therapy may be required for several decades.

## 8.5 TREATMENT OF HIV/AIDS

Even though HIV was first identified in the early 1980s, there is still no effective treatment or vaccine to prevent HIV infection (Anderson et al., 2014). However, there have been significant improvements in treatment, control, and prevention. The morbidity and mortality of HIV/AIDS were significantly decreased in 1996 with the introduction of antiretroviral drugs and highly active antiretroviral therapy (HAART). Currently, ART is advised for all HIV-positive adults. Abacavir with lamivudine or tenofovir disoproxil fumarate with emtricitabine are two Nucleoside Reverse Transcriptase Inhibitors (NRTIs) that are advised for initial regimens, along with an integrase strand transfer inhibitor like dolutegravir, elvitegravir, or raltegravir; a nonnucleoside RT inhibitor (efavirenz or rilpivirine); or a boosted (darunavir or atazanavir) (Maartens et al., 2014) (Figure 8.5).

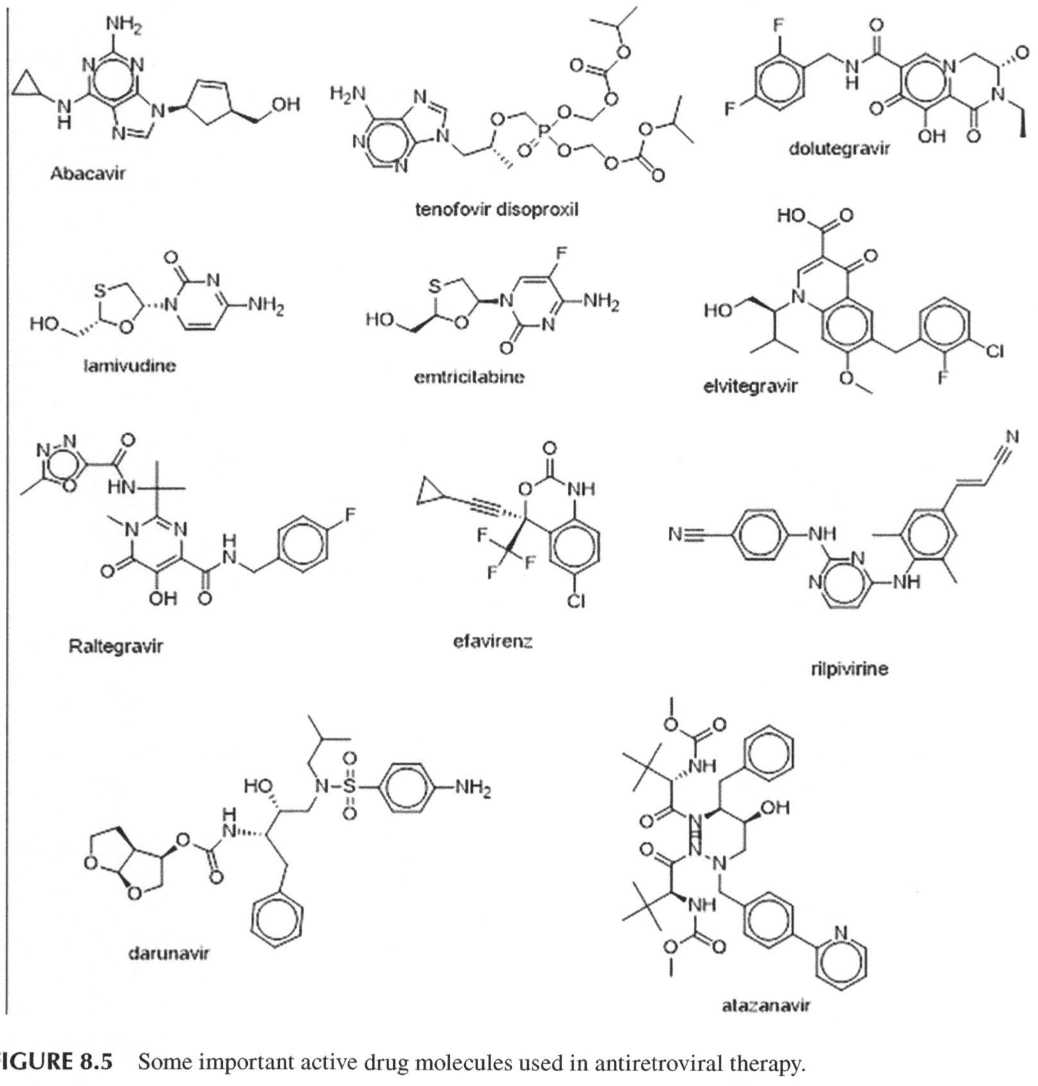


**FIGURE 8.5** Some important active drug molecules used in antiretroviral therapy.

Although NRTI-sparing strategies may be taken into consideration, protease inhibitor monotherapy is typically not advised. When a treatment is suspected to be failing, it must be confirmed quickly, resistance testing must be done while the patient is still on the failing regimen, and the causes of the failure must be assessed before switching the patient's course of treatment is even considered. To avoid jeopardising antiretroviral potency, therapeutic regimen changes brought on by side effects, convenience, or cost savings should be carefully considered. Aside from voluntary male medical circumcision (Saag et al., 2020) and antiretrovirals for the prevention of mother-to-child transmission (Gravatt et al., 2017).

Research on HIV vaccines and antimicrobial agents is still ongoing. However, these measures are not the only significant advancements in HIV prevention. The disease has become a chronic, controllable condition because of the decrease in morbidity and mortality. Intriguingly, the increased survival rate has led to an ageing HIV/AIDS population, which has presented a whole new set of issues. These issues include a higher prevalence of chronic diseases in this population, such as cardiovascular and pulmonary diseases, malignancies, and even a specific set of comorbidities, which are now designated as HIV-associated non-AIDS (HANA) conditions. HIV treatment and prevention still rely heavily on antiretroviral drugs. ART is currently advised for all HIV-infected patients with detectable virus, regardless of their CD4 cell count, to begin as soon as possible after diagnosis. This will help to prevent disease progression, improve clinical outcomes, such as lowering the incidence of AIDS-related events and non-AIDS-related events, as well as all-cause mortality and transmission. All HIV-infected people with detectable plasma virus should receive treatment with recommended initial regimens consisting of an integrase strand transfer inhibitor (InSTI) plus two NRTIs. These recommendations are supported by large randomised controlled clinical trials (NRTIs). Antiretroviral medications suppress HIV and stop the spread of new HIV infections when used properly with these treatment plans. It has been hypothesised that adult HIV-positive patients' survival rates may approach those of adult HIV-negative patients (Keele and Estes, 2011). Drugs with novel mechanisms of action, such as histone deacetylase (HDAC) inhibitors, gene therapies, broadly neutralising anti-HIV antibodies, immune modulation, and drugs with novel mechanisms to block HIV entry, have been the subject of a recent review of HIV therapies in phase 2 clinical trials (Gonzalez et al., 2019). The new treatments are being developed as both supplements to the current combination ART and as agents to be used when the course of treatment is interrupted. Various degrees of success were seen in the early trials for the drugs that are currently being developed. Each of these novel medicines has the potential to close a gap in the current ART regimens, which will ultimately improve the prognosis for HIV-positive people.

## 8.6 CURRENT ONGOING TREATMENT TO COMBAT HIV/AIDS

Following the start of treatment, the HIV RNA level must be pragmatically checked every 4–6 weeks or changed as little as possible, usually aiming for a level below 20–50 copies/mL. Within 24 weeks of starting ART, while the infection is still acute, the viral particles are suppressed. Once the virus has been suppressed, the presence of HIV RNA will be checked every 3 months for the first year, and then at least every 6 months for devoted patients who remain clinically protected. Except in very specific circumstances, monitoring therapeutics is not advised (Bean, 2005). A CD4 cell count is used to determine whether Opportunistic Infection (OI) prophylaxis is necessary. Re-examination is optional every 3–4 months if the pre-treatment CD4 cell count is less than 200/μL until the HIV RNA is consistently suppressed and the CD4 cell count has been over 350/μL for a year. After that, CD4 cell counts will be checked every 6 months until the virus has been dormant for at least 2 years and the CD4 cell count is consistently steady at 500/μL. As a result, going over the track is not required if virologic collapse, concurrent immune-suppressive events, or immune-suppressive therapies are started.

### 8.6.1 Monitoring of Safety

The safety process consists of renal, hepatic, and fasting lipid tasks, which should be tailored based on age, comorbid conditions, and concurrent medications. Presided over with guidelines, local prevalence, and patient threat, seclusion for sexually transmitted infections would be trustworthy.

### 8.6.2 Novel Therapeutical Study against HIV

In a recent review of HIV therapeutics, phase 2 clinical trials of therapeutic drugs to prevent HIV entry through novel mechanisms of action, histone deacetylase (HDAC) inhibitors, and gene therapies to promote deeply neutralising anti-HIV antibodies were described (Sharifi-Rad, 2016). According to a press release from Merk and Co., the US Food and Drug Administration (FDA) has approved Pifeltro and Delstrigo as two new oral treatments for adults with HIV-1 infection. Recently, the USFDA approved Dovato, the first full two-drug HIV treatment regimen, for patients who had not previously been on ART. This medication aids the infected people in removing any additional toxicity and possible drug interactions from a third medication in a single tablet. HIV that is resistant to drugs may initially infect a person or may start to become resistant after taking HIV medications. Additionally, this drug-resistant form of HIV can pass from person to person. The selection of HIV medications to be included in an HIV treatment regimen is aided by the screening of a person's drug-resistance strain. It is important to take HIV medication as directed each day to avoid drug resistance and maintain viral load without running the risk of sexually transmitting HIV to HIV-negative partners (Lucera et al., 2014).

## 8.7 NATURAL PRODUCTS AND MEDICINAL PLANTS: GAME CHANGERS IN THE TREATMENT OF HIV/AIDS

Despite several advantages of ART, some serious side effects are especially noticeable in patients receiving long-term treatment. This includes the development of multidrug resistance which limits the effectiveness of current therapies, thus necessitating the development of new drugs and novel targets to address the problem of the body's HIV reservoirs and to achieve the goal of the total eradication of HIV and AIDS (WHO, 2001). HIV-1 eradication continues to be hampered by latently infected cells. A number of medications that can selectively reactivate latent proviruses without inducing polyclonal T-cell activation have been discovered over the past 10 years as a result of the molecular mechanism by which HIV latency persists (Shaw et al., 2012). It is interesting to note that vorinostat and other histone deacetylase (HDAC) inhibitors can also activate latently infected cells in HIV transcription. According to research, vorinostat increases $CD4^+$ T-cell susceptibility to HIV infection in a dose- and time-dependent manner. It does not promote viral fusion with cells, but it does promote reverse transcription, nuclear import, and integration, as well as viral production in a spreading-infection assay. HDAC inhibitors, in particular vorinostat, are currently being studied in clinical trials as part of a "shock-and-kill" approach to eliminate HIV latent reservoirs (Shaw and Hunter, 2012). The WHO has recommended that ethnomedicines and other natural products should be systematically tested against HIV as they may yield efficient and more cost-effective therapeutic agents (World Health Organization (Nazli et al., 2019). Multiple factors necessitate further investigation and identification of natural remedies for preventing and treating infections related to HIV. First, past efforts to discover a reliable treatment for HIV infection have yielded disappointing results. Second, creating an HIV vaccine seems like an unattainable goal. HAART, which combines nucleoside or nonnucleoside RT inhibitors with protease inhibitors, is the most effective treatment option for HIV-infected people. Only people who continuously follow a therapeutic regimen can control their HIV infection with HAART, and stopping the medication causes the re-emergence of the diseases. Additionally, based on the HIV pathogenesis, efforts have been made to target every

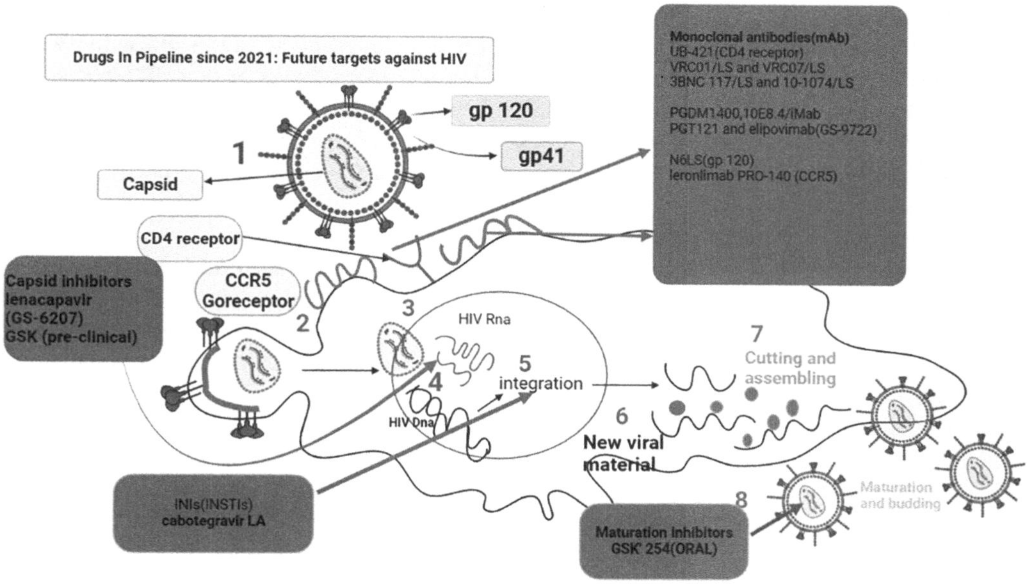


**FIGURE 8.6** HIV life cycle and efficacy of antiviral drugs.

stage of the viral life cycle, from viral entry to viral morphogenesis. Despite this, none of these approaches have produced a cure (Gallay, 2004). This can urge for an alternative requirement for the management of HIV (Figure 8.6).

From microscopic, unicellular plants to enormous, long-lived trees, the plant kingdom is incredibly diverse. It is impossible to screen every single plant or each of its individual parts (stem, leaves, roots, flowers, and bark) for the presence of antiviral components. However, the current knowledge of antiviral abilities of medicinal plants is based on opinion. Additionally, there are instances of plants with antiviral properties or products that have undergone phytochemical screening based on the characteristics of recently discovered phytocomponents. An illustration of such a substance is the 11-kDa protein known as cyanovirin N (CV-N), which was isolated from the cyanobacterium Nostoc ellipsosporum. The CV-N is one of the potentially emerging antiviral compounds (Wang et al., 2004) and exhibits a broad-spectrum antiviral activity in addition to inhibiting HIV-1 infection. The molecular analysis of CV-antiviral N's activity reveals that it interacts very specifically with the viral envelope glycoprotein gp120 (Bessong et al., 2006). The viral envelope mediates interactions with permissive cells' CD4 receptors in conjunction with chemokine receptors CXCR4, CCR5, and plays a significant role in the viral particle's ability to infect permissive cells. Additionally, a group of antiviral substances known as HIV entry inhibitors interact with either the viral envelope or the host cell receptors that are responsible for mediating viral entry.

In addition to CV-N, Baicalin, a natural substance derived from plants, interacts with chemokine receptors to prevent HIV from entering the body (Fu et al., 2010). Currently, efforts are being made to characterise the molecular basis of medicinal plants' anti-HIV activity. In a recent study, the ability of different plants and each of their component parts to inhibit viral RT and integrase—the two crucial enzymes involved in HIV infection—was compared (Chen, 2019). The alcohol-based, n-butanol fraction of the medicinal plant *Bridelia micrantha* (Hochst) demonstrated the highest anti-RT activity when compared to its aqueous and organic solvent fractions. The integration step was not inhibited by any of the six potential antiviral medicinal plants tested in this study. This study also discovered a potential anti-RT element in a medicinal plant that needs more research. Already available anti-HIV medications include RT inhibitors. In the past, several studies have linked the ability of medicinal plants to inhibit RT activity as an indirect measurement of anti-HIV effect (Kurapati et al., 2016). This is because RT plays a significant role in controlling viral infectivity and replication. Additionally, therapeutic plants have demonstrated inhibitory effects on viral protease (PR), an enzyme necessary for the proteolytic conversion of polyprotein precursor into proteins required for the construction of viral particles. Protease inhibitors are important components of the drug cocktail used in HAART, the most effective therapeutic regimen for treating HIV-infected individuals. Phytochemicals, like triterpene derivatives from various medicinal plants, have previously demonstrated inhibitory effects on PR (Deng et al., 2014). The continued development of previously characterised protease inhibitors and the discovery of novel compounds with such activity depend on these efforts. Different medicinal plants that naturally contain RT and protease inhibitors provide an additional source of untapped antiviral compounds.

It is interesting to note that a significant amount of research on investigations of natural products with activities against HIV-1 RT, HIV-1 and -2 proteases, and HIV integrases was carried out in the 1990s by Kurapati and group. It has been suggested that the natural substances such as calanolides (coumarins), ursolic and betulinic acids (triterpenes), baicalin (flavonoid), polycitone A (alkaloid), and lithospermic acid (phenolic compound) are promising candidates for anti-HIV drugs (Bhusal et al., 2020). However, the majority of these studies have been conducted *in vitro*, and not enough in *in vivo* or human studies. In order to assess the effectiveness of traditional Chinese medicines based on clinical data, a meta-analysis of 12 clinical trials involving 881 patients with AIDS was conducted (TCM).

The findings demonstrated that when compared to a placebo, TCM interventions were associated with a significantly lower plasma viral load. The findings demonstrated that when compared to a placebo, TCM interventions were associated with a significantly lower plasma viral load. According to this study, TCM treatments significantly outperformed a placebo in terms of lowering plasma viral loads and raising CD4+ T-lymphocyte counts in AIDS patients. TCM interventions were associated with improved symptoms in a greater number of patients and with fewer adverse events than conventional Western medicine, but they were significantly less effective in reducing viral load when compared to Western medicine (Jasinska et al., 2022). Therefore, natural products and conventional medicines have a lot of potential for managing HIV infections and symptoms, but *in vivo* and human studies are lacking (Figure 8.7).

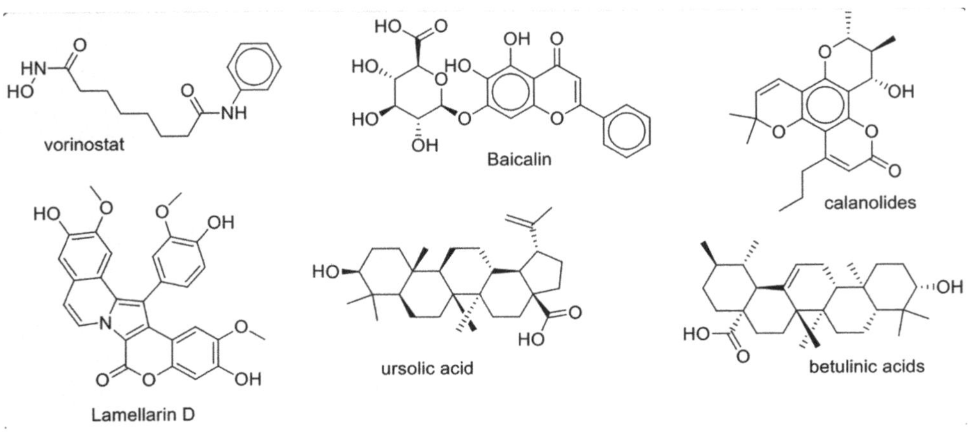


**FIGURE 8.7** Some important antiviral drug used in the treatment of HIV.

**TABLE 8.1**
**Anti-HIV Activity of Few Medical Plants**

| Plant Name | Part Used | Bioactive Phytoconstituent | Anti-HIV Activity (% Inhibition) |
|---|---|---|---|
| *Aegle marmelos* | Fruits | Coumarins | 44.0 (65) |
| *Alstonia scholaris* | Leaves | Alkaloids | 40.4 (66) |
| *Terminalia arjuna* | Stem bark | Terpenes | 42.8 (67) |
| *Rubia cordifolia* | Roots | Quinones | 0.0 (66) |
| *Madhuca indica* | Bark | Saponins | 39.0 (68) |
| *Butea monosperma* | Roots | Flavonoids | 75.7 (68) |
| *Butea monosperma* | Stem bark | Flavonoids | 34.0 (68) |
| *Cassia occidentalis* | Leaves | Flavonoids | 68.6 (69) |
| *Rauwolfia serpentina* | Roots | Alkaloids | 40.8 (70) |
| *Aegle marmelos* | Leaves | Coumarins | 6.4 (71) |
| *Glycyrrhiza glabra* | Roots | Steroids | 51.1 (72) |
| *Tinospora cordifolia* | Stem bark | Triterpene | 51.0 (73) |
| *Withania somnifera* | Roots | Triterpenoids | 47.2 (74) |
| *Argemone mexicana* | Leaves | Alkaloids | 64.3 (75) |

## 8.8 TRADITIONAL INFORMATION ON PLANTS AGAINST HIV

Medicinal plant serves as a promising alternative for several illnesses and conditions. This article reports few of the plant species with their antiviral potency against HIV. In proper identification of plant species, taxonomy plays a vital role. The antiretroviral drugs for HIV infection are classified according to the Food and Drug Administration (FDA or USFDA) into the following types like Protease Inhibitors (PIs), Fusion Inhibitors, Multi-class Combination Products, Nucleoside Reverse Transcriptase Inhibitors (NRTIs), Nonnucleoside Reverse Transcriptase Inhibitors (NNRTIs), Entry Inhibitors—CCR5 coreceptor antagonist, and HIV integrase strand transfer inhibitors. Table 8.1 depicts some of the fruitful plants with anti-HIV activity.

## 8.9 ANTI-HIV ACTIVITY OF PLANT EXTRACTS AND SOME SECONDARY METABOLITES

The use of plant materials to treat illnesses has a long history in the majority of cultures around the world. Natural products are growing in popularity as a source for researchers to discover new and more effective antiviral drugs due to recent advancements in pharmacognosy and technology, as well as current trends of a more health-conscious general public (Harvey, 2000). This is because natural products have comparatively fewer side effects and are more cost-effective when used on a commercial scale. Plants produce a variety of secondary metabolites, including phenolics, glycosides, alkaloids, coumarins, terpenoids, essential oils, and peptides, as evolutionary responses to parasitic infections by nematodes, nematodes, and other organisms, to avoid herbivory, and to compete for light and space. Different biological activities associated with these metabolites have been discovered. Some of them, such as viral infections linked to HIV types 1 and 2 (HIV-1 and HIV-2) as genetic variations, have a significant impact on immune system improvement and have antiviral potential (Vlietinck et al., 1998). A growing number of HIV-positive patients are unable to use the anti-HIV medications currently approved, such as reverse transcriptase and protease inhibitors, because of the side effects of antiretroviral medications, particularly liver diseases. Even the most effective antiretroviral therapy (HAART) has not been able to completely stop HIV replication. As a result, in order to minimise side effects when compared to current synthetic drugs, it is necessary to discover and develop new anti-HIV agents or new mechanisms of activity from medicinal plants (Jadaun et al., 2016).

One of the most crucial strategies for developing an effective treatment for AIDS is the potential use of plant extracts and their secondary metabolites to thwart the development of anti-HIV agents. Novel compounds are systematically sourced from medicinal plants using bioassay-guided fractionation and isolation of secondary metabolites in accordance with their preliminary high throughput screenings. The therapeutic potential of these chemical compounds was confirmed by the *in vitro* and *in vivo* evaluation. Since the isolation of bioactive lead compounds has been the primary focus of early research, traditional medicines can, therefore, be used as sources for potential new drug candidates (Lubbe et al., 2012). Numerous substances with anti-HIV-1 effects have been screened, isolated from natural sources, and found to inhibit HIV at almost all stages of the viral life cycle. The list of these substances is extensive and includes alkaloids, sulphated polysaccharides, polyphenolics, flavonoids, coumarins, phenolics, tannins, triterpenes, lectins, phloroglucinols, lactones, iridoids, depsidones, O-caffeoyl derivatives, lignans, ribosome inactivating proteins, saponins, and many more. Owing to the large structural diversity, natural products offer a significant pool of potential candidates for the screening of anti-HIV agents with novel structures and antiviral mechanisms. Numerous natural substances have been discovered to inhibit specific enzymes and proteins essential to the life cycle of HIV, including effective interference with reverse transcription, virus entry, and inhibitors of integrase and protease (Burack et al., 1996).

Although it is still unknown, many natural products have anti-HIV properties. Some plant extracts have significantly reduced the enzyme activity required for HIV-1 replication, while defending HIV-infected cells. They were also found to be additionally effective against other retroviruses like the *Herpes Simplex Virus* since these extracts possessed anti-HIV activity (HSV). A small number of *in vivo* studies have been conducted using compounds isolated from natural sources, but the majority of studies have used *in vitro* test systems for anti-HIV-1 enzyme assays such as HIV-1 RT colorimetric assay, HIV-1 integrase assay, and HIV-1 protease fluorogenic assay (Kalvatchev et al., 1997). Several medicinal plant extracts, fractions, or isolated compounds with potential anti-HIV properties have been listed.

### 8.9.1 *Artemisia annua* L. (Asteraceae)

Using validated cellular systems, the tea infusion made from the Chinese medicinal plant *Artemisia annua L.* was tested for its anti-HIV properties. Artemisia annua tea was discovered to be extremely potent, with $IC_{50}$ values as low as 2.0 µg/mL. The related species *Artemisia afra* (which does not contain artemisinin) has also demonstrated a similar level of activity. Artemisinin was also found to be inactive at 25 µg/mL (Kashman et al., 1992) (Figure 8.8).

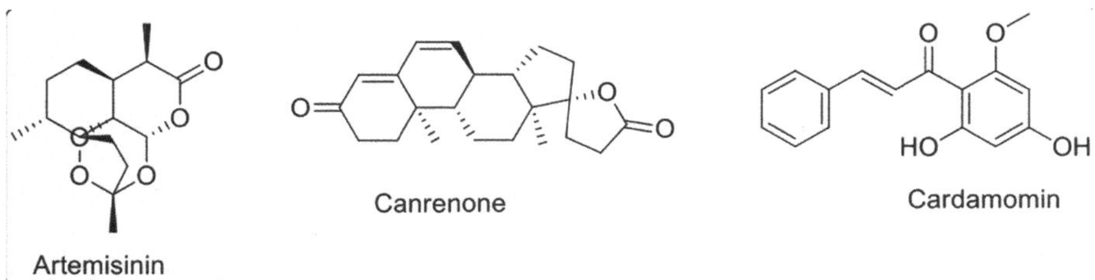


**FIGURE 8.8** Anti-HIV active compounds isolated from *Artemisia*.

### 8.9.2 *Calendula officinalis* L. (Asteraceae)

In India, ointments made from the flowers of the *Calendula officinalis* plant are used to treat wounds, herpes, ulcers, frostbite, skin damage, scars, and blood purification. Varicose veins have traditionally been treated with infusions made from the leaves. Flowers from *Calendula officinalis* showed strong anti-HIV activity in an *in vitro* test using the 3-(4,5-dimethylthiazolyl-2)-2,5-diphenyltetrazolium bromide) (MTT)/tetrazolium-based assay. Dichloromethane-methanol (1:1) extract of the flowers exhibited this activity, which was attributed to the suppression of the HIV mediated fusion at a concentration of 500 μg/mL and the inhibition of HIV1-RT at a concentration of 1,000 μg/mL (Muley et al., 2009). The ability to prevent the reproduction of the human immunodeficiency virus type 1 (HIV-l) was tested for in the organic and aqueous extracts of dried flowers from *Calendula officinalis*. It was found that in an *in vitro* MTT ketrazolium-based assay, only the organic extract showed potent anti-HIV activity, while the other was relatively nontoxic to human lymphocytic Molt-4 cells. Uninfected Molt-4 cells were also completely shielded for up to 24 hours in the presence of the organic extract (500 pg/mL) from fusion and subsequent death induced by cocultivation with persistently infected U-937/HIV-1 cells. A significant dose- and time-dependent reduction of HIV-l reverse transcription (RT) activity was also discovered to be induced by the organic extract from *Calendula officinalis* flowers. After 30 minutes of treatment with a partially purified enzyme in a cell-free system, 85% RT inhibition was attained. These findings suggested that an organic extract of *Calendula officinalis* flowers contained therapeutic-grade anti-HIV properties (Pengsuparp et al., 1996).

### 8.9.3 *Astragalus membranaceus* Bunge (Fabaceae)

*Astragalus membranaceus* is a well-known immunostimulant in Chinese traditional medicine. Studies on immune-suppressed and immune-competent human patients have shown that astragalus extracts can restore or enhance local graft versus host rejection. In patients with HIV, these extracts have improved symptomology. These findings indicate that *Astragalus* extracts are safe, but mutagenicity testing is still needed (Leteane et al., 2012).

### 8.9.4 *Calophyllum lanigerum* Miq. var. austrocoriaceum (T.C. Whitmore) P.F. Stevens (Clusiaceae)

It has been discovered that *Calophyllum lanigerum* var. austrocoriaceum inhibits the cytopathic effects of *in vitro* HIV infection. The biological characterization of the active constituents as coumarin derivatives and bioassay-guided fractionation of the extract and chemical have both been reported (Jiratchariyakul et al., 2001). Soulattrolide, a coumarin isolated from *Calophyllum teysmanni* latex, has demonstrated that the latex of this plant is active against HIV-1 RT (Jiratchariyakul et al., 2001) (Figure 8.9).

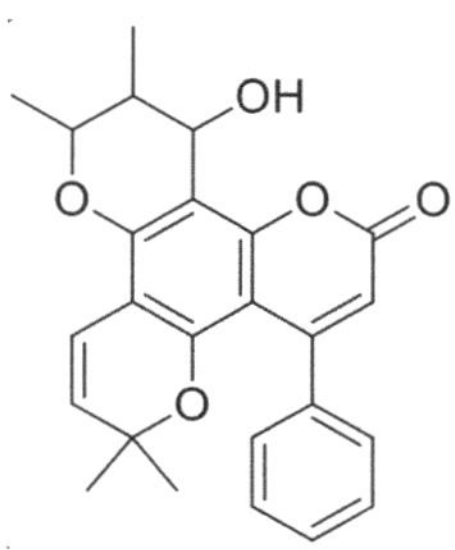


**FIGURE 8.9** Soulattrolide, a coumarin derivative anti-HIV activity isolated from *Calophyllum teysmanni*.

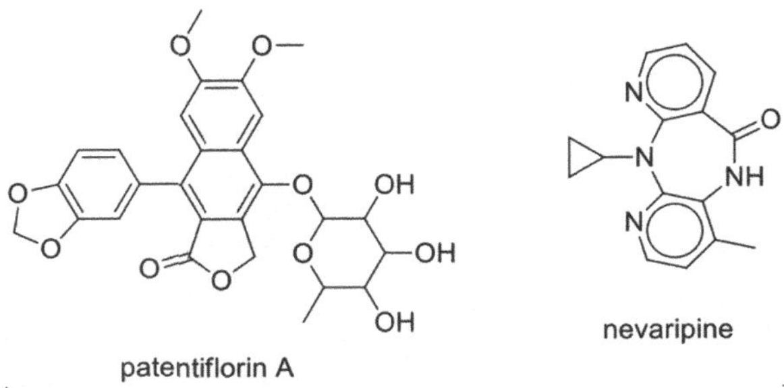


**FIGURE 8.10** Anti-HIV compounds isolated from *Justicia gendarussa.*

### 8.9.5 *Justicia gendarussa* Burm. f. (syn: Gendarussa vulgaris Nees) (Acanthaceae)

Through the evaluation of over 4,500 plant species found in Vietnam and Laos, *Justicia gendarussa* was found to be a powerful anti-HIV-1 active lead. At a concentration of 20 µg/mL, it completely inhibited HIV replication. Justiprocumins A and B were isolated as new arylnaphthalide lignan glycosides by using bioassay-guided isolation from the methanol extract of the plant's stems and bark. Justiprocumin B exhibits potent activity against a variety of HIV strains, with $IC_{50}$ values ranging between 15 and 21 nM (AZT, $IC_{50}$ 77–95 nM, as positive control). The NRTI-resistant isolate (HIV-11617-1) of the analogue (AZT) and the NNRTI (nonnucleoside reverse transcriptase inhibitor)-resistant isolate (HIV-1N119) of the analogue (nevaripine) both showed significant inhibitory activity in response to justiprocumin B (Gerencer et al., 2006). At a concentration of 20 µg/mL, the dichloromethane plant extract completely inhibited HIV replication. The evaluation of the MeOH extract made from a fresh sample of the same plant, which inhibited HIV-1 replication at an $IC_{50}$ value of 40 ng/mL, provided further evidence of this bioactivity. The stem and root extracts of this plant were separated under the direction of bioassays to yield patentiflorin A, an anti-HIV arylnaphthalene lignan (ANL) glycoside. The drug was tested against both M- and T-tropic HIV-1 isolates and demonstrated a significantly higher level of inhibition than the currently prescribed nucleotide analogue (AZT) and non-nucleotide analogue anti-HIV medications (nevaripine). Because of this, patentiflorin A has the potential to be developed into a cutting-edge HIV medication (Asres and Bucar, 2005). Patentiflorin A demonstrated anti-HIV-1 activity in the flawed HIV-based pseudo typed assay with an $IC_{50}$ value of 26.9 nM. The findings unambiguously demonstrated that patentiflorin A exhibits a broad-spectrum activity against both M-tropic and T-tropic HIV-1 isolates, with $IC_{50}$ values lower than those of AZT, the first anti-HIV medication created and still in use for the treatment of HIV/AIDS. It effectively and dose-dependently inhibited the particle production of all four HIV-1 isolates, like AZT. Compared to 77–95 nM for AZT, the $IC_{50}$ value for patentiflorin A was 24–37 nM (Figure 8.10).

### 8.9.6 *Cassia abbreviata* Oliv. Oliv., C. sieberiana D.C. (Fabaceae)

The Botswana plant *Cassia abbreviata*, which is used by traditional healers to treat HIV/AIDS, was tested for its inhibitory effects on HIV replication against the HIV-1c (MJ4) clone, measuring the protection from the cytopathic effect and the levels of viral p24 antigen in infected PBMCs. Extracts from *Cassia sieberiana* and *Cassia abbreviata* have demonstrated a notable inhibition of HIV-1c (MJ4) replication. Using effective concentrations ($EC_{50}$) of 65.1, 85.3, and 102.8 µg/mL, respectively, *Cassia sieberiana* root and bark extracts, and *Cassia abbreviata* root extracts demonstrated concentration-dependent anti-HIV activity (Burack et al., 1996).

### 8.9.7 *Chelidonium majus* L. (Papaveraceae)

Freshly made crude extract of *Chelidonium majus L.* was tested for anti-retroviral activity, and a low-sulphated poly-glycosaminoglycan moiety with a molecular weight of about 3,800 Da was isolated from the extract (Kalvatchev et al., 1997).

Measurements of RT activity and p24 content in cell cultures were used to determine the substance's ability to prevent HIV-1 infection of human $CD4^+$ T-cell lines AA2 and H9 at a concentration of 25 g/mL as well as the spread of the virus from cell to cell in H9 cells that were continuously infected with HIV-1. In addition, it was reported that in a murine AIDS model, treatment with a purified substance significantly reduced splenomegaly and cervical lymph node enlargement in C57Bl/6 mice chronically infected with a pool of murine leukaemia retroviruses.

### 8.9.8 *Terminalia paniculata* (Combretaceae)

Acetone and methanol extracts made from *Terminalia paniculata* fruits were tested for their *in vitro* anti-HIV1 activity. The $EC_{50}$ values for the *T. paniculata* acetone and methanol extracts were 10.3 g/mL. The anti-HIV1 activity may be caused by the inhibition of RT (77.7% inhibition) and protease (69.9% inhibition), according to the enzymatic assays that were carried out to ascertain the mechanism of action (Kashman et al., 1992).

### 8.9.9 *Hoodia gordonii* (Masson) Sweet ex Decne (Apocynaceae)

The ethanol and ethylacetate extracts of *Hoodia gordonii* were tested for their *in vitro* anti-HIV potential. HIV-1 RT was effectively inhibited by both extracts in a dose-dependent manner, with $IC_{50}$ values of 73.55±0.04 and 69.81±9.45 g/mL, respectively. As a positive control, doxorubicin, a well-known RT inhibitor, inhibited HIV RT by 68% at 25 μg/mL ($IC_{50}$ 25 μg/mL). Additionally, both extracts showed inhibitory activity against HIV protease (PR), with $IC_{50}$ values for ethanol and ethyl acetate extracts of 97.29±0.01 and 63.76±9.01 μg/mL, respectively. As a well-known PR inhibitor, acetyl pepstatin inhibited HIV PR by as much as 82% at 50 μg/mL ($IC_{50}$ 50 μg/mL). Additionally, at the highest tested concentration of 400 μg/mL, neither the ethanol nor the ethyl acetate extracts significantly inhibited HIV-1 integrase (IN), showing only a 50% inhibition. As a positive control substance for IN inhibition, sodium azide was used (Muley et al., 2009).

In the same study, phenolics, alkaloids, terpenes, steroids, cardiac glycosides, and tannins were all detected in the ethanolic extract of *Hoodia gordonii*, but only phenolics, cardiac glycosides, and steroids were detected in the ethyl acetate extract (Figure 8.11).

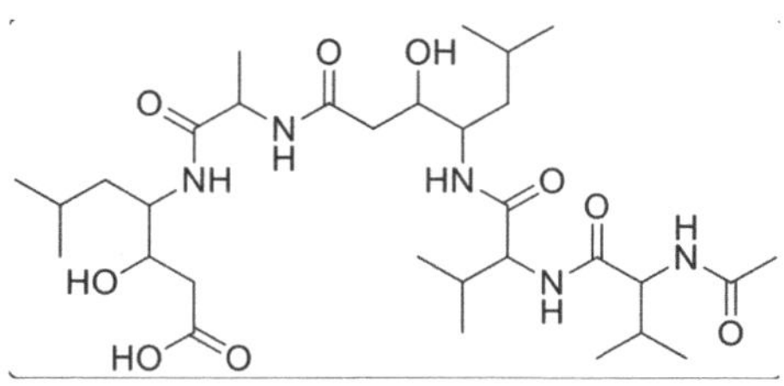


**FIGURE 8.11** Acetyl pepstatin, a peptide inhibitor against HIV protease isolated from *Hoodia gordonii*.

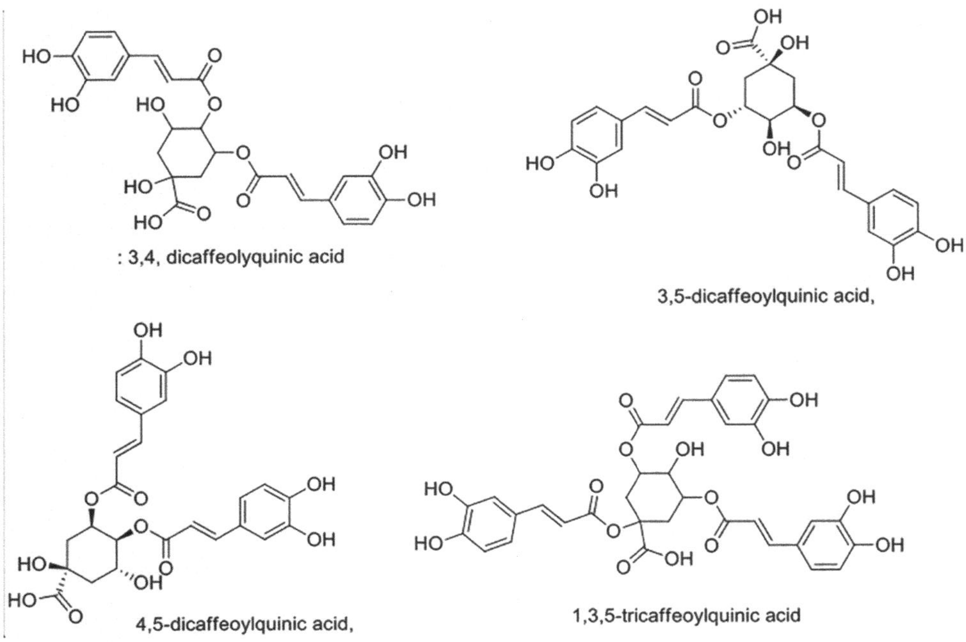


**FIGURE 8.12** Caffeolyquinic acid derivatives with anti-HIV activity isolated from in *H. populifolium*.

### 8.9.10 *Helichrysum populifolium* (Asteraceae)

HeLa-SXR5 cells that express the CD4 receptor and the CXCR4/CCR5 chemokine receptors were used to test the anti-HIV activity of a methanol: water (1:1) extract of the aerial parts of *Helichrysum populifolium* growing in South Africa (Pengsuparp et al., 1996). The extract was found to be active ($IC_{50}$ value of 12 μg/mL). Three dicaffeolyquinic acid derivatives, 3,4-dicaffeolyquinic acid, 3,5-dicaffeolyquinic acid, and 4,5-dicaffeoylquinic acid, as well as two tricaffeoylquinic acid derivatives, 1,3,5-tricaffeoylquinic acid and either 5-malonyl-1,3,4-tricaffeoylquinic or 3-malonyl-1,4,5-tricaffeoylquinic acid, were the anti-HIV compounds found in *H. populifolium* (Figure 8.12).

### 8.9.11 *Garcinia edulis* Exell (Clusiaceae)

The1,4,6-trihydroxy-3-methoxy-2-(3-methyl-2-butenyl)-5-(1,1-dimethyl-prop-2enyl)xanthone derivative was isolated from ethanolic extract of the root bark of *Garcinia edulis*. It demonstrated anti-HIV-1 protease activity with an $IC_{50}$ value of 11.3 μg/mL *in vitro*, while acetyl pepstatin was used as a positive control and had anti-HIV-1 PR activity with an $IC_{50}$ value of 2.2 μg/mL (Leteane et al., 2012). With an *in vitro* $LC_{50}$ value of 2.36 μg/mL, this substance has also demonstrated potent lethality toward brine shrimp larvae (Figure 8.13).

### 8.9.12 *Pachymahoelen* Rumph (Polyporaceae)

In comparison to the other extracts tested, *Pachymahoelen* Rumph's hexane extract, which is used in folk medicine in Korea, was found to have the strongest anti-HIV-1 activity. The highest value for this extract was 37.3 μg/mL ($EC_{50}$) on the p24 antigen assay and 36.8% on the RT activity test (at 200 μg/mL). Additionally, this extract had demonstrated protective effects on infected MT-4 cells; the highest observed level of protection was 58.2%. The hexane extract of this plant species' 50% cytotoxic concentration ($CC_{50}$) was discovered to be 100.6 μg/mL (Jiratchariyakul et al., 2001).

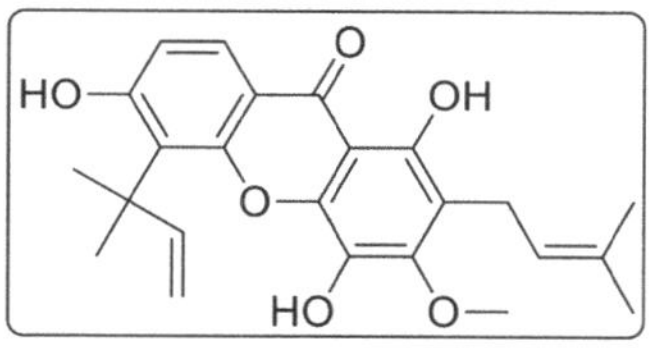


**FIGURE 8.13** Xanthone derivative with anti-HIV-1 protease activity, isolated from *Garcinia edulis*.

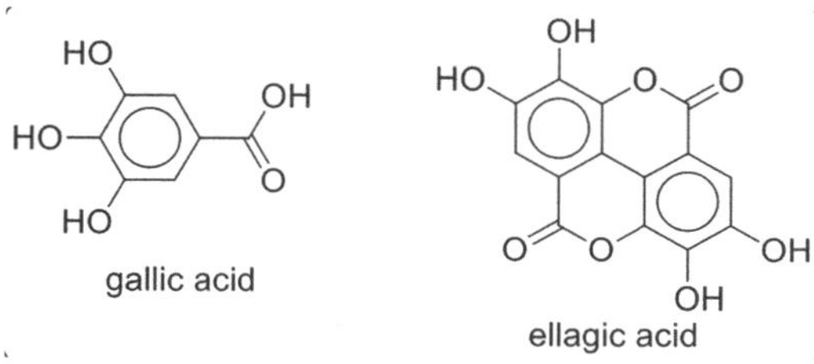


**FIGURE 8.14** Anti-HIV active compounds isolated from *Diospyros* lotus fruit.

### 8.9.13 *Diospyros lotus* L. (Ebenaceae)

*Diospyros* lotus fruit methanol extract was tested for anti-HIV-1 activity. With a Therapeutic Index (TI) value of >32.84, gallic acid was discovered to be the compound that was most effective against HIV-1. The other compounds were less effective. Lotus fruits from the *Diospyros* genus may serve as an anti-HIV chemical reserve. All identified substances underwent cytotoxicity and anti-HIV-1 activity testing. The commercially available medication azido-thymidine (AZT) was examined using the same procedures as a positive control. The percentages of cytotoxicity ($CC_{50}$), effectiveness ($EC_{50}$), and TI, which is the ratio of $CC_{50}/EC_{50}$, were used to describe the activity data. Compared to ellagic acid ($CC_{50}$=35.84 μg/mL), seven isolated phenolic compounds ($CC_{50}$>200 μg/mL) were less toxic to C8166 cells. More than any other compound, gallic acid inhibited HIV-1 replication with an $EC_{50}$ of 6.09 μg/mL and a TI value of>32.84. The syncytia assay was used to measure the anti-HIV-1 activity. The seven phenolic compounds demonstrated effective anti-HIV-1 activity, with gallic acid, a simple tannin compound, being the most active and having the highest TI value (Gerencer et al., 2006) (Figure 8.14).

### 8.9.14 *Tuberaria lignosa* (Sweet) Sampaio (Asteraceae)

The ethanolic and aqueous extracts of *Tuberaria lignosa* were tested for their anti-HIV activity by inhibiting HIV replication. *Tuberaria lignosa* was widely used in folk medicine to treat diseases of viral origin in the Iberian Peninsula. Additionally, the extracts' toxicity to MT-2 cells was examined. The evaluation of their possible antiviral effects at higher concentrations was not possible because the ethanolic extract was particularly toxic. The aqueous extract of *T. lignosa*, however, demonstrated anti-HIV activity at concentrations ranging from 12.5 to 50 g/mL, while being relatively nontoxic to human lymphocytic MT-2 cells (Asres and Bucar, 2005). In conclusion, secondary metabolites produced by terrestrial plants for chemical defence have played crucial roles in human health and have unique chemical structures. In order to effectively treat diseases, it is constantly necessary to develop new drug candidates. The process of discovering new drugs can be accomplished by combining both traditional and contemporary research methodologies. Numerous avenues are there for research to ascertain the biological activities of some medicinal plants because they are still relatively unexplored.

The anti-HIV activity of a few plant extracts and their potential application as anti-HIV agents have already been outlined. In the process of creating new candidates with anti-HIV properties of therapeutic interest, *Calendula officinalis*, *Justicia gendarussa*, and *Sceletium tortuosum* may prove to be useful potential sources for new lead compounds. These studies are regarded as one of the most significant steps toward an AIDS treatment.

## 8.10 NEWLY DISCOVERED NATURAL PRODUCTS OBTAINED FROM PLANT SOURCES WITH ANTI-HIV ACTIVITY

### 8.10.1 Recently Discovered Anti-HIV Alkaloid Compounds

The leaves and twigs of the Chinese native *Glycosmis montana* were used to isolate the diprenylated indole alkaloid (E)-3-(3-hydroxymethyl-2-butenyl)-7-(3-methyl2-butenyl)-1H-indole. The alkaloid exhibited potent anti-HIV activity with an $IC_{50}$ value of 1.17 µg/mL and an SI of 11.68. It was suggested that these diprenylated indole alkaloids might be candidates for further study as potential anti-HIV agents (Jayasuriya et al., 2004).

Tripfordine A, a sesquiterpene-pyridine alkaloid, was discovered in an ethanolic root extract of the deadly liana *Tripterygium wilfordii*. In Chinese folklore, the extremely poisonous roots are referred to as "lei gong teng" and are employed as an insecticide. This compound has a peculiar structure that consists of a macrocycle connecting a sesquiterpene skeleton to a pyridine ring. This substance was tested for HIV-1 along with other sesquiterpene pyridine alkaloids. While the other alkaloids displayed more potent activity, with $EC_{50}$ values of 2.54 to <0.1 µg/mL, tripfordine A demonstrated moderate anti-HIV activity with an EC50 of 13.4 µg/mL and TI of 1.9. The SAR hypothesised that the C-5 acetyl group might be crucial for the activity because compounds with a 5-acetyl group were more active than those with 5-hydroxy, 5-benzyloxy, and 5-furanoyl groups. The activity ($EC_{50}$<0.10 µg/mL) was unaffected by the position of the carboxypropyl or carboxybutyl side chain (C-20 or C-40) on the nicotinic moiety. Additionally, the compounds that contained a tertiary hydroxy group in either the propionic (C-80) or butyric (C-90) acid side chain exhibited anti-HIV activity, whereas those that did not contain an OH were inactive or less active. However, one alkaloid, hyponine B, which lacks an OH on C-80, was also very active, indicating that other structural factors might also be crucial for the activity (Li et al., 2009 a,b).

The methanolic extract of the stems of an *Aniba* species was used to isolate anibamine, a brand-new pyridine quaternary alkaloid, and it was then purified using TFA salt. Anibamine had an $IC_{50}$ of 1 µM (Hsieh et al., 2004) and competed with 125I-gp120 for binding to human CCR5. The binding site of anibamine shared some characteristics with the other CCR5 antagonists, according to molecular docking studies of anibamine in two different CCR5 homology models (human β2-adrenergic receptors and bovine rhodopsin). This suggests that this compound may bind in similar binding sites and/or modes (Saklani and Kutty, 2008).

Drymaritin, a newly discovered canthin-4-one type alkaloid, was isolated from *Drymaria diandra*, also known as *Drymaria cordata* subsp. *diandra*, and it demonstrated anti-HIV activity in H9 lymphocytes with an $EC_{50}$ value of 0.699 mg/mL and a TI of 20.6. Alkaloids of the canthin-6-one type have already been linked to anti-HIV activity (Chen et al., 2006) (Figure 8.15).

### 8.10.2 Recently Discovered Anti-HIV Phenolic Compounds

The phenolic compounds like calanolide A, DCK, and DCQA are under clinical trials for anti-HIV profile (Figure 8.16).

The fruits of the southern Chinese native *Schisandra rubriflora* were used to isolate an inseparable 2 : 1 mixture of new dibenzo cyclooctadiene lignans called rubrisandrins A. $^{1}H$,$^{13}C$, and

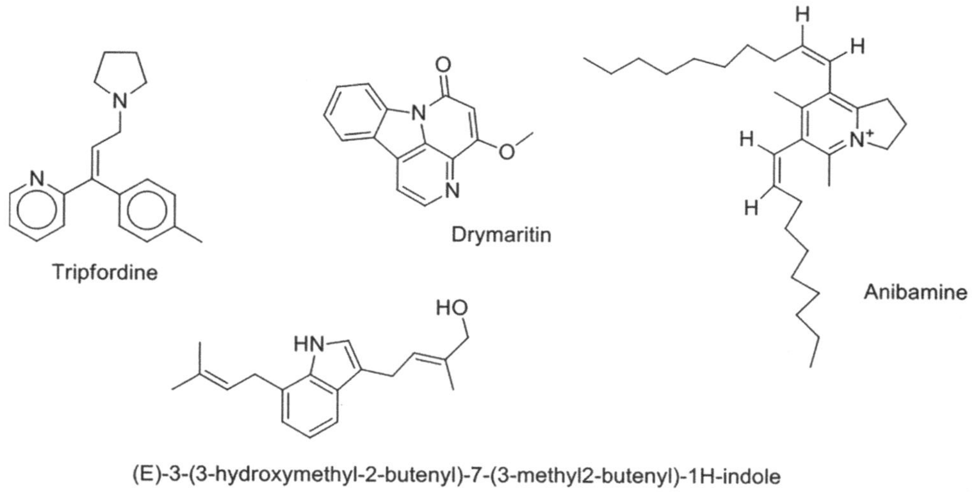


**FIGURE 8.15** Some potential anti-HIV alkaloid molecules.

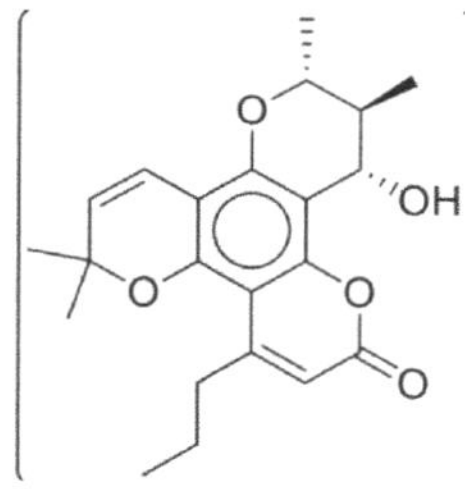

**FIGURE 8.16** Calanolide A, a phenolic compound under clinical trials.

2D-NMR data were used to distinguish between the two regioisomers. With an $EC_{50}$ of 11.3 μM (Li et al., 2009 a,b), the mixture had *in vitro* inhibitory effects against HIV replication in H9 lymphocytes.

*Kadsura longipedunculata*'s roots and stems were used to make longipedunin A, which has an $IC_{50}$ of 50 mg/mL against HIV-1 PR (Appendino et al., 2007). An important functional group for the anti-HIV PR activity may be a trans-cinnamic acid ester group, as evidenced by the higher activity of 28 compared to other dibenzocyclooctadiene lignans.

A 70% aqueous acetone extract of the aerial parts of *Schisandra propinqua* var. sinensis, more commonly known as "tie-gusan" in China, yielded Tiegusanin G, 1 of 14 new lignans. In the C8166 cell line, the compound displayed anti-HIV-1 activity with an $EC_{50}$ value of 7.9 mM and TI of greater than 25 (Dat et al., 2007). Various flavonoid glycoside component like 6(γ-γ-dimethylally) dihydrokaempferol-O-β-D-glucoside, 6-(γ-γ-dimethylallyl) dihydroquercetin-7-O-β-D-glucoside, 6-(3-hydroxy-3-methylbutyl) taxifolin-7-O-β-D-glucoside, and 6-(3-hydroxy-3-methylbutyl)-quercetin-7-O-β-D-glucoside were isolated for anti-HIV activity from the methanolic extraction of *Ochna integerrima*. In the syncytium assay, these substances demonstrated anti-HIV-1 activity with an $EC_{50}$ of 29.1, 14.0, 102.4, and 45.8 mg/mL, respectively. The same investigation revealed that vitexin, which has a sugar unit on ring A, is inactive. Previously, McKee et al. reported that compounds with 30-OH were effective against HIV, whereas Meragelman et al. reported that 6-isoprenylated flavonoids with two hydroxyls at C-5 and C-7 were inactive (Kim et al., 2006). It was concluded from this that the isoprenyl and sugar unit on ring A and 30-OH (present in the compounds mentioned above) are necessary for anti-HIV activity.

Three phenolic compounds, 2,3-dihydroxy-4-methoxy-6,6,9-trimethyl-6H-dibenzo[b,d] pyran, 8-methoxy-2-methyl-2-(4-methyl-3-pentenyl)-2H-1-benzopyran-6-ol, and

4-methoxy-3-(3-ethyl-2-butenyl)benzoic acid, were isolated from *Wigandia urens*. The structure of homoarenol was altered as a result of the structural clarification of arzanol. While arzanol failed to inhibit luciferase activity in the HeLa Tet-On-Luc assay, it did inhibit TNFa-induced HIV-LTR transactivation in a T-cell line with an $IC_{50}$ of 5 mM. With $IC_{50}$ values of 1.5 and 1.1 mM, respectively, the dimeric lactone ardimerindigallate isolated from the whole plants of *Ardisia japonica* exhibited inhibitory activity against both HIV-1 and HIV-2 RNase H *in vitro*. At 200 mM, it had no effect on either human or *E. coli* RNase H, proving that dimeric lactone ardimerindigallate is selective for inhibiting the retroviral enzyme (Pu et al., 2008).

Ardimerin, a comparable substance devoid of a galloyl moiety, failed to inhibit RNase H, indicating that the galloyl group is crucial for anti-HIV activity. The ethyl acetate extract of *Gardenia jasminoides* was fractionated using bioactivity to isolate three quinic acid derivatives: methyl-5-O-caffeoyl-3-O-sinapoylquinate, methyl-5-O-caffeoyl-4-O-sinapoylquinate, and methyl-3,5-di-O-caffeoyl-4-O-(3-hydroxy-3-methyl) (*Gardeniae fructus*). With an $IC_{50}$ of 20 mg/mL, these compounds demonstrated HIV-1 IN inhibitory activity. (Chen et al., 2006).

Southern Chinese native *Kadsura heteroclita* stems were used to isolate 16 well-known substances, including lignans of the dibenzocyclooctadiene type. Acetoxy oxokadsurane, heteroclitin F, and lignans kadsurin all were effective anti-HIV agents. Interiorin and interiorin B, which have $EC_{50}$ values of 1.6 and 1.4 mg/mL and TI values of 52.9 and 65.9, respectively, showed anti-HIV activity. The αα'ββ'-dienone may play a significant role in the anti-HIV activity of lignans 47 and 48 when compared to other dibenzocyclooctadiene lignans with comparable skeletons (Xiao et al., 2008).

*Schisandra rubriflora* fruits were used to isolate the dibenzocyclooctadiene lignans gomisin J, (±)-gomisin M1, (+)-gomisinM2, and schisanhenol. The strongest anti-HIV activity was demonstrated by (±)-Gomisin M1, with $EC_{50}$ and TI values of <0.65 μM and >68, respectively. The anti-HIV activity of dibenzocyclooctadiene lignans was found to depend heavily on the presence of aromatic hydroxyl groups, which were found to be more active than those lacking aromatic hydroxyls. As shown by the fact that (±)-gomisin M1 (with a 3′-OH) was more effective than (+)-gomisinM2 (with a 3-OH) (Gao et al., 2008), the position of the hydroxyl was also crucial for enhancing the anti-HIV activity (with a 3-OH).

From the leaves and stems of *Schisandra sphenanthera*, four well-known lignans were isolated that included gomisin J, schizandrin, schisantherin A, and schisantherin D (Yu et al., 2007a, b). The inhibition assay for the cytopathic effects of HIV-1IIIB was used to assess the anti-HIV-1 activity of dibenzocyclooctadiene lignans gomisin J and schizandrin, schisantherin A, and schisantherin D. The $EC_{50}$ values for these compounds were 17.2, 15.5, 19.0, and 20.5 mg/mL, respectively, demonstrating their anti-HIV-1 activity (Wu et al., 2004). Gomisin J is also known for promoting glucose uptake into peripheral tissue and inhibiting L-type calcium channels. Along with 11 new lignans, binakadsurin A was isolated from a 70% aqueous acetone extract of *Kadsura angustifolia* stems. In the C8166 cell line, this compound exhibited anti-HIV-1 activity with $EC_{50}$ values of 3.86 mM, $CC_{50}$ values of 227.16 mM, and SI values of 58.92 (Lee et al., 2009) (Figure 8.17).

Flavonoids derived from EtOAc extract of *Alnus firma* leaf material were quercetin, quercitrin, and myricetin 3-O beta-D-galactopyranoside. 5,7-Dihydroxychromone and (-)-catechin, isolated from *Begonia nantoensis*, endemic to the mountains of central Taiwan, demonstrated activity against HIV replication in H9 lymphocyte cells with $EC_{50}$ values of 18.7 and 14.3 mg/mL, and TI values of 1.3 and 1.8, respectively (Ma et al., 2009). The compounds demonstrated inhibition against HIV-1 RT with an $IC_{50}$ of 60 mM, whereas they were inactive against HIV-1 PR (Reutrakul et al., 2007).

A DCM extract of the Cameroonian medicinal plant Erythrina senegalensis yielded eight prenylisoflavonoids. 8-Prenylluteone, auriculatin, erysenegalensein O, erysenegalensein D, erysenegalensein N, derrone, alpinumisoflavone, and 6,8-diprenylgenistein all demonstrated HIV-1 PR inhibitory activity with $IC_{50}$ values of 4.0, 3.5, 5.0, 2.5, and 4.5. It was discovered that the two prenyl groups at the 6- and 8-positions, as well as the hydroxy at the 4′-position, were crucial for the activity. It has been discovered that the roots of *Astragalus membranaceus* var. mongholicus contain calycosin-7-O-beta-D glucopyranoside (Kuo et al., 2009). This plant has been used in traditional

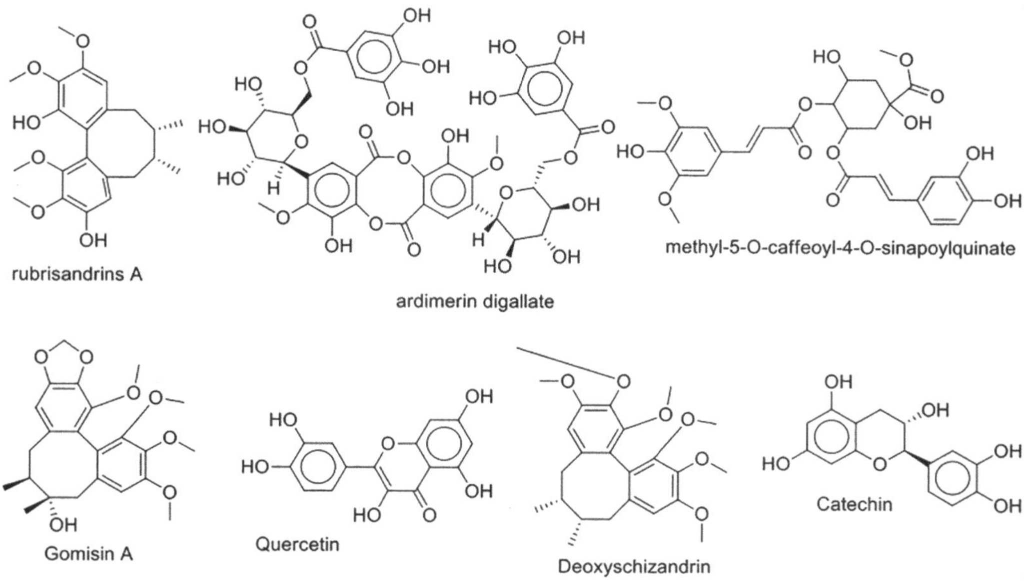


**FIGURE 8.17** Some potential anti-HIV phenolic compounds.

Chinese medicine to treat a variety of illnesses, including diabetes, oedema, anaemia, abnormal uterine bleeding, diarrhoea, and chronic nephritis albuminuria. Compound Calycosin-7-O-beta-D glucopyranoside had an $EC_{50}$ of 15.7 mM and a TI of 28.5 that prevented cytopathic effects (Zhang et al., 2000). In the syncytium assay, the bifunctional compounds ochnaflavone 7″-O-methyl ether and 2″,3″-dihydroochnaflavone 7″-O-methyl ether, which were isolated from *Ochna integerrima*, were found to be active with $EC_{50}$ values of 2.0 and 0.9 mg/mL, respectively. These substances inhibited HIV-1 RT with $IC_{50}$ values of 2.0 and 2.4 mg/mL, respectively, in a mechanism-based assay (Fujioka et al., 1994).

### 8.10.3 Anti-HIV Effects of Triterpenoids

Betulinic acid (1, BA), which has an $EC_{50}$ value of 1.4 µM and exhibits an inhibitory effect against HIV-1IIIB replication, was first identified by the Lee et al. Natural Products Research Laboratories in 1994(111). Since then, betulinic acid has been a key molecular building block in the creation of triterpene analogues that are effective HIV medications. Numerous effective anti-HIV agents were discovered as a result of systematic structural modifications made to the C-3 hydroxyl, A-ring, C-19 isopropenyl group, and C-28 carboxyl moiety (Kashiwada et al., 1996).

A wide range of anhydrides and acid chlorides can easily acylate the C-3 hydroxy group. The identification of bevirimat (3-O-3′,3′-dimethylsuccinyl BA), which had a significant anti-HIV $EC_{50}$ value of 0.35 nM, was the outcome of this modification (Qian et al., 2007). The SAR study also demonstrated that the enhanced antiviral potency of bevirimat depends on the presence of an ester functionality at position C-3, an appropriate side chain length at position C-3, a terminal carboxylic acid moiety, and a dimethyl substitution at position C-3′. Additionally, SAR showed that the 3'S-methyl moiety within the 3′-dimethyl moiety contributed more to the activity because 3'S-monomethylsuccinyl-substituted BA (3'S-MSB) and bevirimat both exhibited similar antiviral potency ($EC_{50}$: 8.7 and 1.3 nM, respectively), while 3'R-MSB exhibited significantly lower activity ($EC_{50}$: 0.12 µM) (Li et al., 2003). Studies on the mechanisms of action showed that bevirimat maintained its nanomolar antiviral potency against HIV-1 isolates that were resistant to RT and PR inhibitors. Scientists from Panacos Pharmaceuticals Inc. and the National Institutes of Health characterised the virus generated by cells treated with the compound to define the target of bevirimat.

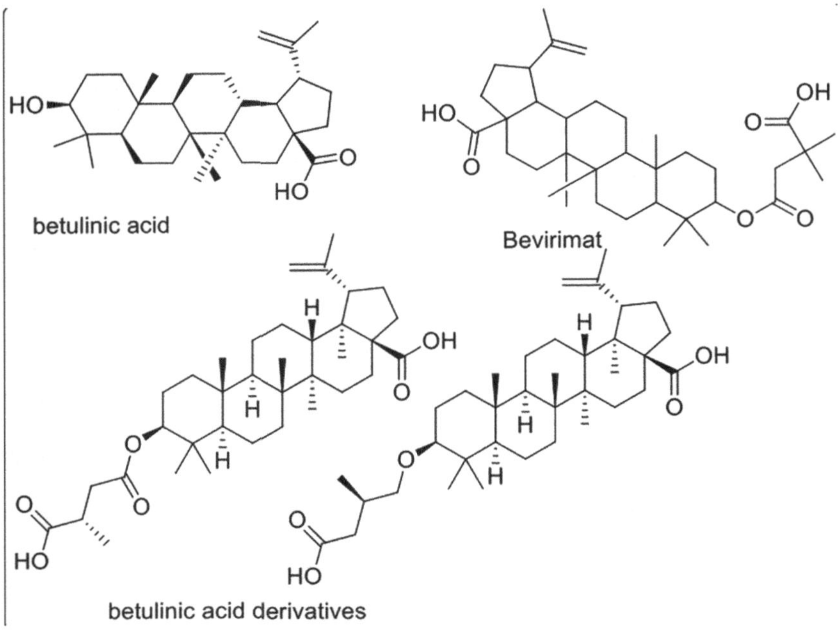


**FIGURE 8.18** Betulinic acid and its analogues with potential anti-HIV activity.

The research revealed that bevirimat inhibited, in a dose-dependent manner, the maturation of capsid precursor p25 to mature capsid (CA) p24, a specific step in the processing of the viral Gag polyprotein (Soler et al., 1996) (Figure 8.18).

Due to the poor cleavage of p25 to p24, immature, morphologically flawed viral particles were created that were not contagious. Bevirimat was consequently classified as a maturation inhibitor (MI), a new class of antiretroviral medication. It is currently undergoing Phase IIb clinical trials and advanced to Phase III trials in 2009. On the other hand, it was found that betulinic acid derivatives containing C-28ω-aminoalkanoic acid work by preventing viral entry into the host cells. The introduction of a second aminoalkanoic acid at the end of the C-28 side chain may further modulate the antiviral potency (Mayaux et al., 1994; Sun et al., 2002). Incremental chain lengthening significantly affected the anti-HIV potency of the derivatives. This research resulted in the identification of two new compounds: IC9564, a stereoisomer of RPR103611, which demonstrated equipotency in antiviral testing, and RPR103611, a statine derivative that inhibits the infectivity of several HIV-1 strains with $EC_{50}$ values in the range of 0.05–2 mM.

A43-D was created with a C-28 substituent that has been slightly simplified. Additionally, compared to IC9564 and RPR103611, it displayed comparable anti-HIV activity (Huang et al., 2006; Lai et al., 2008) (Figure 8.19).

A study on the mechanism of action found that C-28-modified BA analogues work at a post-binding stage required for virus entry into the host cell. A thorough investigation revealed that IC9564 caused an unproductive conformation of gp120, which was unable to cause the conformational rearrangement of gp41 necessary for membrane fusion. There is evidence that chemokine receptor binding occurs at the V3 loop of the gp120 protein, which is a target of IC9564 (Qian et al.). Investigations are still ongoing to determine the precise binding target of triterpene derivatives. The C-19 isopropenyl group modification had little impact on the anti-HIV activity. The anti-HIV activity of the betulinic acid derivatives was not significantly affected by allylic substitution at C-30 through an ether or thioether linkage, saturation of the 20 double bond, or either. More investigation revealed that the C-30 position is a good location to include water-soluble moieties and boost the hydrophilicity of this compound class. It is interesting to note that depending on where the

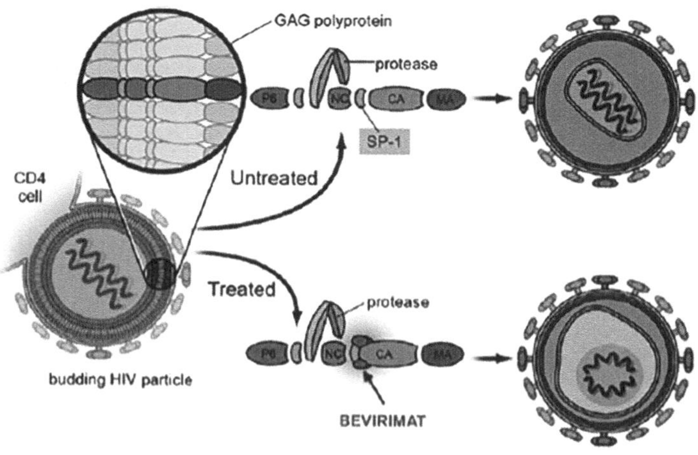


**FIGURE 8.19** Bevirimat targets Gag at clevagesitr CA-SP1 resulting in the generation of morphologically flawed non-infectious viral particles.

side chain modifications are made, different triterpene analogues have different anti-HIV-1 targets. Betulinic acid derivatives modified with C-3 or C-28 are potent HIV entry inhibitors, preventing virus maturation as well. A12-2, a dual-purpose HIV inhibitor that prevents both viral entry and maturation, was discovered as a result of these two modifications (Gerrish et al., 2008). In a mixture of human liver microsomes, compounds 167 and 168 that have a secondary amine forming the C-28 amide bond displayed noticeably higher metabolic stability (Li et al., 2007).

The anti-HIV activity of compound A12-2 was extremely strong and marginally superior to that of bevirimat. Another series of 3,28-disubstituted betulinic acid analogues were created by Gerrish et al. These compounds had different substituted benzyl and phenethyl amides at C-28 and a dimethylsuccinyl ester side chain at C-3. A number of the compounds in this series demonstrated strong anti-HIV activity, and bevirimat derivatives displayed the best pharmacokinetic/pharmacodynamic (PK/PD) characteristics. The best activity was typically produced by substitutions at the ortho-position of the benzyl or phenyl ring. A new generation of betulinic acid-derived HIV-1 inhibitors as potential clinical trial candidates may be developed using the 3,28-dimodified derivatives as attractive, promising leads (Wei et al., 2008). A lupane-like triterpene with a rearranged five-membered A-ring was isolated from the thorns of Gleditsia sinensis and is known as 2β-Carboxyl-3β-hydroxyl-norlupA(1)-20(29)-en-28-oic acid. It had a significant $EC_{50}$ value <0.13 μM, indicating anti-HIV activity. A-ring functionalities are likely significant pharmacophores for the antiviral activity of the lupane-type triterpenes, as evidenced by the fact that 2β-Carboxyl-3β-hydroxyl-norlupA(1)-20(29)-en-28-oic acid differs from betulinic acid only in the A-ring configuration (Parra et al., 2009). During a search for antiviral compounds in the stems of *Stauntonia obovatifoliola*, a novel 2,3-seco-lupane triterpene, 16β-hydroxy-2,3-seco-lup-20(29)-ene-2,3-dioic acid, was discovered. When compared to the control substance 30S-MSB, a well-known HIV-1 protease inhibitor, compound 16β-hydroxy-2,3-seco-lup-20(29)-ene-2,3-dioic acid exhibited good anti-HIV-1 protease activity ($EC_{50}$: 17.8 μM) (Huang et al., 2007). This finding indicated that triterpenes' antiviral protease activity might be enhanced by a seco A-ring with dual polar moieties. During testing, triterpenes of the lupane, 30-nor-oleanane, and oleanane types all exhibited comparable anti-HIV protease activity. A closer look at the structures revealed that the 16β-OH group is crucial for the increased antiviral activity of lupane triterpenes, as evidenced by substances like 16β-hydroxy-2,3-seco-lup-20(29)-ene-2,3-dioic acid (Huang et al., 2007) (Figure 8.20).

Maslinic acid, an oleanane-type triterpene, inhibits the HIV-1 protease. Several different α- and ω-amino acids (single, dipeptide, or tripeptide) were coupled at the C-28 position of maslinic acid by

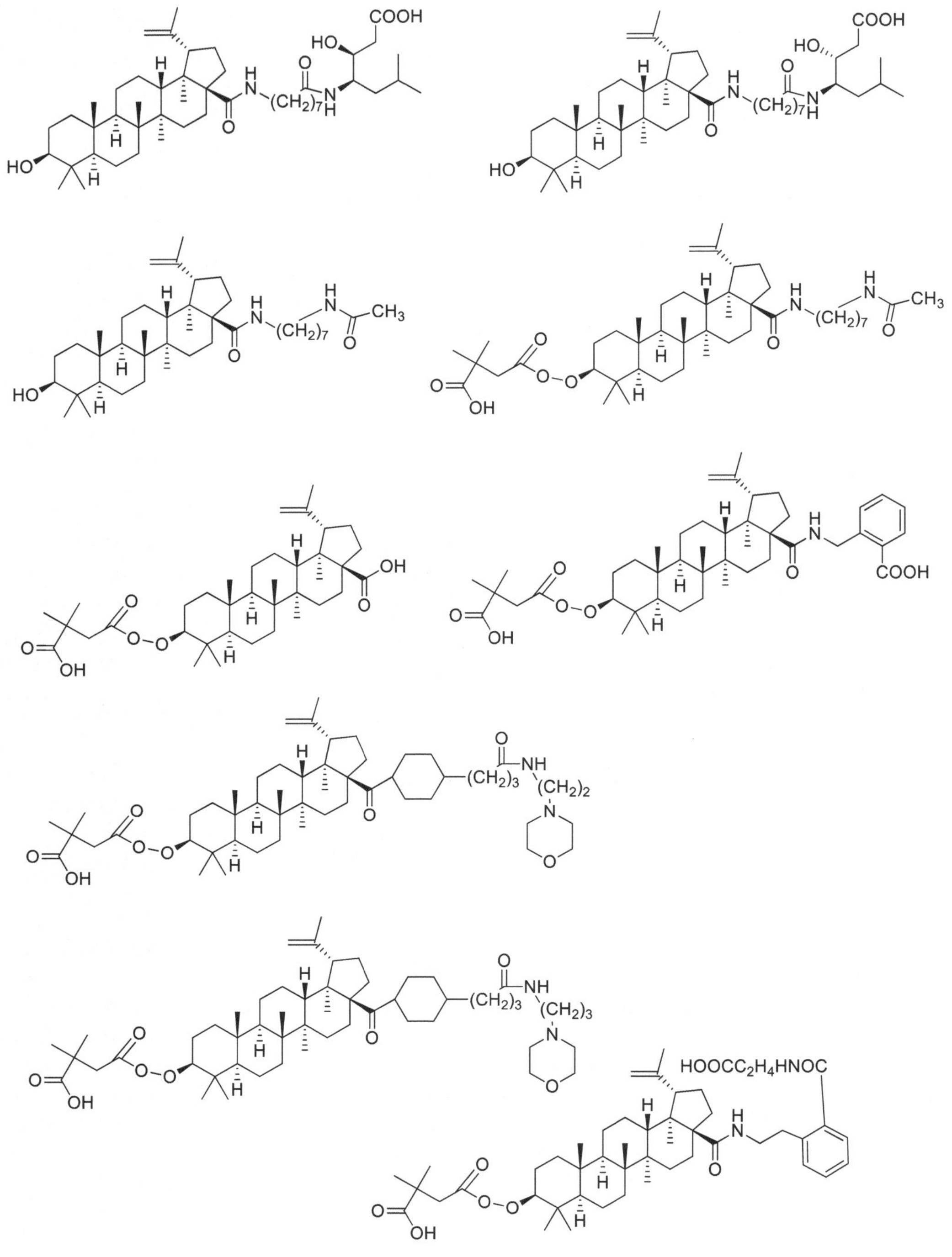

**FIGURE 8.20** Structures of some betulinic acid derivatives with potential anti-HIV activity.

using both solution- and solid-phase synthetic techniques in order to further investigate the antiviral activity. Only two substances, however—those [with 11-aminoundecanoic acid (11AUA) at C-28] and [with a C-28 conjugate dipeptide of 11AUA and Lvaline]—exhibited about 50% inhibition activity against virus replication at 10 μM. Although maslinic acid itself exhibited both cytotoxic

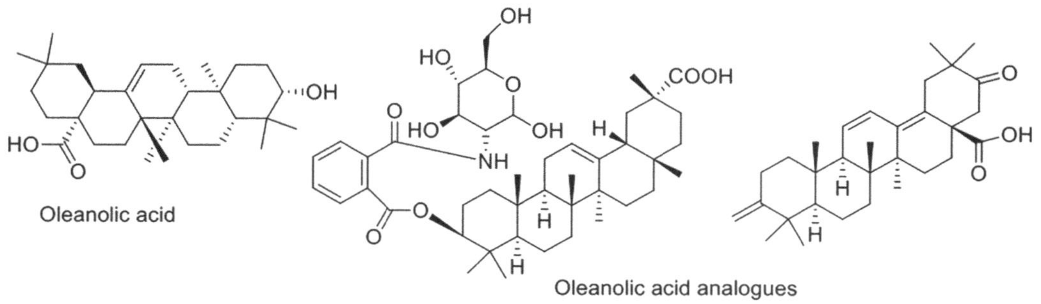


**FIGURE 8.21** Oleanolic acid and its potential anti-HIV derivatives.

and antiviral activities, these derivatives did not exhibit any cytotoxicity (Yu et al., 2006). The majority of earlier studies on glycyrrhetic acid and its synthetic derivatives produced unfavourable findings. The changes mainly involved replacing the C-3 hydroxy and C-30 carboxylic acid groups. The skeleton may be crucial to the anti-HIV and other bioactivities since the rigid triterpene scaffold controls the relative position of the functional side chains. As a result, modelling software was used to predict the distance between the hydroxy group and the carboxylic moiety as well as the dihedral angles of the D/E rings. The findings offered a potential explanation for why betulinic or moronic acid analogues exhibited stronger anti-HIV effects than glycyrrhetic acid analogues (Baltina et al., 2008).

Oleanolic acid (OA) and its biological activities of derivatives against HIV and other diseases were examined by Sultana et al. in 2007. Since OA inhibited HIV-1 replication with an $EC_{50}$ value of 3.7 μM and a TI of 12.8, it was basically determined that it was an anti-HIV principle. 3-O-3′,3′-dimethylsuccinyl-OA, which had significant antiviral $EC_{50}$ (0.86 nM) and TI (22 400) values, was produced by altering OA at the C-3 position. However, the reported activity of oleanolic acid may differ in different publications due to variations in the screening system. The mechanism of action of 3-O-3′,3′-dimethylsuccinyl-OA is comparable to that of bevirimat, a maturation inhibitor that prevents the cleavage of p25 to p24 and causes the production of morphologically flawed, non-infectious HIV particles. When compared to 3-O-acyl BA analogues, the SAR of 3-O-acyl OAs with various succinyl or glutaryl substitutions showed similar trends, indicating that the C-3 side chain is a very significant pharmacophore and that the triterpene backbone may act as a docker to enable the side chains to interact with the targets. The ideal C-3 and C-28 side chains in bi-functional betulinic acid (BA) analogues have previously been found. Therefore, some 3,28-disubstituted OA derivatives were also prepared to broaden the research's focus on replacing triterpenoid docker. With an EC50 value of 0.95 μM, OA derivative of compound, with 1,7-diaminoheptane linked at the C-28 position and 3′,3′-dimethylsuccinyl linked at the C-3 position, exhibited anti-HIV activity. Indicating that BA is a better-fit docker than OA, this activity is lower than that of the identically substituted BA analogue, A12-2 ($EC_{50}$: 0.0026 μM) (Figure 8.21).

The lanostane group plants of the genus *Schisandra* are known to contain dibenzocyclooctadiene lignans, as well as lanostane and cycloartane triterpenes. In 2003, Sun's group began a phytochemical study of several species of the *Schisandra* genus collected in the Yunnan province of China, which led to the discovery of a series of norcycloartane triterpenoids. During 2006–2008, Sun's group also identified several nortriterpenoids and bisnortriterpenoids (Huang et al., 2007).

Various species of Schisandra possessed weak to moderate anti-HIV-1 activity. Rubriflordilactones A and B were isolated from the leaves and stems of *Schisandra rubriflora*. They are highly unsaturated rearranged bisnortriterpenoids, comprising a biosynthetically modified aromatic D-ring. The anti-HIV activity of both compounds was tested in HIV-1IIIB-infected C8166 cell lines, and their cytotoxicity was evaluated in C8166 and MT-4 cells (Kashiwada et al., 1998) (Figure 8.22).

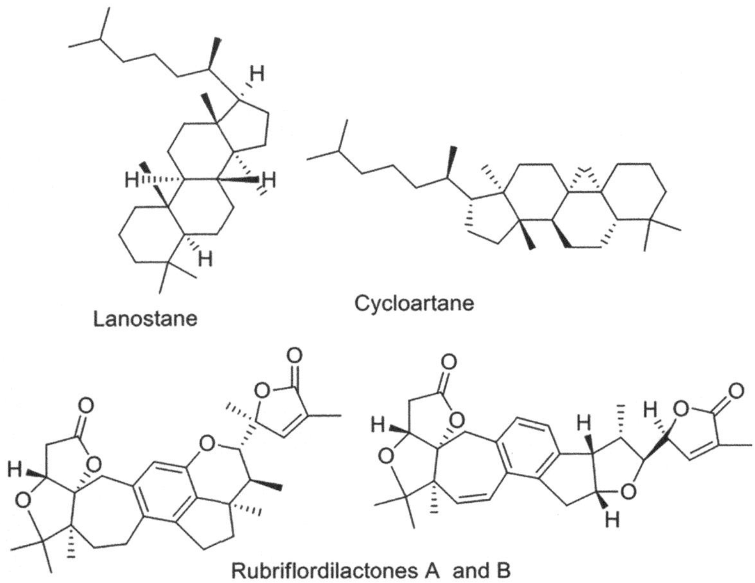


**FIGURE 8.22** Bioactive anti-HIV compounds from genus *Schisandra*.

Sphenadilactones A and B were isolated from the leaves and stems of *Schisandra sphenanthera*. Their structural elucidation was accomplished by extensive NMR and single-crystal X-ray crystallography (Ho et al., 2007). However, an erratum was later published to correct the stereochemistry of Me-27, which should have an a- rather than b-orientation. Sphenadilactones A exhibited very weak anti-HIV-1 activity, with an $EC_{50}$ of 237.8 mM in HIV1IIIB-infected C8166 cells. No obvious cytotoxicity was observed when both compounds were tested against three human tumour cell lines, K562, A549, and HT-29. A minor nortriterpenoid, sphenadilactone C, was also isolated in a subsequent study of *Schisandra sphenanthera*. It is characterised by the presence of a partial enol structure, and is the first example of an acetamide bearing highly oxygenated nortriterpenoid from a cycloartane. Sphenadilactone C also exhibited very weak antiviral activity against HIV-1IIIB, with an $EC_{50}$ of 47.8 mM (Kashiwada et al., 2000). Re-investigation of *Schisandra sphenanthera* resulted in the isolation of sphenalactones A–D, which have an unusual 27-carbon skeleton derived from cycloartane, due to the loss of C-2, C-3, and CH3–28. This is the first class in the Schisandra species and the first case in the cycloartane triterpenoid family in which C-2 and C-3 are degraded in the molecule. Other cycloartane triterpenoid compounds demonstrated weak anti-HIV-1 replication activity, with $EC_{50}$ values of 185.1, 153.7, 105.4, and 76.2 mM, respectively isolated from *Schisandra chinensis*. Both compounds have a unique carbon skeleton, with an unstable stereogeniccenter at C-20. The two diastereomers slowly interconvert through an enol intermediate. Nortriterpenoids schindilactones D–G and bisnortriterpenoidswuweizidilactones G and H were also isolated from the stems and leaves of *Schisandra chinensis*. However, they demonstrated only weak antiviral activity, with $EC_{50}$ values ranging from 32 to 181 mM. Wilsonianadilactones A–C were isolated from the stems and leaves of *Schisandra wilsoniana*, a climbing plant distributed in the Yunnan province of China. All three compounds showed weak anti-HIV-1 activity, with $EC_{50}$ values of 38.3, 99.1, and 118.6 mM in an *in vitro* antiviral assay. Lancifodilactone H, a trinorcycloartane triterpenoid with a biosynthetically modified seven-membered lactone ring, and lancifoic acid A, a new A ring-secocycloartane triterpenoid, were isolated from the stems and leaves of *Schisandra lancifolia*. They exhibited moderate anti-HIV activity, with $EC_{50}$ values of 37.4 and 33.2 mM, respectively. The genus *Kadsura* belongs to the family Schisandraceae. In an investigation of *Kadsura heteroclita*, a new nortriterpenoid, longipedlactone J, was isolated. When tested in HIV-1IIIB infected

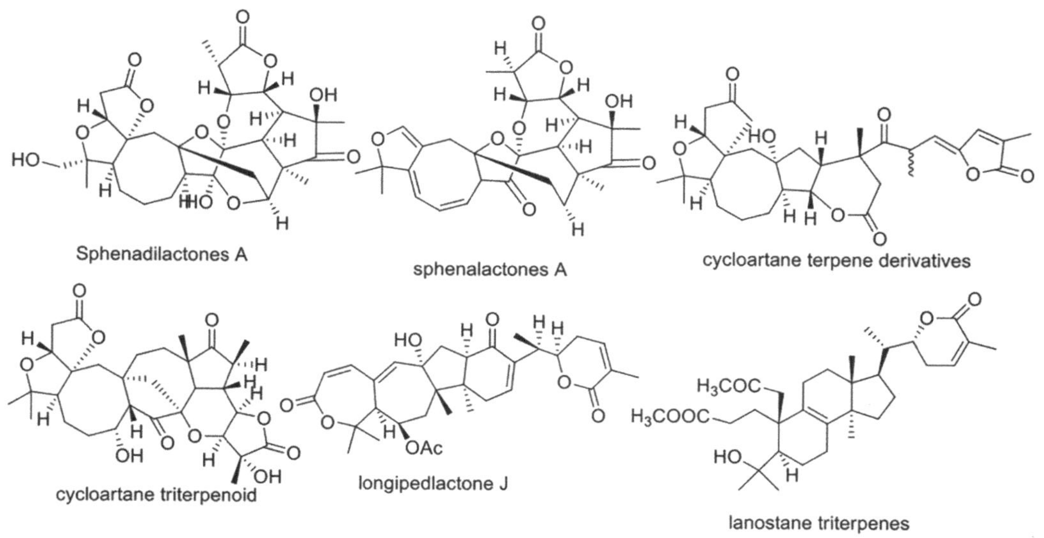


**FIGURE 8.23** Some other bioactive anti-HIV compounds from genus *Schisandra*.

C8166 cells, longipedlactone J showed a moderate anti-HIV $EC_{50}$ of 7.1 mM. Four new lanostane triterpenes, colossolactones V, VI, VII, and VIII, were isolated from the fruiting bodies of the Vietnamese mushroom *Ganoderma colossum*. Lanostane triterpenes V have seco A- and E-rings, which include a seven-membered lactone A-ring and six-membered lactone E-ring (Deng et al., 2007; Tshibangu et al., 2004) (Figure 8.23).

Brazilian propolis, an herbal remedy used all over the world, was used to isolate moronic acid (MA). With an $EC_{50}$ value of 0.2 μM and a TI of 186, it demonstrated significant anti-HIV activity. The authors' team announced in 2006 that they had found a number of MA derivatives with exceptional anti-HIV potency. With $EC_{50}$ values of 0.017, 0.0156, and 0.007 mM, respectively, derivatives of Compounds Moronic Acid with C-3 3′,3′-dimethylsuccinyl substitution and various C-28 amide side chains demonstrated strong anti-HIV activity in H9 lymphocytes. These activity results are comparable to those of the drug candidate bevirimat (EC50: 0.007 μM), which is undergoing Phase IIb anti-AIDS clinical trials. The antiviral $EC_{50}$ value of another 3,28-disubstituted MA derivative created was 0.055 μM. 3,28-disubstituted betulinic acid or moronic acid analogues exhibit both anti-fusion and anti-maturation activity, according to a mechanism of action study (Mills et al., 2005).

## 8.11 ONGOING CLINICAL TRIALS OF PLANTS WITH ANTI-HIV ACTIVITY

The use of herbal medicine in clinical research and HIV/AIDS treatment is rarely reported. Herbal remedies are the main form of treatment for HIV, AIDS, and diseases associated with HIV, which are the most common problems in Africa. In China, three different types of treatment systems are involved with HAART. One of them is the traditional Chinese medicine, which is given by certified Chinese herbalists. There are a number of randomised studies examining the beneficial effects of conventional medicinal plants on patients with HIV or AIDS and comparing them to a control group (without treatment and placebo). The effects on the promotion of $CD4^{+}$ cells were monitored. Approximately 11 different Chinese traditional medicinal plants, including *Panax ginseng*, *Astragalus membranaceus*, *Lycium barbarum*, *Trichosanthis kirilowii*, and *Viola mandshurica*, have been tested in about 1,000 patients across various studies. Traditional medical plant treatments increased CD4 cells when compared to placebo, but further research is the utmost requirement (Tani et al., 2002). During a 24-week study, a Chinese herbal supplement made of 14 plants (*Coptis chinensis*, *Jasminum officinale*, *Wolfiporia extensa*, *Sparganium stoloniferum*,

*Polygonatum odoratum*, and *Scrophularia buergeriana*) was found to increase plasma CD4 count and also exhibited HIV growth inhibition (Djohan et al., 2009). In one US study, it was discovered that 26% of HIV-positive individuals also take herbal medications. A study conducted in Europe found that 25% of HIV-positive individuals use herbal medications (Tani et al., 2002). The study, which included 366 HIV-positive African-American women enrolled in herbal medicine therapy, revealed that these patients had antiretroviral effects that were 1.69 times stronger than those of women not using the therapy based on medicinal plants (Djohan et al., 2009). In South Africa, the effectiveness of *Calendula officinalis* was tested on 33 HIV-positive volunteers (7 men and 26 women, aged 22–43). Viral loads and CD4 T-cell counts both markedly decreased. *Hypoxis hemerocallidea* and *Sutherlandia frutescens* are two plant remedies that have been used for HIV/AIDS treatment, according to the South African Ministry of Health, which is actively promoting their use with antiretroviral medications. Additionally, it was observed in Romania that children with AIDS who received natural herbal treatments had lower mortality rates. Furthermore, a significant increase in CD4 and CD8 lymphocytes was seen in blood samples taken from 30 adults who used an *Alternanthera pungens* extract. Twenty-five practitioners of traditional medicine were interviewed for the study that was carried out to show the use of medicinal plants in various districts of Uganda, where this disease was first described and 1 million people are infected. The doctors saw an average of 29 patients annually (range: 2–250). They mentioned that 145 members belong to the *Euphorbiaceae, Fabaceae*, and *Asteraceae* families. It was also noted that *Aloe* spp., *Erythrina abyssinica*, *Sarcocephalus latifolius*, *Psorospermum febrifugum*, *Mangifera indica*, and *Warburgia salutaris* were among the most commonly used plants. A progressive loss of CD4positive T-cell lymphocytes in the blood was seen in patients receiving treatment with herbal medicine.

## 8.12 CONCLUSION AND FUTURE ASPECTS

From an overall review of literature, it is observed that numerous plant-derived molecules were useful against AIDS, but the extensive and fruitful mechanistic biological activity evaluation posed one of the biggest challenges to the researchers. The significant clinical retorts of natural plant derivatives often encourage putting endeavour towards better research. This chapter's study highlights plant-derived compounds and plant extracts or fractions as an advantageous and privileged moiety in the field of pharmaceutical chemistry with its broad range of anti-HIV profile, including numerous targets to throw enormous scientific knowledge for development and designing of newer analogues. In some areas, various semi-synthetic analogues were also described. The in-detail explanation of different architectures of natural products and their essential pharmacophore delivered a better understanding of both design and position of substituents on this exceptional potent drug molecule deemed to be liable for its efficiency.

Natural products have been significant contributors to the pharmaceutical industry for era considerable period. These derivatives offer ideal possibilities for interacting with a broad variety of biomacromolecules as it is structurally identical to that of many naturally occurring nucleotides and are found in various naturally existing compounds. Additionally, this nucleus, owing to its electron-rich character, has been employed in making metal complexes. In order to bring novelty in AIDS medication, a large number of intriguing natural pharmacophore moiety have emerged in recent years. Despite these developments, there are ample opportunities to further study this intriguing moiety by extensive and detailed mechanistic biological activity evaluation, which may yield beneficial results in the field of AIDS drug development. In response to the growing need to treat the fatal maladies, researchers have put their focus in developing novel therapies derived from natural bioactive molecules with a reduced potential for adverse effects and an enhanced pharmacodynamic as well as pharmacokinetic profile.

## REFERENCES

Alkhatib G, Berger EA, HIV coreceptors: from discovery and designation to new paradigms and promise. *Eur J Med Res 2007*; 12:375–380.

Anderson S-J, Cherutich P, Kilonzo N, Cremin I, Fecht D, Kimanga D, Harper M, Masha RL, Ngongo PB, Maina W, Maximising the effect of combination HIV prevention through prioritisation of the people and places in greatest need: a modelling study. *Lancet* 2014; 384:249–256.

Appendino G, Ottino M, Marquez N, Bianchi F, Giana A, Ballero M, Sterner O, Fiebich BL, Munoz E, Arzanol, an anti-inflammatory and anti-HIV-1 phloroglucinol α-pyrone from Helichrysum italicum ssp. microphyllum. *J Nat Prod* 2007; 70:608–612.

Arrildt KT, LaBranche CC, Joseph SB, Dukhovlinova EN, Graham WD, Ping L-H, … Swanstrom R, Phenotypic correlates of HIV-1 macrophage tropism. *J Virol* 2015; 89(22):11294–11311.

Asres K, Bucar F, Anti-HIV activity against immunodeficiency virus type 1 (HIV-I) and type II (HIV-II) of compounds isolated from the stem bark of Combtetummolle. *Ethiop Med J* 2005; 43:15–20.

Baltina Jr. LA, Kondratenko RM, Plyasunova OA, Baltina LA, Pokrovskii AG, Khalilov LM, Galin FZ, Tolstikov GA, *Pharm Chem J* 2008; 42:64–67.

Barak V, Halperin T, Kalickman I, Barak T, Halperin IK, The effect of Sambucol, a black elderberry-based, natural product, on the production of human cytokines : I. Inflammatory cytokines Healthy donors blood monocyte cultures Cytokine assays Evaluation of Sambucol's effect. *Eur Cytokine Netw* 2012; 12(2):4–7.

Bean P, New drug targets for HIV. *Clin Infect Dis*. 2005; 41:96–100.

Bessong PO, Rojas LB, Obi LC, Tshisikawe PM, Igunbor EO, Further screening of Venda medicinal plants for activity against HIV type 1 reverse transcriptase and integrase. *African J Biotechnol* 2006; 5:526–528.

Bhusal RP, Foster SR, Stone MJ, Structural basis of chemokine and receptor interactions: key regulators of leukocyte recruitment in inflammatory responses. *Protein Sci* 2020; 29(2):420–432.

Bonolo PF, Gomes RRFM, Guimarães MDC, Adesão à terapia anti-retroviral (HIV/aids): fatoresassociados e medidas da adesão. *EpidemiolServSaúde* 2007; 16(4):261–278. https://dx.doi.org/10.5123/S1679-49742007000400005

Brenchley JM, Hill BJ, Ambrozak DR, Price DA, Guenaga FJ, Casazza JP, et al., T-cell subsets that harbor human immunodeficiency virus (HIV) in vivo: implications for HIV pathogenesis, *J Virol* 2004; 78(3):1160–1168.

Burack JH, Cohen MR, Hahn JA, Abrams DI, Pilot randomized controlled trial of Chinese herbal treatment for HIV-associated symptoms. *J Acquir Immune Defic Syndr* 1996; 12:386–393.

Cardozo T, Kimura T, Philpott S, Weiser B, Burger H, Zolla-Pazner S, Structural basis for coreceptor selectivity by the HIV type 1 V3 loop. *AIDS Res Hum Retrovirus* 2007; 23(3):415–426.

Centers for Disease Control and Prevention (CDC), *HIV Surveillance Report*. 2019; 26(1):1–81.

Chen B, Molecular mechanism of HIV-1 entry. *Trends Microbiol* 2019; 27(10):878–891. Available from: https://doi.org/10.1016/j.tim.2019.06.002

Chen M, Kilgore N, Lee KH, Chen DF, Rubrisandrins A and B, lignans and related anti-HIV compounds from Schisandra rubriflora. *J Nat Prod* 2006; 69:1697–1701.

Chiang LC, Chiang W, Chang MY, Lin CC, In vitro cytotoxic, antiviral and immunomodulatory effects of Plantago major and Plantago asiatica. *Am J Chin Med* 2003; 31(2):225–234.

Chinsembu K, Hijarunguru A, Mbangu A, Ethnomedicinal plant used by traditional healers in the management of HIV/AIDS opportunistic diseases in Rundu, Kavango East Region, Namibia. *S Afr J Bot* 2015; 100:33–42.

Dat NT, Bae K, Wamiru A, McMahon JB, Le Grice SF, Bona M, Beutler JA, Kim YH, A dimeric lactone from Ardisia japonica with inhibitory activity for HIV-1 and HIV-2 ribonuclease H. *J Nat Prod* 2007; 70:839–841.

Deng S-L, Baglin I, Nour M, Flekhter O, Vita C, Cave C, Synthesis of ursolic phosphonate derivatives as potential anti-HIV agents. *Phosphorus, Sulfur Silicon Relat Elem* 2007; 182:951–967.

Deng X, Jiang M, Zhao X, Liang J, Efficacy and safety of traditional Chinese medicine for the treatment of acquired immunodeficiency syndrome: a systematic review. *J Tradit Chinese Med* 2014; 34:1–9. https://doi.org/10.1016/s0254-6272(14)60046-7.

Dhonde SM, Review on Role of Medicinal Plants in HIV/AIDS. *WORLD JOURNAL OF PHARMACY AND PHARMACEUTICAL SCIENCES* 2022; 11(8): 230–239.

Djohan Y, Camara C, Monde A, Koffi G, Niamké G, Déré L, Tiahou G, Djessou P, Sess D, Interest of antioxidants in the care of the patients infected by the HIV: the experience of long term administration of alternanthera pungens herb tea. *Ann Biol Clin* 2009; 67:563–568.

Fanales-Belasio E, Raimondo M, Suligoi B, Butto S, HIV virology and pathogenetic mechanisms of infection: a brief overview. *Ann Ist Super Sanita* 2010; 46:5–14.

Fu M, Ng TB, Jiang Y, Pi ZF, Liu ZK, Li L, Liu F, Compounds from rose (Rosa rugosa) flowers with human immunodeficiency virus type 1 reverse transcriptase inhibitory activity. *J Pharm Pharmacol* 2010; 58:1275–1280. https://doi.org/10.1211/jpp.58.9.0015.

Fujioka T, Kashiwada Y, Kilkuskie RE, Cosentino LM, Ballas LM, Jiang JB, Janzen WP, Chen IS, Lee KH, Anti-AIDS agents, 11. Betulinic acid and platanic acid as anti-HIV principles from Syzigium claviflorum, and the anti-HIV activity of structurally related triterpenoids. *J Nat Prod* 1994; 57:243–247.

Gallay P, Syndecans and HIV-1 pathogenesis. *Microbes Infect* 2004; 6(6):617–622.

Gao XM, Pu JX, Huang SX, Yang LM, Huang H, Xiao WL, Zheng YT, Sun HD, Lignans from Kadsura angustifolia. *J Nat Prod* 2008; 71:558–563.

Garzino-Demo A, Chemokines and defensins as HIV suppressivefactors: an evolving story. *Curr Pharm Des* 2007; 13:163–172.

Gerencer M, Turecek PL, Kistner O, Mitterer A, Savidis-Dacho H, Barrett NP, In vitro and in vivo anti-retroviral activity of the substance purified from the aqueous extract of Chelidonium majus L. *Antivir Res* 2006; 72:153–156.

Gerrish D, Kim C, Kumar Dange V, Austin H, Garrus Jennifer E, Baichwal V, Saunders M, McKinnon Rena S, Anderson Mark B, Carlson R, Arranz-Plaza E, Yager Kraig M, Triterpene based compounds with potent anti-maturation activity against HIV-1. *Bioorg Med Chem Lett* 2008; 18:6377–6380.

Gonzalez SM, Aguilar-Jimenez W, Su RC, Rugeles MT, Mucosa: key interactions determining sexual transmission of the HIV infection. *Front Immunol* 2019; 10. https://doi.org/10.3389/fimmu.2019.00144.

Gouws E, White PJ, Stover J, Brown T, Short term estimates of adult HIV incidence by mode of transmission: Kenya and Thailand as examples. *Sex Transm Infect* 2006; 82:51–55.

Gravatt LAH, Leibrand CR, Patel S, McRae MP, New drugs in the pipeline for the treatment of HIV: a review. *Curr Infect Dis Rep* 2017; 19:1–9. https://doi.org/10.1007/s11908-017-0601-x.

Harvey A, Strategies for discovering drugs from previously unexplored natural products. *Drug Discov Today* 2000; 5:294–300.

Ho DD, Perspectives series: host/pathogen interactions. Dynamicsof HIV-1 replication in vivo. *J Clin Invest* 1997; 99:2565–2567.

Ho JC, Chen CM, Row LC, Oleanane-type triterpenes from the flowers, pith, leaves, and fruit of Tetrapanax papyriferus. *Phytochemistry* 2007; 68:631–635.

Hsieh PW, Chang FR, Lee KH, Hwang TL, Chang SM, Wu YC, A new anti-HIV alkaloid, drymaritin, and a new C-Glycoside flavonoid, diandraflavone, from Drymaria d iandra. *J Nat Prod* 2004; 67:1175–1177.

Huang L, Ho P, Lee KH, Chen CH, Synthesis and anti-HIV activity of bi-functional betulinic acid derivatives. *Bioorg Med Chem* 2006; 14:2279–2289.

Huang L, Yu D, Ho P, Lee K-H, Chen C-H, Synthesis and anti-HIV activity of bi-functional triterpene derivatives. *Lett Drug Des Discovery*. 2007; 4:471–478.

Jadaun P, Khopkar P, Kulkarni S, Repurposing phytochemicals as anti-HIV agents. *J Antivir Antiretrovir* 2016; 8:139–141.

Jasinska AJ, Pandrea I, Apetrei C, CCR5 as a coreceptor for human immunodeficiency virus and simian immunodeficiency viruses: a prototypic love-hate affair. *Front Immunol 2022*; 13:835994. https://doi.org/10.3389/fimmu.2022.835994.

Jayasuriya H, Herath KB, Ondeyka JG, Polishook JD, Bills GF, Dombrowski AW, Springer MS, Siciliano S, Malkowitz L, Sanchez M, Guan Z, Tiwari S, Stevenson DW, Borris RP, Singh SB, Isolation and structure of antagonists of chemokine receptor (CCR5). *J Nat Prod* 2004; 67:1036–1038.

Jiratchariyakul W, Wiwat C, Vongsakul M, Somanabandhu A, Leelamanit W, Fujii I, Suwannaroj N, Ebizuka Y, HIV inhibitor from Thai bitter gourd. *Planta Med* 2001; 67:350–353.

Kalvatchev Z, Walder R, Garzaro D, Anti-HIV activity of extracts from Calendula officinalis flowers. *Biomed Pharmacother* 1997; 51:176–180.

Kanta V, Unnati S, Ritu M, A review on: aids and herbal remedies. *Int J Res Ayurveda Pharm* 2011; 2:1709–1713.

Kashiwada Y, Hashimoto F, Cosentino LM, Chen CH, Garrett PE, Lee KH, Betulinic acid and dihydrobetulinic acid derivatives as potent anti-HIV agents. *J Med Chem* 1996; 39:1016–1017.

Kashiwada Y, Wang HK, Nagao T, Kitanaka S, Yasuda I, Fujioka T, Yamagishi T, Cosentino LM, Kozuka M, Okabe H, Ikeshiro Y, Hu CQ, Yeh E, Lee KH, Anti-AIDS agents. 30. Anti-HIV activity of oleanolic acid, pomolic acid, and structurally related triterpenoids. *J Nat Prod* 1998; 61:1090–1095.

Kashiwada Y, Nagao T, Hashimoto A, Ikeshiro Y, Okabe H, Cosentino LM, Lee KH, Anti-AIDS agents 38. Anti-HIV activity of 3-O-acyl ursolic acid derivatives. *J Nat Prod* 2000; 63:1619–1622.

Kashman Y, Gustafson KR, Fuller R, Cardellina J, 2nd, McMahon J, Currens M, Buckheit R, Jr, Hughes S, Cragg G, Boyd M, The calanolides, a novel HIV-inhibitory class of coumarin derivatives from the tropical rainforest tree, Calophyllumlanigerum. *J Med Chem* 1992; 35:2735–2743.

Kaur R, Sharma P, Gupta GK, Ntie-Kang F, Kumar D, Structure-activity-relationship and mechanistic insights for anti-HIV natural products. *Molecules* 2020; 25(9):1–48.

Keele BF, Estes JD, Barriers to mucosal transmission of immunodeficiency viruses. *Blood* 2011; 118:839–846. https://doi.org/10.1182/blood-2010-12-325860.

Keller PW, Morrison O, Vassell R, Weiss CD, HIV-1 gp41 residues modulate CD4-induced conformational changes in the envelope glycoprotein and evolution of a relaxed conformation of gp120. *J Virol* 2018; 92(16):e00583-18.

Kim HJ, Kim EJ, Seo SH, Shin CG, Jin C, Lee YS, Vanillic acid glycoside and quinic acid derivatives from Gardeniae Fructus. *J Nat Prod* 2006; 69:600–603.

Klos M, Van M, Milne P, Traore H, Meyer D, Oosthuizen V, In vitro anti-HIV activity of five selected South African medicinal plant extracts. *J Ethnopharmacol* 2009; 124:182–188.

Kuo RY, Qian K, Morris-Natschke SL, Lee KH, Plant-derived triterpenoids and analogues as antitumor and anti-HIV agents. *Nat Prod Rep* 2009; 26(10):1321–1344.

Kurapati KRV, Atluri VS, Samikkannu T, Garcia G, Nair MPN, Natural products as anti-HIV agents and role in HIV-associated neurocognitive disorders (HAND): a brief overview. *Front Microbiol* 2016; 6:1–14. https://doi.org/10.3389/fmicb.2015.01444.

Lackner AA, Mohan M, Veazey RS, The gastrointestinal tract and AIDS pathogenesis, *Gastroenterology* 2009; 136(6):1966–1978.

Lai W, Huang L, Ho P, Li Z, Montefiori D, Chen CH, Betulinic acid derivatives that target gp120 and inhibit multiple genetic subtypes of human immunodeficiency virus type 1. *Antimicrob Agents Chemother* 2008; 52:128–136.

Lee J, Oh WK, Ahn JS, Kim YH, Mbafor JT, Wandji J, Fomum ZT, Prenylisoflavonoids from Erythrina senegalensis as novel HIV-1 protease inhibitors. *Planta Med* 2009; 75:268–270.

Leteane MM, Ngwenya BN, Muzila M, Namushe A, Mwinga J, Musonda R, Moyo S, Mengestu YB, Abegaz BM, Andrae-Marobela K, Old plants newly discovered: Cassia sieberiana DC and Cassia abbreviataOliv. Oliv. root extracts inhibit in vitro HIV-1C replication in peripheral blood mononuclear cells (pbmcs) by different modes of action. *J Ethnopharmacol* 2012; 141:48–56.

Levy JA, HIV pathogenesis: 25 years of progress and persistent challenges. *Aids* 2009; 23(2):147–160.

Li F, Goila-Gaur R, Salzwedel K, Kilgore NR, Reddick M, Matallana C, Castillo A, Zoumplis D, Martin DE, Orenstein JM, Allaway GP, Freed EO, Wild CT, *Proc. Natl Acad Sci USA* 2003; 100:13555–13560.

Li WH, Zhang XM, Tian RR, Zheng YT, Zhao WM, Qiu MH, A new anti-HIV lupane acid from Gleditsia sinensis Lam. *J Asian Nat Prod Res* 2007; 9:551–555.

Li G, Haney KM, Kellogg GE, Zhang Y, Comparative docking study of anibamine as the first natural product CCR5 antagonist in CCR5 homology models. *J Chem Inf Model* 2009a; 49:120–132.

Li XN, Pu JX, Du X, Yang LM, An HM, Lei C, He F, Luo X, Zheng YT, Lu Y, Xiao WL, Sun HD, Lignans with anti-HIV activity from Schisandra propinqua var. sinensis. *J Nat Prod* 2009b; 72:1133–1141.

Lindegaard B, Hansen A, Pilegaard H, Keller P, Gerstoft J, Pedersen BK, Adipose tissue expression of IL-18 and HIV-associated lipodystrophy. *AIDS* 2004; 18:1956–1958.

Lubbe A, Seibert I, Klimkait T, Van der Kooy F, Ethnopharmacology in overdrive: the remarkable anti-HIV activity of Artemisia annua. *J Ethnopharmacol* 2012; 141:854–859.

Lucas S, Nelson AM, HIV and the spectrum of human disease. *J Pathol* 2015; 235(2):229–241.

Lucera MB, Tilton CA, Mao H, Dobrowolski C, Tabler CO, Haqqani AA, Karn J, Tilton JC, The histone deacetylase inhibitor vorinostat (SAHA) increases the susceptibility of uninfected CD4 + T cells to HIV by increasing the kinetics and efficiency of postentry viral events. *J Virol* 2014; 88:10803–10812. https://doi.org/10.1128/jvi.00320-14.

Ma CH, Wang RR, Tian RR, Ye G, Fan MS, Zheng YT, Huang CG, Calycosin 7-O-β-D-glucopyranoside, an anti-HIV agent from the roots of Astragalus membranaceus var. mongholicus. *Chem Nat Compd* 2009; 45:282–285.

Maartens G, Celum C, Lewin SR, HIV infection: epidemiology, pathogenesis, treatment, and prevention. *Lancet* 2014; 384:258–271.

Magérus-Chatinet A, Yu H, Garcia S, Ducloux E, Terris B, Bomsel M, Galactosyl ceramide expressed on dendritic cells can mediate HIV-1 transfer from monocyte derived dendritic cells to autologous T cells. *Virology* 2007; 362(1):67–74.

Mayaux JF, Bousseau A, Pauwels R, Huet T, Henin Y, Dereu N, Evers M, Soler F, Poujade C, De Clercq E, Le Pecq JB, Triterpene derivatives that block entry of human immunodeficiency virus type 1 into cells. *Proc Natl Acad Sci USA* 1994; 91:3564–3568.

Miceli MC, Parnes JR, Role of CD4 and CD8 in T cell activation and differentiation. *Adv Immunol* 1993; 53:59–122.

Mills E, Cooper C, Seely D, Kanfer I, African herbal medicines in the treatment of HIV: hypoxis and Sutherlandia. An overview of evidence and pharmacology. *Nutr J* 2005; 4:1–6. https://doi.org/10.1186/1475-2891-4-19.

Muley BP, Khadabadi SS, Banarase NB, Phytochemical constituents and pharmacological activities of Calendula officinalis Linn (Asteraceae): a review. *Trop J Pharm Res* 2009; 8:455–465.

Nazli A, Dizzell S, Zahoor MA, Ferreira VH, Kafka J, Woods MW, Ouellet M, Ashkar AA, Tremblay MJ, Bowdish DME, Kaushic C, Interferon-β induced in female genital epithelium by HIV-1 glycoprotein 120 via Toll-like-receptor 2 pathway acts to protect the mucosal barrier. *Cell Mol Immunol* 2019; 16:178–194.

Ooi LSM, Sun SSM, Ooi VEC, Purification and characterization of a new antiviral protein from the leaves of Pandanus amaryllifolius (Pandanaceae). *Int J Biochem Cell Biol* 2004; 36(8):1440–1446.

Parra A, Rivas F, Lopez PE, Garcia-Granados A, Martinez A, Albericio F, Marquez N, Munoz E, Solution-and solid-phase synthesis and anti-HIV activity of maslinic acid derivatives containing amino acids and peptides. *Bioorg Med Chem* 2009; 17:1139–1145.

Pastorino G, Cornara L, Soares S, Rodrigues F, Oliveira MBPP, Liquorice (Glycyrrhiza glabra): a phytochemical and pharmacological review. *Phyther Res* 2018; 32(12):2323–2339.

Pengsuparp T, Serit M, Hughes SH, Soejarto DD, Pezzuto JM, Specific inhibition of human immunodeficiency virus type 1 reverse transcriptase mediated by soulattrolide, a coumarin isolated from the latex of Calophyllumteysmannii. *J Nat Prod* 1996; 59:839–842.

Pu JX, Yang LM, Xiao WL, Li RT, Lei C, Gao XM, Huang SX, Li SH, Zheng YT, Huang H, Sun HD, Compounds from Kadsura heteroclita and related anti-HIV activity. *Phytochemistry* 2008; 69:1266–1272.

Qian K, Nakagawa-Goto K, Yu D, Morris-Natschke SL, Nitz TJ, Kilgore N, Allaway GP, Lee KH, Anti-AIDS agents 73: Structure–activity relationship study and asymmetric synthesis of 3-O-monomethylsuccinyl-betulinic acid derivatives. *Bioorg Med Chem Lett* 2007; 17:6553–6557.

Qian K, Morris-Natschke SL, Lee KH, HIV entry inhibitors and their potential in HIV therapy. *Med Res Rev* 2009a; 29:369–393.

Qian K, Yu D, Chen CH, Huang L, Morris-Natschke SL, Nitz TJ, Salzwedel K, Reddick M, Allaway GP, Lee KH, Design, synthesis, metabolic stability assessment, and antiviral evaluation of novel betulinic acid derivatives as potent anti-human immunodeficiency virus (HIV) agents. *J Med Chem* 2009b; 52:3248–3258.

Reutrakul V, Ningnuek N, Pohmakotr M, Yoosook C, Napaswad C, Kasisit J, Santisuk T, Tuchinda P, Anti HIV-1 flavonoid glycosides from Ochna integerrima. *Planta Med* 2007; 73:683–688.

Saag MS, Gandhi RT, Hoy JF, Landovitz RJ, Thompson MA, Sax PE, Smith DM, Benson CA, Buchbinder SP, Del Rio C, Eron JJ, Fätkenheuer G, Günthard HF, Molina JM, Jacobsen DM, Volberding PA, Antiretroviral drugs for treatment and prevention of HIV infection in adults: 2020 recommendations of the international antiviral society-USA Panel. *JAMA - J Am Med Assoc* 2020; 324:1651–1669. https://doi.org/10.1001/jama.2020.17025.

Saklani A, Kutty SK, Plant-derived compounds in clinical trials. *Drug Discovery Today* 2008, 13:161–171.

Sarafianos SG, Marchand B, Das K, Himmel DM, Parniak MA, Hughes SH, Arnold E, Structure and function ofHIV-1 reverse transcriptase: molecular mechanisms of polymerization and inhibition. *J Mol Biol* 2009; 385:693–713.

Sharifi-Rad J, Herbal antibiotics: moving back into the mainstream as an alternative for "superbugs". *Cell Mol Biol* 2016; 62:1–2.

Shaw GM, Hunter E, HIV transmission, Cold Spring Harb. *Perspect Med* 2012; 2:a006965.

Sierra S, Kupfer B, Kaiser R, Basics of the virology of HIV-1 and its replication. *J Clin Virol* 2005; 34(4):233–244.

Soler F, Poujade C, Evers M, Carry JC, Henin Y, Bousseau A, Huet T, Pauwels R, De Clercq E, Mayaux JF, Le Pecq JB, Dereu N, *J Med Chem* 1996; 39:1069–1083.

Stebbing J, Gazzard B, Douek DC, Where does HIV live? *N Engl J Med.* 2004; 350(18):1872–1880.

Sun IC, Chen CH, Kashiwada Y, Wu JH, Wang HK, Lee KH, Synthesis, anti-HIV, and anti-fusion activities of IC9564 analogues based on betulinic acid. *J Med Chem* 2002; 45:4271–4275.

Sundquist WI, Kräusslich HG, HIV-1 assembly, budding, and maturation. *Cold Spring Harb Perspect Med* 2012; 2(7).

Tani M, Nagase M, Nishiyama T, Yamamoto T, Matusa R, The effects of long-term herbal treatment for pediatric AIDS. *Am J Chin Med* 2002; 30:51–64. https://doi.org/10.1142/S0192415X02000065.

Tshibangu K, Worku Z, De Jongh M, Van Wyk A, Mokwena S, Peranovic V, Assessment of effectiveness of traditional herbal medicine in managing HIV/AIDS patients in south Africa. *East Afr Med J* 2004; 81:499–504.

Villela WV, Barbosa RM, Trajetórias de mulheresvivendo com HIV/aids no Brasil. Avanços e permanências da resposta à Epidemia. *CiêncSaúdecoletiva* 2017; 22(1):87–96. https://doi.org/10.1590/1413-81232017221.14222016

Vlietinck A, De Bruyne T, Apers S, Pieters L, Plant-derived leading compounds for chemotherapy of human immunodeficiency virus (HIV) infection. *Planta Med* 1998; 64:97–109.

Wang Q, Wang YT, Pu SP, Zheng YT, Zinc coupling potentiates anti-HIV-1 activity of baicalin, *Biochem Biophys Res Commun* 2004; 324:605–610. https://doi.org/10.1016/j.bbrc.2004.09.093.

Wang Z, Shang H, Jiang Y, Chemokines and chemokine receptors: accomplices for human immunodeficiency virus infection and latency. *Front Immunol* 2017; 8(OCT):1–12.

Webster D, Taschereau P, Lee TDG, Jurgens T, Immunostimulant properties of Heracleum maximum Bartr. *J Ethnopharmacol* 2006; 106(3):360–363.

Wei Y, Ma CM, Chen DY, Hattori M, *Phytochemistry.* 2008; 69:1875–1879.

Worku ED, Asemahagn MA, Endalifer ML, Epidemiology of hiv infection in the amhara region of ethiopia, 2015 to 2018 surveillance data analysis. *HIV/AIDS - Res Palliat Care* 2020; 12:307–314.

World Health Organization, Report of an informal WHO consultation on traditional medicine and AIDS: in vitro screening for anti-HIV activities. 2001.

Wu PL, Lin FW, Wu TS, Kuoh CS, Lee KH, Lee SJ, Cytotoxic and anti-HIV principles from the rhizomes of Begonia nantoensis. *Chem Pharm Bull* 2004; 52:345–349.

Xiao WL, Huang SX, Wang RR, Zhong JL, Gao XM, He F, Pu JX, Lu Y, Zheng YT, Zheng QT, Sun HD, Nortriterpenoids and lignans from Schisandra sphenanthera. *Phytochemistry* 2008; 69:2862–2866.

Yu D, Sakurai Y, Chen CH, Chang FR, Huang L, Kashiwada Y, Lee KH, Moronic acid and other triterpene derivatives as novel potent anti-HIV agents. *J Med Chem* 2006; 49:5462–5469.

Yu D, Morris-Natschke SL, Lee KH, New developments in natural products-based anti-AIDS research. *Med Res Rev* 2007a; 27:108–132.

Yu YB, Miyashiro H, Nakamura N, Hattori M, Park JC, Effects of triterpenoids and flavonoids isolated from Alnus firma on HIV-1 viral enzymes. *Arch Pharmacal Res* 2007b; 30:820–826.

Zhang L, Yuc ST, Xue YX, Attele AS, Yuan CS, Effects of qian-kun-nin, a Chinese herbal medicine formulation, on HIV positive subjects: a pilot study. *Am J Chin Med* 2000; 28:305–312. https://doi.org/10.1142/s0192415x00000362.

# 9 Medicinal Plants/Herbs and Influenza Virus

*Abhijit Kumar and Pooja Bhadrecha*
University Institute of Biotechnology, Chandigarh University

## 9.1 INTRODUCTION

Viruses are the root cause of a wide variety of human pathogeneses, including some forms of cancer. They have been linked to a wide range of incurable, complex diseases, including Alzheimer's disease, type 1 diabetes, and hepatocellular carcinoma (Bal et al., 2013; Hober et al., 2012; Morgan et al., 2013). The threat to public health posed by pandemic outbreaks driven by emerging and re-emerging viruses has only increased with the development of international travel and fast urbanization, especially in the absence of effective prophylactic measures such as vaccination and antiviral medication. Recent examples include a slew of new viruses like Dengue, influenza, Measles, SARS, and West Nile Virus (Christou, 2011; Cascio et al., 2011; Grais et al., 2011). Unfortunately, only a limited number of antiviral medications have received clinical approval, leaving many viruses without effective vaccines. The problem is made much worse by the fact that drug-resistant mutants can emerge, which is especially problematic when employing viral enzyme-specific inhibitors (Sheu et al., 2008; Geretti et al., 2012; Locarnini and Yuen, 2010; Wyles, 2013). Since vaccines or traditional treatments are unavailable for viral infections, there is an urgent need to discover new antivirals that are both highly effective and cost-effective. Many herbal medicines and naturally derived compounds have shown the potential in the development of new antiviral drugs. The antiviral mechanisms of these natural agents have shed light on how viruses interact with their hosts and where these interactions happen in the life cycle of viruses. In this short report, we delve into how different natural products and herbal medicines combat different viruses. These viruses include the common cold, the flu, the measles, and the respiratory syncytial virus (RSV). In this chapter, we specifically focus on the influenza virus, the infections it causes, the reason it is concerned as a threat, along with the promising potentials of plants like geranium, green tea, and echinacea in combating the influenza virus due to the bioactive compounds present in various parts of these plants.

## 9.2 INFLUENZA VIRUS

Influenza viruses, specifically influenza A, B, and C (IAVA, IAVB, and IAVC), belong to the *Orthomyxoviridae* family. They are enveloped, negative sense ssRNA viruses. Infection with these viruses can lead to respiratory infections causing fever, headache, sore throat, sneezing, and problems with your muscles and joints. In severe cases, this can lead to more serious and potentially fatal illnesses like pneumonia (Eccles, 2005; Rello and Pop-Vicas, 2009). Even though Infuenza Virus B (IFB) seems to naturally infect people and Influenza virus C (IFC) is seen less often, it may be isolated from humans and pigs. However, Influenza Virus A (IFA), which is the most widespread, can infect birds, humans, and other mammals (Pleschka, 2013). Influenza virus infections have caused a lot of sickness and death among people. About 250,000–500,000 people die each year because of seasonal epidemics. Large pandemics, like the H1N1 Spanish Flu in 1918, have been linked to an additional 20–40 million deaths (WHO). The envelope proteins of influenza viruses, hemagglutinin (HA) and neuraminidase (NA), are known to undergo continual evolution, despite the availability of vaccinations based on expected circulating strains (Yamada et al., 2006; Van der Vries et al., 2012).

 DOI: 10.1201/9781003329169-11

Due to this shift, the host is no longer protected by antibodies that were previously generated in response to previous exposure or vaccination. Concerns have also been voiced about the possibility of extremely pathogenic strains of influenza developing from cross-species transmission and host adaptation (Mak et al., 2012). An additional problem is the emergence of resistance to first-generation anti-influenza drugs. "This is especially true for the M2 ion channel blockers amantadine and rimantadine (Fiore et al., 2011). Some strains of influenza have already developed resistance to neuraminidase inhibitors like oseltamivir and zanamivir, which are currently in widespread use (Samson et al., 2013). Complex antiviral techniques are urgently needed to prevent and control possible pandemics with developing influenza strains in light of drug resistance issues, the fast evolution of influenza viruses, and the emergence of multiple recent outbreaks (e.g., H5N1, H1N1, H7N9)".

"Several natural products have been examined for their effects against influenza. Standardized elderberry (*Sambucus nigra*) liquid extract exerts *in vitro* antiviral effects against IFA, IFB, as well as respiratory bacterial pathogens (Krawitz et al., 2011). A licensed commercial extract from *Pelargonium sidoides* roots inhibits the entry of IFA, impairs viral hemagglutination as well as neuraminidase activity, and improves the symptoms of influenza-infected mice (Theisen and Muller, 2012). The aqueous extract from dandelion (*Taraxacum officinale*) impedes IFA infection and decreases its polymerase activity as well as the nucleoprotein (NP) RNA level (He et al., 2011). Spirooliganone B from the roots of *Illicium oligandrum* exhibits potent anti-IFA activities (Ma et al., 2013). A multitude of secondary plant metabolites have also been identified as potential influenza NA inhibitors, (Grienke et al., 2012) and more recent ones include chalcones from *Glycyrrhiza inflata*, (Dao et al., 2011) xanthones from *Polygala karensium*, (Dao et al., 2012) and homoisoflavonoids from *Caesalpinia sappan*" (Jeong et al., 2012).

The influenza virus is extremely common; it has circulated on Earth for hundreds of years and will likely continue to do so for the foreseeable future. They cause widespread illness and death each year, and occasionally a pandemic that can have far-reaching effects on human and animal health as well as the global economy (Suzuki, 2009). The influenza A, B, and C viruses all cause seasonal illness, but only in humans. Seasonal influenza C virus infections tend to be milder in humans than in other animals. "There have been a number of notable human pandemics in the recent century linked to the influenza A virus, and there have also been an increasing number of epidemics (epizootics) in domestic birds (Cannell et al., 2008; Michaelis et al., 2009; Neumann et al., 2009) (Table 9.1). Many different subtypes (strains) of influenza A virus have been identified, and it is generally agreed that they all originated in wild birds, presumably waterfowl like ducks and geese (Boyce et al., 2009)".

## 9.3 INFLUENZA INFECTIONS

Most healthy people recover quickly from the traditional flu symptoms of cough, malaise, and fever, along with other symptoms such as sore throat, nasal blockage, and sputum production. However, those with impaired immune systems or who are elderly are more likely to be affected. Bronchitis, pneumonia, and the worsening of asthma and COPD are all possible complications (Johnston, 2002). Dangerous overreactions of the innate immune response, including the emission of dangerously high amounts of cytokines ("cytokine storms") and other inflammatory mediators, are sometimes observed in healthy persons with more severe disease, especially during pandemics (Baskin et al., 2009). Furthermore, a concomitant bacterial infection must not be disregarded due to the severity of its potential effects (Brundage and Shanks, 2008). As a result, an effective control agent would inhibit the overproduction of inflammatory mediators while also stopping the transmission of the virus and reducing the likelihood of any dangerous microorganisms infecting the host. Many other "influenza-like" infections could also benefit from a multi-targeted strategy.

Though H3N2 and H1N1 are the most common subtypes of influenza A virus in humans and H5N1 and various H7 viruses are the most common subtypes of influenza A virus in domestic birds,

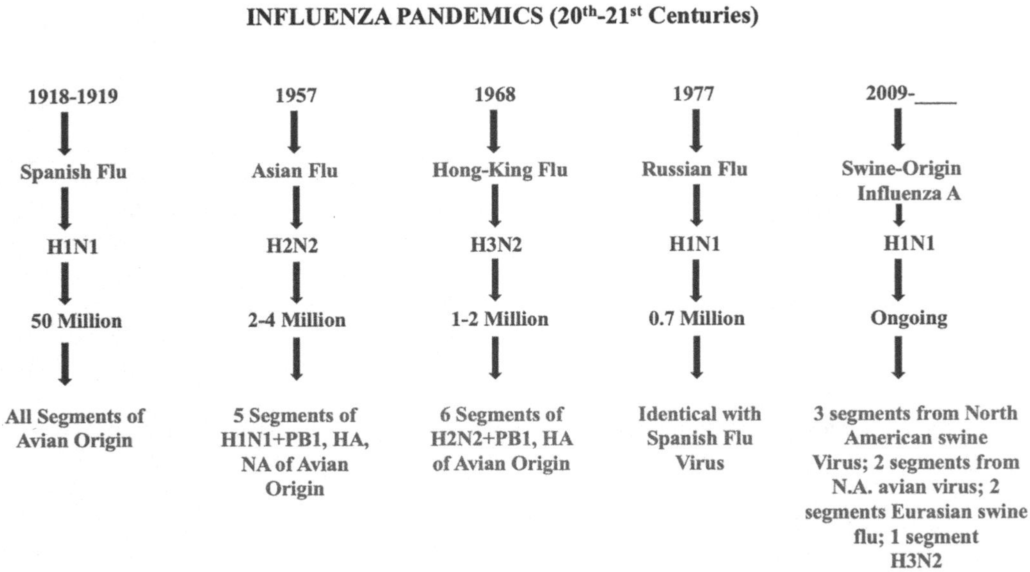


**FIGURE 9.1** Summary of influenza pandemics in 20th–21st centuries.

new dangerous strains could emerge at any time from the avian reservoir (Figure 9.1) (Michaelis et al., 2009). As many nations enter a new winter flu season, the pandemic H1N1 (or more accurately known as swine-origin flu virus (S-OIV)), continues to spread and inflict illness and mortality, as reported by Suzuki (Suzuki, 2009).

## 9.4 INFLUENZA A VIRUS AS A THREAT?

Influenza viruses, being RNA viruses, undergo rapid mutation, a phenomenon known as genetic or antigenic "drift." The eight separate portions of the RNA genome also make it susceptible to reassortment, allowing for the creation of new viral subtypes, some of which may be more virulent (pathogenic) than the parental viruses. This phenomenon is known as genetic or antigenic "shift," and it occurs entirely at random. The virus can be easily passed from one species to another, even from birds to mammals, aiding its spread. Hemagglutinin (HA) and neuraminidase (NA) are the two viral exterior proteins that facilitate virus entry into cells and subsequent spread to other cells. The host immune system and vaccines may stop recognizing them if they undergo significant alterations in their genes (HA or H gene and NA or N gene). All 16 HA gene subtypes (numbered H1 through H16) and 9 N/NA gene subtypes (numbered A through N) have been discovered in free-living avian species (Boyce et al., 2009; Suzuki, 2009).

## 9.5 CONTROL OF INFLUENZA: THE SCARCITY OF EXISTING STRATEGIES

Vaccines are the first line of defence against many different types of infectious agents, and they are commonly recommended for widespread use at the start of each influenza season, based on the virus strain that predominated the previous year. There is no assurance that community protection by a given vaccine will be adequate, especially if the virus becomes progressively more virulent, but it becomes plausible to produce a more appropriate vaccine once a novel virus strain has been isolated and characterized, a process that could take up to 6 months. Several medical professionals have even questioned the benefits of mandatory vaccination programs (Cannell et al., 2008; Jefferson et al., 2009; Fedson, 2009).

There have been several attempts to develop effective antiviral medications for use in infected patients, but so far, none have met with any success. Neuraminidase inhibitors, such as oseltamivir (Tamiflu®) and zanamivir (Relenza®), are the newest synthetic drugs. However, just as antibiotic-resistant bacteria are a concern with any new medicine, so is the possibility that resistant mutants will be selected during treatment. Current S-OIV is just one example of a growing trend of drug-resistant strains of human and avian influenza viruses (Jefferson et al., 2006; Cheng et al., 2009).

## 9.6 ACTIVITIES OF MEDICINAL PLANTS AS ANTI-INFLUENZA: HERBAL APPROACH

The search for natural inhibitors of the influenza virus is centuries old. Throughout the last century, researchers have worked hard to pinpoint phytochemicals with the potential to impede the spread of the influenza virus (Cochran et al., 1966; May and Willuhn, 1978). Numerous natural compounds extracted from different plants have been shown to inhibit influenza virus *in vitro* (Mothana et al., 2006; Pantev et al., 2006; Prajoubklang et al., 2005) and *in vivo* (Ivanova et al., 2005; Prahoveanu et al., 1986). Curiously, many different phytochemicals, including alkaloids, flavonoids, glucosides, polyphenols, and saponins, have been shown to have anti-influenza action (Wang et al., 2006). It is envisaged that a phytochemical capable of inhibiting the influenza virus will become available in the not-too-distant future.

"Some new molecular approaches have been suggested based on how the virus communicates with cells when it replicates (Ludwig, 2009; Fedson, 2009). Traditional Chinese Medicine (TCM) and other herbal treatments have also been suggested as alternatives (Wang et al., 2006). Because they have more than one use, they may be safer than chemical drugs and viruses are less likely to become resistant to them. Also, some herbal remedies can work on both the virus and the flu symptoms, many of which are caused by the overproduction of inflammatory mediators like cytokines that the virus infection causes (Baskin et al., 2009; Sharma et al., 2009a). Some of them can also kill bacteria and fight free radicals."

### 9.6.1 Geranium (*Geranium sanguineum* L.)

Dry geranium aerial roots are used medicinally in Bulgaria (and perhaps elsewhere) for a variety of gastrointestinal problems, infections, and inflammations; these remedies are extracted either in water or alcohol (Serkedjieva and Hay, 1998). Extensive antiviral actions have been observed in cell culture and animal models against a variety of human, avian, and equine influenza A virus strains, including amantadine-resistant virus (Serkedjieva et al., 2008). At concentrations as low as 10 g/mL, the extracts appear to inhibit a critical step in viral propagation. Above 50 g/mL, they exhibit virucidal and cytotoxic effects. But the authors found that the extract seemed to be much stronger *in vivo*, in a mouse influenza model, than in cell cultures. This is probably due to the extra bioactivities that were seen in mice with the virus, such as the return of oxidative stress and the activation of macrophage functions that the virus had stopped (Murzakhmetova et al., 2008).

Intranasal application of the extract, especially in the form of an aerosol, has shown to be the most effective way to reduce lung virus load and pathology while increasing survival time. This approach is beneficial both before and after virus infection. Flavonoids, phenolic acids, catechins, and gallotannins are just some of the polyphenols found in abundance in the alcoholic extracts; it would be tempting to attribute their antiviral and antioxidative bioactivities to these molecules. Recent studies on the anti-influenza effects, however, found that separated individual components are no more effective than the crude ethanol extract considering their weight. These findings point to the synergistic effect of multiple phenolic compounds rather than the importance of any one of them alone. Additionally, as-yet-unidentified ingredients may also have a role in the antiviral effect. These findings suggest that the Geranium extract has a wide range of therapeutic properties and may be useful as an additional treatment for influenza infections.

### 9.6.2 Green Tea (*Camellia sinensis*)

Green tea and other aqueous extracts of *C. sinensis* have been found to contain numerous bio-activities, including antiviral, antibacterial, antifungal, and antioxidant properties. Catechins, such as Hudson 1191 and their chemically related theaflavins, can make up as much as 10% of the dry weight of the leaves, and they appear to be responsible for most of the plant's health benefits (Friedman, 2007). "Multiple influenza A and B virus subtypes and a subset of respiratory virus subtypes are vulnerable. Inhibition of viral hemagglutinin, neuraminidase activity, and viral RNA synthesis were all seen in cell culture experiments, suggesting that the antiviral effects against influenza are multifaceted. When the extract was administered to cells just before virus attachment, the effect was maximized. Yet, these may have resulted from the active chemicals' interaction with viral components at an earlier stage (Imanishi et al., 2002; Song et al., 2005). EGC [(-) epigallo-catechin] and theaflavin digallate were substantially more active than their structural equivalents, probably due to variations in binding affinities with the viral target proteins, however not all the individual polyphenols were effective. Interestingly, as compared to the full extract or a mixture of the polyphenol components, the active catechins were no more potent." This finding suggests that the catechins and polyphenols work together synergistically.

The restricted bioavailability and pharmacodynamics of tea polyphenols have been examined by some writers, particularly in relation to the intestinal tissues that typically receive these compounds after ingestion ((Friedman, 2007). To combat upper respiratory viruses and bacteria, a mouth wash or rinse would be more effective *in vivo* than drinking tea due to the organisms' initial encounter occurring in mucosal tissues. Here, a gargled solution of green tea catechins was tested in an early study for its capacity to prevent influenza infection. Although the results were encouraging, the sample sizes were too low to draw any firm conclusions (Yamada et al., 2006). Subjective assessments of "cold and flu" symptoms were used in another trial in which people took either a placebo or a capsule containing a standardized green tea rich in polyphenols. About a third fewer symptoms were reported in the green tea group (Rowe et al., 2007). Green tea consumption appears to be advantageous during flu season, but without microbial/viral characterization, a direct link to influenza cannot be made.

### 9.6.3 Echinacea (*Echinacea purpurea* and Other Species)

Extracts made from the aerial portions and roots of many species of Echinacea were commonly used in North America to treat respiratory infections, wounds, and other inflammatory disorders in the past (Barrett, 2003). As "cold and flu" cures, they have exploded in popularity over the past few decades. Although other species are occasionally employed, most commercial medicines are based on pressed juice or ethanol tinctures of aerial portions of *E. purpurea* or root extracts of *E. angustifolia*. Although polysaccharides were at first thought to be the primary bioactive component, newer studies have instead concentrated on alkylamides and caffeic acid derivatives. It has not been able to correlate the many *in vitro* and *in vivo* reported antiviral, antibacterial, antifungal, antioxidant, and immune-modulatory actions with particular chemicals (Barnes et al., 2005; Hudson et al., 2005; Vimalanathan et al., 2009). The necessity for standardized extracts has been highlighted, however, because of the obvious variations between extracts made from different plant species and sections (Vohra et al., 2009). "Direct contact with *E. purpurea* preparations is highly effective against influenza virus and other membrane containing viruses (more than 3 log 10 inactivation at dilutions of 1:100 or more, equivalent to one tenth of the recommended oral dose), while rhinoviruses (common cold viruses) and adenoviruses (more resistant) lack membranes. It is possible that the polyphenolic components of Echinacea are responsible for its antiviral effect, and that they operate at a very early stage in the virus replication cycle. There is strong bactericidal action in echinacea extracts against various bacteria that can cause 'cold and flu' symptoms, including strep throat (*Streptococcus pyogenes*) and respiratory problems (*Hemophilus influenzae*, *Legionella pneumophila*). Nonetheless,

**TABLE 9.1**
**Medicinal Plants with Potential Against influenza**

| Plant | Plant Part | Bioactive Compounds | Potential | Mode of Action against Influenza |
|---|---|---|---|---|
| Geranium (*Geranium sanguineum* L.) | Aerial roots | Flavonoids, phenolic acids, catechins, gallotannins | Antiviral, antioxidative | Inhibits viral propagation and is viricidal |
| Green Tea (*Camellia sinensis*) | Leaves | Theaflavins | Antiviral, antibacterial, antifungal, antioxidants | Inhibits viral hemagglutinin, neuraminidase activity and viral RNA synthesis |
| Echinacea (*Echinacea purpurea*) | Roots and aerial parts | Alkyl amides, caffeic acid derivatives | Antiviral, antibacterial, antifungal, antioxidants, immune-modulatory | Inhibits viral replication |

the antibiotic impact is selective, as evidenced by the fact that other bacteria are comparatively resistant. Numerous 'friendly' bacteria (those normally found in the mouth) would be spared, therefore this is a good thing. Epithelial cell models of nasal, mucosal, and tracheo-bronchial tissues showed that several respiratory viruses and bacteria, including those resistant to direct assault by Echinacea, stimulated the release of pro-inflammatory cytokines and associated mediators, as well as mucin secretion." In each of these animal studies, an orally administered standardized strong tincture of *E. purpurea* (Echinaforce®) successfully halted the inflammatory reactions (Sharma et al., 2008, 2009a, 2009b).

Many transcription factors, including Nuclear factor kappa-light-chain-enhancer of activated B cells (NFkB) and others, were involved in mediating the responses to the infectious agents, which ultimately resulted in enhanced expression of inflammatory genes. However, Echinacea suppressed cytokine and mucin release by inhibiting these transcription factors (Altamirano-Dimas et al., 2009). Although Echinacea extracts have been the subject of certain clinical trials, these studies have always focused on the possible prevention or therapy of "common colds," not influenza. Furthermore, not many of them conducted sufficient statistical analysis or used Echinacea preparations that were well-characterized. Although the effects of Echinacea were inconsistent, there was a general positive trend, and the preparations' safety was established (Schoop et al., 2006). Table 9.1 summarizes the potential of some medicinal plants against influenza virus, and mode of action by their bioactive compounds.

## 9.7 ISOLATION AND CHARACTERIZATION OF PLANT COMPONENTS DISPLAYING MEDICINAL PROPERTIES

Preparing a crude cellular lysate of the plant matrix and then extracting different components with potential medicinal value is the first and most fundamental stage in assessing a medicinal plant's therapeutic potential. Standardized extraction processes from medicinal plants are described in a number of books and reviews (Wagner et al., 1996; Ahmad et al., 2007; Houghton and Raman, 1998). The vast majority of published works on phytomedicine focus on traditional methods of isolation. The reliability and quality of some of these traditional methods are low, jeopardizing the reliability and efficacy of phytomedicinal products. Therefore, there is a pressing need to improve upon traditional methods in order to achieve methodological uniformity and obtain highly pure plant components displaying therapeutic significance. An uptick in curiosity about herbal remedies has coincided with the development of more thorough quality assurance measures and the standardization of processes for isolating phytochemicals (Atta-ur-Rahman and Choudhary, 1998; Ong, 2004). Efforts are being made to replace conventional techniques with contemporary sample preparation and extraction processes in order to guarantee the highest possible quality of herbal products. Standardization of approaches that may be utilized to research the pharmacokinetics and pharmacodynamics of phytomedicinal products is also being pursued as a means of easing the process of discovering new drugs derived from plants (Lin et al., 2005).

## 9.8 FUTURE DIRECTIONS AS ANTIVIRAL POTENTIALS

An estimated 80% of the world's population relies on phytomedicinal sources to meet their healthcare needs, highlighting the significance of these plants, according to the World Health Organization (WHO) (Anon, 2005a). In nations where the use of medicinal plants and their products is prevalent, government funding should support initiatives that promote this practice. Countries that rely heavily on alternative medicine should fund research into developing contemporary pharmaceuticals from plant compounds. Screening medicinal plants for their anti-infectious and antiviral properties may become possible with the development of improved separation technologies. Problems that were once deemed to be insurmountable for the antiviral drug discovery program based on medicinal plants have been resolved. Accidental infection of personnel was always a possibility, for instance, when testing the antiviral properties of plant extracts. Overcoming these obstacles has been greatly aided by vector-based testing approaches. Many antivirals have been screened using recombinant viral vectors that replicate the infection by expressing the firefly luciferase marker gene (Esimone et al., 2005).

The creation of vaccines and protein-based therapies using plants is a shrewd application. Several studies have indicated that this approach may be useful for producing pharmaceutical-grade proteins and peptides (Glenz and Warzecha, 2006; Koprowski and Yusibov, 2001). "Over the years following the successful expression of the first subunit vaccine for HBV surface antigen in 1992, various other vaccine antigens have been effectively expressed in plants and their safety has been evaluated in animal and human studies (Glenz and Warzecha, 2006; Ma et al., 2003; Thanavala et al., 2005). The bulk of viral vaccines are made up of attenuated or inactivated viral particles, so caution must be exercised while evaluating this element of plants for treating human viral infections. These constraints have motivated research into producing viral coat proteins, which are thought to self-assemble into virus-like particles (VLP) in plants and to have an immunogenic effect. There are a number of other factors that must be taken into account, such as the correct protein processing for plant expression. Though herbal (medicinal) preparations are commonly utilized in various regions of the world, either singly or in combination, there is a lack of information regarding the effects of medicinal plant interactions on biological systems." Only the indigenous people's own knowledge of how a certain plant or phytochemical product works to treat a disease can be relied upon. Providers of traditional medicine should have access to clinical results, such as the link between the use of kava-kava and St. John's wort that resulted in hepatotoxicity (Musch et al., 2006). Traditional herbal medicines have a reputation for being safe, but there have been concerns that some of them can be toxic to the liver (Pak et al., 2004; Schiano, 2003). It's important to promote and include into a centralized database system the publication of scientific findings concerning the cytotoxicity of medicinal plants' applications. Additionally, before incorporating a specific herbal cure into the treatment of humans, bigger randomized, double-blind, placebo-controlled multicenter clinical trials should be conducted.

## 9.9 FUTURE PROSPECTS AND CONCLUSION

Since many viruses still lack both preventative immunizations and efficient antiviral therapies, their complete eradication is unlikely. However, natural products offer promising avenue for discovering new antivirals, elucidating structure-activity connections, and developing efficient preventative and treatment measures against viral infections. Discoveries of the potent antiviral activity of many plant compounds have led to the creation of new pharmaceuticals and drug candidates. To aid in the development of effective antiviral medicines, it is recommended that more research be conducted to characterize the bioactive substances, define the underlying mechanisms, and evaluate the efficacy and possible applicability *in vivo*. As multitarget therapies may lessen the likelihood of drug-resistant viruses, more research is needed to investigate the possibilities of combination therapies with other natural agents or with traditional medications. Natural ingredients have played an essential role and contributed to the creation of antiviral drugs, and we expect this trend to continue.

## REFERENCES


Ahmad, I.; Aqil, F.; Owais, M. *Modern Phytomedicine: Turning Medicinal Plants into Drugs*. 1st ed. Wiley-VCH, Verlag GmbH & Co, KGaA: Weinheim, 2007.

Altamirano-Dimas, M.; Sharma, M.; Hudson, J.B. *Echinacea* and anti-inflammatory cytokine responses: Results of a gene and protein array analysis. *Pharm. Biol.* 2009, 47, 500–508.

Anon. Medicinal plants—time to harness the potential of neglected genetic resources? Plant Genetic Resources: Characterization and Utilization 2005a. http://www.cabi-publishing.org/pra

Atta-ur-Rahman; Choudhary, M.I. New natural products from medicinal plants of Pakistan. *Pure Appl. Chem.* 1998, 70, 385–389.

Bal, M.J.; Lukiw, W.J., Kammerman, E.M.; Hill, J.M. Intracerebral propagation of Alzheimer's disease: Strengthening evidence of a herpes simplex virus etiology. *Alzheimers Dement.* 2013, 9, 169–175.

Barnes, J.; Anderson, L.A.; Gibbons, S.; Phillipson, J. D. *Echinacea* species (*Echinacea angustifolia* (DC.) Hell. *Echinacea pallida* (Nutt.) Nutt., *Echinacea purpurea* (L.) Moench: A review of their chemistry, pharmacology and clinical properties. *J. Pharm. Pharmacol.* 2005, 57, 929–954.

Barrett, B. Medicinal properties of *Echinacea*: A critical review. *Phytomedicine*. 2003, 10, 66–86.

Baskin, C.R.; Carole, R.B.; Helle, B-O.; et al. Early and sustained innate immune response defines pathology and death in nonhuman primates infected by highly pathogenic influenza virus. *PNAS*. 2009, 106, 3455–3460.

Boyce, W.M.; Sandrock, C.; Kreuder-Johnson, C.; Kelly, T.; Cardona, C. Avian influenza viruses in wild birds: A moving target. *Comp. Immunol. Microbiol. Infec. Dis.* 2009, 32, 275–286.

Brundage, J.F.; Shanks, G.D. Deaths from bacterial pneumonia during 1918-19 influenza pandemic. *Emerg. Inf. Dis.* 2008, 14, 1193–1196.

Cannell, J.J.; Zasloff, M.; Garland, C.F.; Scragg, R.; Giovannucci, E. On the epidemiology of influenza. *Virol. J.* 2008, 5, 29.

Cascio, A.; Bosilkovski, M.; Rodriguez-Morales, A.J.; Pappas, G. The socio-ecology of zoonotic infections. *Clin. Microbiol. Infect.* 2011, 17, 336–342.

Cheng, P.K.C.; Leung, T.W.C.; Ho, E.C.M.; et al. Oseltamivir- and amantadine-resistant influenza viruses A (H1N1). *Emerg. Infec. Dis.* 2009, 15, 966–968

Christou, L. The global burden of bacterial and viral zoonotic infections. *Clin. Microbiol. Infect.* 2011, 17, 326–330.

Cochran, K.W.; Nishikawa, T.; Beneke, E.S. Botanical sources of influenza inhibitors. *Antimicrob. Agents Chemother.* (Bethesda) 1966, 6, 515–520.

Dao, T.T.; Dang, T.T.; Nguyen, P.H.; Kim, E.; Thuong, P.T.; Oh, W.K. Xanthones from Polygala karensium inhibit neuraminidases from influenza A viruses. *Bioorg. Med. Chem. Lett.* 2012, 22, 3688–3692.

Dao, T.T.; Nguyen, P.H.; Lee, H.S.; et al. Chalcones as novel influenza A (H1N1) neuraminidase inhibitors from *Glycyrrhiza inflata*. *Bioorg. Med. Chem. Lett.* 2011, 21, 294–298.

Eccles, R. Understanding the symptoms of the common cold and influenza. *Lancet Infect. Dis.* 2005, 5, 718–725.

Esimone, C.O.; Grunwald, T.; Wildner, O.; et al. *In vitro* pharmacodynamic evaluation of antiviral medicinal plants using a vector-based assay technique. *J. Appl. Microbiol.* 2005, 99(6), 1346–1355.

Fedson, D.S. Confronting the next influenza pandemic with anti-inflammatory and immunomodulatory agents: Why they are needed and how they might work. *Influenza Other Resp. Viruses*. 2009, 3, 129–142.

Fiore, A.E.; Fry, A.; Shay, D.; Gubareva, L.; Bresee, J.S.; Uyeki, T.M. Antiviral agents for the treatment and chemoprophylaxis of influenza - recommendations of the Advisory Committee on Immunization Practices (ACIP). *MMWR Recomm. Rep.* 2011, 60, 1–24.

Friedman, M. Overview of antibacterial, antitoxin, antiviral, antifungal activities of tea flavonoids and teas. *Mol. Nutr. Food Res.* 2007, 51, 116–134.

Geretti, A.M.; Armenia, D.; Ceccherini-Silberstein, F. Emerging patterns and implications of HIV-1 integrase inhibitor resistance. *Curr. Opin. Infect. Dis.* 2012, 25, 677–686.

Glenz, K.; Warzecha, H. New medicinal plants for the production of vaccines. *J. Verbr. Lebensm.* 2006, 1, 126–130.

Grais, R.F.; Strebel, P.; Mala, P.; Watson, J.; Nandy, R.; Gayer, M. Measles vaccination in humanitarian emergencies: A review of recent practice. *Confl. Health.* 2011, 5, 21.

Grienke, U.; Schmidtke, M.; von Grafenstein, S.; Kirchmair, J.; Liedl, K.R.; Rollinger, J.M. Influenza neuraminidase: A druggable target for natural products. *Nat. Prod. Rep.* 2012, 29, 11–36.

He, W.; Han, H.; Wang, W.; Gao, B. Anti-influenza virus effect of aqueous extracts from dandelion. *Virol. J.* 2011, 8, 538.

Hober, D.; Sane, F.; Jaidane, H.; Riedweg, K.; Goffard, A.; Desailloud, R. Immunology in the clinic review series; focus on type 1 diabetes and viruses: Role of antibodies enhancing the infection with Coxsackievirus-B in the pathogenesis of type 1 diabetes. *Clin. Exp. Immunol.* 2012, 168, 47–51.

Houghton, P.J.; Raman, *A. Laboratory Handbook for the Fractionation of Natural Extracts.* Chapman and Hall: New York, 1998.

Hudson, J.; Vimalanathan, S.; Kang, L.; Treyvaud Amiguet, V.; Livesey, J.; Arnason, J.T. Characterization of antiviral activities in *Echinacea* root preparations. *Pharm. Biol.* 2005, 43, 790–796.

Imanishi, N.; Tuji, Y.; Katada, Y.; et al. Additional inhibitory effect of tea extract on the growth of influenza A and B viruses in MDCK cells. *Microbiol. Immunol.* 2002, 46, 491–494.

Ivanova, E.; Toshkova, R.; Serkedjieva, J. A plant polyphenol-rich extract restores the suppressed functions of phagocytes in influenza virus-infected mice. *Microbes Infect.* 2005, 7 (3), 391–398.

Jefferson, T.; Demicheli, V.; Jones, M.; Di Pietrantonj, C.; Rivetti, A. Antivirals for influenza in healthy adults: Systematic review. *Lancet.* 2006, 367, 303–313.

Jefferson, T.; Di Pietrantonj, C.; Debalini, M.J.; Rivetti, A.; Demicheli, V. Inactivated influenza vaccines: Methods, policies, and politics. *J. Clin Epidem.* 2009, 62, 677–686.

Jeong, H.J.; Kim, Y.M.; Kim, J.H.; et al. Homoisoflavonoids from Caesalpinia sappan displaying viral neuraminidases inhibition. *Biol. Pharm. Bull.* 2012, 35, 786-90.

Johnston, S.L. Anti-influenza therapies. *Virus Res.* 2002, 82, 147–152.

Koprowski, H.; Yusibov, V. The green revolution: Plants as heterologous expression vectors. *Vaccine.* 2001, 19, 2735–2741.

Krawitz, C.; Mraheil, M.A.; Stein, M.; et al. Inhibitory activity of a standardized elderberry liquid extract against clinically-relevant human respiratory bacterial pathogens and influenza A and B viruses. *BMC Complement. Altern. Med.* 2011, 11, 16.

Lin, Z.J.; Qiu, S.X.; Wufuer, A.; Shum, L. Simultaneous determination of glycyrrhizin, a marker component in radix Glycyrrhizae, and its major metabolite glycyrrhetic acid in human plasma by LC-MS/MS. *J. Chromatogr. B Analyt. Technol. Biomed. Life Sci.* 2005, 814 (2), 201–207.

Locarnini, S.A.; Yuen, L. Molecular genesis of drug-resistant and vaccine-escape HBV mutants. *Antivir. Ther.* 2010, 15, 451–461.

Ludwig, S. Targeting cell signaling pathways to fight the flu: Towards a paradigm change in anti-influenza therapy. *J. Antimic. Ther.* 2009, 64, 1–4.

Ma, J.K.; Drake, P.M.; Christou, P. The production of recombinant pharmaceutical proteins in plants. *Nat. Rev. Genet.* 2003, 4, 794–805.

Ma, S.G.; Gao, R.M.; Li, Y.H.; et al. Antiviral spirooliganones A and B with unprecedented skeletons from the roots of *Illicium oligandrum. Org. Lett.* 2013, 15, 4450–4453.

Mak, P.W.; Jayawardena, S.; Poon, L.L. The evolving threat of influenza viruses of animal origin and the challenges in developing appropriate diagnostics. *Clin. Chem.* 2012, 58, 1527–1533.

May, G.; Willuhn, G. Antiviral effect of aqueous plant extracts in tissue culture. *Arzneimittelforschung.* 1978, 28 (1), 1–7.

Michaelis, M.; Doerr, H.W.; Cinatl, J. Novel swine-origin influenza A virus in humans: Another pandemic knocking at the door Med. *Microbiol. Immunol.* 2009, 198, 175–183.

Morgan, R.L.; Baack, B.; Smith, B.D.; Yartel, A.; Pitasi, M.; Falck-Ytter, Y. Eradication of hepatitis C virus infection and the development of hepatocellular carcinoma: A meta-analysis of observational studies. *Ann. Intern. Med.* 2013, 158, 329–337.

Mothana, R.A.; Mentel, R.; Reiss, C.; Lindequist, U. Phytochemical screening and antiviral activity of some medicinal plants from the island Soqotra. *Phytother. Res.* 2006, 20 (4), 298–302.

Murzakhmetova, M.; Moldakarimov, S.; Tancheva, L.; Abarova, S.; Serkedjieva, J. Antioxidant and prooxidant properties of a polyphenol-rich extract from Geranium sanguineum L. *in vitro* and *in vivo*. *Phytother. Res.* 2008, 22, 746–751.

Musch, E.; Chrissafidou, A.; Malek, M. Acute hepatitis due to kavakava and St. John's Wort: An immune-mediated mechanism? *Dtsch. Med. Wochenschr.* 2006, 131 (21), 1214–1217.

Neumann, G.; Noda, T.; Kawaoka, Y. Emergence and pandemic potential of swine-origin H1N1 influenza virus. *Nature.* 2009, 459, 931–939.

Ong, E.S. Extraction methods and chemical standardization of botanicals and herbal preparations. *J. Chromatogr. B Analyt. Technol. Biomed. Life Sci.* 2004, 812 (1/2), 23–33.
Pak, E.; Esrason, K.T.; Wu, V.H. Hepatotoxicity of herbal remedies: An emerging dilemma. *Prog. Transplant.* 2004, 14 (2), 91–96.
Pantev, A.; Ivancheva, S.; Staneva, L.; Serkedjieva, J. Biologically active constituents of a polyphenol extract from *Geranium sanguineum L.* with anti influenza activity. *Z Naturforsch.* 2006, 61(7/8), 508–516.
Pleschka, S. Overview of influenza viruses. *Curr. Top Microbiol. Immunol.* 2013, 370, 1–20.
Prahoveanu, E.; Esanu, V.; Anton, G.; Frunzulica, S. Prophylactic effect of a *Beta vulgaris* extract on experimental influenza infection in mice. *Virologie.* 1986, 37 (2), 121–123.
Prajoubklang, A.; Sirithunyalug, B.; Charoenchai, P.; et al. Bioactive deoxypreussomerins and dimeric naphthoquinones from *Diospyros ehretioides* fruits: Deoxypreussomerins may not be plant metabolites but may be from fungal epiphytes or endophytes. *Chem. Biodivers.* 2005, 2 (10), 1358–1367.
Rello, J.; Pop-Vicas, A. Clinical review: Primary influenza viral pneumonia. *Crit. Care.* 2009, 13, 235.
Rowe, C.A.; Nantz, M.P.; Bukowski, J.F.; Percival, S.S. Specific formulation of *Camellia sinensis* prevents cold and flu symptoms and enhances gd T cell function: A randomized, double-blind, placebo-controlled study. *J. Am. Coll. Nutr.* 2007, 26, 445–452.
Samson, M.; Pizzorno, A.; Abed, Y.; Boivin, G. Influenza virus resistance to neuraminidase inhibitors. *Antiviral Res.* 2013, 98, 174–185.
Schiano, T.D. Hepatotoxicity and complementary and alternative medicines. *Clin. Liver Dis.* 2003, 7 (2), 453–473.
Schoop, R.; Klein, P.; Suter, A.; Johnston, S.L. *Echinacea* in the prevention of induced rhinovirus colds: A meta-analysis. *Clin. Therapeutics.* 2006, 28, 174–183.
Serkedjieva, J.; Gegova, G.; Mladenov, K. Protective efficacy of an aerosol preparation, obtained from *Geranium sanguineum* L., in experimental influenza infection. *Pharmazie.* 2008, 63, 160–163.
Serkedjieva, J.; Hay, A.J. In vitro anti-influenza virus activity of a plant preparation from Geranium sanguineum L. *Antivir. Res.* 1998, 37, 121–130
Sharma, M.; Anderson, S.A.; Schoop, R.; Hudson, J.B. Induction of pro-inflammatory cytokines by respiratory viruses and reversal by standardized *Echinacea*, a potent antiviral herbal extract. *Antiviral Res.* 2009a, 83, 165–170.
Sharma, M.; Schoop, R.; Hudson, J.B. Echinacea as an anti-inflammatory agent: The influence of physiologically relevant parameters. *Phytother. Res.* 2008, 23, 863–867.
Sharma, M.; Schoop, R.; Hudson, J.B. The efficacy of *Echinacea* in a 3-D tissue model of human airway epithelium. *Phytother. Res.* 2009b. DOI:10.1002/ptr.3051
Sheu, T.G..; Deyde, V.M.; Okomo-Adhiambo, M.; et al. Surveillance for neuraminidase inhibitor resistance among human influenza A and B viruses circulating worldwide from 2004 to 2008. *Antimicrob. Agents Chemother.* 2008, 52, 3284–3292.
Song, J-M.; Lee, K-H.; Seong, B-L. Antiviral effect of catechins in green tea on influenza virus. *Antivir. Res.* 2005, 68, 66–74.
Suzuki, Y. The highly pathogenic avian influenza H5N1-initial molecular signals for the next influenza pandemic. *Chang. Gung. Med. J.* 2009, 32, 258–263.
Thanavala, Y.; Mahoney, M.; Pal, S.; et al. Immunogenicity in humans of an edible vaccine for hepatitis B. *Proc. Natl. Acad. Sci. U.S.A.* 2005, 102, 3378–3382.
Theisen, L.L.; Muller, C.P. EPs (R) 7630 (Umckaloabo (R)), an extract from Pelargonium sidoides roots, exerts anti-influenza virus activity *in vitro* and *in vivo*. *Antiviral Res.* 2012, 94, 147–156.
Van der Vries, E.; Collins, P.J.; Vachieri, S.G.; et al. H1N1 2012 pandemic influenza virus: Resistance of the I223R neuraminidase mutant explained by kinetic and structural analysis. *PLoS Pathog.* 2012, 8, e1002914.
Vimalanathan, S.; Arnason, J.T.; Hudson, J.B. Anti-inflammatory activities of *Echinacea* extracts do not correlate with traditional marker components. *Pharm. Biol.* 2009, 47, 430–435.
Vohra, S.; Adams, D.; Hudson, J.B.; et al. Selection of natural health products for clinical trials: A preclinical template. *Can. J. Physiol. Pharmacol.* 2009, 87, 371–378.
Wagner, H.; Bladt, S.; Ricki, V. *Plant Drug Analysis: A Thin Layer Chromatography*, 2nd ed. Springer: New York, 1996.
Wang, X.; Jia, W.; Zhao, A.; Wang, X. Anti-influenza agents from plants and traditional Chinese medicine. *Phytother. Res.* 2006, 20 (5), 335–341.
Wyles, D.L. Antiviral resistance and the future landscape of hepatitis C virus infection therapy. *J. Infect. Dis.* 2013, 207 Suppl 1, S33–S39.
Yamada, S.; Suzuki, Y.; Suzuki, T.; et al. Haemagglutinin mutations responsible for the binding of H5N1 influenza A viruses to human-type receptors. *Nature.* 2006, 444, 378–382.

# 10 Medicinal Plants/Herbs in Respiratory Syncytial Virus and Viral Pneumonia

*Gursharan Kaur, Palak Rana, and Manisha*
University Institute of Biotechnology, Chandigarh University

*Aleena James Chirayimmel*
Vellore Institute of Technology

*Santosh Kumar Rath*
DIT University

*Vagish Dwibedi*
University Institute of Biotechnology, Chandigarh University
Volcani Research Center, Agricultural Research Organization

## 10.1 INTRODUCTION

Respiratory syncytial virus (RSV) continues to pose a predominant challenge to worldwide healthcare. Seasonal endemic outbreaks, particularly in winter, have been observed in most parts of the world. Even though RSV causes infection in all age groups, it is found to be more prevalent in infants (0–1 years). Almost every child gets affected by this virus before reaching the age of two, making it the leading cause of acute respiratory tract infections. Under severe conditions, it can lead to bronchiolitis, pneumonia, and, in some cases, may result in asthma with a wheezy cough (Efstathiou et al., 2020). Other symptoms that are observed during RSV infection are congestion, fever, loss of appetite, and difficulty in breathing (Carvajal et al., 2019). The historical background of RSV starts with its isolation from chimpanzees in 1955 at the Walter Reed Army Institute of Research in the United States, which led to further investigations on this virus (Battles and McLellan, 2019).

RSV is an enveloped, pleomorphic, filamentary virus that contains single-stranded negatively sensed RNA (15.2 kb) as its genomic material, encoding for 10 genes producing 11 proteins belonging to the Pneumoviridae family and Orthopneumovirus genus. Basically, the viral envelope of RSV consists of three different integral membrane proteins: the fusion protein (F) responsible for the entry of the virus inside the host, the glycoprotein (G) for endocytic attachment, and small hydrophobic protein (SH), which causes delayed apoptosis of infected cell (Carvajal et al., 2019). It also contains other structural proteins found beneath the viral envelope: matrix protein (M), nucleoprotein (N), phosphoprotein (P), RNA polymerase (L), and transcription factors (M2-1 and M2-2). An overview of the structure of RSV is depicted in Figure 10.1.

The transmission of the virus occurs via the air or direct/indirect contact with mucous droplets released by an RSV-infected person. The spread of the infection usually begins in the nasal epithelial cells. The glycoprotein (G) interacts with the cell surface factors and facilitates the attachment

 DOI: 10.1201/9781003329169-12

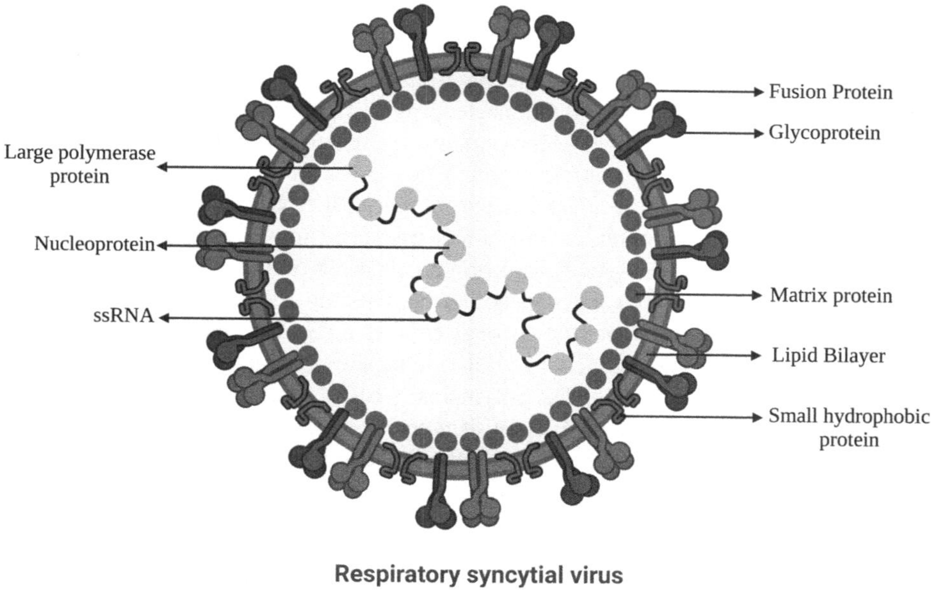


**FIGURE 10.1** Structure of respiratory syncytial virus (RSV).

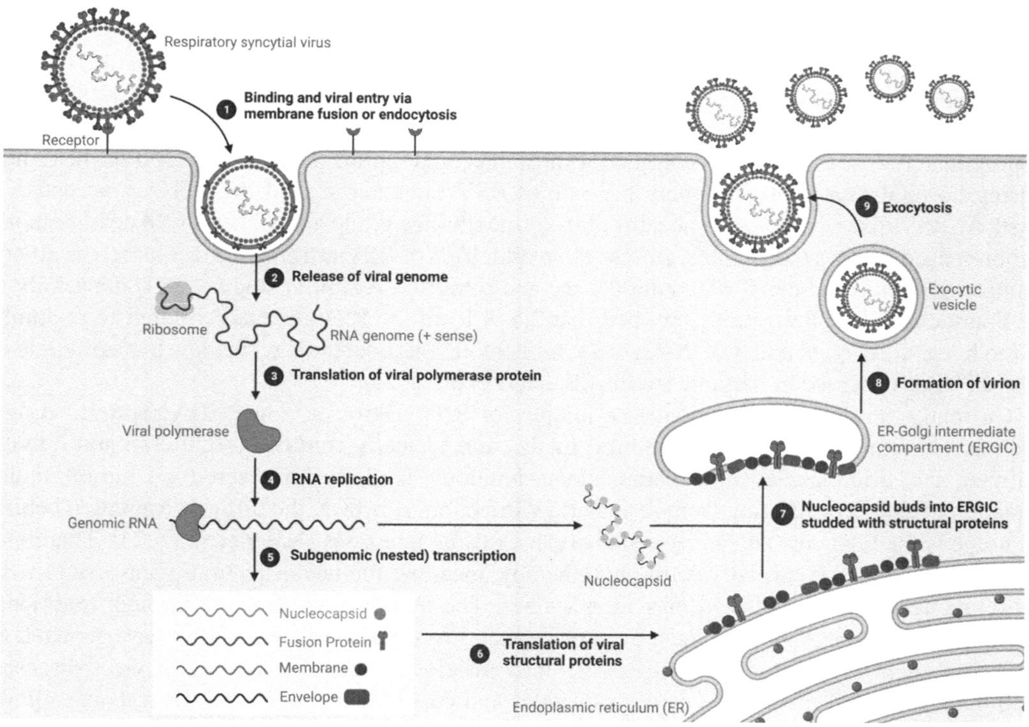


**FIGURE 10.2** Schematic overview of RSV life cycle.

of virion to epithelial cells (Figure 10.2). Then it binds with glycosaminoglycans (GAGs) present on the surface of mammalian cells. The heparin-binding domain, a segment of positively charged amino acids between the two mucin-like domains of G protein, is the major mediator of this interaction (Kaler et al., 2023). The viral fusion F glycoprotein is a trimeric class 1 fusion protein that causes the fusion of viral and host cell membranes by undergoing significant conformational changes, facilitating the entry of the viral genome into the host cell. When the virus fuses with epithelial cell membranes RNP (helical ribonucleoprotein complex) assembled by L Polymerase and viral RNA wrapped by N, P, and M2-1 proteins are released into the cytoplasm of the host cell. Replication of viral genome and transcription take place in the cell cytoplasm (Battles and McLellan, 2019). The transcription is carried out by the viral RNA-dependent RNA polymerase (RdRp) complex. After transcription, the mRNA is translated into various components required to form viral progeny called IB (cytoplasmic protein inclusion body) (Shang et al., 2021). After the process is completed, the virus starts invading the adjacent cells and spreading the infection. RSV binds to the host membrane facilitated by glycoprotein (G) and enters via membrane fusion (F protein) or endocytosis. Then, a helical viral ribonucleic genome (RNP) is released into the cytoplasm. The viral RNA-dependent RNA polymerase (RdRp) complex begins the transcription of viral mRNA and synthesis of positive sense anti-genome needed for the replication of new negative sense genome for the packaging of virions. Assembly of RSV virions starts after the translation of viral structural proteins. Nucleocapsid buds into the ERGIC (ER-Golgi intermediate compartment) and the viral particle buds out through exocytosis (Figure 10.2). The immunological response is controlled by the nuclear factor κB (NF-κB) pathway, which is initiated by the M2-1 protein. It also leads to a strong neutrophilic response along with the release of a high number of interleukins (IL)-6, IL-8, thymic stromal lymphopoietin (TSLP), and tumor necrosis factor (TNF-α). In addition, RSV infection triggers the activation of protein kinase R (PKR), which plays an important role in restricting viral replication. PKR also regulates the activation of the eukaryotic translation initiation factor 2-alpha (eIF-2α) (Efstathiou et al., 2020).

Recent regional-level meta-analyses by Li et al. (2022) state that around 33 million RSV infection episodes were observed globally in children aged between 0 and 60 months in 2019. Out of these, 3.6 million got hospitalized with hypoxemia, and overall, 1,01,400 died. In the case of infants, approximately 6.6 million were admitted to hospitals, and 45,700 died, with 13,300 of them died during hospital treatment. The mortality rate of RSV infection around the globe came out to 1 death by RSV disease in every 50 deaths of 0–60 months age group and 1 in every 28 deaths among children aged 28 days to 6 months. Across the world, 97% of RSV-attributable deaths across all age groups were found to be in low and middle-income countries. A similar study of 58 countries about RSV-associated hospitalizations revealed that 2.5–4.1 million RSV-infected children were admitted to hospitals (Cohen and Zar, 2022). RSV bronchiolitis is associated with a low mortality rate of 0%–0.3%, as observed in multiple studies (Bahabri et al., 2023).

Currently, no decisive treatment is available for RSV. However, some FDA-approved drugs, such as ribavirin, are used as an antiviral medication typically confined to children and infants. Palivizumab, a humanized mouse monoclonal antibody, is also administered via intramuscular injection to children suffering from serious RSV infection. However, the primary limitation behind its usage is the high cost and adverse side effects, raising questions (Kaler et al., 2023). Therefore, nowadays attention is turned toward phytotherapy, meaning the use of natural plant-based medication to treat diseases, instead of synthetic ones. The treatment of diseases through traditional medicines has been going on since the era of yoga and Ayurveda. Various studies have shown how medicinal plant extracts are giving positive antiviral effects against a number of virus diseases. Additionally, herbal medication serves as a safer and cost-effective alternative and its use allows a reduction in the occurrence of drug resistance, helping improve the immune system (Yasmin et al., 2020). The antiviral activities are mainly attributed due to the presence of various bioactive compounds such as flavonoids, terpenoids, saponins, lignans, etc. that have direct interference with different steps of the viral cycle including viral entry, replication, and release from the host cell for the spread of infection (Ponticelli et al., 2023).

Preliminary investigations into the benefits of medicinal plants have revealed the presence of the antiviral potential of various plant species, such as aloe vera, olive, turmeric, and java grass. These plants contain specific compounds that hold great therapeutic value, such as 7-O-Methylaloresin, oleuropein, curcumin, and cyperen (3,14) (Sharma et al., 2023).

Giloy (Guduchi), a medicinal plant native to tropical parts of India, has been found to prevent various fatal infections like swine flu, H1N1flu, and also the recent Covid-19 disease (Yasmin et al., 2020; Chowdhury, 2021). Similarly, Golden Bell (*Fructus forsythia*) has been found to have anti-inflammatory and antiviral activity against the influenza virus and shows therapeutic activity for skin infections. The major drawback of targeting a particular viral protein arises due to the property of rapid mutation of viruses. Therefore, each aspect of medicinal herb needs to be analyzed, including the identification of the correct part of the plant to be used and the preparation of the extract. The most potent antiviral activity is achieved by acetone or methanol fractions. Therefore, the correct methodology should be employed for the exploration of potent medicinal plant-derived drugs.

## 10.2 DIFFERENT APPROACHES FOR MANAGEMENT OF RESPIRATORY SYNCYTIAL VIRUS AND VIRAL PNEUMONIA (CHEMICAL DRUG BASED)

For the management of respiratory syncytial virus and viral pneumonia, specific RSV therapy, immunization therapy, and inhibitors are three basic different approaches that are employed (Table 10.1).

**TABLE 10.1**
**Role of Chemical Drugs on Various Respiratory Syncytial Virus**

| S. No. | Chemical Drug | Routes of Administration | Mode of Action | References |
|---|---|---|---|---|
| 1. | 2B11 | Intravenous | RSV Glycoprotein inhibitor | Rocca et al. (2021) |
| 2. | ALN-RSV01 | Intranasal | siRNA directed against the mRNA of the respiratory syncytial virus and nucleocapsid protein | Xing and Proesmans (2019) |
| 3. | ALS-008176 | Oral | RSV polymerase inhibitor | Behzadi and Leyva-Grado (2019) |
| 4. | ALX-0171 | Nebulization or Inhalation | Trivalent nanobody inhibits fusion protein | Verwey and Madhi (2023) |
| 5. | Benzoazepinequnoline | Intravenous | RSV glycoprotein F inhibitor | Zheng et al. (2018) |
| 6. | Clesrovimab | Intravenous | Humanized monoclonal antibody binding F protein site Ø | Verwey and Madhi (2023) |
| 7. | Favipiravir | Oral | Induces RSV virus RdRp backtracking | Stevaert et al. (2022) |
| 8. | Galidesivir | Oral, Intravenous, Intramuscular | Inhibits RNA synthesis | Stevaert et al. (2022) |
| 9. | GS-5806 | Oral | Targeting RSV fusion protein | Guo et al. (2023) |
| 10. | Lumicitabine | Oral | Nucleoside analog inhibits the replication of RSV by polymerase selectivity | Stevaert et al. (2022) |
| 11. | MDT-637 | Nebulization or Inhalation | Targeting RSV fusion protein | Xing and Proesmans (2019) |
| 12. | MK-1654 | Intravenous | Inhibits the site III of the F glycoprotein | Rocca et al. (2021) |
| 13. | Motavizumab | Intravenous | Humanized monoclonal antibody inhibits Glycoprotein F | Sun et al. (2023) |

(*Continued*)

**TABLE 10.1 (*Continued*)**
**Role of Chemical Drugs on Various Respiratory Syncytial Virus**

| S. No. | Chemical Drug | Routes of Administration | Mode of Action | References |
|---|---|---|---|---|
| 14. | Nirsevimab | Intravenous | Humanized monoclonal antibody binding F protein site Ø | Bergeron and Tripp (2022) |
| 15. | Palivizumab | Intramuscular | Inhibits fusion protein epitope II | Sun et al. (2023) |
| 16. | Pimodivir | Oral | Non-nucleoside polymerase inhibitor that targets the PB2 subunit | Behzadi and Leyva-Grado (2019) |
| 17. | RI-001 | Intravenous | Immune RSV infection | Xing and Proesmans (2019) |
| 18. | Ribavirin | Aerosolized, Oral, and Intravenous | Targets and metabolize in the liver and intracellularly | Tejada et al. (2022) |
| 19. | Rilematovir | Oral | Targeting RSV fusion protein | Mackman et al. (2021) |
| 20. | RSM01 | Intravenous | Humanized monoclonal antibody binding F protein site Ø | Verwey and Madhi (2023) |
| 21. | RSV604 | Mucosal inoculation | Inhibits RNA synthesis and the infectivity of released virus | Gao et al. (2021) |
| 22. | Sisunatovir | Oral | RSV fusion inhibitor | Cockerill et al. (2021) |
| 23. | Suptavumab | Intravenous | Humanized monoclonal antibody binding F protein site IV | Verwey and Madhi (2023) |
| 24. | TMC353121 | Intravenous | Targeting RSV fusion protein | Behzadi and Leyva-Grado (2019) |
| 25. | Ziresovir | Oral | Targeting RSV fusion protein | Gao et al. (2021) |

### 10.2.1 Specific RSV Therapy

Diagnosing RSV is unlikely to be possible based solely on clinical symptoms. Therefore, to differentiate it from other respiratory viruses, laboratory diagnostics tests are essential. Several selective potent therapeutic antiviral compounds have been identified for the management of respiratory syncytial virus and viral pneumonia, including both chemical drugs and bioactive compounds extracted from medicinal plants (Russell et al., 2021). Among chemical drugs, ribavirin, corticosteroids, palivizumab, and motavizumab are used as specific treatment as they have shown significant results. Ribavirin is a broad-spectrum guanosine analog that inhibits the replication of both RNA and DNA viruses. It is utilized in life-threatening conditions of RSV patients. However, ribavirin alone is not sufficient for termination of virus chain. The effectiveness of ribavirin is increased by combining it with rIFN-α2a, rIFN-α2b, and rIFB-β1a (Arabi et al., 2020). In RSV patients, ribavirin is administrated by orally, aerosolized, and intravenous formulations. In aerosolized formulation, a 2 g of dose is administrated after every 8 hours for 15 total doses, while oral/intravenous formulations are administrated 15–20 mg/kg in three doses over 10 days (Foolad et al., 2019). Clinical studies were conducted on different patients to differentiate the effectiveness of oral and intravenous ribavirin. Intravenous ribavirin was administrated in 52 patients in three divided doses, while oral ribavirin was given to 21 patients in two divided doses with an interval of 11 days. Studies showed that oral ribavirin is more effective than intravenous ribavirin, as it showed a reduction in viral load and proved to be cost-effective and safe alternative (Gatt et al., 2023).

Palivizumab is humanized monoclonal immunoglobulin G1 prepared by utilizing Recombinant DNA technology. It was licensed in 1998 by Food and Drug Administration for infants, who were at a higher risk of RSV. Palivizumab targets antigenic site II of fusion protein. Controlled trial period conducted in various hospitals reported the reduction of hospitalization rate from 9.7% to 5.3% among infants. Another trial was conducted to study the effect of palivizumab administrated at different age period of children. The children who received palivizumab in the first year of their life showed 61% reduction rate of RSV bronchitis (Eichinger et al., 2021). Motavizumab is monoclonal immunoglobulin G1, having greater affinity for F glycoprotein developed from palivizumab. To understand the therapeutic effect of motavizumab, clinical trials was conducted on pediatric patients with RSV positive. The results from studies showed that palivizumab reduced the chance of hospitalization to 26% more than palivizumab (Sun et al., 2023).

## 10.2.2 Immunization Strategy

RSV penetrates epithelial mucus layer and activates innate immunity, which fails to clear the virus load to control the infection and protection against RSV, adaptive immunity plays an essential part. Immunization strategy is basically categorized into five categories named as recombinant vector-based vaccines, nucleic acid-based vaccines, protein-based vaccines, live attenuated vaccines, and monoclonal antibodies (Figure 10.3).

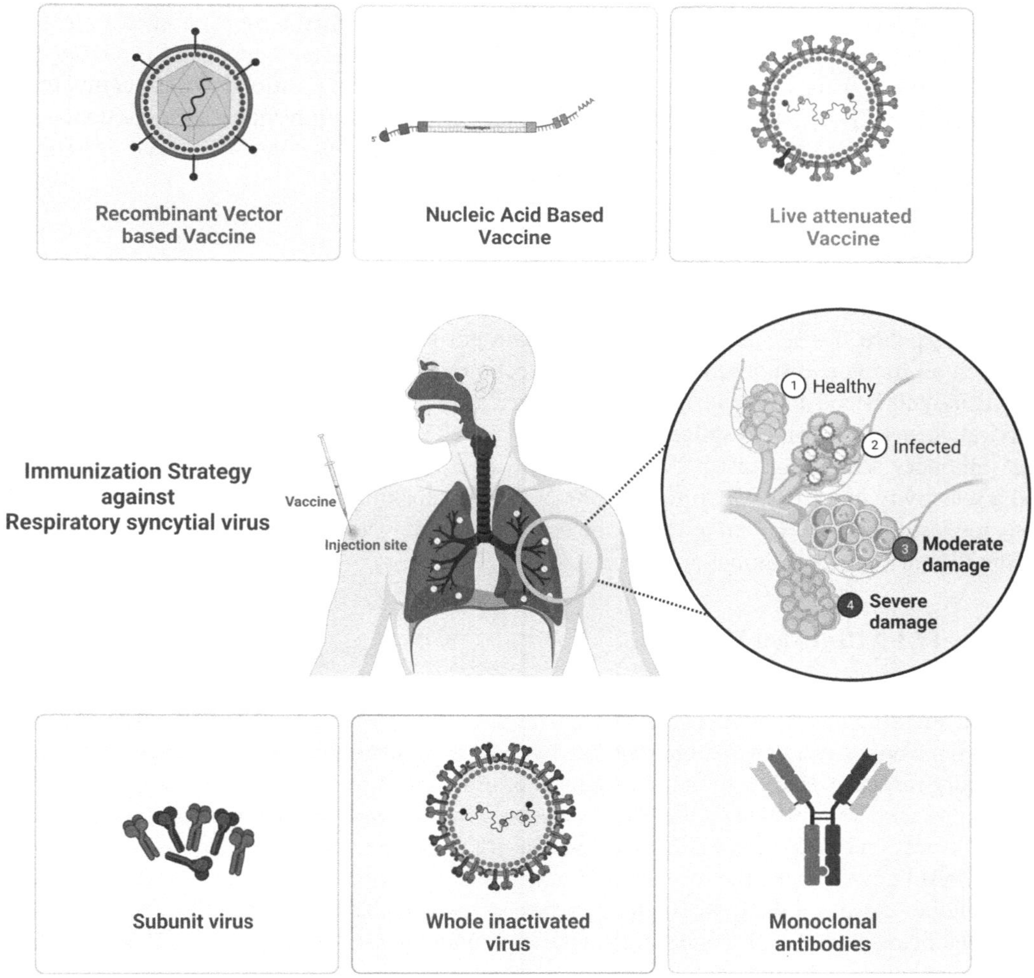


**FIGURE 10.3** Immunization strategy against respiratory syncytial virus.

### 10.2.3 Recombinant Vector-Based Vaccines

Recombinant vector-based vaccines for respiratory syncytial virus (RSV) are a type of vaccine that utilizes modified viruses or viral vectors to deliver genetic material from RSV into the cells of the vaccinated individual. They can induce both antibody and cellular immunity, potentially offering long-lasting protection against RSV (Russell et al., 2021). In a study, a codon-optimized rNDV-BRSV Fopt vaccine was tested on colostrum-deprived calves to protect them from BRSV infection. The vaccine was well-tolerated, showed no signs of vaccine-enhanced disease, and reduced the severity of lung lesions and viral load in the lungs. The vaccinated calves also developed neutralizing antibodies and exhibited lower levels of lung CXC chemokines after the BRSV challenge, demonstrating the safety and effectiveness of the rNDV-BRSV Fopt vaccine in protecting against BRSV infection in neonatal calves (Sacco et al., 2022).

### 10.2.4 Nucleic Acid-Based Vaccines

Nucleic acid-based vaccines are a type of vaccine that uses genetic material, such as DNA or RNA, to stimulate an immune response against a specific pathogen. They work by introducing the genetic material into cells, which then produce antigens that trigger an immune response, providing protection against the targeted disease (Aliprantis et al., 2021). Several clinical trials are underway for mRNA vaccines targeting RSV infection. In phase I trials, mRNA-1777 demonstrated the ability to stimulate RSV neutralizing antibodies and elicit a robust humoral response. Through further optimization and engineering, mRNA-1345 exhibited enhanced immunogenicity, with a phase I trial showing seven times higher neutralizing antibody titers compared to mRNA-1777. mRNA-1345 has received FDA approval for RSV prevention in individuals aged 60 and above, and is now being evaluated in a multicenter, placebo-controlled phase 2/3 clinical trial (Qiu et al., 2022).

### 10.2.5 Protein-Based Vaccines

Protein-based vaccines are a type of vaccine that use specific proteins from RSV to stimulate an immune response. These vaccines typically contain purified or recombinant RSV proteins, which can elicit an immune response and provide protection against RSV infection (Biagi et al., 2020). Scientists synthesized and evaluated a series of monoterpene-substituted arylcoumarins as potential antiviral agents against the respiratory syncytial virus (RSV). These compounds exhibited significant inhibitory activity against RSV replication, with compound 14e showing the highest potency and a selectivity index of approximately 200. Molecular docking and molecular dynamics simulations suggest that the F protein of RSV may be a potential target for these compounds, particularly during the early stages of infection (Khomenko et al., 2023).

### 10.2.6 Live Attenuated Vaccines

Live attenuated vaccines are a type of vaccine that use weakened or modified forms of the virus to stimulate an immune response. These vaccines contain live RSV that has been attenuated or weakened in the laboratory, allowing it to replicate but not cause severe illness. The live virus stimulates an immune response, leading to the development of immunity against RSV (Mueller et al., 2020). In a study, researchers investigated the T-cell responses in the lungs of mice following immunization with influenza and RSV viruses. They found that RSV epitopes did not impact influenza-specific T-cell levels but influenced the formation of lung-resident memory T cells (TRM). The addition of RSV epitopes enhanced the CD8 T-cell responses, suggesting that the local environment created by RSV-specific memory T cells contributed to the augmentation of influenza-specific T-cell immunity (Matyushenko et al., 2020).

### 10.2.7 Monoclonal Antibodies

Monoclonal antibodies are laboratory-produced molecules that mimic the natural antibodies produced by the immune system and are designed to target specific proteins or cells in the body, such as those associated with diseases like cancer or autoimmune disorders (Rodriguez-Fernandez et al., 2021). Nirsevimab, a monoclonal antibody, targets a specific antigenic site and is effective in preventing respiratory infections caused by RSV in infants. A single dose of nirsevimab significantly lowers the incidence of medically attended RSV-associated lower respiratory tract infection compared to a placebo. It shows greater potency and longer serum half-life than palivizumab, potentially leading to more effective antibody-mediated protection (Viguria et al., 2021).

## 10.3 INHIBITORS

Inhibitors are classified into five categories, namely, fusion inhibitors, nucleoprotein inhibitors, nucleoside analog and non-nucleoside inhibitors, virus polymerase inhibitors, and host targeting (Figure 10.4).

### 10.3.1 Fusion Inhibitors

Fusion inhibitors are a class of antiviral drugs that block the fusion process between a virus and host cell, preventing viral entry and infection. They work by targeting specific proteins on the viral envelope or cell membrane, inhibiting the fusion of these membranes, and blocking viral replication (Pribut et al., 2020). In a recent Phase II study, the ResVax maternal RSV vaccine candidate showed favorable tolerability and no severe adverse events in 50 healthy pregnant women. Despite variations in baseline RSV antibody levels due to small sample sizes and natural exposure, immunized

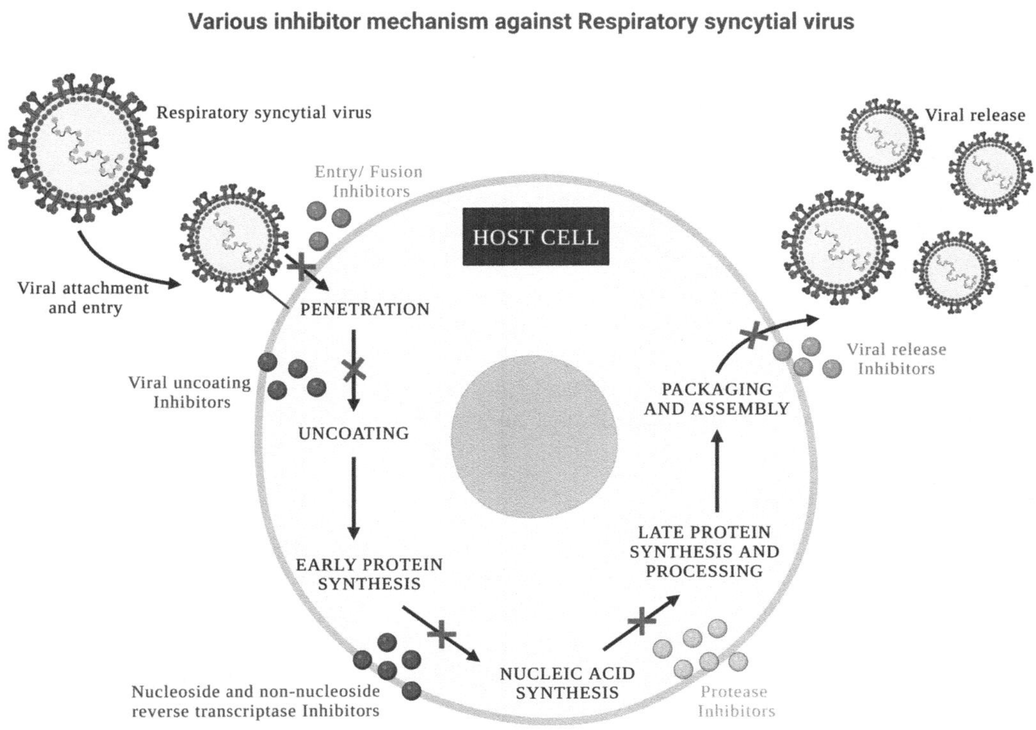


**FIGURE 10.4** Various inhibitor mechanism against respiratory syncytial virus.

women produced antibodies within 14 days, leading to higher RSV-specific antibody levels in cord blood of their infants compared to the placebo group (Novavax, Inc., 2019).

### 10.3.2 Nucleoprotein Inhibitors

Nucleoprotein inhibitors are a type of antiviral drugs that target the nucleoprotein of a virus, which is essential for viral replication and transcription. By binding to the nucleoprotein, these inhibitors disrupt the normal functioning of the virus, inhibiting its ability to replicate and spread (Khodjoyan et al., 2022). EDP-938, a novel non-fusion replication inhibitor, demonstrates potent activity against various RSV strains *in vitro*, including clinical isolates. It targets a post-entry replication step and appears to act through the viral nucleoprotein (N). EDP-938 exhibits a high barrier to resistance, shows no cross-resistance, and displays synergistic antiviral effects when combined with other RSV inhibitors. Additionally, it has demonstrated efficacy in a non-human primate model of RSV infection (Rhodin et al., 2021).

### 10.3.3 Nucleoside Analog and Non-nucleoside Inhibitors

Nucleoside analog inhibitors are antiviral drugs that mimic the structure of natural nucleosides and are incorporated into viral DNA or RNA during replication. By doing so, they disrupt viral replication and inhibit further spread of the virus (Stevaert et al., 2022). Non-nucleoside inhibitors, on the other hand, target specific viral enzymes or proteins, blocking their function and interfering with viral replication (Denel-Bobrowska and Olejniczak, 2022). Scientists presented findings on VV116, an oral derivative of RDV, which exhibited strong effectiveness against SARS-CoV-2 in mice. VV116, a prodrug of the C7-deuterated GS-441524 analog (X1), is currently undergoing Phase II/III clinical trials for COVID-19 treatment. Our study reveals VV116's potential as a promising oral nucleoside antiviral for RSV infection (Zhang et al., 2022).

### 10.3.4 Virus Polymerase Inhibitors

Virus polymerase inhibitors are a class of antiviral drugs that specifically target the viral polymerase enzymes responsible for replicating the genetic material of the virus. By inhibiting the activity of these polymerases, the inhibitors disrupt viral replication and prevent the production of new infectious particles (Ludeke and Fearns, 2020). In the study, scientists examined the effects of template variations on RNA synthesis by the RSV polymerase using *in vitro* assays. Their findings confirmed the previously observed back-priming activity of the RSV polymerase, which is influenced by the secondary structure of the RNA template. Expanding the hairpin loop size of the template eliminated the back-priming activity without affecting *de novo* RNA synthesis. They discovered a new primer-based terminal extension activity of the RSV polymerase that adds nucleotides to the template and primer in a nonspecific manner. These insights contribute to our understanding of RSV polymerase RNA synthesis and may aid in the development of effective antiviral drugs for RSV treatment (Xu et al., 2023).

### 10.3.5 Host Targeting

Host targeting refers to a therapeutic approach that focuses on modulating or manipulating host factors rather than directly targeting the pathogen itself. This strategy aims to disrupt the essential processes or interactions that viruses rely on within host cells, effectively inhibiting viral replication or reducing the severity of the infection. By targeting host factors, such as cellular receptors or signaling pathways, this approach offers a potential broad-spectrum antiviral activity against multiple viral strains (Hu et al., 2020). Thapsigargin (TG) inhibits the ER $Ca^{2+}$ ATPase pump and activates the UPR, inducing a strong antiviral response that blocks the replication of influenza A and RSV viruses. TG is effective in human cells, stable in low pH conditions, and provides therapeutic protection in mice against lethal influenza infection when administered orally post-infection (Al-Beltagi et al., 2021).

## 10.4 ROLE OF MEDICINAL PLANTS/HERBS AS ANTIVIRAL DRUGS DURING RESPIRATORY SYNCYTIAL VIRUS AND VIRAL PNEUMONIA INFECTION

Medicinal plants represent a valuable resource for discovering antiviral drugs. Currently, almost 40% of the drugs are either directly or indirectly derived from natural resources. Since the ancient times in the field of medical sciences, various diseases, including viral infections, have been widely treated by medicinal plants in the Asia-Pacific region, which have demonstrated potent therapeutic effects. However, the current approach of using synthetically derived antiviral drugs has limited efficacy and often leads to severe side effects. Classic examples of synthetic antiviral drugs, such as ribavirin and interferon, exhibit only *in vitro* effectiveness against most viruses but fail to yield the same results in patients (Ben-Shabat et al., 2020). Additionally, the high cost and increasing antiviral drug resistance (ADR) in viruses like herpes, hepatitis B, influenza, and others have raised a major demand to exploring alternative sources for safe and effective antiviral drugs. Medicinal plant extracts or bioactive compounds have become a possible substitute for the development of antiviral drugs responsible for restricting the replication of viral particles in hosts (Adhikari et al., 2021). Approximately around 2,500 plant species have been discovered worldwide for the treatment of various diseases caused by viruses and bacteria (Kapoor et al., 2017). Considerable research is underway in the field of medicinal plants to assess the antiviral activity against various human viruses including respiratory syncytial virus (RSV), a major cause of viral pneumonia in young children (under 5 years of age) and elderly people (over 65 years of age) (Table 10.2) (Ali et al., 2021). However, nowadays, many people are becoming increasingly aware of the side effects of using synthetic drugs and have started adapting themselves to Ayurveda, which is completely based upon the use of nature-based herbal medicines.

Various naturally occurring compounds in medicinal plants like polyphenols, alkaloids, flavonoids, glucosides, and others, have been isolated and observed to show antiviral activity against RSV. *Eleutherococcus senticosus*, also known as Siberian ginesang, is a medicinal plant belonging to the *Araliaceae* family. It is said to regulate immune responses due to the presence of adaptogenic substances. Extracts from the roots and bark of this plant cause the inhibition of RNA replication of the RSV virus (Adhikari et al., 2021). Similarly, *Terminalia chebula* (chebulic myrobalan), native to southern Asia and known for its extraordinary nutritional value and antiviral activities (Sultan et al., 2023), contains bioactive compounds like chebulagic acid and punicalagin. These compounds extracted from *Terminalia chebula* are responsible for the inactivation of the free RSV particles and also inhibiting the early entry, attachment, and penetration of the virus (Ali et al., 2021). In severe cases of RSV infection resulting in bronchiolitis which is inflammation of the bronchioles (Carvajal et al., 2019), it has been stated that the ethanol extract of plant *Lophatherum gracile* (bamboo leaf) inhibits RSV infection and RSV-induced inflammation (Ali et al., 2021).

## 10.5 MECHANISM OF ACTION

The extraction and estimation of therapeutically important properties of small biomolecules through the examination of their interaction with a target protein are critical steps in the discovery of a drug. The selection of a particular bioactive compound to be extracted from the medicinal plant begins with virtual screening, which includes the estimation of the drug affinity, pharmacokinetics, and lipophilicity properties (Chowdhary & Pan, 2020). While designing an antiviral drug, three major pathways of the virus cycle are to be considered: entry, which includes the first attachment of the virus on the host cell, replication of the viral genome, and transcription (Hu et al., 2021). Based on a

**TABLE 10.2**
**Role of Medicinal Plants on Various Respiratory Syncytial Virus and Viral Pneumonia**

| S. No. | Species of Plant | Source of Extraction | Compound Isolated | Virus | References |
|---|---|---|---|---|---|
| 1. | *Burkea Africana hook.* | Stembark | Oleanane-type triterpene saponins 7 | Influenza | Mair et al. (2018) |
| 2. | *Bletilla striata(thunb.) Rchb.f.* | Tubers | Phenanthrenes | Influenza | Xiao et al. (2017) |
| 3. | *Cimicifuga foetida* L. | Roots | Acetin, Cimilactone A | RSV | Fang et al. (2019) |
| 4. | *Cleistoclyx operculatus (Roxb.).* | Flower bud | Cleistocaltones A Cleistocaltones B | RSV | Le et al. (2021) |
| 5. | *Clerodendrum trichotomum thumb.* | Leaves and Twigs | Acetoside | RSV | Z Hu et al. (2021) |
| 6. | *Sambucus nigra* L. | Fruit | Flavonoids | Influenza | Roschek et al. (2009) |
| 7. | *Canarium album* | Leaf | Isocorilagin | Influenza A Virus | Chen et al. (2020) |
| 8. | *Portulaca oleracea* L. | Edible potherb, aerial part | Arabinoglucomannan | Influenza | Li et al. (2019) |
| 9. | *Salvia plebeian R.Br.* | Aerial part | Nepetin | Influenza | Bang et al. (2016) |
| 10. | *Glycine max* (L.) | Aerial part | Daidzein and Glycitein | Influenza | Nagai et al. (2019) |
| 11. | *Cedrus deodara (Roxb.)* | Leave and Twig | Polysaccharide | H1N1 | Sinha (2019) |
| 12. | *Houttuynia Cordata thunb.* | Dried whole plant | Polysaccharide | Influenza | Zhu et al. (2018) |
| 13. | *Aloevera (L.) Burm.f* | Leaves | Aloin | Influenza, SARS-CoV | Piana et al. (2020) |
| 14. | *Curcuma longa* L. | Roots | Curcumin | Influenza | Chen et al. (2020) |
| 15. | *Camellia sinensis* (L.) | Leaves | Theaflavin-3 gallate | Influenza | Hu et al. (2021) |
| 16. | *Allium cepa* | Steam Skin | Quercetin & isorhamnetin | Influenza | Kwak et al. (2017) |
| 17. | *Allium sativum* L. | Stem | Allicin | Influenza B/H1N1 | Venu and Austin (2020) |
| 18. | *Glycine max* (L.) | Leaves | Daidzein glycitein | Influenza | Seo et al. (2017) |
| 19. | *Anacardium occidentale* L. | Leaves | Pentagalloylglucose | Influenza/H1N1/H3N2 | Zhang et al. (2020) |
| 20. | *Lens culinaris Medik* | Aerial part | Kaempferol | Influenza | Zandavar and Babazad (2023) |
| 21. | *Eupatorium perfoliatum* L. | Leaves | Flavonoids | H1N1/influenza virus | Venu and Austin (2020) |
| 22. | *Justicia adhatoda* L. | Leaves | Flavonoids, Alkanoids | Influenza | Venu and Austin (2020) |
| 23. | *Glycyrrhiza glabra* L. | Root | Glycyrrhizin | HIV-1 | Venu and Austin (2020) |
| 24. | *Ficus religiosa* L. | Bark,leaves | Glycoside, Flavonoid | H2N2 | Perera et al. (2021) |
| 25. | *Pyrrosia lingua (Thunb.)* | Whole plant | Chlorogenic acid, Flavonoids | SARS-CoV | Jang et al. (2022) |
| 26. | *Scutellaria indica* L. | Root | Luteolin, Chrysin | RSV | Ergasheva et al. (2021) |
| 27. | *Justicia adhatoda* L. | Leaves | Alkaloids | Influenza | Dhankar et al. (2011) |
| 28. | *Scutellaria indica* L. | Dry whole herb, Root | Apigenin, Flavonoid | RSV | Yu and Dan (2020) |
| 29. | *Mussaenda pubescens Dryand* | Leaves | Triterpene | RSV | Huong et al. (2021) |
| 30. | *Elephantopus scaber* L. | Whole Herb | Flavonoid | RSV | Bai et al. (2020) |

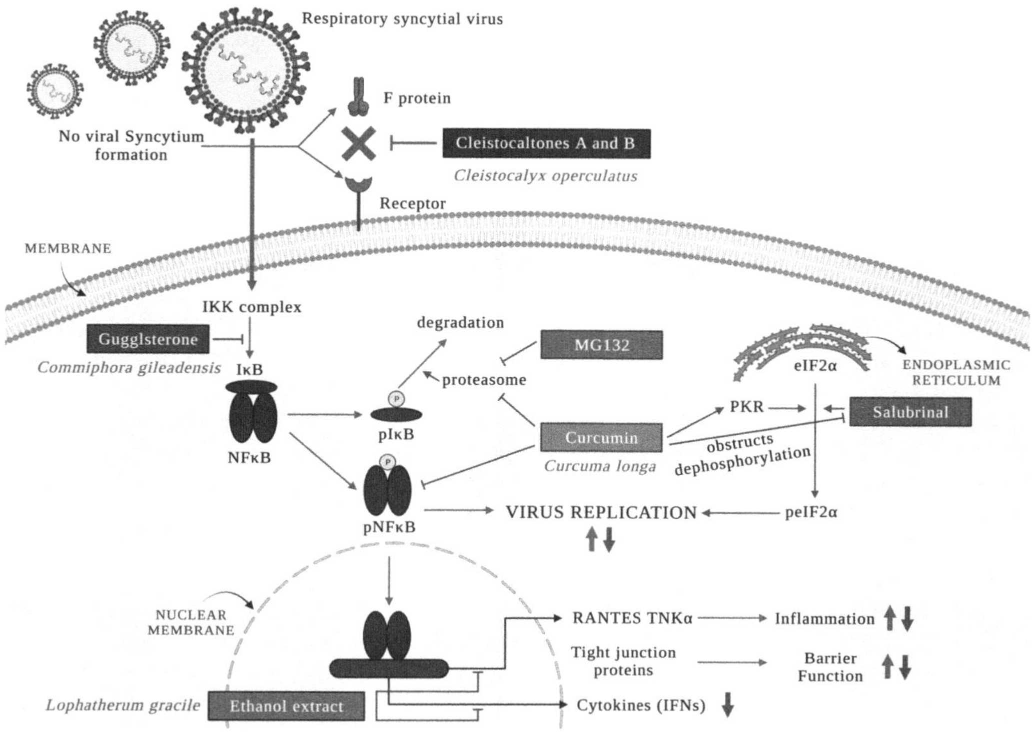


**FIGURE 10.5** Schematic diagram of mechanism of action of bioactive extracted from medicinal plants.

compound's cytotoxicity and antiviral activity, the biological value of phytoconstituent is evaluated. A substance is most likely to be useful as an antiviral medicine if it has a low $IC_{50}$ (half maximal inhibitory concentration) and a high SI (selectivity index) (Hu, et al., 2021). The discussion continues with several examples of the mechanism of action (Figure 10.5).

Curcumin (1,7-bis(4-hydroxy-3-methoxyphenyl)-1,6-heptadiene-3,5-dione), a polyphenolic biocompound, is naturally observed in the rhizomes Curcuma longa, commonly known as turmeric, and is utilized as a medicinal herb in Asian countries (Hewlings and Kalman, 2017). It has been popularly known for its various therapeutic applications including antiviral activity against RSV (Thimmulappa et al., 2021). An experiment by Ben-Shabat et al. (2020) was conducted to show the release of poorly soluble curcumin. In this experiment, they compared a curcumin-loaded solid nano lipid carrier to a similar curcumin-loaded nanolipid carrier modified with the addition of N-acetyl L-cysteine polyethylene glycol (PEG). The goal was to create mucoadhesive carriers that could penetrate mucous effectively. The results showed that after 4 hours of administration, 80% of the drug was released from the curcumin solution in the case of the modified nanolipid carrier, which was lower in comparison to the unmodified one. As soon as it enters the RSV-infected human nasal epithelial cells, it directly alters the NF-κB pathway, which is responsible for the replication and budding of RSV, as well as the release of proinflammatory cytokines in response to immune response without causing any toxicity to living cells. Additionally, curcumin also inhibits eIF-2α dephosphorylation, proteasome activity, and COX2 in the presence of salubrinal, which obstructs the dephosphorylation, MG132 (a proteasome inhibitor), and COX 2 inhibitor (Obata et al., 2013). These results indicate that curcumin is an effective, non-cytotoxic, naturally derived therapeutic agent that can be used for treating lower respiratory tract diseases caused by RSV.

*Lophatherum gracile* is a therapeutically important medicinal plant widely used as traditional Chinese medicine for treating UTIs and fever. The ethanol extract from the leaves of this plant that form a major part of flavonoids was investigated by Chen et al. (2019) to check its anti-RSV activity. The treatment of the ethanol extract on the lungs of the RSV-infected mice did not show any inflammation in the bronchial tubes, interstitial tissues, and blood vessels. The administration of the ethanol extract resulted in a significant decrease in the expression level of IL-1β and TNF-α. Furthermore, the production of interferon-gamma (IFN-γ) in the RSV-infected group was measured at 2,667.2 pg/mL. The treatment also demonstrated a reduction in IFN-γ secretion, which showed a notable difference. The observed function was found to be associated with the control of TNF-α and IL-1β expression, an elevation in the ratio of $CD^{4+}/CD^{8+}$ T cells, and the inhibition of IFN-γ expression originating from Th1 cells within the adaptive immune system. Therefore, these discoveries offer a logical basis and scientific substantiation for the widespread utilization of *L. gracile* in traditional medicine to address ailments that may be linked to RSV.

Another bio-compound guggulsterone ($C_{21}H_{28}O_2$) has been isolated from *Commiphora gileadensis*, a native plant of Saudi Arabia known locally as becham. It is used as an herbal Arabian medicine for treating stomach, liver, and constipation inflammation-related disorders. This compound has been found to show antiviral activity against herpes simplex virus type 2, adenovirus type 5, dengue virus, and RSV type B (Abdulhakim, 2022). Research conducted by Bouslama et al. (2019) has shown the antiviral activity of guggulsterone against RSV-B without any cytotoxic effect. Guggulsterone was isolated from the methanol extract of leaves using various bio-guided assays such as plaque assay, anti-CVB-3, and anti-ADV-5 assay, etc. The results showed approximately 20 μg/mL and >10 of IC50 and SI values of the methanol extract. The mechanism of action behind this antiviral effect of the bioactive compound guggulsterone was observed to involve the suppression of NF-κB signal, which is responsible for the transcription activation and the regulation of inflammation-enhancing factors such as chemokines, cytokines, fibroblasts, and molecules associated with cell adhesion during the RSV life cycle inside the host cell. Therefore, it is a highly effective natural bioactive compound derived from medicinal plants that can be used against RSV infections.

*Cleistocalyx operculatus* is an evergreen plant from the *Myrtaceae* family, widely found in tropical areas of southeast Asia, known as "Voi" in Vietnamese and "Shui Weng" in Chinese. Its leaves and flower buds hold great therapeutical importance and are used as traditional medicine externally to treat inflammation, cold, fever, and various skin-related problems, such as acne and bruises (Pham et al., 2020). In a study conducted by Song et al. (2019) on this plant, two unique phloroglucinol derivatives, cleistocaltones A and B, were found to show antiviral activity against RSV. The cytopathic effect (CPE) reduction assay showed IC50 values of $6.75 \pm 0.75$ and $2.81 \pm 0.31$ μM, which were even higher in comparison to the positive control ribavirin. Electron microscopy revealed that the treatment with cleistocaltones A and B prevented the formation of syncytium in RSV-infected cells. This effect can be attributed to the suppression of RSV F proteins, which are responsible for the formation of viral syncytium or multi-nucleated cell by fusion of cells. These results evidently indicate that these two compounds A and B can be observed as an effective anti-RSV plant-derived bio-compound.

*Celastrus hindsii* Benth, a medicinal plant of genus *Celastraceae*, is commonly found in the forests of Vietnam and is used in Chinese herbal medicine as diuretic or water pills. It is also employed as an emmenagogue, which refers to herbs used for delayed menses. Terpenoids are the bioactive compounds responsible for the various therapeutic effects of this plant such as anti-fungal, anti-tumor, anti-inflammatory, and anti-parasitic (Gundoju et al., 2022). In a study conducted by Luo et al. (2018), the antiviral activity of the terpenoids against the respiratory syncytial virus was shown with the help of cytopathic effect (CPE) reduction assay. They isolated a total number of 24 chemical compounds from ethanol extract of the stems of *C. hindsii*. The results obtained showed that compounds like 2α-hydroxyabietatriene, 3-oxototarol, celahin D,(4bS,7R,8aR)-4b,5,6,7,8,8a,9, 10-octahydro4b,8,8-trimethyl-2-(1-methylethyl)-1,7 phenanthrenediol, and wilforlide B showed significantly higher anti-RSV activities. These compounds displayed IC50 values ranging from 1.55 to 6.25 μM outperforming ribavirin (positive control). Therefore, the discussed medicinal plants and their mechanism of action against RSV offer logical basis and scientific support for the widespread use of medicinal plants in addressing illness specifically associated with RSV.

## 10.6 PLANT-DERIVED COMPOUNDS AGAINST RESPIRATORY SYNCYTIAL VIRUS AND VIRAL PNEUMONIA

### 10.6.1 *Zingiber officinale* Roscoe

Gingerol, which is the primary bioactive compound in ginger, has shown antiviral properties in multiple studies. It exhibits inhibitory effects on the replication of various viruses, including respiratory syncytial virus (RSV), influenza virus, herpes simplex virus (HSV), and human immunodeficiency virus (HIV) (Kiyama, 2020). Gingerol functions by interfering with viral attachment and entry into host cells, inhibiting viral replication, and decreasing the release of new virus particles. Ginger contains more than 400 different compounds chemically, but its pharmacological effects are primarily attributed to the presence of terpene and phenolic compounds (Mao et al., 2019). Based on the molecular docking analyses, it was found that 6-gingerol exhibits a high binding affinity with a number of virus proteins (Goswami et al., 2020). Through the process of dehydration by drying or cooking, ginger undergoes a conversion where 6-gingerol is transformed into 6-shogaol. This conversion enhances the stability of the compound and amplifies its pharmacological effects compared to 6-gingerol (Kiyama, 2020). Shogaol exhibits antioxidant, anti-inflammatory, anticancer, antiemetic, and neuroprotective properties (Kou et al., 2018). Zingerone is not naturally present in fresh ginger, but it can be synthesized through reverse aldolization of gingerols when ginger is dried, heated, or roasted (Choi et al., 2018). Fresh ginger exhibits potent antiviral effects against human respiratory syncytial virus (HRSV) and rhinovirus, making it a promising treatment for respiratory viral infections. The aqueous extract of fresh ginger specifically inhibits the attachment and penetration of HRSV into human larynx epidermoid carcinoma cells and human lung carcinoma cell lines when administered 1–2 hours before virus inoculation, unlike dried ginger. The extract obtained from fresh ginger induces the production of IFN-α and IFN-β from epithelial cells that are infected with HRSV (Jafarzadeh et al., 2021). Ginger compounds have been shown to inhibit RSV replication, reduce viral-induced inflammation, interfere with viral attachment and entry, inhibit viral replication, and modulate the host immune response. Further research is necessary to fully understand the effectiveness of ginger compounds against RSV in humans. Clinical trials are needed to assess safety, optimal dosage, and potential therapeutic applications for RSV treatment (Mustafa et al., 2019).

### 10.6.2 *Terminalia chebula* Retz

*Terminalia chebula* Retz, also known as Haritaki or Myrobalan, is a widely used traditional medicine with a history of treating various human ailments. It is known for its diverse therapeutic properties, including its ability to stop bleeding, aid digestion, treat dysentery, act as a liver tonic, alleviate pain, possess anthelmintic and antibacterial properties, and provide relief for skin disorders (Bulbul et al., 2022). It also contains a variety of bioactive compounds, including tannins, flavonoids, sterols, amino acids, fructose, and resins. Among these compounds, tannins such as chebulinic acid, chebulagic acid, gallic acid, chebulic acid, corilagin, and ellagic acid are believed to contribute to its potential applications (Bulbul et al., 2022). *Terminalia chebula* (*T. chebula*) is a safe herb that does not cause any weakening or toxic effects. This characteristic makes it a promising option for the development of nutraceutical formulations or herbal functional products in the future (Nigam et al., 2020).

### 10.6.3 *Ginkgo biloba*, *Hypericum perforatum*, and *Sambucus canadensis*

Quercetin, with the molecular formula $C_{15}H_{10}O_7$, is a flavonoid found in medicinal plants like *Ginkgo biloba*, *Hypericum perforatum*, and elderberry, as well as in common foods like onions, apples, and tea. Quercetin acts as a natural polar auxin transport inhibitor. It contains a ketocarbonyl group and a basic oxygen atom on the first carbon, forming salts with strong acids. Quercetin has four active groups, including a dihydroxy group, an o-dihydroxy group, a $C_2$–$C_3$ double bond, and a 4-carbonyl group (Yang et al., 2020). It also acts as an anti-HRSV agent. In experiments conducted by Lopes et al., both in the laboratory (*in vitro*) and through computer simulations (*in silico*), the anti-HRSV effects of Quercetin (Q0) and its derivative molecule (Q1) were compared. Acetylating Quercetin enhanced its effectiveness against the HRSV. The acetylated form, Quercetin pentaacetate, exhibited improved interaction with the F protein, resulting in a stronger binding and enhanced stability, which effectively blocked viral adhesion. These findings present a promising alternative strategy for combating hRSV and contribute to the advancement of drug discovery and development in this field. Quercetin's effectiveness against RSV is limited compared to other antiviral compounds. Its bioavailability is a challenge due to poor absorption and rapid metabolism. Further research and optimization are needed to enhance its antiviral activity against RSV (Lopes et al., 2020).

### 10.6.4 *Eucalyptus globulus*

*Eucalyptus globulus*, known as the blue gum tree, is a tall and perennial evergreen tree species. It is widely distributed, with trees in Yunnan Province, China, being notable for their high concentration of 1,8-cineole in the leaves. These trees are commercially utilized for the extraction of essential oils, which find application in the pharmaceutical and cosmetic industries (Dhakad et al., 2018). It also has been utilized in traditional medicine as a natural remedy for alleviating symptoms associated with colds and bronchitis. According to a study, it has been reported that the administration of a high dosage of eucalyptus oil (EO) can have detrimental effects on the immune function of the respiratory tract and overall body immunity. This negative impact is believed to occur through the modulation of B cells, leading to an increase in plasma cells, inflammatory substances, and immune complexes. To promote the enhancement of immune function and minimize potential harm to both the immune system in general and the respiratory tract specifically, it is crucial to ensure the proper dosage and rational utilization of eucalyptus oil (EO). However, it is necessary to conduct further research to investigate and understand the specific mechanisms of action underlying the effects of different dosage levels (Shao et al., 2020).

### 10.6.5 *Mentha piperita* L.

*Mentha*, commonly known as peppermint, belongs to the *Lamiaceae* family, which is a group of plants and it is found in various regions around the world with temperate climates. *Mentha* plants have different components, including peppermint essential oil (PEO) and other non-essential substances. PEO is primarily composed of menthol, menthone, neomenthol, and iso-menthone. It is a combination of volatile compounds that possess several beneficial properties such as anti-inflammatory, antibacterial, antiviral, scolicidal, immunomodulatory, anti-tumor, neuroprotective, antifatigue, and antioxidant activities.

Several methods are available for extracting active compounds from essential oils. These techniques include steam distillation, hydrogenation distillation, microwave-assisted extraction, supercritical fluid extraction, ultrasonic-assisted extraction, and countercurrent extraction (Messaoudi et al., 2021). Studies have shown that peppermint essential oil (PEO) has potent activity against respiratory syncytial virus (RSV). The free form of the virus is highly susceptible to the effects of PEO and it also inhibits the herpes simplex virus (HSV) either prior to or during its entry into host cells. It is believed that PEO reduces the infectivity of the virus by directly interacting with the viral envelope and glycoprotein and it also disrupts the later stages of the HSV-1 life cycle, leading to the suppression of viral replication. Peppermint essential oil (PEO) has been found to provide pharmacological benefits by protecting the gastrointestinal, liver, kidney, skin, respiratory, brain, and nervous systems. It also exhibits effects that help lower blood glucose levels and reduce lipid levels in the body. Currently, PEO is widely used in the treatment of gastrointestinal and dermatological diseases, as well as in postoperative adjuvant therapy, nutraceutical products, the cosmetic industry, and various other fields (Zhao et al., 2022).

### 10.6.6 *Panax ginseng*

Ginseng is a renowned adaptogenic herb used as a traditional herbal remedy in East Asian countries for over 5,000 years. It was first discovered in the mountains of Manchuria, China. Belonging to the *Araliaceae* family, ginseng consists of 8–13 species within the Panax genus. The most commonly known ones are P. ginseng, referred to as "Asian or Chinese ginseng," P. notoginseng, known as "Sanchi," and P. quinquefolius, commonly called "American ginseng." The name "Panax" is derived from "panacea," signifying its reputation as a cure-all. Ginseng has indeed demonstrated its effectiveness in various pharmacological applications. Additionally, it is believed to provide protection against cancer, cardiovascular diseases, inflammatory conditions, microbial infections, and neurological disorders (Ratan et al., 2021a). Respiratory syncytial virus (RSV) is a type of negative single-stranded RNA virus belonging to the *Paramyxoviridae* family. It is a leading cause of severe respiratory disease and pneumonia, particularly in infants. Reports suggest that Korean Red Ginseng (KRG) extracts have the ability to protect human epithelial cells from RSV-induced cell death and viral replication. In a mice model infected with RSV, treatment with RGE (KRG extract) showed improvements in viral clearance from the lungs, enhanced production of IFN-γ (interferon-gamma), and increased levels of $CD^{8+}$ T cells and $CD11c^{+}$ DCs (dendritic cells) (Adhikari et al., 2021). Certain ginsenosides, namely Re, Rf, and $Rg_2$, demonstrated significant antiviral activity against rhinovirus and coxsackievirus, while Rg2 exhibited anti-enterovirus activity. Furthermore, KRG extract, along with ginsenosides Rb1 and Rg1, demonstrated a reduction in feline calicivirus and murine norovirus levels (Ratan et al., 2021b).

### 10.6.7 *Glycyrrhiza glabra*

*Glycyrrhiza glabra L.*, commonly known as liquorice, is a highly valued medicinal plant. Its name originates from the Greek words 'glykos,' meaning sweet, and 'rhiza,' meaning root. The genus *Glycyrrhiza* encompasses over 30 species, widely distributed across various regions, particularly in the Mediterranean regions of Asia. *Glycyrrhiza glabra* and its bioactive phytochemicals possess a diverse range of pharmacological activities. These include properties including being an antidemulcent, expectorant, antiulcer, anti-inflammatory, anticancer, and antidiabetic (Sharifi-Rad et al., 2021). In a separate trial examining the effectiveness of an herbal extract in enhancing the innate immunity of patients infected with COVID-19, researchers evaluated the evidence based on the reduction of viral load by the fourth day of treatment and the early recovery observed in the patients (Rangnekar et al., 2020).

### 10.6.8 *Echinacea purpurea*

*Echinacea purpurea*, commonly known as purple coneflower, is a flowering plant belonging to the *Asteraceae* family. Native to North America, it is widely recognized for its medicinal properties and has been used for centuries in traditional herbal remedies. *Echinacea purpurea* has been traditionally believed to possess immune-stimulating properties and is often used to support immune system function. It is rich in bioactive compounds, such as polysaccharides, flavonoids, and alkamides, which are thought to contribute to its therapeutic effects. These compounds are believed to exhibit anti-inflammatory, antioxidant, and antimicrobial activities. Due to its potential immune-boosting properties, *Echinacea purpurea* is commonly used as a natural remedy for preventing and treating the common cold, coughs, and respiratory infections. It is available in various forms, including teas, tinctures, capsules, and extracts. Additionally, *Echinacea purpurea* is also utilized in skincare products for its purported benefits in promoting healthy skin and reducing inflammation (Ogal et al., 2021). *In vitro*, studies have demonstrated the ability of Echinaforce to inactivate four human coronaviruses (HCoV-229E, MERS-CoV, SARS-CoV-1, and SARS-CoV-2). This finding reinforces the potential of Echinaforce as a preventive treatment against a wide range of respiratory viruses, including those causing severe pulmonary diseases or common colds. Moreover, the availability of a broad-spectrum antiviral compound for long-term prophylaxis could greatly benefit healthcare workers involved in the treatment of severe coronavirus infections, while also potentially reducing the transmission and morbidity of highly pathogenic coronaviruses within the general population (Signer et al., 2020).

### 10.6.9 *Amentoflavone*

Amentoflavone is a naturally occurring compound that belongs to a class of compounds known as flavonoids. It is found in various plants, including *Selaginella heptaphylla*, *Ginkgo biloba*, and *Hypericum perforatum*. Amentoflavone has gained attention in scientific research due to its potential health benefits and medicinal properties. Amentoflavone along with three other flavonoids obtained from the ethanol extract of *S. sinensis* has demonstrated significant activity against RSV, with an IC50 value of 5.5 μ/mL. Notably, among nine species of *Selaginella*, *S. sinensis* exhibited a high content of amentoflavone, with a concentration of 1.13% (Venu & Austin, 2020).

### 10.6.10 *S. heptaphylla*

Heptaphylla is a species of plant that belongs to the genus Selaginella. The plant *S. heptaphylla* has shown potent antiviral activity against respiratory syncytial virus (RSV). Through fractionation and isolation, two triterpenoids, namely 3α-hydroxylup-20(29)-ene-23,28-dioic acid and 3-epi-betulinic acid 3-O-sulfate, were identified as the active compounds. These triterpenoids exhibited a broad-spectrum antiviral activity against RSV, as well as influenza A (H1N1) virus, Coxsackie B3 (Cox B3) virus, and Herpes simplex virus type 1 (HSV-1). The newly discovered caffeoylquinic acid derivatives, specifically 3-O-caffeoylquinic acid, 3,4-di-O-caffeoylquinic acid, and 3,5-di-O-caffeoylquinic acid, also demonstrated effectiveness against RSV. These compounds showed low cytotoxicity and did not inhibit the attachment of RSV to host cells. This suggests that their antiviral effects are exerted by inhibiting the fusion of the virus with host cells during the early stage of infection, as well as inhibiting the cell-to-cell fusion during the replication cycle of RSV. However, they did not exhibit antiviral activity against influenza A (Flu A), Coxsackie B3 (Cox B3), and herpes simplex type one (HSV-1) viruses. These findings highlight the potential of these compounds in combating RSV infection, but further research is needed to fully understand their mechanisms of action and their potential application as antiviral agents (Lakshmi et al., 2020).

### 10.6.11 *Rosmarinus officinalis*

From the methanol extract of Rosmarinus officinalis, a compound called carnosic acid was discovered to have potent antiviral activity against HRSV. Carnosic acid was observed to inhibit the replication of enterovirus and exhibit significant activity against both A-type and B-type hRSV viruses. It demonstrated the ability to suppress viral gene expression and hinder the replication of hRSV without triggering the production of type-I interferon or affecting cell viability. This suggests that carnosic acid directly targets viral factors in a concentration-dependent manner. In a time course analysis, carnosic acid exhibited a notable inhibition of hRSV gene expression within 8 hours of administration, further emphasizing its ability to impede hRSV replication directly. These findings highlight the potential of carnosic acid as a promising therapeutic option for combating hRSV infections. Its strong antiviral activity and selective action on viral factors make it a valuable compound for further investigation and development of antiviral treatments (Lakshmi et al., 2020).

### 10.6.12 Caffeoylquinic Acids from Different Plants

Caffeoylquinic acids (CQAs) are the bioactive metabolites derived from the phenylpropanoid biosynthesis pathway and are known to play a protective role against biotic and abiotic stress (Uleberg et al., 2012). CQAs have a chemical structure where quinic acid is combined with one to four caffeic acid moieties. Quinic acid itself has four specific carbon positions, namely 1, 3, 4, and 5, that serve as stereocenters (Kim et al., 2020). CQAs have been proposed to be involved in the storage of lignin precursor compounds (Schoch et al., 2001). These compounds are abundant in coffee beans (Farah et al., 2005), apple varieties (Mullen et al., 2003), and pears (Galvis-Sánchez et al., 2004). Berries like blueberries, strawberries, and blackberries have been documented to contain CQAs, as mentioned in studies conducted by Li et al. (2013).

Mateus et al. conducted a study in 2002 that specifically investigated the extraction of CQAs from coffee beans using various solvents. In 2013, Oliveira et al. explored the extraction of CQAs from berry fruits using liquid-liquid extraction. Clifford et al., in 2005, discussed the extraction and quantification of CQAs specifically from tea leaves. Additionally, Wang et al. provided an overview of various extraction techniques for CQAs, including both conventional and advanced methods, in their 2015 publication. To extract CQAs from plant materials, solvent extraction is commonly used. Finely ground plant material is mixed with a suitable solvent (e.g., water, methanol, ethanol) and subjected to maceration, sonication, or reflux techniques. The mixture is then filtered to remove solid particles. The resulting filtrate, containing CQAs, can be concentrated through rotary evaporation or freeze-drying. Solid-phase extraction (SPE) selectively adsorbs CQAs onto a solid material, while liquid-liquid extraction (LLE) transfers CQAs between liquid phases. These extraction methods yield CQAs for further analysis or downstream applications (Wang et al., 2021). Several studies have been conducted by scientists to investigate the potential of CQAs (chlorogenic acids) for treating viral infections. These experiments have explored the antiviral properties of CQAs against various types of viruses. For instance, research by Nzekoue et al. (2020) demonstrated that CQAs exhibit antioxidant and antibacterial effects. They found that CQAs can help combat oxidative stress and inhibit the growth of certain bacterial strains. In terms of antiviral activity, a study by Tsunoda et al. (2018) and Matthews et al. (2020) highlighted the potential of CQAs in combating Alzheimer's disease and antiviral effects against specific viruses.

Similarly, Bourgou et al. (2017) found evidence of CQAs' impact on cancer-related activities, suggesting their potential as anticancer agents. Moreover, Liu, Li, Zu, et al. (2020) investigated the effects of CQAs on cancer cells and reported encouraging results in terms of their anticancer properties. Additionally, studies by Sasaki et al. (2019) highlighted the neuroprotective effects of CQAs. Some experiments showed promising results in terms of inhibiting viral replication or reducing viral load, the outcomes were not consistent across all viruses and experimental conditions.

## 10.7 CONCLUSION AND FUTURE PROSPECTIVE

This review chapter provides a comprehensive overview of the current knowledge regarding the use of medicinal plants and herbs in the treatment of respiratory syncytial virus (RSV) and viral pneumonia. However, several challenges and limitations remain in harnessing the full therapeutic potential of medicinal plants and herbs. Standardizing extracts, determining optimal dosage and formulations, and rigorous clinical trials are necessary to establish their safety, efficacy, and therapeutic benefits. Additionally, further research is needed to identify the active compounds responsible for the antiviral effects, elucidate their mechanisms of action, and evaluate their long-term safety. Advances in technology, such as high-throughput screening and metabolomics, can facilitate the identification of novel bioactive compounds from natural sources. Furthermore, integrating traditional knowledge and ethnopharmacology with modern scientific approaches can provide valuable insights into the selection, preparation, and utilization of medicinal plants and herbs. The development of effective and affordable antiviral therapies for RSV and viral pneumonia is of paramount importance, considering the global burden of these respiratory infections. Medicinal plants and herbs offer a sustainable and accessible alternative to synthetic drugs, with the potential for fewer side effects and reduced development of drug resistance.

## REFERENCES

Abdulhakim, J.A. Effect of guggulsterone, a sterol identified in Commiphora gileadensis (Becham), on the dengue virus enzymes: Pharmacokinetics, molecular docking and molecular dynamics simulations studies. *J King Saud. Univ. Sci.* 2022, 34(6), 102140.

Adhikari, B.; Marasini, B. P.; Rayamajhee, B.; Bhattarai, B. R.; Lamichhane, G.; Khadayat, K.; Adhikari, A.; Khanal, S.; Parajuli, N. Potential roles of medicinal plants for the treatment of viral diseases focusing on COVID-19: A review. *Phytother. Res.* 2021, 35(3), 1298–1312.

Ali, SI.; Sheikh, WM.; Rather, MA; Venkatesalu, V; Muzamil, B.; Nabi, SU. Medicinal plants: Treasure for antiviral drug discovery. *Phytother Res.* 2021, Jul, 35(7): 3447–3483.

Al-Beltagi, S.; Preda, C. A.; Goulding, L. V.; James, J.; Pu, J.; Skinner, P.; Jiang, Z.; Wang, B. L.; Yang, J.; Banyard, A. C.; Mellits, K. H.; Gershkovich, P.; Hayes, C. J.; Nguyen-Van-Tam, J.; Brown, I. H.; Liu, J.; Chang, K.-C. Thapsigargin is a broad-spectrum inhibitor of major human respiratory viruses: Coronavirus, respiratory syncytial virus and Influenza A virus. *Viruses* 2021, 13(2), 234. DOI:10.3390/v13020234.

Aliprantis, A. O.; Wolford, D.; Caro, L.; Maas, B. M.; Ma, H.; Montgomery, D. L.; Sterling, L. M.; Hunt, A.; Cox, K. S.; Vora, K. A.; Roadcap, B. A.; Railkar, R. A.; Lee, A. W.; Stoch, S. A.; Lai, E. A Phase 1 randomized, double-blind, placebo-controlled trial to assess the safety, tolerability, and pharmacokinetics of a respiratory syncytial virus neutralizing monoclonal antibody Mk-1654 in healthy adults. *Clinical Pharmacology in Drug Development* 2021, 10(5), 556–566. DOI:10.1002/cpdd.883.

Arabi, Y. M.; Shalhoub, S.; Mandourah, Y.; Al-Hameed, F.; Al-Omari, A.; Al Qasim, E.; Jose, J.; Alraddadi, B.; Almotairi, A.; Al Khatib, K.; Abdulmomen, A.; Qushmaq, I.; Sindi, A. A.; Mady, A.; Solaiman, O.; Al-Raddadi, R.; Maghrabi, K.; Ragab, A.; Al Mekhlafi, G. A.; Balkhy, H. H.; Al Harthy, A.; Kharaba, A.; Gramish, J. A.; Al-Aithan, A. M.; Al-Dawood, A.; Merson, L.; Hayden, F. G.; Fowler, R. Ribavirin and interferon therapy for critically ill patients with Middle East respiratory syndrome: A multicenter observational study. *Clinical Infectious Diseases* 2020, 70(9), 1837–1844.

Bahabri, I.; Abdulaal, A.; Alanazi, T.; Alenazy, S.; Alrumih, Y.; Alqahtani, R.; Al, Johani S.; Bosaeed, M.; Al-Dorzi, HM. Characteristics, management, and outcomes of community-acquired pneumonia due to respiratory syncytial virus: A retrospective study. *Pulm. Med.* 2023, Mar 6, 2023.

Bai, M.; Chen, JJ.; Xu, W.; Dong, SH.; Liu, QB.; Yao, GD.; Lin, B.; Huang, XX.; Song, SJ. Germacranolides from Elephantopus scaber L. and their cytotoxic activities. *Phytochem.* 2020, 178, 112479.

Bang, S.; Ha, T.K.Q.; Lee, C.; Li, W.; Oh, W. K.; Shim, S. H. Antiviral activities of compounds from aerial parts of Salvia plebeia R. *Br. J. Ethnopharmacol.* 2016, 192, 398–405.

Battles, M. B.; McLellan, J. S. Respiratory syncytial virus entry and how to block it. *Nat. Rev. Microbiol.* 2019, 17(4), 233–245.

Behzadi, M. A.; Leyva-Grado, V. H. Overview of current therapeutics and novel candidates against influenza, respiratory syncytial virus, and Middle East respiratory syndrome coronavirus infections. *Front. Microbiol.* 2019, 10, 1327.

Ben-Shabat, S.; Yarmolinsky, L.; Porat, D.; Dahan, A. Antiviral effect of phytochemicals from medicinal plants: Applications and drug delivery strategies. *Drug Deliv. Transl. Res.* 2020, 10, 354–367.

Bergeron, H. C.; Tripp, R. A. Breakthrough therapy designation of nirsevimab for the prevention of lower respiratory tract illness caused by respiratory syncytial virus infections (RSV). *Expert Opin. Investig. Drugs.*, 2022, 31(1), 23–29.

Biagi, C.; Dondi, A.; Scarpini, S.; Rocca, A.; Vandini, S.; Poletti, G.; Lanari, M. Current state and challenges in developing respiratory syncytial virus vaccines. *Vaccines* 2020, 8(4), 672.

Bourgou, S.; Bettaieb Rebey, I.; Mkadmini, K.; Isoda, H.; Ksouri, R.; Ksouri, W. M. LC-ESI-TOF-MS and GC-MS profiling of Artemisia Herba-Alba and evaluation of its bioactive properties. *Food Research International* 2017, 99, 702–712. DOI:10.1016/j.foodres.2017.06.009.

Bouslama, L.; Kouidhi, B.; Alqurashi, Y. M.; Chaieb, K.; Papetti, A. Virucidal effect of guggulsterone isolated from Commiphoragileadensis. *Planta Medica.* 2019, 85(16), 1225–1232.

Bulbul, M.R.H.; Chowdhury, M.N.U.; Naima, T.A.; Sami, S.A.; Imtiaj, M.S.; Huda, N.; Uddin, M.G. A comprehensive review on the diverse pharmacological perspectives of Terminalia chebula Retz. *Heliyon.* 2022, 8(8), e10220.

Carvajal, J. J.; Avellaneda, A. M.; Maya, J. E.; Kalergis, A. M.; Lay, M. K. Host components contributing to respiratory syncytial virus pathogenesis. *Front. Immunol.* 2019, 10, 2152.

Chen, L. F.; Zhong, Y. L.; Luo, D.; Liu, Z.; Tang, W.; Cheng, W.; Xiong, S.; Li, Y.L.; Li, M. M. Antiviral activity of ethanol extract of Lophatherum gracile against respiratory syncytial virus infection. *J. Ethnopharmacol.* 2019, 242, 111575.

Chen, F.; Yang, L.; Huang, Y.; Chen, Y.; Sang, H.; Duan, W.; Yang, J. Isocorilagin, isolated from Canarium album (Lour.) Raeusch, as a potent neuraminidase inhibitor against influenza A virus. *Biochem. Biophys. Res. Commun.* 2020, 523(1), 183–189.

Choi, J.G.; Kim, S.Y.; Jeong, M.; Oh, M.S. Pharmacotherapeutic potential of ginger and its compounds in age-related neurological disorders. *Clin. Pharm. Therap.* 2018, 182, 56–69.

Choudhary, D.; Pan, M.-H. Antiviral effects of anthocyanins and phytochemicals as natural dietary compounds on different virus sources. *Current Research in Nutrition and Food Science Journal* 2020, 8(3), 674–681. DOI:10.12944/crnfsj.8.3.01.

Chowdhury, P. In silico investigation of phytoconstituents from Indian medicinal herb 'Tinosporacordifolia (giloy)' against SARS-CoV-2 (COVID-19) by molecular dynamics approach. *J. Biomol. Struct. Dyn.* 2021, 39(17), 6792–6809.

Cockerill, G. S.; Angell, R. M.; Bedernjak, A.; Chuckowree, I.; Fraser, I.; Gascon-Simorte, J.; Gilman, M.S.; Good, J.A.; Harland, R.; Johnson, S.M.; Powell, K. Discovery of sisunatovir (RV521), an inhibitor of respiratory syncytial virus fusion. *J. Med. Chem.* 2021, 64(7), 3658–3676.

Cohen, C.; Zar, H. J. Deaths from RSV in young infants-the hidden community burden. *Lancet Glob. Health.* 2022, 10(2), e169–e170.

Denel-Bobrowska, M.; Olejniczak, A. B. Non-nucleoside structured compounds with antiviral activity—past 10 years (2010–2020). *European Journal of Medicinal Chemistry* 2022, 231, 114136. DOI:10.1016/j.ejmech.2022.114136.

Dhakad, A.K.; Pandey, V.V.; Beg, S.; Rawat, J.M.; Singh, A. Biological, medicinal and toxicological significance of Eucalyptus leaf essential oil: A review. *J. Sci. Food Agric.* 2018, 98(3), 833–848.

Dhankhar, S.; Kaur, R.; Ruhil, S.; Balhara, M.; Dhankhar, S.; Chhillar, A. K. A review on Justicia adhatoda: A potential source of natural medicine. *Afr. J. Plant Sci.* 2011, 5(11), 620–627.

Efstathiou, C.; Abidi, S.H.; Harker, J.; Stevenson, N.J. Revisiting respiratory syncytial virus's interaction with host immunity, towards novel therapeutics. *Cell. Mol. Life Sci.* 2020, 77, 5045–5058.

Eichinger, K. M.; Kosanovich, J. L.; Lipp, M.; Empey, K. M.; Petrovsky, N. Strategies for active and passive pediatric RSV immunization. *Therapeutic Advances in Vaccines and Immunotherapy* 2021, 9, 251513552098151.

Ergasheva, S. A.; Mamatkhanova, M. A.; Nabiev, A.; Karimov, A. M.; Khalilov, R. M.; Mamatkhanov, A. U. Development of the flow chart for obtaining and studying the antihypoxic activity of dry extracts from the aerial part of scutellaria adenostegia herbs. *Pharm. Chem. J.* 2021, 55, 580–584.

Fang, Z. J.; Zhang, T.; Chen, S. X.; Wang, Y. L.; Zhou, C. X.; Mo, J. X.; Wu, Y.J.; Xu, Y.K.; Lin, L.G.; Gan, L. S. Cycloartane triterpenoids from Actaea vaginata with anti-inflammatory effects in LPS-stimulated RAW264. 7 macrophages. *Phytochem.* 2019, 160, 1–10.

Farah, A.; de Paulis, T.; Trugo, L. C.; Martin, P. R. Effect of roasting on the formation of chlorogenic acid lactones in coffee. *Journal of Agricultural and Food Chemistry* 2005, 53(5), 1505–1513. DOI:10.1021/jf048701t.

Foolad, F.; Aitken, S. L.; Shigle, T. L.; Prayag, A.; Ghantoji, S.; Ariza-Heredia, E.; Chemaly, R. F. Oral versus aerosolized ribavirin for the treatment of respiratory syncytial virus infections in hematopoietic cell transplant recipients. *Clinical Infectious Diseases* 2019, 68(10), 1641–1649.

Galvis-Sánchez, A. C.; Fonseca, S. C.; Morais, A. M.; Malcata, F. X. Sensorial and physicochemical quality responses of pears (CV Rocha) to long-term storage under controlled atmospheres. *Journal of the Science of Food and Agriculture* 2004, 84(13), 1646–1656. DOI:10.1002/jsfa.1798.

Gao, Y.; Cao, J.; Xing, P.; Altmeyer, R.; Zhang, Y. Evaluation of small molecule combinations against respiratory syncytial virus in vitro. *Molecules.* 2021, 26(9), 2607.

Gatt, D.; Martin, I.; AlFouzan, R.; Moraes, T. J. Prevention and treatment strategies for respiratory syncytial virus (RSV). *Pathogens* 2023, 12(2), 154.

Goswami, D.; Kumar, M.; Ghosh, S. K.; Das, A. *Natural product compounds in Alpinia officinarum and ginger are potent SARS-COV-2 papain-like protease inhibitors* 2020. DOI:10.26434/chemrxiv.12071997.

Gundoju, N. R.; Sadasivuni, H.; Krishna, B. R.; Somasekhar, T.; Vangara, R. A review on triterpenoids from plant sources as potential antiviral agents. *Indian J. Chem.* 2022, 61(7), 761–770.

Guo, Qiuli; Jin, Qian; Qingbei, Zeng; Lin, Zhang; Xuehua, Zhu; Jie, Zheng; Kan, Chen; Enmei Liu. Characterization of an orally available respiratory syncytial virus L protein polymerase inhibitor DZ7487. *Am. J. Transl. Res.* 2023, 15(3), 1680.

Hewlings, S. J.; Kalman, D. S. Curcumin: A review of its' effects on human health. *Foods.* 2017, 6(10), 92.

Hu, M.; Bogoyevitch, M. A.; Jans, D. A. Impact of respiratory syncytial virus infection on host functions: Implications for antiviral strategies. *Physiological Reviews* 2020, 100(4), 1527–1594. DOI:10.1152/physrev.00030.2019.

Hu, Z.; Lin, J.; Chen, J.; Cai, T.; Xia, L.; Liu, Y.; Song, X.; He, Z. Overview of viral pneumonia associated with influenza virus, respiratory syncytial virus, and coronavirus, and therapeutics based on natural products of medicinal plants. *Front. Pharmacol.* 2021, 12, 630834.

Huong, L. T.; Thao, T. P.; Trang, D. T.; Lee, J. H.; Tai, B. H.; Kiem, P. V.; Nhiem, N.X.; Dang, N. H. (2021). Anti-osteoclastogenic cycloartane saponins from Mussaenda pubescens. *Nat. Prod. Res.* 2021, 36(18), 4591–4598.

Jafarzadeh, A.; Nemati, M.; Saha, B.; Bansode, Y.D.; Jafarzadeh, S. Protective potentials of type III interferons in COVID-19 patients: Lessons from differential properties of type I- and III interferons. *Viral. Immuno.* 2021, 34(5), 307–320.

Jang, S. A.; Hwang, Y. H.; Yang, H.; Ryuk, J. A.; Gu, D. R.; Ha, H. Ethanolic extract of Pyrrosia lingua (Thunb.) Farw. ameliorates OVX-induced bone loss and RANKL-induced osteoclastogenesis. *Biomed. Pharmacother.* 2022, 147, 112640.

Kaler, J.; Hussain, A.; Patel, K.; Hernandez, T.; Ray, S. Respiratory syncytial virus: A comprehensive review of transmission, pathophysiology, and manifestation. *Cureus.* 2023, 15(3).

Kapoor, R.; Sharma, B.; Kanwar, S. S. Antiviral phytochemicals: An overview. *Biochem. Physiol. Open access.* 2017, 6, 220.

Khodjoyan, S.; Morissette, D.; Hontonnou, F.; Checa Ruano, L.; Richard, C. A.; Sperandio, O.; Sizun, C. Investigation of the Fuzzy Complex between RSV Nucleoprotein and Phosphoprotein to Optimize an Inhibition Assay by Fluorescence Polarization. *International Journal of Molecular Sciences* 2022, 24(1), 569.

Khomenko, T. M.; Shtro, A. A.; Galochkina, A. V.; Nikolaeva, Y. V.; Garshinina, A. V.; Borisevich, S. S.; Korchagina, D. V.; Volcho, K. P.; Salakhutdinov, N. F. New inhibitors of respiratory syncytial virus (RSV) replication based on monoterpene-substituted arylcoumarins. *Molecules* 2023, 28(6), 2673.

Kim, S.; Chen, J.; Cheng, T.; Gindulyte, A.; He, J.; He, S.; Li, Q.; Shoemaker, B. A.; Thiessen, P. A.; Yu, B.; Zaslavsky, L.; Zhang, J.; Bolton, E. E. PubChem in 2021: New data content and improved web interfaces. *Nucleic Acids Research* 2020, 49(D1). DOI:10.1093/nar/gkaa971.

Kiyama, R. Nutritional implications of ginger: chemistry, biological activities and signaling pathways. *The Journal of Nutritional Biochemistry* 2020, 86, 108486. DOI:10.1016/j.jnutbio.2020.108486.

Kou, X. et al. Occurrence, biological activity and metabolism of 6-shogaol. *Food Funct.* 2018, 9(3), 1310–1327.

Kwak, J. H.; Seo, J. M.; Kim, N. H.; Arasu, M. V.; Kim, S.; Yoon, M. K.; Kim, S. J. Variation of quercetin glycoside derivatives in three onion (Allium cepa L.) varieties. *Saudi J. Biol. Sci.* 2017, 24(6), 1387–1391.

Le, K.H.; La, D.D.; Nguyen, P.T.M.; Nguyen, M.D.B.; Vo, A.T.K.; Nguyen, M.T.H.; Tran, D.L.; Chang, S.W.; Nguyen, X.H.; Nguyen, D.D. Fabrication of Cleistocalyx operculatus extracts/chitosan/gum arabic composite as an edible coating for preservation of banana. *Prog. Org. Coat.* 2021, 161, 106550.

Li, C.; Feng, J.; Huang, W.-Y.; An, X.-T. Composition of polyphenols and antioxidant activity of rabbiteye blueberry (*Vaccinium Ashei*) in Nanjing. *Journal of Agricultural and Food Chemistry* 2013, 61(3), 523–531. DOI:10.1021/jf3046158.

Li, Y.; Wang, X.; Blau, D.M.; Caballero, M.T.; Feikin, D.R.; Gill, C.J.; Madhi, S.A.; Omer, S.B.; Simões, E.A.; Campbell, H.; Pariente, A.B. Global, regional, and national disease burden estimates of acute lower respiratory infections due to respiratory syncytial virus in children younger than 5 years in 2019: A systematic analysis. *The Lancet.* 2022, 399(10340), 2047–2064.

Liu, W.; Li, J.; Zhang, X.; Zu, Y.; Yang, Y.; Liu, W.; Xu, Z.; Gao, H.; Sun, X.; Jiang, X.; Zhao, Q. Current advances in naturally occurring caffeoylquinic acids: Structure, bioactivity, and synthesis. *J. Agric. Food Chem.* 2020, 68(39), 10489–10516.

Lopes, B.R.P.; da Costa, M.F.; Ribeiro, A.G.; da Silva, T.F.; Lima, C.S.; Caruso, I.P.; de Araujo, G.C.; Kubo, L.H.; Iacovelli, F.; Falconi, M.; Desideri, A. Quercetin pentaacetate inhibits in vitro human respiratory syncytial virus adhesion. *Virus Res.* 2020, 276, 197805.

Ludeke, B.; Fearns, R. The respiratory syncytial virus polymerase can perform RNA synthesis with modified primers and nucleotide analogs. *Virology* 2020, 540, 66–74. DOI:10.1016/j.virol.2019.11.002.

Luo, D.; Xiong, S.; Li, Q. G.; Jiang, L.; Niu, Q. W.; He, L. J.; Li, Y. L.; Zhang, Y. B.; Wang, G. C. Terpenoids from the stems of Celastrus hindsii and their anti-RSV activities. *Fitoterapia.* 2018, 130, 118–124.

Mackman, R.L.; Hui, H.C.; Perron, M.; Murakami, E.; Palmiotti, C.; Lee, G.; Stray, K.; Zhang, L.; Goyal, B.; Chun, K.; Byun, D. Prodrugs of a 1′-CN-4-aza-7, 9-dideazaadenosine C-nucleoside leading to the discovery of remdesivir (GS-5734) as a potent inhibitor of respiratory syncytial virus with efficacy in the African green monkey model of RSV. *J. Med. Chem.* 2021, 64(8), 5001–5017.

Mair, C. E.; Grienke, U.; Wilhelm, A.; Urban, E.; Zehl, M.; Schmidtke, M.; Rollinger, J. M. Anti-influenza triterpene saponins from the bark of Burkea africana. *J. Nat. Prod.* 2018, 81(3), 515–523.

Mao, Q.Q.; Xu, X.Y.; Cao, S.Y.; Gan, R.Y.; Corke, H.; Beta, T.; Li, H.B. Bioactive compounds and bioactivities of ginger (Zingiber officinale Roscoe). *Foods.* 2019, 8(6), 185.

Matthews, D. G.; Caruso, M.; Alcazar Magana, A.; Wright, K. M.; Maier, C. S.; Stevens, J. F.; Gray, N. E.; Quinn, J. F.; Soumyanath, A. Caffeoylquinic acids in centella asiatica reverse cognitive deficits in male 5xFAD Alzheimer's disease model mice. *Nutrients* 2020, 12(11), 3488. DOI:10.3390/nu12113488.

Matyushenko, V.; Kotomina, T.; Kudryavtsev, I.; Mezhenskaya, D.; Prokopenko, P.; Matushkina, A.; Sivak, K.; Muzhikyan, A.; Rudenko, L.; Isakova-Sivak, I. Conserved T-cell epitopes of respiratory syncytial virus (RSV) delivered by recombinant live attenuated influenza vaccine viruses efficiently induce RSV-specific lung-localized memory T cells and augment influenza-specific resident memory T-cell responses. *Antiviral Research* 2020, 182, 104864. DOI:10.1016/j.antiviral.2020.104864.

Messaoudi, M.; Rebiai, A.; Sawicka, B.; Atanassova, M.; Ouakouak, H.; Larkem, I.; Egbuna, C.; Awuchi, C.G.; Boubekeur, S.; Ferhat, M.A.; Begaa, S. Effect of extraction methods on polyphenols, flavonoids, mineral elements, and biological activities of essential oil and extracts of Mentha pulegium L. *Molecules.* 2021, 27(1), 11.

Mueller, S.; Stauft, C. B.; Kalkeri, R.; Koidei, F.; Kushnir, A.; Tasker, S.; Coleman, J. R. A Codon-Pair deoptimized live-attenuated vaccine against respiratory syncytial virus is immunogenic and efficacious in non-human primates. *Vaccine* 2020, 38(14), 2943–2948.

Mullen, W.; Yokota, T.; Lean, M. E. J.; Crozier, A. Analysis of ellagitannins and conjugates of ellagic acid and quercetin in raspberry fruits by LC–MSN. *Phytochemistry* 2003, 64(2), 617–624. DOI:10.1016/s0031-9422(03)00281-4.

Mustafa, I.; Chin, N.L.; Fakurazi, S.; Palanisamy, A. Comparison of phytochemicals, antioxidant and anti-inflammatory properties of sun-, oven- and freeze-dried ginger extracts. *Foods.* 2019, 8(10), 456.

Nagai, E.; Iwai, M.; Koketsu, R.; Okuno, Y.; Suzuki, Y.; Morimoto, R.; Sumitani, H.; Ohshima, A.; Enomoto, T.; Isegawa, Y. Anti-Influenza virus activity of Adlay tea components. *Plant Foods for Human Nutrition* 2019, 74(4), 538–543. DOI:10.1007/s11130-019-00773-3.

Nigam, M.; Mishra, A.P.; Adhikari-Devkota, A.; Dirar, A.I.; Hassan, M.M.; Adhikari, A.; Belwal, T.; Devkota, H.P. "Fruits of Terminalia chebula Retz.: A review on traditional uses, bioactive chemical constituents and pharmacological activities," *Phytother. Res.* 2020, 34(10), 2518–2533.

Novavax, Inc. Novavax announces topline results from Phase 3 prepare TM trial of ResVax TM for prevention of RSV disease in infants via maternal immunization. *Press Release* 2019, 1–3.

Nzekoue, F. K.; Angeloni, S.; Navarini, L.; Angeloni, C.; Freschi, M.; Hrelia, S.; Vitali, L. A.; Sagratini, G.; Vittori, S.; Caprioli, G. Coffee silverskin extracts: Quantification of 30 bioactive compounds by a new HPLC-MS/MS method and evaluation of their antioxidant and antibacterial activities. *Food Research International* 2020, 133, 109128. DOI:10.1016/j.foodres.2020.109128.

Obata, K.; Kojima, T.; Masaki, T.; Okabayashi, T.; Yokota, S.; Hirakawa, S.; Nomura, K.; Takasawa, A.; Murata, M.; Tanaka, S.; Fuchimoto, J.; Fujii, N.; Tsutsumi, H.; Himi, T.; Sawada, N. Curcumin prevents replication of respiratory syncytial virus and the epithelial responses to it in human nasal epithelial cells. *PLoS One.* 2013, 8(9), e70225.

Ogal, M.; Johnston, S.L.; Klein, P.; Schoop, R. Echinacea reduces antibiotic usage in children through respiratory tract infection prevention: A randomized, blinded, controlled clinical trial. *Eur. J. Med. Res.* 2021, 26(1), 1–9.

Perera, W.P.R.T.; Liyanage, J.A.; Dissanayake, K.G.C.; Gunathilaka, H.; Weerakoon, W.M.T.D.N.; Wanigasekara, D.N.; Fernando, W.S.K.; Rajapaksha, R.M.H.; Liyanage, R.P.; Perera, B.T. Antiviral potential of selected medicinal herbs and their isolated natural products. *Biomed Res. Int.* 2021, 18.

Pham, G. N.; Nguyen, T. T. T.; Nguyen-Ngoc, H. Ethnopharmacology, phytochem.; and pharmacology of Syzygium nervosum. *Evid. Based Complementary Altern. Med.* 2020, 14.

Ponticelli, M.; Bellone, M.L.; Parisi, V.; Iannuzzi, A.; Braca, A.; de Tommasi, N.; Russo, D.; Sileo, A.; Quaranta, P.; Freer, G.; Pistello, M. Specialized metabolites from plants as a source of new multi-target antiviral drugs: A systematic review. *Phytochem Rev.* 2023, 22, 615–693.

Qiu, X.; Xu, S.; Lu, Y.; Luo, Z.; Yan, Y.; Wang, C.; Ji, J. Development of MRNA vaccines against respiratory syncytial virus (RSV). *Cytokine & Growth Factor Reviews* 2022, 68, 37–53.

Rangnekar, H.; Patankar, S.; Suryawanshi, K.; Soni, P. Safety and efficacy of herbal extracts to restore respiratory health and improve innate immunity in COVID-19 positive patients with mild to moderate severity: A structured summary of a study protocol for a randomised controlled trial. *Trials.* 2020, 21(1), 943.

Ratan, Z.A.; Haidere, M.F.; Hong, Y.H.; Park, S.H.; Lee, J.O.; Lee, J.; Cho, J.Y. Pharmacological potential of ginseng and its major component ginsenosides. *J. Ginseng Res.* 2021a, 45(2), 199–210.

Ratan, Z.A.; Youn, S.H.; Kwak, Y.S.; Han, C.K.; Haidere, M.F.; Kim, J.K.; Min, H.; Jung, Y.J.; Hosseinzadeh, H.; Hyun, S.H.; Cho, J.Y. Adaptogenic effects of Panax ginseng on modulation of immune functions. *J. Ginseng Res.* 2021b, 45(1), 32–40.

Rhodin, M. H.; McAllister, N. V.; Castillo, J.; Noton, S. L.; Fearns, R.; Kim, I. J.; Yu, J.; Blaisdell, T. P.; Panarese, J.; Shook, B. C.; Or, Y. S.; Goodwin, B.; Lin, K. EDP-938, a Novel nucleoprotein inhibitor of respiratory syncytial virus, demonstrates potent antiviral activities in vitro and in a non-human primate model. *PLOS Pathogens* 2021, 17 (3).

Rocca, A.; Biagi, C.; Scarpini, S.; Dondi, A.; Vandini, S.; Pierantoni, L.; Lanari, M. Passive immunoprophylaxis against respiratory syncytial virus in children: Where are we now? *Int. J. Mol. Sci.* 2021, 22(7), 3703.

Rodriguez-Fernandez, R.; Mejias, A.; Ramilo, O. Monoclonal antibodies for prevention of respiratory syncytial virus infection. *The Pediatric Infectious Disease Journal* 2021, 40(5S), S35-S39.

Roschek Jr, B.; Fink, R. C.; McMichael, M. D.; Li, D.; Alberte, R. S. Elderberry flavonoids bind to and prevent H1N1 infection in vitro. *Phytochem.* 2009, 70(10), 1255–1261.

Russell, M. S.; Thulasi Raman, S. N.; Gravel, C.; Zhang, W.; Pfeifle, A.; Chen, W.; Van Domselaar, G.; Safronetz, D.; Johnston, M.; Sauve, S.; Wang, L.; Rosu-Myles, M.; Cao, J.; Li, X. Single immunization of a vaccine vectored by a novel recombinant Vaccinia virus affords effective protection against respiratory syncytial virus infection in cotton rats. *Frontiers in Immunology* 2021, 12. DOI:10.3389/fimmu.2021.747866.

Sacco, R. E.; Mena, I.; Palmer, M. V.; Durbin, R. K.; García-Sastre, A.; Durbin, J. E. An intranasal recombinant NDV-BRSV Fopt vaccine is safe and reduces lesion severity in a colostrum-deprived calf model of RSV infection. *Scientific Reports* 2022, 12(1), 22552.

Schoch, G.; Goepfert, S.; Morant, M.; Hehn, A.; Meyer, D.; Ullmann, P.; Werck-Reichhart, D. CYP98A3 from Arabidopsis Thaliana is a 3′-Hydroxylase of phenolic esters, a missing link in the phenylpropanoid pathway. *Journal of Biological Chemistry* 2001, 276(39), 36566–36574. DOI:10.1074/jbc.m104047200.

Sasaki, K.; Davies, J.; Doldán, N. G.; Arao, S.; Ferdousi, F.; Szele, F. G.; Isoda, H. 3,4,5-Tricaffeoylquinic acid induces adult neurogenesis and improves deficit of learning and memory in aging model senescence-accelerated prone 8 mice. *Aging* 2019, 11(2), 401–422. DOI:10.18632/aging.101748.

Seo, W. D.; Kang, J. E.; Choi, S. W.; Lee, K. S.; Lee, M. J.; Park, K. D.; Lee, J. H. Comparison of nutritional components (isoflavone, protein, oil, and fatty acid) and antioxidant properties at the growth stage of different parts of soybean [Glycine max (L.) Merrill]. *Food Biotechnol.* 2017, 26, 339–347.

Shang, Z.; Tan, S.; Ma, D. Respiratory syncytial virus: From pathogenesis to potential therapeutic strategies. *Int. J. Biol. Sci.* 2021, 17(14), 4073–4091.

Shao, J.; Yin, Z.; Wang, Y.; Yang, Y.; Tang, Q.; Zhang, M.; Jiao, J.; Liu, C.; Yang, M.; Zhen, L.; Hassouna, A. Effects of different doses of eucalyptus oil From Eucalyptus globulus labill on respiratory tract immunity and immune function in healthy rats. *Front. Pharmacol.* 2020, 11, 1287.

Sharifi-Rad, J.; Cruz-Martins, N.; López-Jornet, P.; Lopez, E. P.-F.; Harun, N.; Yeskaliyeva, B.; Beyatli, A.; Sytar, O.; Shaheen, S.; Sharopov, F.; Taheri, Y.; Docea, A. O.; Calina, D.; Cho, W. C. Natural coumarins: Exploring the pharmacological complexity and underlying molecular mechanisms. *Oxidative Medicine and Cellular Longevity* 2021, 1–19. DOI:10.1155/2021/6492346.

Sharma, R.; Bhattu, M.; Tripathi, A.; Verma, M.; Acevedo, R.; Kumar, P.; Rajput, V. D.; Singh, J. Potential medicinal plants to combat viral infections: A way forward to environmental biotechnology. *Environ. Res.* 2023, 227, 115725.

Signer, J.; Jonsdottir, H. R.; Albrich, W. C.; Strasser, M.; Züst, R.; Ryter, S.; Ackermann-Gäumann, R.; Lenz, N.; Siegrist, D.; Suter, A.; Schoop, R.; Engler, O. B. *In vitro antiviral activity of Echinaforce®, an echinacea purpurea preparation, against common cold coronavirus 229e and highly pathogenic MERS-COV and SARS-COV* **2020**. DOI:10.21203/rs.2.24724/v2.

Sinha, D. A review on phytochemical, ethnobotanical, pharmacological, and antimicrobial importance of Cedrus deodara (Roxb. Ex D. Don) G. Don. *Int. J. Green Pharm.* 2019, (IJGP), 13(01).

Song, J. G.; Su, J. C.; Song, Q. Y.; Huang, R. L.; Tang, W.; Hu, L. J.; Huang, X.J.; Jiang, R.W.; Li, Y.L.; Ye, W.C.; Wang, Y. Cleistocaltones A and B, antiviral phloroglucinol-terpenoid adducts from Cleistocalyx operculatus. *Org. Lett.* 2019, 21(23), 9579–9583.

Stevaert, A.; Groaz, E.; Naesens, L. Nucleoside analogs for management of respiratory virus infections: Mechanism of action and clinical efficacy. *Curr. Opin. Virol.* 2022, 57, 101279.

Sultan, M.T.; Anwar, M.J.; Imran, M.; Khalil, I.; Saeed, F.; Neelum, S.; Alsagaby, S.A.; Al Abdulmonem, W.; Abdelgawad, M.A.; Hussain, M.; El-Ghorab, A.H. Phytochemicalprofile and pro-healthy properties of Terminaliachebula: A comprehensive review. *Int. J. Food Prop.* 2023, 26(1), 526–551.

Sun, Mingyao; Honghao, Lai; Feiyang, Na; Sheng, Li; Xia, Qiu; Jinhui, Tian; Zhigang, Zhang; Long, Ge. Monoclonal antibody for the prevention of respiratory syncytial virus in infants and children: A systematic review and network meta-analysis. *JAMA Netw. Open.* 2023, 6( 2), e2300233.

Tejada, S.; Martinez-Reviejo, R.; Karakoc, H. N.; Peña-López, Y.; Manuel, O.; Rello, J. Ribavirin for treatment of subjects with respiratory syncytial virus-related infection: A systematic review and meta-analysis. *Adv. Ther.* 2022, 39(9), 4037–4051.

Thimmulappa, R. K.; Mudnakudu-Nagaraju, K.K.; Shivamallu, C.; Subramaniam, J. T.; Radhakrishnan, A.; Bhojraj, S.; Kuppusamy, G. Antiviral and immunomodulatory activity of curcumin: A case for prophylactic therapy for COVID-19. *Heliyon.* 2021, 7(2).

Tsunoda, N.; Hashimoto, M.; Ishikawa, T.; Fukuhara, R.; Yuki, S.; Tanaka, H.; Ikeda, M. Clinical features of auditory hallucinations in patients with dementia with Lewy bodies: a soundtrack of visual hallucinations. *The Journal of clinical psychiatry* 2018, 79(3), 1673.

Uleberg, E.; Rohloff, J.; Jaakola, L.; Trôst, K.; Junttila, O.; Häggman, H.; Martinussen, I. Effects of temperature and photoperiod on yield and chemical composition of northern and southern clones of bilberry (*Vaccinium Myrtillus* L.). *Journal of Agricultural and Food Chemistry* 2012, 60(42), 10406–10414. DOI:10.1021/jf302924m.

Venu, L.N.; Austin, A. Antiviral efficacy of medicinal plants against respiratory viruses: Respiratory syncytial virus (RSV) and coronavirus (COV)/COVID 19. *The Journal of Phytopharmacology* 2020, 9, 281–290. DOI:10.31254/phyto.2020.9412

Verwey, C.; Madhi, S. A. Review and update of active and passive immunization against respiratory syncytial virus. *BioDrugs.* 2023, 1–15.

Wang, D.; Wang, J.; Sun, J.; Qiu, S.; Chu, B.; Fang, R.; Li, L.; Gong, J.; Zheng, F. Degradation kinetics and isomerization of 5-O-Caffeoylquinic acid under ultrasound: influence of epigallocatechin gallate and Vitamin C. *Food Chemistry: X* 2021, 12, 100147. DOI:10.1016/j.fochx.2021.100147.

Xiao, S.; Yuan, F.-M.; Zhang, M.-S.; Yu, S.-Y.; Li, J.-D.; Yi, X.-D.; Xu, D.-L. Three new 1-(*p*-Hydroxybenzyl) Phenanthrenes from *Bletilla Striata. Journal of Asian Natural Products Research* 2017, 19(2), 140–144. DOI:10.1080/10286020.2016.1184254.

Xing, Y.; Proesmans, M. New therapies for acute RSV infections: Where are we? *Eur. J. Pediatr.* 2019, 178(2), 131–138.

Xu, E.; Park, S.; Calderon, J.; Cao, D.; Liang, B. In Silico identification and in vitro validation of repurposed compounds targeting the RSV polymerase. *Microorganisms* 2023, 11(6), 1608. DOI:10.3390/microorganisms11061608

Yang, D.; Wang, T.; Long, M.; Li, P. Quercetin: Its main pharmacological activity and potential application in clinical medicine. *Oxid. Med. Cell. Longev.* 2020, 1–13.

Yasmin, A. R.; Chia, S. L.; Looi, Q. H.; Omar, A. R.; Noordin, M. M.; Ideris, A. Herbal extracts as antiviral agents. *Feed Addit.* 2020, 115–132.

Yu, X. M.; Dan, C. Determination of effective components of Scutellaria indica from different habitats. *IOP Conf. Ser. Earth Environ. Sci.* 2020, 474(5), 052083.

Zandavar, H.; Babazad, M. A. Secondary metabolites: Alkaloids and flavonoids in medicinal plants. *J. Herbs Spices Med. Plants.* 2023, IntechOpen.

Zhang, R.; Zhang, Y.; Zheng, W.; Shang, W.; Wu, Y.; Li, N.; Xiong, J.; Jiang, H.; Shen, J.; Xiao, G.; Xie, Y.; Zhang, L. Oral remdesivir derivative VV116 is a potent inhibitor of respiratory syncytial virus with efficacy in mouse model. *Signal Transduction and Targeted Therapy* 2022, 7(1). DOI:10.1038/s41392-022-00963-7.

Zhang, T.; Lo, C. Y.; Xiao, M.; Cheng, L.; Mok, C. K. P.; Shaw, P. C. Anti-influenza virus phytochemicals from Radix Paeoniae Alba and characterization of their neuraminidase inhibitory activities. *J. Ethnopharmacol.* 2020, 253, 112671.

Zhao, H.; Ren, S.; Yang, H.; Tang, S.; Guo, C.; Liu, M.; Tao, Q.; Ming, T.; Xu, H. Peppermint essential oil: Its Phytochem., biological activity, pharmacological effect and application. *Biomed. Pharmacother.* 2022, 154, 113559.

Zheng, X.; Liang, C.; Wang, L.; Wang, B.; Liu, Y.; Feng, S.; Wu, J.Z.; Gao, L.; Feng, L.; Chen, L.; Guo, T. Discovery of benzoazepinequinoline (BAQ) derivatives as novel, potent, orally bioavailable respiratory syncytial virus fusion inhibitors. *J. Med. Chem.* 2018, 61(22), 10228–10241.

Zhu, H.; Lu, X.; Ling, L.; Li, H.; Ou, Y.; Shi, X.; Lu, Y.; Zhang, Y.; Chen, D. Houttuynia cordata polysaccharides ameliorate pneumonia severity and intestinal injury in mice with influenza virus infection. *J. Ethnopharmacol.* 2018, 218, 90–99.

# 11 Plant Viruses and Nanotechnology Approach to Combat Viral Pathogens

*Davinder Singh, Manpreet Singh, and Zaved Ahmed Khan*
Baba Farid College, Baba Farid Group of Institutions

*Aadil Bathla*
Hanyang University

## 11.1 INTRODUCTION

The world faces numerous challenges due to climatic changes, rapid increase in population, food insecurity, urbanization, and soil degradation, which are detrimental to current agricultural practices as well as human health. The prime objective of the agriculture is to increase its yield without compromising the crop quality in order to fulfill the rising global food demands. Nevertheless, various biotic and abiotic stress factors are adversely affecting the crop production. Among them, the plant viruses are considered as a significant threat to sustainable agriculture. Like other viruses, these obligate intracellular parasites are considered pathogenic to plants and frequently inflict significant economic losses yearly. Recently, it has been reported that 137 diseases and pests are associated with up to 41.10% yield loss in five globally important crops (wheat, rice, maize, potato, and soybean). Overall, a loss of 47% in global food production is mainly attributed to viral diseases.

While bacterial and fungal infections can be treated with antibiotics and fungicides, respectively, plant viral infections currently lack a cure. Therefore, disease management depends on a variety of approaches that must be designed particularly for each virus, host, and/or environment in order to prevent viruses from entering plants or to make plants resilient to viral infections. The methods used to diagnose and identify viruses with precision is crucial for the implementation and assessment of disease management strategies. This chapter discusses the current status and development of various approaches to handle plant viruses. It also includes the various types of plant viruses, their impact on the plant metabolism, and various methods used to combat the viral pathogens. Finally, it explores the role of nanotechnology in efficiently managing viral diseases, which has opened new avenues in disease management against viral pathogens.

## 11.2 PLANT VIRAL DISEASE MANAGEMENT STRATEGIES

Plant viruses exhibit a high genetic variability owing to its rapid replication and generation of large populations. They contain either DNA or RNA much like other viruses. However, most of the plant viruses possess RNA genomes. These RNA genomes are relatively small, typically less than 15 kb, as they tend to move between the plasmodesmata of plant cells. A group of DNA plant viruses have been also reported to cause severe diseases in plants. Disease management caused by plant viruses is limited by two factors, the dynamic nature of virus epidemics and high rate of mutations

DOI: 10.1201/9781003329169-13

in viruses. Therefore, it is important to understand the genetic diversity and evolution of plant viruses to develop specific and rapid diagnostic methods for the efficient control of viruses. Till date, numerous techniques at physiological, biochemical, and molecular levels have been identified to combat plant viruses, with the goal of minimizing food loss and increasing economic benefits to the farmers. These methods include:

### 11.2.1 Immunization

The most common form of immunization is the introduction of resistance genes from domesticated or wild species into the most susceptible varieties through backcrossing. This is mostly achieved by either of active or passive immunization. Active immunization is mediated by resistance proteins encoded by dominant alleles. These proteins have been reported to recognize a specific sequence or conformational pattern of a virus gene (avirulence determinant, Avr) and induce cell death (hypersensitive response), thereby preventing viral movement to adjacent cells and systemic infection (de Ronde et al. 2014; Peiró et al. 2014). Passive immunization has been reported to be imparted by resistant recessive alleles encoding essential host components for viral infection, primarily eukaryotic translation initiation factors (eIF) 4E and 4G, along with their isoforms (Truniger and Aranda 2009; Hashimoto et al. 2016).

Typically, plant breeders aim for complete immunization, where the virus cannot form a systemic infection. Since numerous plants may be tested concurrently, ELISA and molecular hybridization techniques have been employed to evaluate germplasm and resistant cultivars. Breeding for relative or partial resistance or tolerance can be a viable alternative when total resistance is not attainable (Rubio et al. 2003a; Soler et al. 2015). Real-time Quantitative Polymerase Chain Reaction (qPCR) is also used for determining the amount of relative resistance (Soler et al. 2015; Galipienso et al. 2013). Time-course tests can be also used to evaluate relative resistance through ELISA or molecular hybridization and relative tolerance to a pathogen by observing visible symptoms or survival analyses of the host plant (Rubio et al. 2020).

### 11.2.2 Genetic Engineering

Owing to the lack of resistance genes in genetically compatible relatives, breeding resistant cultivars is wholly irrelevant for numerous crops and viruses. Alternately, the specificity or range of known resistance genes could be altered through artificial evolution to confer resistance to novel viruses or virus strains. This strategy is focused on the generation of huge populations of random mutants from a resistance gene by polymerase chain reaction (PCR) with a high error rate, followed by the selection of variations with the required resistance characteristics. *Agrobacterium*-mediated transient co-expression of each mutant resistance gene and the Avr from the challenge virus in *Nicotiana benthamiana* leaves has been utilized to evaluate resistance, with the resistance response detected as a necrotic lesion. This technique has been used to widen the resistance specificity of the potato gene Rx; however, it is not widely employed because most of the mutants are non-functional and screening is time-consuming (Farnham and Baulcombe 2006; Harris et al. 2013).

Genome editing through CRISPR-Cas9 could be applied to crops, both for introducing resistance genes generated through artificial evolution and for directly mutagenizing host genes implicated in recessive resistance to inhibit viral contact (Piron et al. 2010; Chandrasekaran et al. 2016; Pyott et al. 2016). These mutagenized plant genes may be involved in crucial biological functions, therefore, leading to unanticipated negative impacts on plant development or physiology (Rubio et al. 2020).

### 11.2.3 RNA Interference

Utilizing the RNA silencing mechanism is another method for obtaining the virus-resistant plants. RNA silencing is a natural antiviral defense mechanism in eukaryotes. The host RNA silencing machinery targets dsRNA that originates from replicative viral intermediates or secondary structures in the genomic RNA due to internal complementarity. These dsRNAs are detected by

RNases (Dicer-like proteins) and cleaved into small RNA duplexes (siRNA or miRNAs) of 21–24 nucleotides (nt). One of the two strands of the short RNAs is recruited to the RNA-induced silencing complex (RISC), which then sequence-specifically cleaves cognate viral RNAs. These cleaved RNAs are identified by RNA-dependent RNA polymerases, which amplify the dsRNA molecules, hence contributing to the amplification of the host defense mechanism. This process inhibits local and systemic viral infection efficiently (Kaldis et al. 2018).

However, numerous viruses encode RNA silencing suppressors, which further limit this defensive strategy (Voinnet et al. 1999). These suppressors can function at multiple points of the silencing pathway, either by binding to siRNA duplexes or by interacting directly with essential components of the RNA silencing machinery. By introducing into plants DNA constructs that create viral dsRNAs or ssRNAs with some degree of secondary structure to induce RNA silencing, resistance can be established. Since RNA silencing needs a certain nucleotide identity between the targeted virus and the transgene, it is required to evaluate the nucleotide variation of the virus population, as previously mentioned. Real-time quantitative polymerase chain reaction (qPCR) is the optimal method for determining the efficacy of RNA silencing-based resistances.

Mutation and natural selection, as well as co-infection with other viruses, can lead to resistance's breakdown (Syller 2012). Transgenic plants with numerous viral constructions (from the same species and/or distinct species) can be utilized to reduce the likelihood of resistance breakdown (Bucher et al. 2006; Angaji et al. 2010). Another technique involves using a transgene that mimics the secondary structure of endogenous miRNA precursors to create artificial miRNAs (amiRNAs) that target viral sequences (Qu et al. 2007; Liu et al. 2017). The primary benefit is that the short sequence of amiRNAs makes it easier to locate conserved regions that are more challenging to overcome (they are typically subject to significant negative selective pressure) and can be utilized to target a wide variety of organisms (genera and families). However, amiRNA resistance can also be overcome by mutation and natural selection, as well as contact with co-infecting viruses (Simón-Mateo and García 2006; Lin et al. 2009; Lafforgue et al. 2011; Pacheco et al. 2012; Martínez et al. 2013). The expression of several amiRNAs that target distinct sections of a single viral genome is an approach for achieving more robust resistances (Fahim et al. 2012; Kung et al. 2012; Kis et al. 2016; Lafforgue et al. 2013). Synthetic trans-acting small interfering RNAs (syn-tasiRNAs) are an additional type of artificial small RNAs generated in plants that are particularly well-suited to target numerous places within a viral genome or different viruses (Carbonell et al. 2016, 2019; Chen et al. 2016; Carbonell and Daròs 2019). Experimental evolution based on several passages of the virus in resistant plants can be used to determine the persistence of these resistances. qPCR can identify resistance breakdown as an increase in viral accumulation (Carrasco et al. 2007; Pena et al. 2014). PCR followed by cloning, Sanger sequencing, or high-throughput sequencing (HTS) can be used to detect mutations repaired following passes. If infectious clones exist, it is possible to identify the mutations that contribute to resistance breakdown, as each mutation may be checked for an increase in virus accumulation. The exogenous application of *in vitro*-produced dsRNAs from viral sequences to plants provides an alternative to transgenic plants (Kaldis et al. 2018; Niehl et al. 2018). By spraying plants with cell-penetrating peptides or clay nanoparticles stabilizing dsRNAs at high pressure, the efficiency of this vaccination approach can be enhanced (Dalakouras et al. 2016; Numata et al. 2014; Mitter et al. 2017).

### 11.2.4 Cross-Protection

Cross-protection is a form of vaccination based on the inoculation of weak or attenuated viral strains to protect plants from severe strains of the same virus. It has been utilized to other viruses and crops, including PepMV in tomato and CTV in citrus (Pechinger et al. 2019). The mechanism of cross-protection is poorly understood, and several models have been proposed, including prevention of virus entry into cells, competition for host factors for replication, interference with disassembly, translation, or replication, and induction of RNA silencing resulting in sequence-specific degradation of the superinfecting virus (Ziebell and Carr 2010; Folimonova 2013). Cross-protection in certain viruses has been hypothesized to be an active virus-controlled function involving virus-encoded proteins (Folimonova 2013). A recent model suggests that cross-protection is a mechanism that blocks virus offspring from replicating in cells in order to reduce mutation rate, while also targeting highly homologous superinfecting viruses that are indistinguishable from virus progeny (Zhang et al. 2018). To implement cross-protection, it is required to examine the genetic and biological variability of the local virus population and to hunt for mild isolates that are genetically similar to severe ones (Hanssen et al. 2010). Thermal treatment, random mutagenesis with chemicals such as nitrous acids, and selection or guided mutagenesis in viral RNA silencing suppressors can also be used to develop mild or attenuated strains (Ziebell and Carr 2010). Cross-protection experiments can be assessed using real-time qPCR with probes specific to each viral variety and SSCP analysis (Ruiz-Ruiz et al. 2007; Hanssen et al. 2010; Sambade et al. 2002). Through transmission to a different host species, interaction with other viruses in mixed infections, or recombination of divergent strains or viruses, the mild isolate's protective effect can be undone, and a more severe illness can emerge (Fulton 1986). Therefore, cross-protection should be utilized only for deadly diseases when other protection measures fail, and the process should be continuously monitored.

### 11.2.5 Preventive Measures

Quarantine (border control) and sanitary certification of virus-free germplasm (seeds or asexual propagating tissues) are the mandatory steps considered to prevent the introduction and establishment of new viruses in a geographical area. Since discarding healthy material is preferable to the possibility of spreading new diseases, virus identification should be based on sensitive and broad-spectrum approaches. Immediate judgments can be made with the assistance of on-site detecting equipment, preventing importation, and exportation delays. HTS is the most effective detection method since it can detect all known and undiscovered viruses present in a plant without bias. As its price decreases, its usage in quarantine and clean plant initiatives increases (Villamor et al. 2019). Since plants might harbor numerous viruses, phytosanitary certificates should be based on propagative material free of only dangerous viruses (Maree et al. 2018).

Sanitation, or the elimination of viruses from valuable cultivars, is required when there are no healthy plants available. Typically, virus-free plants are created using thermotherapy, chemotherapy, electrotherapy, and tissue culture alone or in combination (Naik and Buckseth 2018). Thermotherapy could render viruses inactive through viral RNA cleavage, viral particle breakup or coat protein rupture, viral replicase inactivation, virus movement suppression, and/or translation decrease (Hull 2002). Chemotherapy is based on antiviral medications that block or disrupt certain parts of the virus's life cycle, such as nucleoside analogues that impede replication and protease inhibitors that prevent protein processing. Antiviral medications are expensive and are only used to regenerate healthy mother plants for vegetative propagation or seed production (Panattoni et al. 2013). Virus-free evaluation and confirmation of the sanitized plants must be performed using highly sensitive techniques, such as real-time qPCR. Nucleoside analogues can also be employed to accelerate the already high mutation rate of RNA viruses, leading to the loss of genetic information and eventual extinction of the virus (lethal mutagenesis or error catastrophe). This strategy was recently tested on a plant virus, tobacco mosaic virus (TMV), resulting in a reduction in viral infectivity (Díaz-Martínez et al. 2018).

Removing virus-infected plants from crops or weeds serving as inoculum sources is an agronomic practice that limits the spread of viruses. This implies the use of quick and specific detection techniques, such as ELISA and molecular hybridization, capable of analyzing a large number of field samples, particularly tissue-prints (Rubio et al. 1999, 2003b). Roguing is only successful if the incidence of the virus is low following a recent introduction or in isolated places. Other approaches involve stopping the cycle of virus transmission. Many seed-borne viruses (carried on the seed coat) can be eliminated with chemical disinfectants such as sodium hypochlorite, trisodium phosphate, hydrochloric acid, and ozone, but other seed-transmitted viruses (infecting the seed embryo) can be eradicated with thermotherapy (Ling 2010; Paylan et al. 2014).

Multiplex (RT)-PCR can be utilized to detect concurrently both seed-transmitted and seed-borne viruses in a crop (Panno et al. 2012). By taking hygienic steps such as using disposable gloves or washing hands with disinfectant, heat sterilizing equipment and debris, and restricting access to crops, the incidence of contact-transmissible plant viruses, such as those of the genus Tobamovirus, can be decreased (Broadbent 1976).

Primarily aphids, whiteflies, and thrips serve as vectors for the majority of plant viruses (Antignus 2012; Fereres and Raccah 2015). Three methods have been employed to prevent the transmission of plant viruses via vectors: Reducing vector populations with pesticides; biological control with natural enemies including arthropod predators and parasitoids (del Mar Tellez et al. 2017) or entomopathogenic fungi, nematodes, bacteria, and viruses (Kalha et al. 2014); and biotechnology-based approaches based on protease inhibitors, neurotoxins. Additionally, preventing the vector from reaching the crop can be achieved through various means, such as using barriers like greenhouses and barrier plants, interfering with the insect's vision with UV-absorbing plastics and reflective surfaces, implementing agronomic practices such as altering planting or sowing dates to avoid high vector populations, or imposing a time gap between crops and/or space gap between plots to break the transmission cycle (Antignus 2012; Fereres and Raccah 2015). Spraying mineral oils, synthetic peptides, or modified proteins to compete with virus-encoded proteins required for virus attachment to insect receptors can also impede transmission.

## 11.3 NANOTECHNOLOGY VS PLANT VIRUSES

Plant viral diseases are becoming increasingly challenging to control due to their fast spread. It is important to adopt an efficient management approach that is mostly based on the ecology of the virus (Dutta et al. 2022). In this regard, nanotechnology has emerged as a boon for the agriculture sector due to its ease of use and low cost and resist various biotic and abiotic stresses (Marwal and Gaur 2019; Singh and Kumar 2020, 2021; Singh et al. 2020, 2021; Farooq et al. 2021) (Figure 11.1).

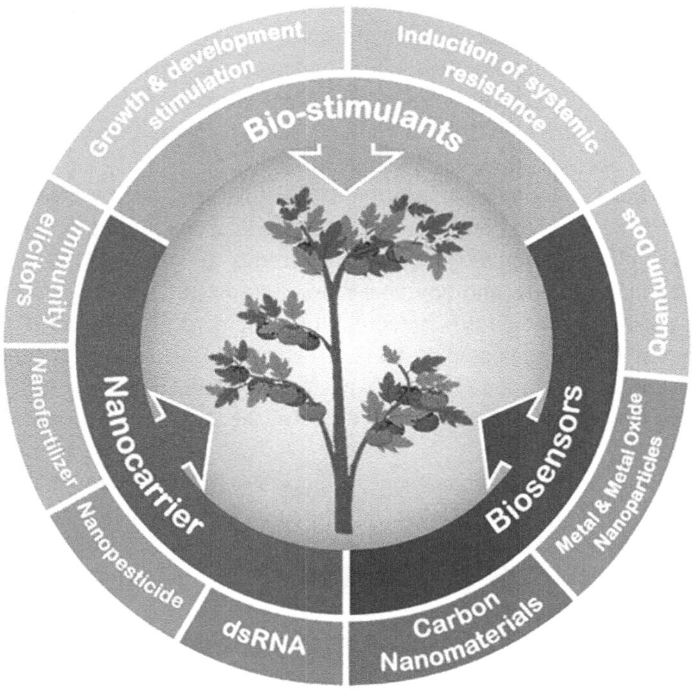


**FIGURE 11.1** The use of nanotechnology-based approaches to combat viral pathogens. Reproduced with permission from Farooq et al. (2021).

The synthesized nanomaterial offers numerous environmental benefits such as low toxicity, improved shelf life, and solubility of poorly soluble pesticides (Worrall et al. 2018). The synthesized nanomaterials have been reported to be either used as protectants or as carriers for pesticides and other compounds. Earlier viruses were stained with fluorescent dyes, but these days, quantum dots (QDs) and nanoparticles (NPs) have been produced to contain detection tags (dyes or antiviral antibodies) for the identification of viruses and reported to be useful for imaging and labeling (Cheng et al. 2009) (Table 11.1). NPs have been also reported to be coupled with probes such as nucleic acids, antibodies, or aptamers for the identification of virus-derived molecules upon binding (Dougan et al. 2007; Wang et al. 2009). The development of nano-biosensors for the identification of viral vectors has been reported earlier (Safarpour et al. 2012; Shojaei et al. 2016). A nano-biosensor has been developed by a group of scientists to detect the plasmodiophoromycete vector of beet necrotic yellow vein virus (BNYVV), i.e., *Polymyxa betae*, an obligate parasite responsible for causing rhizomania disease in sugarbeet roots. The developed immunosensor is made of particular antibodies conjugated with Cadmium-Telluride QDs and glutathione-S-transferase proteins. Fluorescence resonance energy transfer (FRET) is exhibited by the developed nano-biosensor for the detection of the vector along with high sensitivity and 100% specificity (Safarpour et al. 2012). Lavanya and Arun (2021) devised a visual detection approach based on functionalized gold nanoparticles (AuNPs) bearing the clccpi1 probe for the detection of begomoviruses in tomato, chili plant, common bean, green gram, and black gram plants. The AuNPs assay produced superior detection findings (77.7%) than the standard PCR method (49.4%).

Using changes in the refractive index of a metal surface, surface plasmon resonance (SPR) is an additional method for the detection and identification of viruses. This strategy involved conjugating gold nanoparticles with antiviral antibodies adsorbed on a glass substrate and finally detecting them using SPR (Davatgaran Taghipour et al. 2018). Quartz-crystal microbalance (QCM) is a well-known technology for quantitatively measuring the thickness of thin films that is commercially available. The basic mechanism consisted of a nanometer-sized quartz-crystal device coated with antiviral antibodies against the targeted plant viruses. When the gadget comes into contact with a virus, the mass increases, causing the resonance frequency to decrease. Changes in the resonance frequency are then monitored, and the virus is finally detected (Eun et al. 2002). A QCM- and SPR-based nano-biosensor for the detection of different plant viruses, including Tomato mosaic virus (TMV), Cymbidium mosaic virus (CMV), Odontoglossum ringspot virus (ORSV), and numerous orchid viruses, has also been developed (Shojaei et al. 2016).

To detect viral particles, nanowires containing transistors have been created. Several immuno-biosensors have been developed to detect phytoviruses, such as the plum pox virus, in the leaves of plum and tobacco host plants. Using modified gold electrodes containing AuNPs, 1,6-hexanedithiol, anti-PPV IgG polyclonal antibody, and bovine serum albumin, the biosensor was created (Jarocka et al. 2011). When nanowires come into contact with the charged viral capsid of the target virus, a simple change in conductance indicated a change in charge. A similar approach, lithographically patterned nanowire electrodeposition (LPNE), was developed as label-free chemi-resistive sensors based on polypyrrole (PPy) nanoribbon coated with antiviral antibodies to target CMV, an essential phytovirus (Hayden et al. 2003).

In addition, NP-based particle bombardment system-mediated gene cannons can also be utilized for virus identification and infectivity testing (Abubakar et al. 2018). This method also permits the construction of transient or stable gene expression systems for the study of plant-virus interactions at the cellular level (Acanda et al. 2019). The research must be directed toward the improvement of biosensor selectivity, affinity, and efficiency.

Numerous studies have reported the effectual use of nanoparticles in reducing viral infection, as synthesized nanoparticles possess inherent virucidal activity (Table 11.1; Figure 11.2). Jain and Kothari (2014) demonstrated that AgNPs are efficient antiviral agents by achieving the total suppression of sunhemp rosette virus (SHRV) disease on cluster bean after spraying a 50 ppm aqueous solution of AgNPs. Elbeshehy et al. (2015) biosynthesized AgNPs employing newly isolated strains of *Bacillus*

**TABLE 11.1**
**Role of Various Synthesized Nanoparticles in Combating Viral Pathogens in Plants**

| Type of NPs Used | Characteristics of NPs | Target Plant | Target Virus | Key Findings | Reference |
|---|---|---|---|---|---|
| Chitosan/dextran nanoparticles (CDNPs) | 20–160 nm, Spherical | *Nicotiana glutinosa* | Alfalfa mosaic virus | • Reduced disease severity by 70.43%<br>• Accumulation of total carbohydrates and total phenolic contents<br>• Increase in the transcriptional levels of peroxidase, pathogen-related protein-1, and phenylalanine ammonia-lyase | Abdelkhalek et al. (2021) |
| Silver nanoparticles (AgNPs) | 15 nm, Cubic and Hexagonal | Cluster bean leaves | Sunhemp rosette virus (SHV) | • 100% disease suppression<br>• Inhibited viral replication in the host plant | Jain and Kothari (2014) |
| AgNPs | 77–92 nm, Triangular, hexagonal and spherical shapes | Fava Bean | Bean Yellow Mosaic Virus (BYMV) | • Reduction in disease severity by 100% when treated simultaneously with BYMV infection and after viral infection | Elbeshehy et al. (2015) |
| AgNPs | 12 nm, Spherical | Potato | Potato Virus Y (PVY) | • AgNPs at 0.1 μg/μL significantly decreased the viral infection when treated after 24 hour post viral infection<br>• Decreased viral multiplication, accumulation, and entry to the vascular system | El-Shazly et al. (2017) |
| AgNPs | - | Tomato | PVY and Tomato mosaic virus (ToMV) | • Reduction in disease severity and relative concentration of both the viruses when treated with 50 ppm AgNPs by binding to viral coat proteins<br>• Increase in total soluble protein, pigment content, peroxidase, and polyphenol oxidase activities | El-Dougdoug et al. (2018) |
| Silica nanoparticles (SiNPs) | 100 nm | Tomato | Tomato Yellow Leaf Curl Virus (TYLCV) | • Reduction in disease severity up to 40% when treated with 200 μg/mL SiNPs<br>• Increase in peroxidase and polyphenol oxidase activities | El-Sawy et al. (2017a) |
| Zinc oxide nanoparticles (ZnO NPs) | 4.0 nm | Mosambi plants (*Citrus sinensis*) | Huanglongbing (HLB) | • Up to 98% reduction in pathogen growth by the combined application of ZnO NPs and 2S albumin | Ghosh et al. (2018) |
| Titanium oxide nanoparticles ($TiO_2$ NPs) | Hollow spherical | Faba plants | Broad bean stain virus (BBSV) | • Reduction in disease severity when treated with $TiO_2$ NPs either through foliar spray or soil drench method<br>• Increase in expression of pathogenesis related gene (*PR1*) | Elsharkawy and Derbalah (2019) |

**FIGURE 11.2** Effect of synthesized silver nanoparticles from different *Bacillus* species i.e., *Bacillus pumilus, B. persicus*, and *B. licheniformis* on BYMV infected *Vicia faba* L. cv. Giza3 and non-infected plants. The nanoparticles were introduced to the plant by spraying simultaneously with virus inoculation (experimental Group I); before inoculation (Group II) or after inoculation (Group III) (Elbeshehy et al. 2015), Open access journal.

*pumilus, B. persicus*, and *B. licheniformis,* and showed an *in vitro* antiviral efficacy against Bean Yellow Mosaic Virus (BYMV). The authors also reported that AgNP treatment administered after 24 hours post-infection prevented all damaging BYMV symptoms. In addition, a reduction in virus concentration, infection rate, and disease severity was recorded in treated plants as compared to untreated plants. However, no effect was detected when AgNPs were delivered at the pre-viral infection stage (Figure 11.2). The results of another study indicated that AgNP activity was reported to be more pronounced in the early stages of viral replication since it stimulated the plant's systemic resistance to BYMV. On potato plants infected with potato virus Y (PVY), the inhibitory effects of AgNPs at 0.1 μg/μL alone and a mixture of AgNPs at 0.1 μg/μL and salicylic acid (SA) at 200 μmol/L were observed. The application of NPs before 3–7 days of virus infection resulted in a decrease in the virus concentration and percent infection, as well as an increase in plant growth and tuber yield (El-Shazly et al. 2017).

El-Dougdoug et al. (2018) reported the antiviral activity of AgNPs in tomatoes against TMV and Potato virus Y (PVY). By raising the levels of total soluble proteins (TSP), antioxidant enzymes peroxidase (POD), and polyphenol oxidase (PPO) activities, AgNPs at 50 ppm lowered viral disease severity and percent infection generating systemic acquired resistance (SAR) against TMV and PVY in tomato plants. TMV- and PVY-infected tomato plants exhibited a significant drop in photosynthetic pigments and a rise in total soluble phenols and free proline, respectively (El-Dougdoug et al. 2018). A reduction in the severity of the disease caused by tomato yellow leaf curl virus (TYLCV) by delaying the emergence of viral symptoms was also recorded in silica NPs (SiNPs)-treated tomato plants compared to untreated control plants. POD and PPO activity rose in tomato plants treated with SiNPs, showing enhanced systemic resistance to TYLCV (El-Sawy et al. 2017a). Zinc oxide nanoparticles (ZnONPs) were also found to increase resistance to CMV in brinjal plants grown in greenhouse conditions (El-Sawy et al. 2017b). Similarly, gold nanoparticles were found to be effective against plant-pathogenic viruses, including bean mild mosaic virus in beans, tobacco necrosis virus in tobacco, and barley yellow mosaic virus in barely host plants (Marwal and Gaur 2019; Dutta et al. 2022).

Ghosh et al. (2018) first revealed in planta efficacy of two antimicrobial drugs, namely 2S albumin-a plant-based protein, zinc oxide nanoparticles (ZnONPs), and their combinations against

*Candidatus Liberibacter* asiaticus (*CLas*) causing Haunglongbing (HLB) disease in citrus. The application of 2S albumin and ZnONPs formulation alone or in various combinations resulted in a significant reduction in *CLas* titers in citrus plants infected with HLB. Therefore, it can be deduced from the preceding explanation that NPs alone can be used as an efficient antimicrobial agent to combat the threat posed by plant viruses.

Chitosan NPs were also reported to elicit resistance against a few viruses in host plants, including the potato mosaic virus, peanut mosaic virus, cucumber mosaic virus, alfalfa mosaic virus, and snuff mosaic virus (Rendina et al. 2019; Abdelkhalek et al. 2021; Mishra et al. 2022; Murphy et al. 2003). Treatment of *Nicotiana glutinosa* plants with 100 μg/mL chitoson/dextran nanoparticles (CDNPs) significantly reduced the disease severity of alfalfa mosaic virus by 70.43% and reduced the viral accumulation by 61.65% in the host plants. It was also found that increase in total carbohydrates and total phenolic contents, and significant changes in transcriptional levels of peroxidase, pathogen-related protein-1, and phenylalanine ammonia-lyase, were also recorded in CDNPs treated plants (Abdelkhalek et al. 2021). A significant reduction in cucumber mosaic virus infection was observed in chitosan treated tomato plants (Rendina et al. 2019). Murphy et al. (2003) also evaluated the effect of combinations of two PGPR strains formulated with chitosan molecule, and observed a significant reduction in disease severity alongside significantly increased plant height, root and shoot weight, and the number of flowers and fruits in treated plants as compared to the control. Thus, it was confirmed that tomato plants treated with biopreparations demonstrated improved plant development and increased resistance to CMV infection. The antiviral activity of graphene-based silver nanocomposites was validated against tomato bushy stunt virus (TBSV) by lowering the virus concentration and disease severity in lettuce plants following soil and foliar application (El-Shazly et al. 2017). *Cucumis sativus* infected with Papaya ringspot virus (PRSV) showed a significant decrease in symptom generation when treated with $SiO_2$NPs by soil supplementation (Elsharkawy and Mousa 2015). In addition, a reduction in broad bean strain virus (BBSV) infection was found when $TiO_2$ NPs were given through spray and soil drench 4 hours after inoculation (Elsharkawy and Derbalah 2019).

## 11.4 CONCLUSION AND FUTURE CHALLENGES

Numerous synthesized nanoparticles, including metal and their metal oxide nanoparticles (NPs), have gained tremendous attention in the past few decades due to their broad-spectrum antiviral properties. The literature review indicated that NPs mostly interfere with replication cycle of the viruses, inactive and denature capsid proteins, nucleic acids, and other protein components which ultimately resulted in inducing the systemic acquired resistance to the host plants. However, detailed research on viral mobility, reproduction, and transmission is still required. The long-term effects of synthesized NPs on the soil under different environmental conditions are still pending. The impact of combination of various synthesized NPs could be another proposed area of research for the successful management of complicated viral infections under natural environments. Once nanoparticles enter the plant cell, they may affect their life activities in the future as they are slowly released. Therefore, their biological toxicity needs critical attention.

In most cases, foliar application of nanoparticles is used to investigate the role of synthesized nanoparticles against viral pathogens. However, this method is more expensive and time-consuming and can be applied only in the morning and evening. Therefore, there is a need to adopt alternative methods of application of nanoparticles to overcome the limitations. The utilization of plant tissue culture technology can serve as a better alternative as it can be performed throughout the year under controlled conditions and does not cause any harm to the ecosystem.

Several investigations have determined the efficacy of NPs against viral pathogens and elucidated that synthesized nanoparticles inhibited the growth of viral pathogens by improving the photosynthetic machinery of plants and through production of antioxidant compounds. The utilization of omics approach in this direction would pave a new direction to this research and help in elucidating the molecular mechanism of action.

## REFERENCES


Abdelkhalek A, Qari SH, Abu-Saied MAA-R, Khalil AM, Younes HA, Nehela Y, Behiry SI (2021) Chitosan nanoparticles inactivate alfalfa mosaic virus replication and boost innate immunity in *Nicotiana glutinosa* plants. *Plants* 10 (12):2701.

Abubakar A, Abarshi M, Maruthi M (2018) Testing the infectivity of a Begomovirus by particle bombardment method using a gene gun. *Nigerian Journal of Biotechnology* 35 (2):58–65.

Acanda Y, Wang C, Levy A (2019) Gene expression in citrus plant cells using Helios(r) Gene Gun System for particle bombardment. In: Catara A, Bar-Joseph M, Licciardello G (eds.) *Citrus Tristeza Virus: Methods and Protocols*. Springer, pp. 219–228.

Angaji S, Hedayati SS, Hosein R, Samad S, Shiravi S, Madani S (2010) Application of RNA interference in plants. *Plant Omics* 3 (3):77–84.

Antignus Y (2012) Control methods of virus diseases in the Mediterranean basin. In: Bar-Joseph AF, Licciardello M. (eds.), *Advances in Virus Research, Viruses and Virus Diseases of Vegetables in the Mediterranean Basin*, vol 84. Elsevier, pp. 533–553.

Broadbent L (1976) Epidemiology and control of tomato mosaic virus. *Annual Review of Phytopathology* 14 (1):75–96.

Bucher E, Lohuis D, van Poppel PM, Geerts-Dimitriadou C, Goldbach R, Prins M (2006) Multiple virus resistance at a high frequency using a single transgene construct. *Journal of General Virology* 87 (12): 3697–3701.

Carbonell A, Carrington JC, Daròs J-A (2016) Fast-forward generation of effective artificial small RNAs for enhanced antiviral defense in plants. *RNA and Disease* 3 (1): e1130.

Carbonell A, Daròs J-A (2019) Design, synthesis, and functional analysis of highly specific artificial small RNAs with antiviral activity in plants. In: Kobayashi K, Nishiguchi M (eds.) *Antiviral Resistance in Plants*. Springer, pp 231–246.

Carbonell A, Lisón P, Daròs JA (2019) Multi-targeting of viral RNAs with synthetic trans-acting small interfering RNAs enhances plant antiviral resistance. *The Plant Journal* 100 (4): 720–737.

Carrasco P, Daròs J, Agudelo-Romero P, Elena S (2007) A real-time RT-PCR assay for quantifying the fitness of *Tobacco etch virus* in competition experiments. *Journal of Virological Methods* 139 (2):181–188.

Chandrasekaran J, Brumin M, Wolf D, Leibman D, Klap C, Pearlsman M, Sherman A, Arazi T, Gal-On A (2016) Development of broad virus resistance in non-transgenic cucumber using CRISPR/Cas9 technology. *Molecular Plant Pathology* 17 (7):1140–1153.

Chen L, Cheng X, Cai J, Zhan L, Wu X, Liu Q, Wu X (2016) Multiple virus resistance using artificial trans-acting siRNAs. *Journal of Virological Methods* 228:16–20.

Cheng X, Chen G, Rodriguez WR (2009) Micro-and nanotechnology for viral detection. *Analytical and Bioanalytical Chemistry* 393:487–501.

Dalakouras A, Wassenegger M, McMillan JN, Cardoza V, Maegele I, Dadami E, Runne M, Krczal G, Wassenegger M (2016) Induction of silencing in plants by high-pressure spraying of *in vitro*-synthesized small RNAs. *Frontiers in Plant Science* 7:1327.

Davatgaran Taghipour Y, Kharrazi S, Amini SM (2018) Antibody conjugated gold nanoparticles for detection of small amounts of antigen based on surface plasmon resonance (SPR) spectra. *Nanomedicine Research Journal* 3 (2):102–108.

de Ronde D, Butterbach P, Kormelink R (2014) Dominant resistance against plant viruses. *Frontiers in Plant Science* 5:307.

del Mar Tellez M, Simon A, Rodriguez E, Janssen D (2017) Control of *Tomato leaf curl New Delhi virus* in zucchini using the predatory mite *Amblyseius swirskii*. *Biological Control* 114:106–113.

Díaz-Martínez L, Brichette-Mieg I, Pineño-Ramos A, Domínguez-Huerta G, Grande-Pérez A (2018) Lethal mutagenesis of an RNA plant virus via lethal defection. *Scientific Reports* 8 (1):1–13.

Dougan JA, Karlsson C, Smith WE, Graham D (2007) Enhanced oligonucleotide-nanoparticle conjugate stability using thioctic acid modified oligonucleotides. *Nucleic Acids Research* 35 (11):3668–3675.

Dutta P, Kumari A, Mahanta M, Biswas KK, Dudkiewicz A, Thakuria D, Abdelrhim AS, Singh SB, Muthukrishnan G, Sabarinathan K (2022) Advances in nanotechnology as a potential alternative for plant viral disease management. *Frontiers in Microbiology* 13:935193.

El-Dougdoug N, Bondok A, El-Dougdoug K (2018) Evaluation of silver nanoparticles as antiviral agent against ToMV and PVY in tomato plants. *Middle East Journal of Applied Sciences* 8 (01):100–111.

El-Sawy MM, Elsharkawy MM, Abass JM, Hagag ES (2017a) Inhibition of *tomato yellow leaf curl virus* by *Zingiber officinale* and *Mentha longifolia* extracts and silica nanoparticles and its reflection on tomato growth and yield. *International Journal of Antivirals & Antiretrovirology* 1(1):1–6.

El-Sawy MM, Elsharkawy MM, Abass JM, Kasem MH (2017b) Antiviral activity of 2-nitromethyl phenol, zinc nanoparticles and seaweed extract against cucumber mosaic virus (CMV) in Eggplant. *Journal of Virology & Antiviral Research* 6 (2):1–6.

El-Shazly MA, Attia YA, Kabil FF, Anis E, Hazman M (2017) Inhibitory effects of salicylic acid and silver nanoparticles on potato virus Y-infected potato plants in Egypt. *Middle East Journal of Agriculture Research* 6 (3):835–848.

Elbeshehy EK, Elazzazy AM, Aggelis G (2015) Silver nanoparticles synthesis mediated by new isolates of *Bacillus* spp., nanoparticle characterization and their activity against Bean Yellow Mosaic Virus and human pathogens. *Frontiers in Microbiology* 6:453.

Elsharkawy MM, Derbalah A (2019) Antiviral activity of titanium dioxide nanostructures as a control strategy for broad bean strain virus in faba bean. *Pest Management Science* 75 (3):828–834.

Elsharkawy MM, Mousa KM (2015) Induction of systemic resistance against Papaya ring spot virus (PRSV) and its vector Myzus persicae by *Penicillium simplicissimum* GP17-2 and silica ($SiO_2$) nanopowder. *International Journal of Pest Management* 61 (4):353–358.

Eun AJ- C, Huang L, Chew F-T, Li SF- Y, Wong S-M (2002) Detection of two orchid viruses using quartz crystal microbalance (QCM) immunosensors. *Journal of Virological Methods* 99 (1–2):71–79.

Fahim M, Millar AA, Wood CC, Larkin PJ (2012) Resistance to *Wheat streak mosaic virus* generated by expression of an artificial polycistronic microRNA in wheat. *Plant Biotechnology Journal* 10 (2):150–163.

Farnham G, Baulcombe DC (2006) Artificial evolution extends the spectrum of viruses that are targeted by a disease-resistance gene from potato. *Proceedings of the National Academy of Sciences of the United States of America* 103 (49):18828–18833.

Farooq T, Adeel M, He Z, Umar M, Shakoor N, da Silva W, Elmer W, White JC, Rui Y (2021) Nanotechnology and plant viruses: an emerging disease management approach for resistant pathogens. *ACS Nano* 15 (4):6030–6037.

Fereres A, Raccah B (2015) Plant virus transmission by insects. In: *Encyclopedia of Life Sciences*. John Wiley & Sons, Ltd: Chichester, pp. 1–12. DOI: 10.1002/9780470015902.a0000760.pub3

Folimonova SY (2013) Developing an understanding of cross-protection by Citrus tristeza virus. *Frontiers in Microbiology* 4:76.

Fulton RW (1986) Practices and precautions in the use of cross protection for plant virus disease control. *Annual Review of Phytopathology* 24 (1):67–81.

Galipienso L, Janssen D, Rubio L, Aramburu J, Velasco L (2013) *Cucumber vein yellowing virus* isolate-specific expression of symptoms and viral RNA accumulation in susceptible and resistant cucumber cultivars. *Crop Protection* 43:141–145.

Ghosh DK, Kokane S, Kumar P, Ozcan A, Warghane A, Motghare M, Santra S, Sharma AK (2018) Antimicrobial nano-zinc oxide-2S albumin protein formulation significantly inhibits growth of "Candidatus Liberibacter asiaticus" in planta. *PLoS One* 13 (10):e0204702.

Hanssen IM, Gutiérrez-Aguirre I, Paeleman A, Goen K, Wittemans L, Lievens B, Vanachter AC, Ravnikar M, Thomma B (2010) Cross-protection or enhanced symptom display in greenhouse tomato co-infected with different *Pepino mosaic virus* isolates. *Plant Pathology* 59 (1):13–21.

Harris CJ, Slootweg EJ, Goverse A, Baulcombe DC (2013) Stepwise artificial evolution of a plant disease resistance gene. *Proceedings of the National Academy of Sciences of the United States of America* 110 (52):21189–21194.

Hashimoto M, Neriya Y, Yamaji Y, Namba S (2016) Recessive resistance to plant viruses: potential resistance genes beyond translation initiation factors. *Frontiers in Microbiology* 7:1695.

Hayden O, Bindeus R, Haderspöck C, Mann K-J, Wirl B, Dickert FL (2003) Mass-sensitive detection of cells, viruses and enzymes with artificial receptors. *Sensors and Actuators B: Chemical* 91 (1–3):316–319.

Hull R (2002) Genome organization. In: R. Hull (ed.) *Matthew's Plant Virology*, 4th edn. Academic Press. pp. 171–224.

Jain D, Kothari S (2014) Green synthesis of silver nanoparticles and their application in plant virus inhibition. *Journal of Mycology and Plant Pathology* 44 (1):21–24.

Jarocka U, Wąsowicz M, Radecka H, Malinowski T, Michalczuk L, Radecki J (2011) Impedimetric immunosensor for detection of plum pox virus in plant extracts. *Electroanalysis* 23 (9):2197–2204.

Kaldis A, Berbati M, Melita O, Reppa C, Holeva M, Otten P, Voloudakis A (2018) Exogenously applied dsRNA molecules deriving from the *Zucchini yellow mosaic virus* (ZYMV) genome move systemically and protect cucurbits against ZYMV. *Molecular Plant Pathology* 19 (4):883–895.

Kalha C, Singh P, Kang S, Hunjan M, Gupta V, Sharma R (2014) Entomopathogenic viruses and bacteria for insect-pest control. In: Abrol DP (ed.) *Integrated Pest Management*. Elsevier, pp. 225–244.

Kis A, Tholt G, Ivanics M, Várallyay É, Jenes B, Havelda Z (2016) Polycistronic artificial miRNA-mediated resistance to W heat dwarf virus in barley is highly efficient at low temperature. *Molecular Plant Pathology* 17 (3):427–437.

Kung YJ, Lin SS, Huang YL, Chen TC, Harish SS, Chua NH, Yeh SD (2012) Multiple artificial microRNAs targeting conserved motifs of the replicase gene confer robust transgenic resistance to negative-sense single-stranded RNA plant virus. *Molecular Plant Pathology* 13 (3):303–317.

Lafforgue G, Martínez F, Niu Q-W, Chua N-H, Daròs J-A, Elena SF (2013) Improving the effectiveness of artificial microRNA (amiR)-mediated resistance against *Turnip mosaic virus* by combining two amiRs or by targeting highly conserved viral genomic regions. *Journal of Virology* 87 (14):8254–8256.

Lafforgue G, Martínez F, Sardanyés J, De La Iglesia F, Niu Q-W, Lin S-S, Solé RV, Chua N-H, Daròs J-A, Elena SF (2011) Tempo and mode of plant RNA virus escape from RNA interference-mediated resistance. *Journal of Virology* 85 (19):9686–9695.

Lavanya R, Arun V (2021) Detection of Begomovirus in chilli and tomato plants using functionalized gold nanoparticles. *Scientific Reports* 11 (1):14203.

Lin S-S, Wu H-W, Elena SF, Chen K-C, Niu Q-W, Yeh S-D, Chen C-C, Chua N-H (2009) Molecular evolution of a viral non-coding sequence under the selective pressure of amiRNA-mediated silencing. *PLoS Pathogens* 5 (2):e1000312.

Ling K-S (2010) Effectiveness of chemo-and thermotherapeutic treatments on *Pepino mosaic* virus in tomato seed. *Plant Disease* 94 (3):325–328.

Liu S-R, Zhou J-J, Hu C-G, Wei C-L, Zhang J-Z (2017) MicroRNA-mediated gene silencing in plant defense and viral counter-defense. *Frontiers in Microbiology* 8:1801.

Maree HJ, Fox A, Al Rwahnih M, Boonham N, Candresse TJFi PS (2018) Application of HTS for routine plant virus diagnostics: state of the art and challenges. *Frontiers in Plant Science* 9:1082.

Martínez F, Elena SF, Daròs J-A (2013) Fate of artificial microRNA-mediated resistance to plant viruses in mixed infections. *Phytopathology* 103 (8):870–876.

Marwal A, Gaur R (2019) Nanophytovirology: an emerging field for disease management. In: Topolovec-Pintarić S (ed.) *Plant Diseases-Current Threats and Management Trends.* IntechOpen, pp.1–12.

Mishra KK, Subbanna A, Rajashekara H, Paschapur AU, Singh AK, Jeevan B, Maharana C (2022) Chitosan and chitosan-based nanoparticles for eco-friendly management of plant diseases and insect pests: a concentric overview. In: Kumar S, Madihally SV (eds.) *Role of Chitosan and Chitosan-Based Nanomaterials in Plant Sciences.* Elsevier, pp 435–451.

Mitter N, Worrall EA, Robinson KE, Li P, Jain RG, Taochy C, Fletcher SJ, Carroll BJ, Lu G, Xu ZP (2017) Clay nanosheets for topical delivery of RNAi for sustained protection against plant viruses. *Nature Plants* 3 (2):1–10.

Murphy JF, Reddy M, Ryu C-M, Kloepper JW, Li R (2003) Rhizobacteria-mediated growth promotion of tomato leads to protection against *Cucumber mosaic* virus. *Phytopathology* 93 (10):1301–1307.

Naik PS, Buckseth T (2018) Recent advances in virus elimination and tissue culture for quality potato seed production. *Biotechnologies of Crop Improvement* 1:131–158.

Niehl A, Soininen M, Poranen MM, Heinlein M (2018) Synthetic biology approach for plant protection using ds RNA. *Plant Biotechnology Journal* 16 (9):1679–1687.

Numata K, Ohtani M, Yoshizumi T, Demura T, Kodama Y (2014) Local gene silencing in plants via synthetic ds RNA and carrier peptide. *Plant Biotechnology Journal* 12 (8):1027–1034.

Pacheco R, García-Marcos A, Barajas D, Martiáñez J, Tenllado F (2012) PVX-potyvirus synergistic infections differentially alter microRNA accumulation in *Nicotiana benthamiana. Virus Research* 165 (2):231–235.

Panattoni A, Luvisi A, Triolo E (2013) Elimination of viruses in plants: twenty years of progress. *Spanish Journal of Agricultural Research* (1):173–188.

Panno S, Davino S, Rubio L, Rangel E, Davino M, García-Hernández J, Olmos A (2012) Simultaneous detection of the seven main tomato-infecting RNA viruses by two multiplex reverse transcription polymerase chain reactions. *Journal of Virological Methods* 186 (1–2):152–156.

Paylan İC, Erkan S, Cetinkaya N, Ergun M, Pazarlar S (2014) Effects of different treatments on the inactivation of various seedborne viruses in some vegetables. *Ozone: Science and Engineering* 36 (5):422–426.

Pechinger K, Chooi KM, MacDiarmid RM, Harper SJ, Ziebell H (2019) A new era for mild strain cross-protection. *Viruses* 11 (7):670.

Peiró A, Cañizares MC, Rubio L, López C, Moriones E, Aramburu J, Sánchez-Navarro J (2014) The movement protein (NSm) of *Tomato spotted wilt virus* is the avirulence determinant in the tomato *Sw-5* gene-based resistance. *Molecular Plant Pathology* 15 (8):802–813.

Pena EJ, Ferriol I, Sambade A, Buschmann H, Niehl A, Elena SF, Rubio L, Heinlein M (2014) Experimental virus evolution reveals a role of plant microtubule dynamics and TORTIFOLIA1/SPIRAL2 in RNA trafficking. *PLoS One* 9 (8):e105364.

Piron F, Nicolaï M, Minoïa S, Piednoir E, Moretti A, Salgues A, Zamir D, Caranta C, Bendahmane A (2010) An induced mutation in tomato eIF4E leads to immunity to two potyviruses. *PLoS One* 5 (6):e11313.

Pyott DE, Sheehan E, Molnar A (2016) Engineering of CRISPR/Cas9-mediated potyvirus resistance in transgene-free Arabidopsis plants. *Molecular Plant Pathology* 17 (8):1276–1288.

Qu J, Ye J, Fang R (2007) Artificial microRNA-mediated virus resistance in plants. *Journal of Virology* 81 (12):6690–6699.

Rendina N, Nuzzaci M, Scopa A, Cuypers A, Sofo A (2019) Chitosan-elicited defense responses in Cucumber mosaic virus (CMV)-infected tomato plants. *Journal of Plant Physiology* 234:9–17.

Rubio L, Galipienso L, Ferriol I (2020) Detection of plant viruses and disease management: relevance of genetic diversity and evolution. *Frontiers in Plant Science* 11:1092.

Rubio L, Herrero J, Sarrio J, Moreno P, Guerri J (2003a) A new approach to evaluate relative resistance and tolerance of tomato cultivars to begomoviruses causing the tomato yellow leaf curl disease in Spain. *Plant Pathology* 52 (6):763–769.

Rubio L, Janssen D, Cuadrado I, Moreno P, Guerri J (2003b) Rapid detection of cucumber vein yellowing virus by tissue-print hybridisation with digoxigenin-labelled cDNA probes. *Journal of Virological Methods* 114 (1):105–107.

Rubio L, Soong J, Kao J, Falk BW (1999) Geographic distribution and molecular variation of isolates of three whitefly-borne closteroviruses of cucurbits: *Lettuce infectious yellows* virus, *Cucurbit yellow stunting disorder* virus, and *Beet pseudo-yellows* virus. *Phytopathology* 89 (8):707–711.

Ruiz-Ruiz S, Moreno P, Guerri J, Ambrós S (2007) A real-time RT-PCR assay for detection and absolute quantitation of *Citrus tristeza* virus in different plant tissues. *Journal of Virological Methods* 145 (2):96–105.

Safarpour H, Safarnejad R, Tabatabaei M, Mohsenifar A, Rad F, Shahryari F (2012) Detection of *Polymyxa betae*; the transmitting agent of rhizomania disease of sugar beet, with quantum dots FRET-based biosensor. *Canadian Journal of Plant Pathology* 34 (4):507–515.

Sambade A, Rubio L, Garnsey S, Costa N, Müller G, Peyrou M, Guerri J, Moreno P (2002) Comparison of viral RNA populations of pathogenically distinct isolates of Citrus tristeza virus: application to monitoring cross-protection. *Plant Pathology* 51 (3):257–265.

Shojaei TR, Salleh MAM, Sijam K, Rahim RA, Mohsenifar A, Safarnejad R, Tabatabaei M (2016) Fluorometric immunoassay for detecting the plant virus *Citrus tristeza* using carbon nanoparticles acting as quenchers and antibodies labeled with CdTe quantum dots. *Microchimica Acta* 183:2277–2287.

Simón-Mateo C, García JA (2006) MicroRNA-guided processing impairs Plum pox virus replication, but the virus readily evolves to escape this silencing mechanism. *Journal of Virology* 80 (5):2429–2436.

Singh D, Kaur S, Kumar A (2020) *In vitro* drought tolerance in selected elite clones of *Eucalyptus tereticornis* Sm. *Acta Physiologiae Plantarum* 42 (2):17. https://doi.org/10.1007/s11738-019-3009-4.

Singh D, Kumar A (2020) In vitro screening and characterization of selected elite clones of *Eucalyptus tereticornis* Sm. for salt stress. *Journal of Plant Growth Regulation* 40:694–706.

Singh D, Kumar A (2021) Multivariate screening approach indicated adaptive tolerance to salt stress in the seedlings of an agroforestry tree, Eucalyptus tereticornis Sm. *Plant Cell, Tissue and Organ Culture* 145:545–560. https://doi.org/10.1007/s11240-021-02025-2.

Singh D, Sillu D, Kumar A, Agnihotri S (2021) Dual nanozyme characteristics of iron oxide nanoparticles alleviate salinity stress and promote the growth of an agroforestry tree, Eucalyptus tereticornis Sm. *Environmental Science: Nano* 8 (5):1308–1325. https://doi.org/10.1039/D1EN00040C.

Soler S, Debreczeni DE, Vidal E, Aramburu J, López C, Galipienso L, Rubio L (2015) A new *Capsicum baccatum* accession shows tolerance to wild-type and resistance-breaking isolates of *Tomato spotted wilt virus*. *Annals of Applied Biology* 167 (3):343–353.

Syller J (2012) Facilitative and antagonistic interactions between plant viruses in mixed infections. *Molecular Plant Pathology* 13 (2):204–216.

Truniger V, Aranda MA (2009) Recessive resistance to plant viruses. *Advances in Virus Research* 75:119–231.

Villamor D, Ho T, Al Rwahnih M, Martin R, Tzanetakis I (2019) High throughput sequencing for plant virus detection and discovery. *Phytopathology* 109 (5):716–725.

Voinnet O, Pinto YM, Baulcombe DC (1999) Suppression of gene silencing: a general strategy used by diverse DNA and RNA viruses of plants. *Proceedings of the National Academy of Sciences of the United States of America* 96 (24):14147–14152.

Wang X, Liu L-H, Ramström O, Yan M (2009) Engineering nanomaterial surfaces for biomedical applications. *Experimental Biology and Medicine* 234 (10):1128–1139

Worrall EA, Hamid A, Mody KT, Mitter N, Pappu HR (2018) Nanotechnology for plant disease management. *Agronomy* 8 (12):285.

Zhang X-F, Zhang S, Guo Q, Sun R, Wei T, Qu F (2018) A new mechanistic model for viral cross protection and superinfection exclusion. *Frontiers in Plant Science* 9:40.

Ziebell H, Carr JP (2010) Cross-protection: a century of mystery. In: Carr JP Loebenstein G (eds.) *Advances in Virus Research: Natural and Engineered Resistance to Plant Viruses, Part II*, vol 76. Elsevier, pp 211–264.

# *Section C*

## *Novel Strategies to Combat AMR*

# 12 Prospects of Medicinal Plants and Plant Compounds as Anti-Human Herpes Virus Drugs

*Reena Kumari*
Indian Institute of Technology

*Prachi Vaid*
AP Goyal Shimla University

## 12.1 INTRODUCTION

Herpesviruses have been infecting humans for ages and generally, all adults are infected by at least one of the viral strains. Among more than a hundred strains of human herpes viruses (HHV), only eight strains are responsible to cause infection (Grinde, 2013). HHV belongs to the *Herpesviridae* family, which is further divided into three subfamilies: α-herpesvirinae, β-herpesvirinae, and γ-herpesvirinae. The viruses are differentiated by their viral and structural features, as well as by their pathogenicity. Once these viruses infect humans, they remain inside the individual's body (latent stage) throughout the lifetime. HHV are responsible for many diseases, and benign infections, but they can cause clinically severe infections in immunodeficient persons (Sharma et al., 2016). Since there is no definitive treatment for most of the infections caused by HHV, researchers are exploring new therapeutic strategies by harnessing the potential of medicinal plants and their compounds.

## 12.2 HHV INVASION MECHANISM

HHVs are spherical double-stranded DNA viruses which invade humans' body through skin and mucous membrane. The unique structure of HHV is made up of four layers: a core, capsid, tegument, and envelope. The core encloses a linear double-stranded DNA molecule, surrounded by an icosapentahedral capsid. Between the envelope and capsid is a group of tegument proteins, which help to control the biochemical processes and destabilize virion production, and some provide protection against the host immune responses (Whitley, 1996; Heming et al., 2017). The envelope of HHV consists of unique membrane proteins and glycoproteins that play a significant role in the pathogenicity of the virus. The glycoproteins bind with the specific receptors of the host cell membrane and transfer DNA from the viral nuclear capsid into the host cell nucleus (Kim et al., 2013). After entering the host body, the HHVs will reside lifelong without showing clinical symptoms and can cause latent infection and repeated recurrence in individuals with weak immune system. The HHV has four different stages of viral infection: proliferative, latent, integrated, and congenital infection. The initial stage when the virus enters inside the host cell and proliferates is called proliferative infection. During latent infection, the virus stays inactive in the host cell without showing any symptoms of infection. Integral infection occurs when DNA fragments of HHV integrates

DOI: 10.1201/9781003329169-15

with the DNA of the host cell and inhibit its apoptosis, which can lead to the formation of tumor. Congenital infection affects the newborn babies or unborn fetus, premature birth, and congenital abnormalities. Understanding the structure and pathogenesis of HHV can help to stop and manage its infection and transmission and provide new suggestions for the synthesis of new therapies (Dong et al., 2021; Traylen et al., 2011).

## 12.3 ANTI-HHV PLANT EXTRACTS/FRACTIONS AND COMPOUNDS

Plant extracts and their compounds have been widely utilized for many untreatable diseases, including viral infections, since ancient times. Although numerous phytomedicines have been synthesized with promising antiviral effects, abundant medicinal plants and herbs in nature largely remains unexplored for therapeutic applications against different virus strains. Nowadays, many scientists from different areas are looking to develop safe, inexpensive, and potent antiviral phytomedicines by evaluating and exploiting these unexplored medicinal plants (Kitazato et al., 2007; Ruchawapol et al., 2021). Further, in this chapter, the effect of different plants extracts their fractions, and active compounds against the main strains of HHV responsible for infecting humans (Figure 12.1 and Table 12.1) are explained.

## 12.4 HERPES SIMPLEX VIRUS 1 AND 2 (HSV-1, HSV-2)

HSV-1 and HSV-2 belong to α-herpesvirinae subfamily of HHV viruses. Approximately 90% of adult populations are infected by HSV-1 and -2 viruses worldwide. They have the capability to exhibit long-term latent infections in the peripheral nervous system. HSV-1 is an opportunistic virus and does not act on the host cell under normal conditions, but it starts replicating under stress conditions or conditions of fatigue and impaired immune system (Madavaraju et al., 2021; Suzich et al., 2018; Huang et al., 2011; Cliffe et al., 2015).

HSV-1 and HSV-2 are extremely infectious and responsible for numerous oral and genital diseases. HSV-1 generally infects the mouth called herpes labialis, and HSV-2 infects the anal or genital area known as herpes genitalis (Zhu and Viejo-Borbolla, 2021). HSV-1 infection spread from person to person through oral contact and can cause serious diseases like sporadic encephalitis (brain infection), herpatic keratitis (eye infection), and herpes labialis. The eye infections by HSV-1 usually heal without damaging the eye, but more severe infections can lead to cornea damage or blindness (Richards et al., 2003). It also has the ability to cause cervical cancer (Skinner et al., 1997;

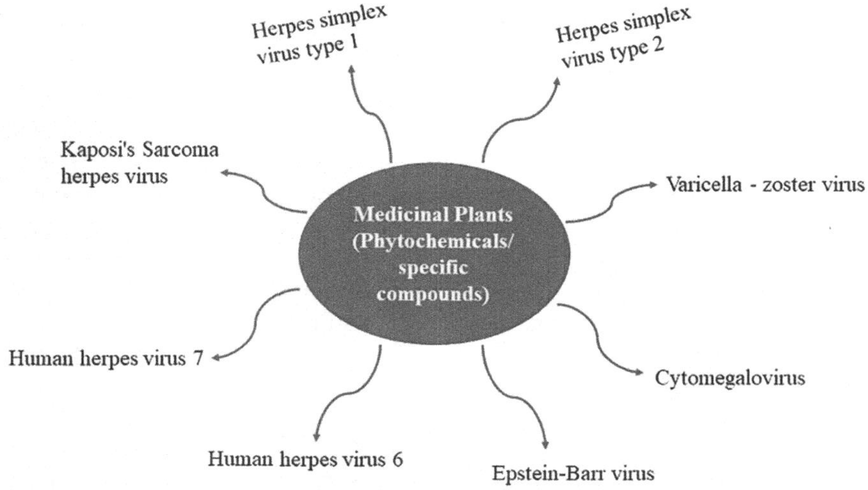


**FIGURE 12.1** Human herpes virus strains that cause infection in humans.

**TABLE 12.1**
**Medicinal Plants and Compounds Exploited for Anti-Herpetic Activities**

| Viral Strain | Plant | Extract/Compounds | References |
|---|---|---|---|
| **HSV1/2** | *Peganum harmala* (L.) | Methanolic extract (seeds) | Huleihel and Mahmoud (2017) |
| | *Eucalyptus camaldulensis* | Ethanolic extract (leaves) | Okba et al. (2017) |
| | *Eucalyptus sideroxylon* | Methanolic extract (leaves) | Roner et al. (2007) |
| | *Quillaja saponaria* | Aqueous extract | Sabouri Ghannad et al. (2014) |
| | *Glycyrrhiza glabra* | Aqueous extract (roots) | Deepika et al. (2014) |
| | *Camellia Sinensis* (Green tea) | Aqueous and methanolic extracts | Lopes et al. (2013) |
| | *Caesalpinia ferrea* | Galactomannans | Lee et al. (2015a) |
| | *Acanthopanax sciadophylloides* | Acetylated galactoglucomannan and arabinogalactan | Chrestani et al. (2009) |
| | *Mimosa scabrella* | Sulfated galactomannan | Zen et al. (1998) |
| | *Polygonatum cyrtonema* | Fructan | Hongmei et al. (2003); Jin et al. (2015) |
| | *Eucheuma gelatin* | Galactose | Sangkitporn et al. (1995) |
| **Varicella-zoster virus** | *Eucalyptus camaldulensis* | Ethanolic extract (Leaves) | Okba et al. (2017) |
| | *Clinacanthus nutans* | Extract | Tolo et al. (2006) |
| | *Carissa edulis, Prunus africana* and *Melia azedarach* | Aqueous extract | Salem and Hossain (2000) |
| | *Nigella sativa* | Black seed oil | Tsang et al. (2013) |
| **EBV/HHV-4** | *Berberis vulgaris* | Berberine | Lee et al. (2015b) |
| | *Glycyrrhiza uralensis* | Quercetin and isoliquiritigenin | Hu et al. (2016) |
| | *Hypericum japonicum* | Hyperjaponols (Aerial parts) | Banno et al. (2004) |
| | *Perilla frutescens* | Ethanolic extract (leaves) | Gruffaz et al. (2018) |
| **KSHV/HHV-8** | *Cryptotheca crypta* | Cytarabine | Long et al. (2016) |
| | *Tripterygium wilfordii* | Triptolide | Hu et al. (2018) |
| | *Hypericum japonicum* | Acylphloroglucinol-based meroterpenoid | Xie et al. (2020) |
| | ***Allium sativum*** | **Allicin** | Xie et al. (2020) |

Szpara et al., 2010) and Alzheimer's disease (Carbone et al., 2014; Harris et al., 2015; Steel and Eslick, 2015), while HSV-2 is known to cause a sexually transmitted disease in humans known as genital herpes. It is also a major contributor to genital ulcer disease and can be transmitted directly from mother to neonatal during birth, resulting in neonatal herpes (Kimberlin, 2004). HSV-1 and HSV-2 viruses can infect HIV patients and can lead to their death. HIV patients have compromised immunity which gives HSV-1 and HSV-2 viruses the opportunity to adhere and multiply within the host, suggesting that HSV viruses and HIV can coexist in the host cell (Crisci et al., 2019).

Medicinal plants and their derivatives are being explored diligently for the development of novel anti-HSV drugs. Nowadays, medicinal plants have been used worldwide, but their extensive use has been limited to the Asian countries such as India, Japan, Sri Lanka, Thailand, and China along with some African regions. However, developed nations have also started using plant-based medicinal products in their healthcare practices (Perera et al., 2021). Plant extracts and metabolites such as polysaccharides and phenols, etc. have shown anti-HSV activity and less toxicity to normal cells of the body (Khan et al., 2005; Arakawa et al., 2009; Li et al., 2018).

Polysaccharides isolated from the medicinal plants have shown promising inhibition of both HSV-1 and HSV-2. Polysaccharides such as glucose, galactose, galacturonic acid, homogalacturonan, pectic arabinogalactan, galactomannans isolated from *Prunella*, *Portulaca oleracea* L., *Basella rubra* L., *Stevia rebaudiana* and *Adenanthera pavonina*, respectively, have exhibited significant activity against HSV-1 and HSV-2 at the distinct stages of viral infection. (Liu et al., 2020). Table 12.1 presents many other plant polysaccharides which have shown antiviral activities. Ansari et al. analyzed the anti-HSV 1 effect of three medicinal plants from the *Lamiaceae* family*:* *Teucrium polium* L., *Ziziphora clinopoides* Lam., and *Salvia rhytidea* Benth. They selected these plants due to the presence of rosmarinic acid, a natural phenol found in the *Lamiaceae* family plants, which are widely used for their antioxidant, antimicrobial, antiallergic, anti-cancer, and anti-HIV-1 properties. The toxic effect of these plant extracts was analyzed on Vero cells. *T. polium* showed minimum toxicity, while *S. rhytidea* had highest toxicity and *Z. clinopoides* showed relatively average toxicity. The rosmarinic acid content was 1.8% in *polium* extract, 1.50% *Z. clinopoides*, and 1.40% *S. rhytidea*. Therefore, they concluded that not only rosmarinic acid is responsible for antiviral activities but flavonoids, tannins, essential oils, and caffeic acid derivatives can also inhibit HSV infection by blocking host cell receptors or viral surface ligands (Ansari et al., 2014). *Achyranthes aspera*, also called Apamarg or Apang, is an indigenous folk remedy. The stem, leaves, and fruits of *A. aspera* are utilized to cure asthma, boils, bronchitis, dysentery, pneumonia, skin diseases, fever, typhoid, piles, pneumonia, cough, kidney stone, and gonorrhea. The methanolic extract of *A. aspera* roots and oleanolic acid (pure compound) showed significant anti-HSV-activity. The study revealed the inactivation of HSV-1 and HSV-2 at early stage of infection with enhanced level of pro-inflammatory cytokines (IL6 and IL12) of the affected host cell, and therefore, can be a utilized as an alternative therapy to treat HSV infection (Mukherjee et al., 2013). *Phyllanthus urinaria* Linnea (Euphorbiaceae), a medicinal plant, is commonly used by Asian people to cure many illnesses. The acetone extract and excoecarianin, compound isolated from the whole plant extract of *P. urinaria*, are specific to inactivate HSV-2 infection (Cheng et al., 2011).

## 12.5 VARICELLA-ZOSTER VIRUS (VZV)

VZV is an omnipresent human virus belonging to the α-herpesvirus family and causes herpes zoster commonly called shingles and chickenpox. Like HSV, these viruses stay inactive (latent phase) in cranial nerve and dorsal root ganglia and recur later. To eradicate VZV related infections, children are vaccinated on a regular basis. However, numerous individuals continue to be infected by VZV and are found sensitive to developing shingles and its dreadful complication. VZV reactivation causes shingles, Ramsay-Hunt syndrome, ophthalmic herpes zoster, and postherpetic neuralgia which are not life-threatening but can cause pain, paralysis, chronic or permanent neuropathies, depression, and other morbid conditions (Mueller et al., 2008).

*Elaeocarpus sylvestris* (ES) is a native plant of Japan, Taiwan, and Jeju Island in Korea. It is utilized to cure many diseases such as anxiety, asthma, arthritis, nerve pain, epilepsy, migraine, liver infections, malaria, etc. Cho et al. evaluated the antiviral efficacy of the ethanolic extract of *E. sylvestris* (ESE) against VZV. ESE increased the cell viability of Vero cells infected by VZV by inhibiting the expression of glycoprotein B (responsible for VZV cell invasion) and a transcription-promoting gene, open reading frame (ORF) 62 of VZV. Furthermore, ESE reduced the phosphorylation of IκB and NF-κB in LPS-induced Raw 264.7 cells and expression of COX-2, iNOS, IL-1β, IL-6, and TNF-α. ESE showed a direct inhibitory effect on VZV and control the pain and inflammatory responses initiated by VZV infection. ESE can be utilized as a potential source of medicine to treat VZV infection (Cho et al., 2022). Bae and Song investigated the antiviral efficacy of *Lysimachia mauritiana* ethanolic extract (LME) against both laboratory and clinical strains of VZV. Antiviral activity of LME was checked on VZV-infected MRC-5 cells, and results showed strong inhibition of the viral replication in both VZV strains (laboratory and clinical) without affecting the MRC-5 cells viability. LME inhibits the expression of immediate-early 62 proteins (IE62), which is an essential viral transactivator. In addition, LME also suppresses the expression of lytic genes in the later stages of viral replication. The results showed the potent inhibitory effect of LME on VZV gene expression and replication and can be utilized as a therapeutic anti-VZV agent (Bae & Song, 2017). *Eucalyptus camaldulensis* (Ec) is believed to be a traditional medicinal plant with beneficial therapeutic properties. The 80% MeOH fraction of Ec displayed significant inhibition of VZV infection. Ec extract showed VZV inhibition at various phases, including the early stage of viral infection (viral adsorption/penetration) and also prevented viral replication after entering into the host cells (Abu-Jafar & Huleihel, 2017). Glycyrrhiza glabra, commonly known as Licorice, is generally used as traditional medicine for its anti-inflammatory, mucoprotectant, and antiviral properties. One of its constituents, named glycyrrhizin, can inhibit viral growth and may also inactivate viral particles (Shebl et al., 2012). Baba and Shigeta analyzed the anti-VZV action of glycyrrhizin (GL). Human embryonic fibroblast (HEF) cells inoculated with Zoster virus and then exposed to GL showed 50%-inhibition at 0.71 mM (ID50) against five VZV strains. GL was also found to be an effective antiviral agent against VZV replication when HEF cells were treated with GL for 24 hours before the inoculation of the virus. More than 99% of viral particles were inhibited by 2.4 mM GL concentration within 30 minutes. A synergistic impact of GL has been seen with anti-herpes medicines (acyclovir, bromovinyldeoxyuridine, adenine arabinoside, and phosphonoformate) or human beta-interferon on the replication of VZV (Baba and Shigeta, 1987).

## 12.6 HUMAN CYTOMEGALOVIRUS (HCMV)

HCMV is a universal herpes virus which infects 50%–90% of the human population by body secretions globally. It belongs to the beta-subfamily of herpes viruses, and is one of the main agents involved in infectious complications after transplantation, in individuals with HIV/AIDS and in neonates because of their weak immunity (Gandhi and Khanna, 2004). After primary infection, HCMV, like other HHVs establishes long-term latent infection inside blood monocytes and myelomonocytic precursors inside bone marrow with regular recurrence. In immunocompromised persons, HCMV infection or reactivation of the virus in latent stage causes acute disorders, such as pneumonitis, encephalitis, retinitis, hepatitis, and gastroenteritis (Azevedo et al., 2015; To et al., 2014). In order to find a potent antiviral agent for HCMV infection, To et al. screened ethanolic extract of 662 plants. Only crude ethanolic extracts (70%) of *Elaeocarpus sylvestris* leaves (ESE) displayed potential inhibitory activity against HCMV. ESE when tested on HCMV infected human foreskin fibroblasts (HFF) cells showed inhibition of immediate-early (IE) gene expression and replication without affecting the HEF viability. Therefore, ESE can be utilized as a potent candidate to develop new drug molecules to cure HCMV-linked infections (Alvarez et al., 2009). Alvarez et al. studied the *in vitro* antiviral activity of Phyllanthus orbicularis H.B.K. against HSV-1, HSV-2, and HCMV. The aqueous extract of *P. orbicularis* showed anti-HCMV activity by inhibiting viral replication at initial phases of infection (Zhen et al., 2006). *Allium sativum,* popularly known as garlic, has been reported to inhibit HCMV infection. Allitridin, a compound obtained from the *A. sativum* extraction, has shown significant effect on the HCMV replication and the expression of immediate-early genes. It displayed dose-dependent inhibition of HCMV with 4.2 μg/mL EC (50) value (selective index, SI = 16.7). Further studies have suggested that allitridin inhibits viral replication in the initial stages of viral cycle. It suppresses the expression of both immediate-early gene (iel) transcription and IEA (IE (1)72 and IE (2)86), although, IE (2)86 was strongly inhibited compared to IE (1)72. Viral DNA is also decreased in infected cells treated with allitridin, possibly due to the reduction in ie gene transcription. This study mechanistically revealed the anti-HCMV potential of allitridin with the suppression of ie gene transcription (Yukawa et al., 1996). Yukawa et al. investigated the antiviral activity of *Geum japonicum*, *Syzygium aromaticum*, *Terminalia chebula*, and *Rhus javanica* aqueous extracts. These medicinal plants inhibit the replication of HCMV and MCMV (Murine Cytomegalovirus) and may be beneficial for the prevention of CMV infection in immunodeficient individuals (Jha et al., 2016).

## 12.7 EPSTEIN-BARR VIRUS (EBV/HHV-4)

EBV belongs to the human γ-herpesvirus subfamily, which infects individuals in adolescence. EBV usually spreads through body fluids, mostly saliva. EBV causes infectious mononucleosis which is a benign acute lymphoproliferative disease. It causes various human cancers such as Hodgkin's lymphoma, Burkitt's lymphoma, gastric cancer, and nasopharyngeal carcinoma. EBV mainly infects circulating B-cells and is maintained in a latent phase. Memory B-cells are responsible for EBV recurrence and virus-linked malignancies. Cytotoxic T cells help to prevent EBV recurrence but under immunosuppressive conditions the immune system was unable to control viral replication, leading to development of malignancies (Cui & Snapper, 2021; Ben-Amor et al., 2021).

Olive leaf (*Olea europaea* L.) is a traditional herb used by Mediterranean people to treat some disorders. In this study, olive leaves extract displayed an antiviral activity against the EBV lytic cycle stimulation. The antioxidant property of the leaves helps to fight against the oxidative stress produced during viral replication which promotes disease development and progression. Therefore, olive leaf extract could be utilized as a supplement to relieve oxidative stress and as an adjuvant therapy for viral infections (Tsai et al., 2020). The whole plant extract/fractions and pure compounds of *Lindernia crustacean* exhibited significant inhibition of EBV in lytic stage of infection. Anti-EBV compounds isolated from *L. crustacea* are phytol, aloe-emodin, byzantionoside B, trans and cis-martynoside, trans and cis-isomartynoside, luteolin-7-O-β-D- glucopyranoside, and apigenin-7-O- [β-D-apiofuranosyl (1 → 6)-β-D-glucopyranoside]. EBV- containing Burkitt's lymphoma cell line (P3HR1) was utilized to evaluate the effect of *L. crustacea* on the expression of EBV lytic proteins and based on the outcome recommended its traditional use for its antiviral effects (Manzano-Santana et al., 2021). A few examples of the antiviral compounds isolated from medicinal plants are curcumin isolated from turmeric (*Curcuma longa* L.) (Imran et al., 2016), Niaziminin, a compound present in *Moringa oleifera* leaves (Perez, 2003), afromosin, and formononetin from *Wisteria brachybotrys* Sieb (Leguminosae), magnolol, honokiol, and monoterpenylmagnolol, derived from the bark of *Magnolia officinalis* Rehd. et Wils (Magnoliaceae), and oleanene-type triterpene from the rhizome of *Cochlospermun tinctorium* A. (Jiwajinda et al., 2002), and 15β-dihydroxyklaineanone extracted from the leaves of *Eurycoma longifolia* Jack (Wolz et al., 2012).

## 12.8 HUMAN HERPES VIRUS-6 (HHV-6) AND HUMAN HERPES VIRUS-7 (HHV-7)

HHV-6 and HHV-7, also known as Roseola virus, belong to the β-herpesvirus family. These are quite common strains that infect kids between the age of 6 months to 6 years. It affects only people with weak immune systems. HHV-6 infection spread through coughing and sneezing via saliva or mucus. Like the other types of human herpesvirus, HHV-6 also causes primary and latent infection inside the host cell, where it can be stimulated by anxiety or weak immunity. It causes roseola infantum (exanthema subitem), a common childhood condition having symptoms such as high fever, irritability, decreased appetite, and rashes (Goncalves et al., 2017; Naimo et al., 2021).

The role of HHV-7 is still not known for human diseases. However, it might cause a roseola infantum, acute febrile respiratory disorder, and seizures in children. HHV-7 also stays in a latent condition like HHV-6 persistent in the lymphocytes and reactivated in hosts with the weak immune system, particularly in transplant recipients when their immune system is suppressed clinically (Ruchawapol et al., 2021). Availability of plant extracts/fractions and active phytocomponents to treat HHV-6 and HHV-7 is not reported in the published literature so far. Hence, it can be concluded that no active phytocomponents are identified for HHV-6 and HHV-7, and this area can be explored for medicinal plants against these two HHVs.

## 12.9 KAPOSI'S SARCOMA HERPES VIRUS (KSHV/HHV-8)

KSHV belongs to the gamma 2-herpesvirus family and causes Kaposi sarcoma (blood vessel cancer), primary effusion lymphoma, and multicentric Castleman's disease (lymph node enlargement) in humans. KSHV can also be responsible for causing a pathological condition named KSHV inflammatory cytokine syndrome and unusual cases of post-transplant polyclonal lymphoproliferative disorders (Gruffas et al., 2018; Goncalves et al., 2017; Mekni-Toujani et al., 2022). Currently, very few effective therapies are present to prevent or manage KSHV infectivity. Thus, the development of new and alternative approaches using plant extracts and their compounds is needed. M. Mekni-Toujani et al. studied the antiviral potential of ethanolic/aqueous extracts, essential oils, and some flavonoids obtained from *Thymus capitatus* (generally called thyme). Only hesperidin, eupafolin, and vicenin (flavonoids) isolated from the aqueous extract of *T. capitatus* displayed potent anti- KSHV activity by inhibiting infectious phases of the virus inside the host cells (Xie et al., 2020). Some phytocompounds such as glycyrrhizic acid (GA), epigallocatechin-3-gallate (EGCG), and allicin inhibit KSHV in latent and lytic phases of infection and proved to be excellent anti-KSHV agents. (Cho et al., 2013). Cho et al. (2013) isolated angelicin (furocoumarin compound) from *Psoralea corylifolia* seeds. Angelicin inhibits the lytic replication of KSHV in the early phases of infection and stop its recurrence (Coen et al., 2014) Some plant-derived compounds such as resveratrol (non-flavonoid polyphenol) isolated from *Polygonum cuspidatum* and delta-9 tetrahydrocannabinol from marijuana help to inhibit the KSHV infection by targeting viral replication and transcription mechanisms (Benzekri et al., 2018).

## 12.10 CONCLUSIONS

Human herpesviruses are well known for their capability to establish long-term infection. So far, there is a lack of vaccines or drugs that can effectively eradicate the HHV infection. In this chapter, we have briefly explained the utilization of numerous medicinal plants and compounds to treat herpesvirus infections. Medicinal plants contain several bioactive compounds which act synergistically to inhibit HHV infection at different stages of viral replication. As the plant extracts and natural compounds have the potential to inhibit different types of HHV infections without showing any toxic effects, further evaluation and exploitation of various anti-herpetic plants and compounds against HHV infections in humans would be required.

## REFERENCES

Abu-Jafar, A.; Huleihel, M. Antiviral activity of Eucalyptus camaldulensis leaves ethanolic extract on herpes viruses infection. *Int J Clin Virol.* 2017, 1, 001–009. doi:10.29328/journal.ijcv.1001001.

Alvarez, A.L.; del Barrio, G.; Kourí, V.; Martínez, P.A.; Suárez, B.; Parra, F. In vitro anti-herpetic activity of an aqueous extract from the plant Phyllanthus orbicularis. *Phytomedicine.* 2009, 16, 960–966. doi: 10.1016/j.phymed.2009.03.023.

Ansari, M.; Sharififar, F.; Arabzadeh, A.M.; Mehni, F.; Mirtadzadini, M.; Iranmanesh, Z.; Nikpour, N. In vitro evaluation of anti-herpes simplex-1 activity of three standardized medicinal plants from Lamiaceae. *Anc Sci Life.* 2014, 34, 33–38. doi: 10.4103/0257-7941.150777.

Arakawa, T.; Yamasaki, H.; Ikeda, K.; Ejima, D.; Naito, T.; Koyama, A.H. Antiviral and virucidal activities of natural products. *Curr Med Chem.* 2009, 16, 2485–2497. doi: 10.2174/092986709788682065.

Azevedo, L.S.; Pierrotti, L.C.; Abdala, E.; Costa, S.F.; Strabelli, T.M.; Campos, S.V.; Ramos, J.F.; Latif, A.Z.; Litvinov, N.; Maluf, N.Z.; Pannuti, C.S. Cytomegalovirus infection in transplant recipients. *Clinics.* 2015, 70, 515–23. doi: 10.6061/clinics/2015(07)09.

Baba, M.; Shigeta, S. Antiviral activity of glycyrrhizin against varicella-zoster virus in vitro. *Antiviral Res.* 1987, 7(2), 99–107.

Bae, S.; Song, Y.J. Inhibition of varicella-zoster virus replication by an ethanol extract of Lysimachia mauritiana. *Mol Med Rep.* 2017, 15, 3847–3851. doi: 10.3892/mmr.2017.6444.

Banno, N.; Akihisa, T.; Tokuda, H.; Yasukawa, K.; Higashihara, H.; Ukiya, M.; Watanabe, K.; Kimura, Y.; Hasegawa, J.; Nishino, H. Triterpene acids from the leaves of Perilla frutescens and their anti-inflammatory and antitumor-promoting effects. *Biosci Biotechnol Biochem.* 2004, 68, 85–90. doi: 10.1271/bbb.68.85.

Ben-Amor, I.; Gargouri, B.; Attia, H.; Tlili, K.; Kallel, I.; Musarra-Pizzo, M.; Sciortino, M.T.; Pennisi, R. In vitro anti-epstein barr virus activity of Olea europaea L. Leaf extracts. *Plants (Basel).* 2021, 10, 2445. doi: 10.3390/plants10112445.

Benzekri, R.; Bouslama, L.; Papetti, A.; Hammami, M.; Smaoui, A.; Limam, F. Anti HSV-2 activity of Peganum harmala (L.) and isolation of the active compound. *Microb Pathog.* 2018, 114, 291–298. doi: 10.1016/j.micpath.2017.12.017.

Carbone, I.; Lazzarotto, T.; Ianni, M.; Porcellini, E.; Forti, P.; Masliah, E.; Gabrielli, L.; Licastro, F. Herpes virus in Alzheimer's disease: relation to progression of the disease. *Neurobiol Aging.* 2014, 35(1), 122–129.

Cheng, H.Y.; Yang, C.M.; Lin, T.C.; Lin, L.T.; Chiang, L.C.; Lin, C.C. Excoecarianin, isolated from phyllanthus urinaria linnea, inhibits herpes simplex virus type 2 infection through inactivation of viral particles. *Evid Based Complement Alternat Med.* 2011, 2011, 259103. doi: 10.1093/ecam/nep157.

Cho, H.J.; Jeong, S.G; Park, J.E.; Han, J.A.; Kang, H.R.; Lee, D.; Song, M.J. Antiviral activity of angelicin against gammaherpesviruses. *Antiviral Res.* 2013, 100, 75–83. doi: 10.1016/j.antiviral.2013.07.009.

Cho, S.M.; Park, D.W.; Lee, Y.G.; Jeong, Y.J.; Jeon, H.; Seo, Y.J.; Kim, H.; Kang, S.C. Anti-varicella zoster virus and related anti-inflammation effects of ethanolic extract of Elaeocarpus sylvestris. *J Ethnopharmacol* 2022, 287, 114951.

Chrestani, F.; Sierakowski, M.R.; de Andrade Uchoa, D.E.; Nozawa, C.; Sassaki, G.L.; Gorin, P.A.J.; Ono, L. In vitro antiherpetic and antirotaviral activities of a sulfate prepared from Mimosa scabrella galactomannan. *Int J Biol Macromol.* 2009, 45(5), 453–457.

Cliffe, A.R.; Arbuckle, J.H.; Vogel, J.L.; Geden, M.J.; Rothbart, S.B.; Cusack, C.L.; Strahl, B.D.; Kristie, T.M.; Deshmukh, M. Neuronal stress pathway mediating a histone methyl/phospho switch is required for herpes simplex virus reactivation. *Cell Host Microbe.* 2015, 18(6), 649–658.

Coen, N.; Duraffour, S.; Snoeck, R.; Andrei, G. KSHV targeted therapy: an update on inhibitors of viral lytic replication. *Viruses.* 2014, 6, 4731–4759. doi: 10.3390/v6114731.

Crisci, E.; Svanberg, C.; Ellegård, R.; Khalid, M.; Hellblom, J.; Okuyama, K.; Bhattacharya, P.; Nystrom, S.; Shankar, E.M.; Eriksson, K.; Larsson, M. HSV-2 cellular programming enables productive HIV infection in dendritic cells. *Front Immunol.* 2019, 10, 2889.

Cui, X.; Snapper, C.M. Epstein barr virus: development of vaccines and immune cell therapy for EBV-associated diseases. *Front Immunol.* 2021, 12, 734471.

Deepika, G.; Durgadevi, H.; Narayan, R.; Sudanthira, M.; Manickan, E. Anti-herpes simplex viruses activity of Camellia sinensis, member of the family Theaceae (green tea). *BMC Infect Dis.* 2014, 14, P47. doi: 10.1186/1471-2334-14-S3-P47.

Dong, L.; Zhongjun, L.; Shuai, Z. The pathogenesis and advances in human herpesvirus associated central nervous system diseases. *Clin Surg.* 2021, 6, 3054.

Gandhi, M.K.; Khanna, R. Human cytomegalovirus: clinical aspects, immune regulation, and emerging treatments. *Lancet Infect Dis.* 2004, 4,725–738. doi: 10.1016/S1473-3099(04)01202-2.

Goncalves, P.H.; Ziegelbauer, J.; Uldrick, T.S.; Yarchoan, R. Kaposi sarcoma herpesvirus-associated cancers and related diseases. *Curr Opin HIV AIDS.* 2017, 12, 47.

Grinde, B. Herpesviruses: latency and reactivation - viral strategies and host response. *J Oral Microbiol.* 2013, 5. doi: 10.3402/jom.v5i0.22766.

Gruffaz, M.; Zhou, S,; Vasan, K,; Rushing, T.; Michael, Q.L.; Lu, C.; Jung, J.U.; Gao, S.J. Repurposing Cytarabine for Treating Primary effusion lymphoma by targeting Kaposi's Sarcoma-associated herpesvirus latent and lytic replications. M. *Bio.* 2018, 9(3), 10–128. doi: 10.1128/mbio.00756-18.

Gruffaz, M.; Zhou, S.; Vasan, K.; Rushing, T.; Michael, Q.L.; Lu, C.; Jung, J.U.; Gao, S.-J. Repurposing cytarabine for treating primary effusion lymphoma by targeting Kaposi's sarcoma-associated herpesvirus latent and lytic replications. *mBio.* 2018, 9, e00756–18. doi: 10.1128/mBio.00756-18.

Heming, J.D.; Conway, J.F.; Homa, F.L. Herpesvirus capsid assembly and DNA packaging. *Adv Anat Embryol Cell Biol.* 2017, 223, 119–142. doi: 10.1007/978-3-319-53168-7_6.

Hongmei, G.U.; Yiwen, M.; Qiang, P.U. Polysaccharide from Polygonatum cyrtonema Hua against herpes simplex virus in virto. *Chin J Appl Environ Biol (China).* 2003, 9, 21–23.

Huang, W.; Xie, P.; Xu, M.; Li, P.; Zao, G. The influence of stress factors on the reactivation of latent herpes simplex virus type 1 in infected mice. *Cell Bioch Biophys.* 2011, 61(1), 115–122. doi: 10.1007/s12013-011-9167-7.

Hu, L.; Zhang, Y.; Zhu, H.; Liu, J.; Li, H.; Li, X.-N.; Sun, W.; Zeng, J.; Xue, Y.; Zhang, Y. Filicinic acid based meroterpenoids with anti-epstein-barr virus activities from Hypericum japonicum. *Org Lett.* 2016, 18, 2272–2275. doi: 10.1021/acs.orglett.6b00906.

Hu, L.; Liu, Y.; Wang, Y.; Wang, Z.; Huang, J.; Xue, Y.; Liu, J.; Liu, Z.; Chen, Y.; Zhang, Y. Discovery of acylphloroglucinol-based meroterpenoid enantiomers as KSHV inhibitors from Hypericum japonicum. *RSC Adv.* 2018, 8, 24101–24109. doi: 10.1039/c8ra04073g.

Imran, I.; Imran, I.; Ashraf, M.; Javeed, A.; Munir, N.; Bashir, R. In vitro evaluation of antiviral activity of leaf extracts of Azadirachta indica, Moringa oleifera, and Morus alba against the foot and mouth disease virus on BHK-21 cell line. *ScienceAsia.* 2016, 42, 392–396. doi: 10.2306/scienceasia1513-1874.2016.42.392.

Jha, H.C.; Pei, Y.; Robertson, E.S. Epstein-barr virus: diseases linked to infection and transformation. *Front Microbiol.* 2016, 7, 1602. doi: 10.3389/fmicb.2016.01602.

Jin, F.; Zhuo, C.; He, Z.; Wang, H.; Liu, W.; Zhang, R.; Wang, Y. Anti-herpes simplex virus activity of polysaccharides from Eucheuma gelatinae. *World J Microbiol Biotechnol.* 2015, 31, 453–460.

Jiwajinda, S.; Santisopasri, V.; Murakami, A.; Sugiyama, H.; Gasquet, M.; Riad, E.; Balansard, G.; Ohigashi, H. In vitro anti-tumor promoting and anti-parasitic activities of the quassinoids from Eurycoma longifolia, a medicinal plant in Southeast Asia. *J Ethnopharmacol.* 2002, 82, 55–58. doi: 10.1016/s0378-8741(02)00160-5.

Khan, M.T.H.; Ather, A.; Thompson, K.D.; Gambari, R. Extracts and molecules from medicinal plants against herpes simplex viruses. *Antiviral Res.* 2005, 67, 107.

Kim, I.J.; Chouljenko, V.N.; Walker, J.D.; Kousoulas, K.G. Herpes simplex virus 1 glycoprotein M and the membrane-associated protein UL11 are required for virus- induced cell fusion and efficient virus entry. *J Virol.* 2013, 87, 8029–8037. doi: 10.1128/JVI.01181-13.

Kimberlin, D.W. Neonatal herpes simplex infection. *Clin Microbiol Rev.* 2004, 17, 1–13. doi: 10.1128/CMR.17.1.1-13.2004.

Kitazato, K.; Wang, Y.; Kobayashi, N. Viral infectious disease and natural products with antiviral activity. *Drug Discov Ther.* 2007, 1, 14–22.

Lee, J.B.; Tanikawa, T.; Hayashi, K.; Asagi, M.; Kasahara, Y.; Hayashi, T. Characterization and biological effects of two polysaccharides isolated from Acanthopanax sciadophylloides. *Carbohydr Polym.* 2015a, 116, 159–166.

Lee, M.; Son, M.; Ryu, E.; Shin, Y.S.; Kim, J.G.; Kang, B.W.; Cho, H.; Kang, H. Quercetin-induced apoptosis prevents EBV infection. *Oncotarget.* 2015b, 6, 12603–12624. doi: 10.18632/oncotarget.3687.

Li, G.J.; Hu, H.Y.; Wang, Y.X.; Du, Y.J.; Zhang, Y.Y.; Chang, C.H.; Lin, Y.Z. In vitro anti-HSV-1 activity of extracellular polysaccharides from Paecilomyces lilacinuson isolated from Mangrove. *China J Modern Med (China).* 2018, 28, 19–23.

Liu, Z.H.; Niu, F.J.; Xie, Y.X.; Xie, S.M.; Liu, Y.N.; Yang, Y.Y.; Zhou, C.Z.; Wan, X.H. A review: natural polysaccharides from medicinal plants and microorganisms and their anti-herpetic mechanism. *Biomed Pharmacother.* 2020, 129, 110469.

Long, C.; Guo, W.; Zhou, H.; Wang, J.; Wang, H.; Sun, X. Triptolide decreases expression of latency-associated nuclear antigen 1 and reduces viral titers in Kaposi's sarcoma-associated and herpesvirus-related primary effusion lymphoma cells. *Int J Oncol.* 2016, 48, 1519–1530. doi: 10.3892/ijo.2016.3353.

Lopes, N.; Faccin-Galhardi, L.C.; Espada, S.F.; Pacheco, A.C.; Ricardo, N.M.P.S.; Linhares, R.E.C.; Nozawa, C. Sulfated polysaccharide of Caesalpinia ferrea inhibits herpes simplex virus and poliovirus. *Int J Biol Macromol.* 2013, 60, 93–99.

Madavaraju, K.; Koganti, R.; Volety, I.; Yadavalli, T.; Shukla, D. Herpes simplex virus cell entry mechanisms: an update. *Front Cell Infect Microbiol.* 2021, 10, 617578. doi: 10.3389/fcimb.2020.617578.

Manzano-Santana, P.I.; Tivillin, J.P.P.; Choez-Guaranda, I.A.; Lucas, A.D.B.; Katherine, A.; Orellana-Manzano, A.K.; Rastrelli, L. Potential bioactive compounds of medicinal plants against new Coronavirus (SARS-CoV-2): a review. *Bionatural Latin Am J Biotech Life Sci.* 2021, 6, 1653–1658.

Mekni-Toujani, M.; Mousavizadeh, L.; Gallo, A.; Ghram, A. Thymus capitatus flavonoids inhibit infection of Kaposi's sarcoma-associated herpesvirus. *FEBS Open Bio.* 2022,12, 1166–1177. doi: 10.1002/2211-5463.13407.

Mueller, N.H.; Gilden, D.H.; Cohrs, R.J.; Mahalingam, R.; Nagel, M.A. Varicella zoster virus infection: clinical features, molecular pathogenesis of disease, and latency. *Neurol Clin.* 2008, 26, 675–697, viii. doi: 10.1016/j.ncl.2008.03.011.

Mukherjee, H.; Ojha, D.; Bag, P.; Chandel, H.S.; Bhattacharyya, S.; Chatterjee, T.K.; Mukherjee, P.K.; Chakraborti, S.; Chattopadhyay, D. Anti-herpes virus activities of Achyranthes aspera: an indian ethnomedicine, and its triterpene acid. *Microbiol Res.* 2013, 168, 238–244. doi: 10.1016/j.micres.2012.11.002.

Naimo, E.; Zischke, J.; Schulz, T.F. Recent advances in developing treatments of Kaposi's sarcoma herpesvirus-related diseases. *Viruses.* 2021, 13, 1797.

Okba, M.M.; El Gedaily, R.A.; Ashour, R.M. UPLC-PDA-ESI-qTOF-MS profiling and potent anti-HSV-II activity of Eucalyptus sideroxylon leaves. *J Chromatogr B Analyt Technol Biomed Life Sci.* 2017, 1068–1069, 335–342.

Perera, W.P.R.T.; Liyanage, J.A.; Dissanayake, K.G.C.; Gunathilaka, H.; Weerakoon, W.M.T..D.N.; Wanigasekara, D.N.; Fernando, W.S.K.; Rajapaksha, R.M.H.; Liyanage, R.P.; Perera, B.T. Antiviral potential of selected medicinal herbs and their isolated natural products. *Biomed Res Int.* 2021, 2021, 7872406. doi: 10.1155/2021/7872406.

Perez, R.M. Antiviral activity of compounds isolated from plants. *Pharm Biol.* 2003, 41, 107–157. doi: 10.1076/phbi.41.2.107.14240.

Richards, C.M.; Case, R.; Hirst, T.R.; Hill, T.J.; Williams, N.A. Protection against recurrent ocular herpes simplex virus type 1 disease after therapeutic vaccination of latently infected mice. *J Virol.* 2003, 77(12), 6692–6699.

Roner, M.R.; Sprayberry, J.; Spinks, M.; Dhanji, S. Antiviral activity obtained from aqueous extracts of the Chilean soapbark tree (Quillaja saponaria Molina). *J Gen Virol.* 2007, 88, 275–285. doi: 10.1099/vir.0.82321-0.

Ruchawapol, C.; Yuan, M.; Wang, S.M.; Fu, W.W.; Xu, H.X. Natural products and their derivatives against human herpesvirus infection. *Molecules.* 2021, 26, 6290. doi: 10.3390/molecules26206290.

Sabouri Ghannad, M.; Mohammadi, A.; Safiallahy, S.; Faradmal, J.; Azizi, M.; Ahmadvand, Z. The effect of aqueous extract of Glycyrrhiza glabra on herpes simplex virus 1. *Jundishapur J Microbiol.* 2014, 7, e11616. doi: 10.5812/jjm.11616.

Salem, M.L.; Hossain, M.S. Protective effect of black seed oil from Nigella sativa against murine cytomegalovirus infection. *Int J Immunopharmacol.* 2000, 22, 729–740. doi: 10.1016/s0192-0561(00)00036-9.

Sangkitporn, S.; Chaiwat, S.; Balachandra, K.; Na-Ayudhaya, TD.; Bunjob, M.; Jayavasu, C. Treatment of herpes zoster with Clinacanthus nutans (bi phaya yaw) extract. *J Med Assoc Thai.* 1995, 78, 624–627.

Sharma, V.; Mobeen, F.; Prakash, T. Comparative genomics of herpesviridae family to look for potential signatures of human infecting strains. *Int J Genomics.* 2016. Doi:10.1155/2016/9543274.

Shebl, R.I.; Amin, M.A.; Emad-Eldin, A.; Bin Dajem, S.M.; Mostafa, A.S.; Ibrahim, E.H.; Mohamed, A.F. Antiviral activity of liquorice powder extract against varicella zoster virus isolated from Egyptian patients. *Chang Gung Med J.* 2012, 35, 231–239. doi: 10.4103/2319-4170.106149.

Skinner, G.R.B.; Turyk, M.E.; Benson, C.A.; Wilbanks, G.D.; Heseltine, P.; Galpin, J.; Kaufmann, R.; Goldberg, L.; Hartley, C.E.; Buchan, A. The efficacy and safety of Skinner herpes simplex vaccine towards modulation of herpes genitalis; report of a prospective double-blind placebo-controlled trial. *Med Microbiol Immunol.* 1997, 186(1), 31–36.

Steel, A.J.; Eslick, G.D. Herpes viruses increase the risk of Alzheimer's disease: a meta-analysis. *J Alzheimer's Dis.* 2015, 47(2), 351–364.

Suzich, J.B.; Cliffe, A.R. Strength in diversity: understanding the pathways to herpes simplex virus reactivation. *Virology.* 2018, 522, 81–91. doi: 10.1016/j.virol.2018.07.011.

Szpara, M.L.; Parsons, L.; Enquist, L.W. Sequence variability in clinical and laboratory isolates of herpes simplex virus 1 reveals new mutations. *J Virol.* 2010, 84(10), 5303–5313.

To, K.P.; Kang, S.C.; Song, Y.J. The extract of Elaeocarpus sylvestris inhibits human cytomegalovirus immediate early gene expression and replication in vitro. *Mol Med Rep.* 2014, 9, 744–748. doi: 10.3892/mmr.2013.1824.

Tolo, F.M.; Rukunga, G.M.; Muli, F.W.; Njagi, E.N.; Njue, W.; Kumon, K.; Mungai, G.M.; Muthaura, C.N.; Muli, J.M.; Keter, L.K.; Oishi, E.; Kofi-Tsekpo, M.W. Anti-viral activity of the extracts of a Kenyan medicinal plant Carissa edulis against herpes simplex virus. *J Ethnopharmacol.* 2006, 104, 92–99.

Traylen, C.M.; Patel, H.R.; Fondaw, W.; Mahatme, S.; Williams, J.F.; Walker, L.R.; Dyson, O.F.; Arce, S.; Akula, S.M. Virus reactivation: a panoramic view in human infections. *Future Virol.* 2011, 6, 451–463. doi: 10.2217/fvl.11.21.

Tsai, Y.C.; Hohmann, J.; El-Shazly, M.; Chang, L.K.; Danko, B.; Kusz, N.; Hsieh, C.T.; Hunyadi, A.; Chang, F.R. Bioactive constituents of Lindernia crustacea and its anti-EBV effect via Rta expression inhibition in the viral lytic cycle. *J Ethnopharmacol.* 2020, 250, 112493. doi: 10.1016/j.jep.2019.112493.

Tsang, C.M.; Cheung, Y.C.; Lui, V.W.; Yip, Y.L.; Zhang, G.; Lin, V.W.; Cheung, K.C.; Feng, Y.; Tsao, S.W. Berberine suppresses tumorigenicity and growth of nasopharyngeal carcinom, a cells by inhibiting STAT3 activation induced by tumor associated fibroblasts. *BMC Cancer.* 2013, 13, 619. doi: 10.1186/1471-2407-13-619.

Whitley, R.J. Herpesviruses. In: Baron S, editor. *Medical Microbiology*. 4th edition. Galveston (TX): University of Texas Medical Branch at Galveston; 1996. Chapter 68. Available from: https://www.ncbi.nlm.nih.gov/books/NBK8157/

Wolz, M.M.; Sciallis, G.F.; Pittelkow, M.R. Human herpesviruses 6, 7, and 8 from a dermatologic perspective. *Mayo Clin Proc*. 2012, 87, 1004–1014. doi: 10.1016/j.mayocp.2012.04.010.

Xie, Y.; Chen, Y.; Guo, Y.; Huang, Y.; Zhu, B. Allicin and glycyrrhizic acid display antiviral activity against latent and lytic kaposi sarcoma-associated herpesvirus. *Infect Microbes Dis*. 2020, 2, 30–34. doi: 10.1097/IM9.0000000000000016.

Yukawa, T.A.; Kurokawa, M.; Sato, H.; Yoshida, Y.; Kageyama, S.; Hasegawa, T.; Namba, T.; Imakita, M.; Hozumi, T.; Shiraki, K. Prophylactic treatment of cytomegalovirus infection with traditional herbs. *Antiviral Res*. 1996, 32, 63–70. doi: 10.1016/0166-3542(95)00978-7.

Zen, Q.H.; Yu, X.L.; Li, X.; Dong, D.Q. A study on the efficacy of Polygonatum polysaccharide eye drops in the treatment of experimental rabbits with herpes simplex keratitis. *Chin J Tradit Chin Ophthalmol (China)*. 1998, 8, 7–10.

Zhen, H.; Fang, F.; Ye, D.Y.; Shu, S.N.; Zhou, Y.F.; Dong, Y.S.; Nie, X.C.; Li, G. Experimental study on the action of allitridin against human cytomegalovirus in vitro: Inhibitory effects on immediate-early genes. *Antiviral Res*. 2006, 72, 68–74

Zhu, S.; Viejo-Borbolla, A. Pathogenesis and virulence of herpes simplex virus. *Virulence*. 2021, 12, 2670–2702. doi: 10.1080/21505594.2021.1982373.

# 13 Rotavirus and Medicinal Plants/Herbs

*Yadveer Kaur and Gurpreet Kaur*
Thapar Institute of Engineering and Technology

*Gursharan Kaur*
University Institute of Biotechnology, Chandigarh University

## 13.1 INTRODUCTION

Viruses, bacteria, fungus, and helminths are the most common human pathogens. Viruses are metabolically inactive until they find a host cell to become activated. They principally includes proteins and nucleic acid. A unique covering made up of proteins called capsid gives them their distinctive shape (Andersson 2010). More than two-thirds of novel human diseases are caused by viruses. Yellow fever virus was the first 'filterable agent' discovered in humans in 1901 (Levine and Enquist 2007). Presently, there are 219 different viral species that can infect humans (Woolhouse et al. 2012). The virus families are classified based on physiochemical properties, structure of genome, size of virion and/or genome, capsid morphology, envelope, and molecular processes. Viruses implicated in human disease are included under 26 families on the basis of the type of genome, namely dsDNA, ssDNA(−), ssRNA(−), ssRNA(+), and partially ssDNA. Understanding virus taxonomy can lead to key generalizations about how to prevent and cure viral infection, as well as insights into how viruses spread and evolve (Siegel 2018).

Virus transmission to humans can occur through multiple routes, such as respiratory route, fecal-oral route, direct contact, sexual, urine-associated, parenteral route, vertical route, arthropod-borne route, rodent-associated transmission, bat-associated transmission, monkey-associated transmission, and other zoonotic associations. The virus multiplication is due to various sequential processes: virus adhesion to the cell membrane, adsorption and administration to the cell, replication and expression of viral genome using host machinery, and finally, the release of virus particles from the cell. To gain access to the cells, viruses need cell surface protein receptors such as claudin-1, occludin, and tetraspanins (Zambon 2001). Furthermore, RNA viruses need reverse transcriptase and integrase enzymes to survive in host cell. Former enzyme is required for transcribing viral RNA to DNA at the time of replication, whereas the latter one is vital to integrate viral genome to host DNA and for uncoating of virus-core proteins. Therefore, for effective infection, virus needs various enzymatic and non-enzymatic proteins (Appel et al. 2006).

Ruth Bishop saw the virus for the first time in a duodenal biopsy of a baby with acute gastroenteritis in 1973, marking the beginning of the virus's discovery. It belongs to the *Reoviridae* family. Other representative members of this family are Colorado tick fever virus, human reovirus, Kemerovo virus. These viruses are naked with icosahedral capsid morphology having double stranded ribonucleic acid (dsRNA). The genome contains 11 gene segments encoding five non-structural and six structural proteins (Christensen 1989). Rotavirus (RV) has been found to infect a wide range of mammals, including humans. It is the leading cause of serious gastroenteritis in babies and small children around the world. They contract the virus through direct fecal-oral route or fecally contaminated food and water, and respiratory droplets (Payne et al. 2011). It has been

 DOI: 10.1201/9781003329169-16

reported that RV infection effect the lungs and large intestine. RV-mediated infection in the large intestine damage may result in substantial release of antibodies and cytokines to bronchoalveolar lavage fluid (BALF) and intestinal mucus. Diarrhea remains the second major reason for death under the age of five around the world. In developing nations, rotavirus alone causes two million hospitalizations and 500 deaths each year (Atherly et al. 2009). According to several reports, it can either be a single cause of diarrhea or can also co-infect along with other viruses such as astrovirus (Liu et al. 2006). Two licensed rotavirus vaccines (pentavalent Rotateq [RV5]; Merck, Whitehouse Station, NJ, and monovalent Rotarix [RV1]; GlaxoSmithKline, Brentwood, UK) have been introduced to manage rotavirus-mediated diarrheal illness worldwide (Tate et al. 2013). Vaccine efficacy, on the other hand, remains a significant problem in low-income areas, while illness burden is highest (Babji and Kang 2012). Recently, studies have reported that human rotavirus (HRVs) bind human histo-blood group antigens (HBGAs) during initial infection.

As an alternative to conventional treatments, traditional methods offer various benefits. Naturally-derived products like dietary supplements, nutraceuticals, and plant-derived extracts of specific plant organs namely roots, stem, bark, flowers, fruits, and seeds have a wide range of applications in treating common to rare infectious and non-infectious ailments (Rates 2001). The World Health Organization (WHO) estimates that 80% of the population in underdeveloped nations relies on herbal medicines to survive. Plant-derived substances, namely salicylates (willow bark), digitalis (from foxglove), ergotamine (from contaminated rye), and quinine (from cinchona) have been used as sources of alternate medicines. Various factors influence the process of selecting a plant for medicinal purposes such as tradition, efficacy, abundance, accessibility, doctrine of signatures, as well as taxonomic affiliation (Phumthum et al. 2019). Null model given by a statistical tool states a linear relationship between species diversity and the number of species that are useful for medicinal purpose (Bennett and Husby 2008). It is interesting to note that some plant families contain more medicinally important species than would be predicted by the null hypothesis. A study revealed that *Asteraceae*, *Apiaceae*, *Ericaceae*, *Rosaceae* and *Ranunculaceae* included more numbers of medicinal species as compared to their species richness in North America (Moerman 1996). In Italy, families such as *Orchidaceae*, *Caryophyllaceae*, *Poaceae*, and *Leguminosae* were reported to have fewer medicinally important plant families than expected based on their total number of species. While *Lamiaceae*, *Rosaceae*, and *Malvaceae* included a disproportionately large number of medicinal species (Weckerle et al. 2011), *Leguminosae*, *Lamiaceae*, *Euphorbiaceae*, *Apocynaceae*, *Malvaceae*, *Apiaceae*, and *Ranunculaceae* have been recently reported as the most important plant families in terms of medicinal plants (Kew 2017). Medicinal plants are the sources of many chemical drugs used for antimalarial and anticancer therapies. Botanical extracts contribute to one-third of clinical medications. These extracts might be the secondary metabolites chemicals produced by plants. These compounds are not involved in the growth and development of a plant. The main categories of secondary metabolites are alkaloids, phenolics, and terpenoids.

Due to developing increased resistance of microorganisms (bacteria, viruses, and parasites), medicinal plants have been re-explored as an alternative to conventional antimicrobial therapies. Various plant species have been shown to be active against a wide range of bacterial strains. For instance, *Helicobacter pylori* is bacterium responsible for peptic and duodenal ulcers. Plant products including quercetin, apigenin, carotenoid-rich algae, tea product, garlic extract, polyphenol from apple peel, and fingerroot extract have been reported to be effective against Anti-H. *pylori*-induced gastric inflammation (Wang 2014). Komape et al. (2017) reported compounds extracted from plant species, such as *C. lemon*, *C. heroroense*, and *A. dimidiata*, conferring antioxidant activities against free radicals generated due to diseased condition. Multidrug-resistant (MDR) bacteria has become a serious risk to the public health. Recently, Voukeng et al. (2017) reported that *Euphorbia prostrata*, commonly known as prostrate spurge, has a potential to cure patients infected with MDR bacteria.

Currently, the available interferon and vaccine therapies are used to manage various viral infections. Due to the high rates of mutation and recurrence, current therapeutic approaches are insufficient to treat patients. Medicinal plants have been reported to possess the therapeutic potential in response to various viral diseases, such as influenza, human immunodeficiency virus (HIV), herpes simplex virus (HSV), hepatitis, and coxsackievirus infections (Naithani et al. 2008). For instance, extract from the herb *Astragalus* has been found to show antiviral activity against different *Flaviviridae* viruses. King (2008) filed a patent for methods and compositions for therapeutically treating viral diseases. They reported a naturally occurring substance conferring Tubercin and/or SSM activity against viral diseases. Table 13.1 describes the antiviral properties of plant-derived compounds.

## 13.2 DIFFERENT APPROACHES FOR MANAGEMENT OF ROTAVIRUS INFECTION

Among the eight known species (A, B, C, D, E, F, G, and H) of RV, humans are reported to be affected by species A, B, and C. Species A is responsible for 90% of diarrheal infections occurring worldwide (Rodrigo et al. 2010). Brief overview of RV infection resulting in diarrhea has been depicted in Figure 13.1. RV infects children under the age of 5. It is evident that boys are twice as likely to be hospitalized as compared to the girls (Rheingans et al. 2012). Cool and dry seasons favor the RV infection. Diarrhea from rotavirus infection is caused by climate variability that spikes in the winter. RV infection is moderately reported in temperate regions like United States, Japan, northern Asian regions, Australia, and Europe (Hossain et al. 2016).

Most of the patients recover within 8 days of RV infection. Still more severe complications result in 2–4 million hospitalizations per year globally. Bishop's proof of the existence of a protective immunity to the virus paved the way for early attempts to develop vaccines. These vaccines are currently approved for use in more than 100 nations. They are included in the standard childhood immunization programs of more than 47 nations on the recommendation of World Health Organization. The introduction of RV vaccines has reduced the severity and incidence of this disease worldwide. Apart from vaccination, researchers have investigated various antiviral drugs for developing effective antirotaviral therapy. Viral infection is a multistep process occurring in a cell. The synthetic drugs development is a laborious work focusing on target identification and screening, optimization, clinical trials, and the drug registration. Since the first drug idoxuridine in 1963 year, various drugs have been developed. Some specific antiviral drugs for specific virus has been listed in Table 13.2.

Moreover, various antirotaviral drugs have been investigated till date. Frazia et al. (2013) studied the antirotaviral activity of the thiazolide. Nitazoxanide has been shown to reduce the rotavirus diarrhea. They revealed that this class of drugs were able to inhibit the replication of simian A/SA11-G3P[2] and human Wa-G1P[8] in various cells with 50% effective concentrations ($EC_{50}$s) ranging from 0.3 to 2 µg/mL, and 50% cytotoxic concentrations ($CC_{50}$s) higher than 50 µg/mL. This drug reduces the rotavirus dsRNA formation by altering the viral morphogenesis. A widely used anticancer drug, namely gemcitabine has been shown to possess potent antirotaviral properties. This drug inhibits pyrimidine biosynthesis in cells (Chen et al. 2020). Methyl-β-cyclodextrin, a cholesterol-sequestering agent, has potent antirotaviral activity. This drug inhibits RV receptor-mediated endocytosis (Dou et al. 2018). Benzekri et al. (2018) revealed that drebrin interacts with the VP4 and VP5 protein and inhibits RV entry into cells. Furthermore, Rapamycin has been shown to sustain rotavirus infection by activating autophagy cascade in the infected cells (Yin et al. 2017). Various synthetic drugs namely ribavirin, cimetidine, famotidine, dipyridamole, nifedipine, and isoprinosine have been also reported to manage the retrovirus-induced gastroenteritis (Gandhi et al. 2017).

Specifically, Rotarix (GSK) and Rotateq (Merck) are commercially available and WHO recommended vaccines (Patel et al. 2011). Rotarix has been found to be derived from a single common human strain of rotavirus, serotype G1 [P8], which was isolated from Cincinnati. This strain has

**TABLE 13.1**
**Effect of Medicinal Plant-Derived Products on Various Viruses**

| S.No. | Plant | Source of Extraction | Compound Isolated | Virus | Mode of Action | References |
|---|---|---|---|---|---|---|
| 1. | *Piper nigrum* L. | Fruits | Piperamides | Coxsackie virus type B3 | Cytopathic effect inhibition assay | Mair et al. (2016) |
| 2. | *Artocarpus integrifolia* | Bark | Dihydromorin Cynomacurin, Isoartocarpin, Cyloartocarpin, Artocarpesin, Artocarpetin, Artocarpanone | (SA-11) and human (HCR3) rotaviruses | Anti-rotavirus activity | Ben-Shabat et al. (2020) |
| 3. | *Tinospora cordifolia* | Leaf extract | Cardiofolioside B, tembetarine and berberine | Chikungunya | Alcoholic leaves extract significantly decreases intracellular reactive oxygen species (ROS) | Sharma et al. (2019) |
| 4. | *Withania somnifera* | Leaves | Withanone | Coronavirus | Blockage of SARS-CoV–2 entry and also its subsequent infection by interrupting electrostatic interactions between the RBD and ACE2 | Balkrishna et al. (2021) |
| 5. | *Adathoda vasica* | Leaf extract | Oseltamivir and Amantadine | Influenza A (H1N1) | Selective suppression for antiviral activity | El-Saadony et al. (2021) |
| 6. | *Aspalathus linearis* | Organic red bush leaf extract | Aspalathin and Nothofagin | Rotavirus<br>HIV-1 | Anti-rotavirus activity<br>Intervene with the connecting of HIV-1 to MT-4 cells | Hong et al. (2014) |
| 7. | *Acacia nilotica* | Pods | Water and methanol extracts | HIV-1 | Suppress HIV-1 replication | El-Saadony et al. (2021) |
| 8. | *Achillea fragrantissima* | Dried plant material | | Poliomyelitis-1 virus (POLIO) | Cytopathogenic effect (CPE) and was expressed as reduction factor (Rf) of the viral titer | Ben-Shabat et al. (2020) |
| 9. | *Azardirachta indica* | Dried bark of neem plant | Aqueous extracts (NBE) | Herpes simplex virus (HSV-1) | Entry inhibitor against HSV-1 infection into natural target cells<br>Locked the viral spread during cell-to-cell fusion assay and polykaryocytes formation | Álvarez et al. (2020) |

*(Continued)*

**TABLE 13.1 (*Continued*)**
**Effect of Medicinal Plant-Derived Products on Various Viruses**

| S.No. | Plant | Source of Extraction | Compound Isolated | Virus | Mode of Action | References |
|---|---|---|---|---|---|---|
| 10. | *Andrographis paniculata* | Plant's tissues | 14-deoxy-11,12-dehydroandrographolide | Influenza A virus, HSV-1, and HIV | Inhibits the influenza A virus, HSV-1, and HIV replication<br>Downregulate CXCR4 and CCR | Adiguna et al. (2021) |
| 11. | *Camellia sinensis* | Leaves and leaf buds | Epigallocathechin-3-gallate | Hepatitis B virus | Reduction in HBV genome and HBeAg from the replicon.\<br>Inhibits the reactive oxygen species (ROS) | Álvarez et al. (2020) |
| 12. | *Scutellaria baicalensis Georgi* | Hairy root<br>Aerial part | Baicalein<br>Isoscutellarein | Influenza virus | To inhibit the influenza virus | Adhikari et al. (2021) |
| 13. | *Kaempferia parviflora Wall. ex Baker* | Rhizome | 5,7-Dimethoxyflavone,<br>5-Hydroxy-7-methoxy flavone,<br>Tetramethyl lueteonin,<br>Tri-methyl apigenin | Avian influenza virus (H5N1) | Upregulation of TNF-α and IFN-β mRNA expressions, suggesting their roles in the inhibition of H5N1 virus replication.\ | Sornpet et al. (2017) |
| 14. | *Capparis spinosa* | Flower buds | Quercetin and kaempferol glycosides | Herpes simplex virus (Type HSV-2) | Inhibit the replication of HSV-2 and modulate the induction kinetics of IL-12, TNF-alpha IFN-gamma | Arena et al. (2008) |
| 15. | *Cassine xylocarpa* | Stem | Pentacyclic lupane-type triterpenoids | HIV | Inhibitory effects of human immunodeficiency virus type 1 replication with $IC_{50}$ values in the micromolar range | Ben-Shabat et al. (2020) |
| 16. | *Curcuma longa* L. | Rhizome | Curcumin(diferuloylmethane) | H5N1 | Direct effect on viral particle infectivity that was reflected by the inhibition of hemagglutination | Ben-Shabat et al. (2020) |
| 17. | *Cistus incanus* | Leaves | CYSTUS052 | H7N7 Avian influenza virus, H1N1, HRV14 | Showed antiviral activity | Musarra-Pizzo et al. (2021) |
| 18. | *Lagerstroemia speciosa* L. | Air dried leaves | Orobol 7-O-d-glucoside (O7G)<br>Quercetin-7-glucoside (Q7G) | Human rhinovirus 2, Human rhinovirus A, Human rhinovirus B | Anti-HRV activity toward HRV of groups A and B | Musarra-Pizzo et al. (2021) |

*(Continued)*

**TABLE 13.1 (*Continued*)**
**Effect of Medicinal Plant-Derived Products on Various Viruses**

| S.No. | Plant | Source of Extraction | Compound Isolated | Virus | Mode of Action | References |
|---|---|---|---|---|---|---|
| 19. | *Ocimum basilicum* | Leaves | Ursolic acid | Coxsackievirus B1, Enterovirus 71,HSV-1 | Showed the strongest activity against HSV-1 | Ali et al. (2021) |
| 20. | *Woodfordia fruticosa* | Flower | Gallic acid | Enterovirus 71 | Strongest anti-EV71 activity | Ali et al. (2021) |
| 21. | *Raoulia australis* | Crushed after drying | Raoulic acid | Human rhinovirus 2, 3, Coxsackievirus B3, B4, Enterovirus 71 | Broad spectrum antiviral activity | Albulescu et al. (2017) |
| 22. | *Sylibum marianum* | Extracted from milk thistle | Silymarin | Enterovirus 71 | Exert direct extracellular virucidal effects against EV-A71 at 50% inhibitory concentration ($IC_{50}$) | Lalani et al. (2020) |
| 23. | *Salvia miltiorrhiza* | Isolated from the ethyl acetate soluble fraction of the methanolic extract | Rosmarinic acid | Enterovirus A71 | Targets viral structures and inhibits viral infection and replication process | Musarra-Pizzo et al. (2021) |
| 24. | *Ornithogalum saundersiae* | Bulbs | Orsaponin (OSW-1) | Enterovirus 71, Coxsackievirus A21, Human rhinovirus 2 | Targets host cellular factors | Ali et al. (2021) |
| 25. | *Panax ginseng* | Root | Ginsenosides | Hepatitis virus A, Rhinovirus, Respiratory syncytial virus, Norovirus | Inhibiting viral attachment, membrane penetration, and replication, the foremost antiviral activities of ginseng are attributed to the enhancement of host immunity | Im et al. (2016) |
| 26. | *Swietenia macrophylla* | Stems | 3-hydroxy caruilignan C | HCV | Inhibition of HCV NS3/4 protease | Lin et al. (2014) |
| 27. | *Bupleurum* species | Root | Saikosaponins (C, D) | Hepatitis B virus | Saikosaponin C inhibits HBeAg expression and HBV DNA replication | Lin et al. (2014) |

*(Continued)*

**TABLE 13.1 (*Continued*)**
**Effect of Medicinal Plant-Derived Products on Various Viruses**

| S.No. | Plant | Source of Extraction | Compound Isolated | Virus | Mode of Action | References |
|---|---|---|---|---|---|---|
| 28. | *Terminalia chebula Retz.* | Dried fruits | Chebulagic acid and punicalagin | Respiratory syncytial virus | Inactivate free virus particles and inhibit early viral entry including attachment and penetration phases; do not affect viral cell-to-cell transmission | Lin et al. (2014) |
| 29. | *Glycyrrhiza uralensis* | Roots | Glycyrrhizin | Rotavirus diarrhea | Inactivate rotaviruses, especially G5 and G8P.\ | Khan and Al-Balushi (2021) |
| 30. | *Magnolia officinalis* | Methanol extract | Honokiol | Dengue virus Type 2 | Suppress the viral production both in DENV-infected BHK and Huh7 cells | Khan and Al-Balushi (2021) |
| 31. | *Peganum harmala* | Methanol seeds extract | Harmine | HSV-2 | Inhibition of viral entry | Bezekri et al. (2018) |
| 32. | *Dryopteris crassirhizoma* | Rhizome | Lapachol, Limonin, Camptothecin, Rhein, Reynoutrin, Eriodictyol, Sakuranetin, Quercetin, Robinetin, Rutin, Poncirin, Gedunin, Shikonin, Vitexin, Hesperetin, Hesperidin, Puerarin, Myricetin | Dengue virus | Antiviral effect against all four DENV serotypes against conserved target | Maryam et al. (2020) |
| 33. | *Morus alba* | Leaves | Naringin, Hesperidin, Quercetin, Kaempferitrin, Robinin, Poncirin, Rutin, Kaempferol, Gedunin, Rhein, Shikonin, Hesperetin, Nobiletin, Rottlerin, Gedunin, Quercetin, Genistein, Rhoifolin | Dengue virus | Antiviral effect against all four DENV serotypes against conserved target | Maryam et al. (2020) |
| 34. | *Zingiber officinale Roscoe.* | Root | Gingerol, shogaol, paradols | Human respiratory syncytial virus (HRSV) and Rhinovirus | Lock viral attachment and penetration into host cells via interacting with G and F proteins | Jafarzadeh et al. (2021) |
| 35. | *Glycyrrhiza glabra* L. | root extract | Glycyrrhizin | hepatitis C virus | The risk to develop hepatocellular carcinoma was reduced in those infected patients | Julsrigival et al. (2021) |

(*Continued*)

**TABLE 13.1 (*Continued*)**
**Effect of Medicinal Plant-Derived Products on Various Viruses**

| S.No. | Plant | Source of Extraction | Compound Isolated | Virus | Mode of Action | References |
|---|---|---|---|---|---|---|
| 36. | *Careya sphaerica Roxb.* | DCM- extract | 1-methyl-2-phenyl-1H-indole | Herpes simplex virus type 1 | Antiviral effect with $IC_{50}$ of dichloromethane–methanol and methanolic extracts were 24, and 30 μg/mL | Julsrigival et al. (2021) |
| 37. | *Caesalpinia sappan* | Hearthwood | Brazilein, brazilin, protosappanin A, 3-deoxysappanchalcone, sappanchalcone and rhamnetin | Influenza virus | Highest activity against influenza virus | Julsrigival et al. (2021) |
| 38. | *Selaginella uncinala* | Dried whole plants | Uncinodide A and B | Respiratory Syncytial Virus | Inhibit RSV infection | Amen et al. (2021) |
| 39. | *Calophyllum brasiliense* | Stem bark | coumarin | HIV | Inhibition of HIV replication *in vitro* model by suppressing nuclear factor-kappa [NF –KB] activation | Julsrigival et al. (2021) |
| 40. | *Alo barbadnesis* | Leaves | Quercetin | Herpes simplex virus (Type 2) | Exhibits promising antiviral effect | Roy et al. (2022) |
| 41. | *Aframomum melegueta* | Seed | Caryophyllene, humulene, 5α-androstan-16-one, benzodioxolo[5,6-c] phenanthridine, and d-norandrostane | Measles virus | Showed antiviral activity | Roy et al. (2022) |
| 42. | *Allium sativum* | Fresh garlic bulbs | Ajoene | HIV-1 | Prevention of HIV-induced destruction of CD + cells and enhance cellular immunity | Rouf et al. (2020) |
| 43. | *Punica Granatum* | Peel | Kaempferol, luteolin, naringin, pelargonidin, prodelphindin, quercetin, and rutin | Influenza virus | Inhibits the growth of the virus | Moradi et al. (2020) |
| 44. | *Artocarpus Heterophyllus* | Leaves | Dichloromethane extract | Hepatitis C (HCV) | Provides immunity against virus | More-Adate et al. (2022) |
| 45. | *Bauhinia variegata* | Leaves | Methanolic extract | Rotavirus | Shows antiviral activity | More-Adate et al. (2022) |
| 46. | *Psidium guajava* | leaves | Quercetin,Tannin, Leukocyanidins, gallic acid | Influenza Virus | Growth inhibition assay with ST6Gal I-expressing MDCK cells (AX4 cells) | Naseer et al. (2018) |

(*Continued*)

**TABLE 13.1 (*Continued*)**
**Effect of Medicinal Plant-Derived Products on Various Viruses**

| S.No. | Plant | Source of Extraction | Compound Isolated | Virus | Mode of Action | References |
|---|---|---|---|---|---|---|
| 47. | *Clerodendrum serratum* | Whole plant | Clerodendiod A and B | Yellow fever virus | Inhibition of the NF-κB pathway and showed potent activity in inhibiting NF-κB | Wang et al. (2012) |
| 48. | *Echinacea purpurea* | Leaves & Roots | Cichoric acid and echinacoside, caffeic acid | Rhinovirus | Reduction in the pro-inflammatory cytokines | Aucoin et al. (2020) |
| 49. | *Pedilanthus tithymaloides* | Leaves | Coumarin | Chikungunya virus | Modulation of NF-κB signaling mediated anti-inflammatory response | Joshi et al. (2020) |
| 50. | *Hypericum perforatum* | Aerial parts | Hypericin and pseudohypericin | Herpes simplex virus | Inhibitory effect against vesicular stomatitis | Mullaicharam and Halligudi (2019) |
| 51. | *Kalanchoe pinnata* | Leaves | Baicalein, fisetin, and quercetagetin | Chikungunya virus | Inhibition on BHK-21 cells | Septembre-Malaterre et al. (2022) |
| 52. | *Lawsonia inermis* | Fruits | Quercetin. Apigenin, apigenin-7-glucoside, apigenin-4-glycoside | Sembiki forest virus | Even after 10–25 days it showed 60%–100% inhibitory properties | Abdulfatai and Ayotunde (2022) |
| 53. | *Origanum vulgare* | Whole plant | 2,2-diphenyl-1-picrylhydrazyl | Respiratory syncytial virus (RSV), Coxsackie virus B3 (CVB3) and herpes simplex virus type 1 (HSV-1) | Compounds showed antiviral activity | Kryvtsova et al. (2020) |
| 54. | *Justicia adhatoda* | Whole plant | Asicoline, Vasicolinone, Vasicinone, Vasicine, Adhatodine, Anisotine | Influenza Virus | Improving the bronchial function with bronchodilatory, expectorant and mucolytic properties | Ghosh et al. (2021) |
| 55. | *Mangifera indica* | Seed coat | Catechins, quercetin, kaempferol, rhamnetin, anthocyanins, and tannic acid | Influenza | Targets rotavirus proteins | Chassagne et al. (2022) |

(*Continued*)

**TABLE 13.1 (*Continued*)**
**Effect of Medicinal Plant-Derived Products on Various Viruses**

| S.No. | Plant | Source of Extraction | Compound Isolated | Virus | Mode of Action | References |
|---|---|---|---|---|---|---|
| 56. | *Nerium Oleander* | New shoots | β-Eudesmol, digitoxigenin | Parainflunza-3 | Shows antiviral activity | Mandal et al. (2022) |
| 57. | *Prunus dulcis* | Skin of the nut | Catechin, epicatechin, kaempferol, and isorhamnetin, the latter as 3-O-rutinoside or 3-O-glucoside | Herpes simplex virus | Interferes viral attachment | Musarra-Pizzo et al. (2019) |
| 58. | *Pinus koraiensis* | Seed shell | α-pinene | Rotavirus | Seed shell found to be interfering the virus adsorption | Kim et al. (2020) |
| 59. | *Camellia thea* | Leaves | Epigallocatechin gallate | Rotavirus | Virus adsorption is interfered | Kim et al. (2020) |
| 60. | *Oryza sativa* | Whole plant | γ-Oryzanol | Rotavirus | Produced antibodies and induces T-cells responses | Kim et al. (2020) |
| 61. | *Tylosema esculentum* | grains | Protocatechuic acid, Caffeic acid, Gallic acid | Rotavirus | Viral replication is hindered and barrier functions are strengthened | Omotayo and Aremu (2021) |
| 62. | *Spongias lutea* | leaves | Saponins, alkaloids, avonoids, tannins, steroids, phenolics and glycosides | Rotavirus | Antirotaviral activity | Omotayo and Aremu (2021) |
| 63. | *Myristica fragrans* | Flower and seeds | Myristicin, trimyristin | Rotavirus | Inhibited the propagation of virus into MA-104 cells | Omotayo and Aremu (2021) |
| 64. | *Theobroma cacao* | husk | Cyanidin-3-arabinoside and cyanidin-3-galactoside | Rotavirus | Interfered adsorption of rotavirus into cells | Hu et al. (2020) |
| 65. | *Thalictrum cirrhosum* | Whole plants | Cirrhobenzazepine A, B,C | Rotavirus | Viral titers are found to be reduced in MA-104 cells | Hu et al. (2020) |
| 66. | *Vitis labrusca* | Seeds | Trans-3,5,4′-trihydroxystilbene | Rotavirus | Shows antiviral activity | Hu et al. (2020) |

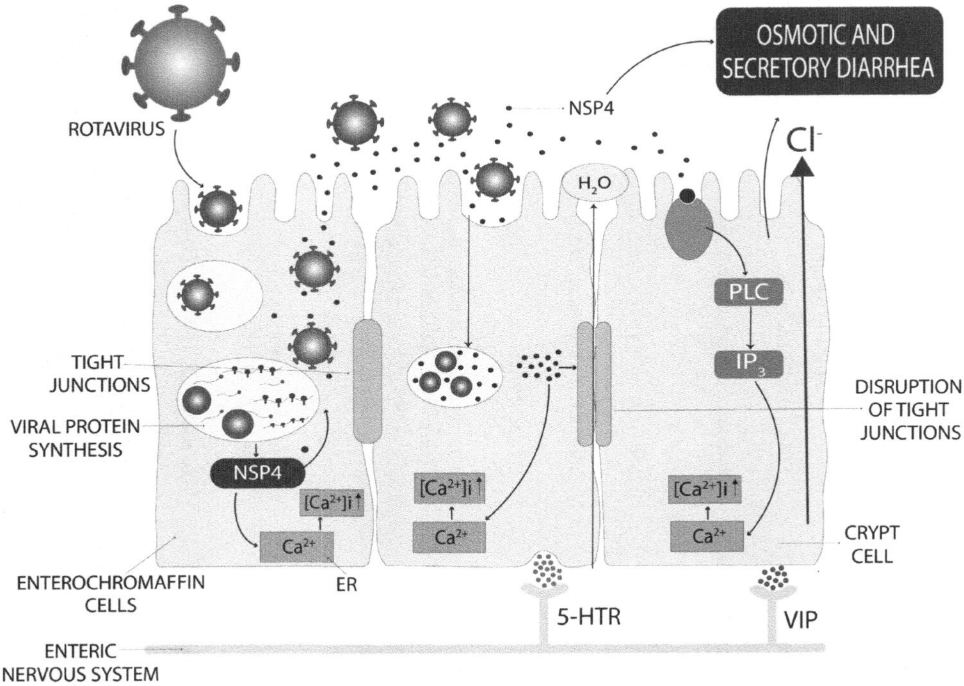


**FIGURE 13.1** RV infects the enterocytes, followed by the release of NSP4 through a Golgi-independent pathway. Intracellular NSP4 results in the calcium release. Additionally, NSP4 also disrupts the tight junctions resulting in paracellular water flow. Extracellular NSP4 released from other infected cells binds to the receptor and triggers a signaling cascade through PLC and IP3 that could also lead to $Ca^2$+ release and loss of water. RV infection can also result in accumulation of 5-HT into secretory granules and its release directly results in osmotic and secretory diarrhea. VIP could also result in the release of electrolytes including chloride.

been attenuated with multiple blind passaging to avoid any chances of infection in infants. During the first dose, replication of this strain in the intestine takes place. Post the dose, symptoms show shedding of this virus through stool. On the other hand, if the infant has been completely immunized, second dose results in very less shedding. Immunologically, this vaccine induce effective IgA response as compared to the IgG1 specific responses. Additionally, this vaccine provides immunity after two doses, as the second dose is responsible for filling gaps in immunity for the ones who failed to show response for first time (Vesikari et al. 2004). As for the other Rotateq vaccine, the bovine rotavirus strain WC3 has been used. Subsequently, reassortment with human G1, G2, G3 and G3 rotaviruses showed more efficiency than WC3 alone. For instance, this vaccine resulted in 74% efficacy against every strain with 100% efficacy to the most severe strains of rotavirus (Vesikari et al. 2006). This vaccine requires three doses.

Clinical trials have suggested that in underdeveloped countries, efficacy of these vaccines is from 40% to 70% with an average of 50%–60% (Jiang et al. 2010). This may be due to several reasons including reduction in the titer of the vaccine during passage of antibodies through placenta or breast milk, environmental enteropathy, and intrusion of other viral vaccines (Glass et al. 2014). A cost-saving equilibrium is not favored by various countries when comparing direct expenditures to the disease's entire societal costs, which even includes indirect costs to the family. For such countries, where vaccination has not been proven as successful as anticipated, several approaches can be adopted. One approach is to adopt and explore plant extracts as an alternative to the commercially available vaccines that might work better with low cost and would be easily available.

The medicinal herbs are helpful in treating the viral infections as either they contain antiviral compounds that can kill the virus by interfering with their replication or they can enhance the

**TABLE 13.2**
**Role of Chemical Drugs Used to Treat Viral Disease**

| S No. | Chemical Drug | Viral Infection Causative | Mode of Action | References |
|---|---|---|---|---|
| 1. | Abacavir | Human Immunodeficiency Virus (HIV) | Inhibits the conversion of viral RNA to DNA | Ramesh et al. (2021) |
| 2. | Acyclovir | Herpes Simplex Viruses-1 and -2 | Chain terminator | Sadowski et al. (2021) |
| 3. | Asunaprevir | Hepatitis C virus (HCV) genotype 1b | NS3/4A protease inhibitor | Wei et al. (2018) |
| 4. | Bromocriptine | Dengue virus | Dopamine receptors D2 and D3 agonist<br>Targeting Dopamine receptors D2 and D3 agonist | Troost et al. (2020) |
| 5. | Cidofovir | Herpes Simplex Viruses-1 and -2 | Selectively inhibiting viral DNA polymerase | Sadowski et al. (2021) |
| 6. | Danoprevir | Hepatitis C virus (HCV) genotype 1b | NS3/4A protease inhibitor | Miao et al. (2020) |
| 7. | Delavirdine | Corona Virus, HIV-1 | Inhibit transcription | Tarcsai et al. (2022) |
| 8. | Didanosine | Human immunodeficiency virus (HIV) | Inhibiting HIV reverse transcriptase by targeting PNP gene | Alakwaa and F. M. (2020) |
| 9. | Dolutegravir | Human immunodeficiency virus | Integrase strand transfer inhibitor | Cahn et al. (2020) |
| 10. | Doravirine | HIV-1 infection | Non-nucleoside reverse transcriptase inhibitor | Deeks and E. D. (2018) |
| 11. | Duramycin | Dengue virus | Targeting TIM1 receptor by Inhibiting entry | Troost et al. (2020) |
| 12. | Efavirenz | HIV, AIDS, COVID-19 | Non-nucleoside reverse transcriptase inhibitor | |
| 13. | Ezetimibe | Rotavirus | Preventive and interference effects by lowering cholesterol | Ding et al. (2021) |
| 14. | Favipiravir | Ebola virus, Lassa virus | Inhibits viral RdRp by terminating elongation at the incorporation site as a chain terminator | Shiraki and Daikoku (2020) |
| 15. | Fluoxetine | Coxsackievirus B1 | Selective serotonin reuptake inhibitor/CYP2D6 inhibitor | Honkimaa et al. (2020) |
| 16. | Foscarnet | Herpes Simplex Viruses-1 and -2 | Reversibly binds to the viral DNA polymerase | Sadowski et al. (2021) |
| 17. | Fucoidan | Dengue virus | Targeting E protein by Inhibiting virus binding and entry | Troost et al. (2020) |
| 18. | Glecaprevir | Hepatitis C virus | NS3/4A protease inhibitor | Cotter et al. (2019) |
| 19. | Grazoprevir | Chronic hepatitis C | NS5A and NS3/4A inhibitors | Wang et al. (2012) |
| 20. | Lamivudine | Human immunodeficiency virus, Hepatitis B virus | Nucleoside reverse transcriptase inhibitor | Cahn et al. (2020) |
| 21. | Nevirapine | HIV | Non-nucleoside reverse transcriptase inhibitor | Tiwari et al. (2022) |
| 22. | Oseltamivir | Avian influenza virus (H5N1) | Neuraminidase inhibitor | Abou Baker et al. (2021) |
| 23. | Pleconaril | Coxsackievirus B1 | Capsid inhibitor | Honkimaa et al. (2020) |

*(Continued)*

**TABLE 13.2 (*Continued*)**
**Role of Chemical Drugs Used to Treat Viral Disease**

| S No. | Chemical Drug | Viral Infection Causative | Mode of Action | References |
|---|---|---|---|---|
| 24. | Posaconazole | Chikungunya virus, Semliki forest virus, Dengue virus | Inhibit replication | Varghese et al. (2019) |
| 25. | Ribavirin | Respiratory syncytial virus, hepatitis C | Inhibitor of the capping guanylation of viral mRNA | Elawar et al. (2021) |
| 26. | Ribavirin | Respiratory syncytial virus, hepatitis C, nairovirus, hantavirus and Lassa virus | Interfere with the synthesis of viral mRNA | Walther et al. (2020); Casanova et al. (2018) |
| 27. | Rupintrivir | Enterovirus-71 and Coxsackievirus | Inhibitor of viral 3C protease | Walther et al. (2020); Casanova et al. (2018) |
| 28. | Tenofovir Alafenamide | HBV, Human immunodeficiency virus (HIV) | Reverse transcriptase incorporates TFV-DP into DNA resulting in Virus DNA chain termination | Byrne et al. (2018) |
| 29. | Zanamivir | Avian influenza virus (H5N1) | Neuraminidase inhibitor | Abou Baker et al. (2021) |
| 30. | Sofosbuvir | Yellow fever virus | Selective activity against YFV in Huh-7 cells | Mendes et al. (2019) |

immune response of the host against the virus (AL-Ballawi et al. 2017). For instance, *Tylosema esculentum*, a legume crop grown in Africa has been reported to inhibit rotavirus (Chingwaru et al. 2011). Moreover, a traditional herbal extract used for the treatment of acute diarrhea has been found to exert antiviral properties against RV (Taherkhani et al. 2013). Recently, AL-Ballawi et al. (2017) reported that plant extracts from *Allium sativum*, *Zingiber officinale*, *Punica granatum*, and *Cymbopogon citrates* exhibited inherent capacity against rotavirus SA-11. Moreover, Knipping et al. (2012) tested nutritionally important plant species, namely *Nelumbo nucifera*, *Aspalathus linearis*, *Urtica dioica*, *Glycyrrhiza glabra*, and *Olea europaea*, against the rotavirus infection. The extracts from these plants were shown to possess antiviral properties. These extracts, when used separately at the concentration with $IC_{50}$, resulted in a 50% inhibition of infection. On the other hand, some of these extracts in combination resulted in infection inhibition up to 89% (Knipping et al. 2012). Peppermint oregano (*O. vulgare*) extract was found to possess antiviral properties (Zhang et al. 2014; Alhazmi, et al. 2021). Hydro-alcoholic *A. kellalensis* extracts are known to contain various phytochemicals such as flavonoids, phenolic acids, terpenoids, and sterols. Taherkhani et al. (2015) evaluated these extracts for toxicity against RV infection. They revealed that the extract was able to prevent replication but it did not affect virus adsorption. Henceforth, the extract exhibited strong antirotaviral activity.

Many extracts have been found to exert antirotaviral activities. Some of these extracts help in managing the symptoms rather than interfering with viral growth cycle. For instance, *Cordia Africana* has been long used for the treatment of liver disease, amebiasis, stomachache, and diarrhea. RV is a well-known causative agent for diarrhea. Asrie et al. (2016) demonstrated that different organ extracts containing phenols, flavonoids, terpenoids, and saponins have the ability to reduce diarrhea by 24.41%, 39.89%, and 51.66% in mice model. Sabo and Knezevic (2019) investigated the effect of *Eucalyptus camaldulensis* extracts against rotavirus Wa. They revealed that the extracts had capacity to inhibit rotavirus multiplication. *Patallus mollis* sea cucumber aqueous extract was used to evaluate the antiviral effect against RV in cell culture. The result analysis revealed 99% inhibition of the virus at the time of absorption and viral inactivation phase (Garcia-Candela et al.

2021). An aglycone called genipin, isolated from the fruit of *Gardenia jasminoides*, has been shown to possess antiviral effect against the human rotavirus Wa and SA-11 strains *in vitro (*Kim et al. 2020). However, none of these herbal medications have yet been used in clinical trials properly as there is a lack of knowledge about mechanism of action behind these individual medicinal drugs.

## 13.3 MECHANISM OF ACTION

As discussed earlier, RV infection has been reported to effect lungs and intestine severely. During infection, there is a loss of transepithelial resistance (TER) due to disruption of tight junction proteins mainly claudin-1, occludin, and ZO-1. It is also associated with many metabolic changes in cell, such as overproduction of lactate, decreased mitochondrial oxygen consumption, and reduced cellular ATP (Chingwaru et al. 2011).

To develop effective treatment, it is crucial to investigate the underlying pathological changes related to RV infection. Phytochemicals are reported to regulate immune responses, restore cell signaling, and repair intestinal damage due to rotavirus infections. They are known to regulate rotavirus-induced reactive oxygen species formation by controlling mRNA concentrations of proinflammatory, Th1, and Th2 cytokines (Alfajaro et al. 2012). *Tylosema esculentum* (marama) cotyledon and seed coat extracts promoted cell survival of RV infected cells by at least 100%. These extracts also enhanced nitric oxide release and transepithelial resistance in the intestinal enterocyte cells of human infant (H4). Moreover, the accumulated nitric oxide interfered with the replication of viral particles, significantly reducing the infectivity (Chingwaru et al. 2011). Some plant-derived products, namely Qiwei Baizhu powder (QWBZP), *Glycyrrhiza uralensis* Fisch (Fabaceae), and 18β-Glycyrrhetinic acid have been found to improve pathological alterations in small intestinal, reduce diarrhea incidences, and minimize virus shedding. Additionally, various plant extracts have been shown to improve signaling pathways in intestine by increasing the expression of signaling molecules such as IL-2, IL-4, IL-8, and IL-10, IFN-γ, IFN-β, and TNF-α. These cytokines are known to protect intestinal epithelial cells from rotavirus-induced reactive oxygen species formation (Gandhi et al. 2017). In contrast, some plant-derived extracts interfere with virus attachment to the cell. For instance, Quillaja Saponaria Molina extract at 1.0 μg/mL concentration has been found to possess strong immune-enhancing activity by blocking the rotavirus attachment and subsequent infection (Anand et al. 2019). Moreover, licorice root metabolite 18β-glycyrrhetinic acid (GRA) has been shown to possess antiviral properties against various viruses including rotavirus. They are reported to inhibit viral attachment and internalization, as well as stimulate IFN secretion to improve immune signaling pathways (Hardy et al. 2012). RV infection causes pathological changes in lungs and intestines such as the release of pro-inflammatory cytokines and antibodies, alterations in $CD4^+$ and $CD8^+$ cells morphology, and dysregulation of STAT3 and STAT1 signaling pathways. A Japanese medicinal plant, *Magnolia obovata* Thunb., leaves have been long used as a crude herbal drug known as "Wakoboku" as a natural remedy for oral infection. Kawahara et al. (2014) reported the therapeutic effects of hot-water extract of *M. obovata* leaves on RV strain SA11. They noticed a reduced incidence of diarrhea, inhibited viral mRNAs encoding viral proteins and improved cytopathic effects in SA11-infected MA104 cells.

Shen et al. (2019) reported the baicalin, a bioactive compound isolated from *Scutellaria baicalensis* Georgi, exerts therapeutic effects against RV infection by restoring the pathological changes that occur in the lungs and intestine. According to a recent study, *Macleaya cordata* extract has been reported to possess antiviral properties. The effect of this extract was also investigated in rotavirus SA11induced diarrhea in mice model. They revealed dose-dependent decrease in viral RNA levels. Also, there was a reduction in proinflammatory cytokines and increase in antiinflammatory cytokine levels. This extract also improved intestinal inflammation. Henceforth, *M. cordata* can be used as potential treatment drug for rotavirusinduced diarrhea (Jiang et al. 2020). Kim et al. (2020) revealed that genipin affect two stages of viral replication cycle, i.e., attachment and penetration followed by assembly and release. In addition, this compound also possesses anti-inflammatory effects. *Rubia cordifolia,* a perennial climber mostly found in China and India, is widely used for its aerial parts to treat diarrhea in children. Sun et al. (2016) reported the antiviral properties of the aerial parts of this plant. MA-104 cells infected with virus were used for this study. They noticed virus-induced apoptosis in the cells, which in turn inhibited rotavirus multiplication.

## 13.4 SECONDARY METABOLITES AS ANTIROTAVIRAL AGENTS

Secondary metabolites are synthesized by plants during later stages of growth. In contrast to primary metabolite, they do not play a direct role in the growth and metabolism of plant life cycle. They are of great importance for the human health and economics. Plants synthesize a wide variety of metabolites with complex chemical composition, namely polyphenols, alkaloids, terpenoids, and sulfur-containing compounds (Guerriero et al. 2018).

Numerous studies on phytochemicals with antiviral properties have been published, focusing on viruses like the human papillomavirus (HPV), human immunodeficiency virus (HIV), human hepatitis virus, the rotavirus, chikungunya virus, Japanese encephalitis virus, herpes, Zika virus, and hepatitis C virus. For instance, flavonoids, alkaloids, lignan, and coumarins have been reported to inhibit growth of Dengue virus (Kaushik et al. 2018). Pinocembrin, a flavanone displayed potent inhibition of ZIKV in human placental JEG-3 cells (Lee et al. 2019). Genistein, an isoflavones alone or in combination with acyclovir and ganciclovir inhibit herpes B virus replication (LeCher et al. 2019). The influenza virus has been reported to be inhibited by alkaloids at several phases of replication. They either prevent the synthesis of viral proteins or impair other stages of viral replication (Moradi et al. 2018). Lin et al. (2014) analyzed saikoponins against hepatitis C virus. These compounds targeted virus at early stages like viral entry/fusion rather than later stages, including transcription and translation. The use of medicinal plants as an antirotaviral agents has been briefly depicted in Figure 13.2.

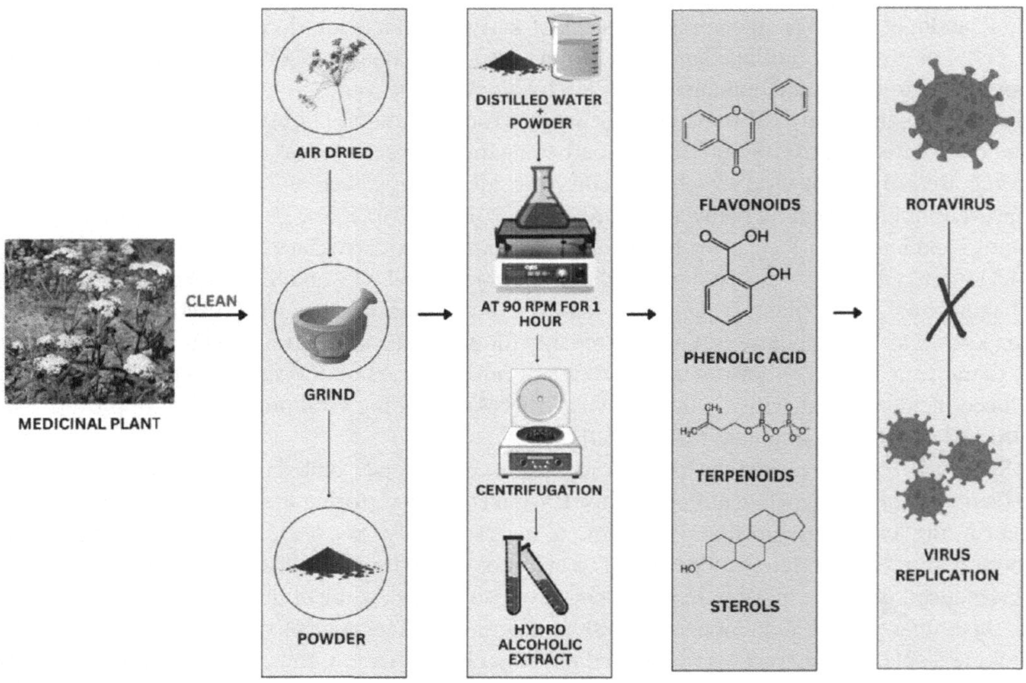


**FIGURE 13.2** Various plant organs are collected and washed and grinded to fine powder. Using various solvent various metabolites and compound acts as antirotaviral by inhibiting virus replication.

### 13.4.1 Polyphenols

Polyphenols are a group of secondary metabolites constituting approximately 8,000 structural variants based upon the number of rings and type of elements linking these rings. They are further classified into subgroups, including flavonoids, phenolic acids, stilbenes, or lignans (Giovinazzo et al. 2020). Continuous research is being conducted on polyphenols for potential antirotaviral properties (Jacob and Thomas 2019). Polyphenols isolated from various plants have been reported to alter viral morphology and act as antiviral compounds. As per early reports, *Citrus aurantium*-derived hesperidin and neohesperidin inhibited rotavirus infectivity (Bae et al. 2000).

Plant-derived phenolic compounds known as kaempferol and chrysin are known to exert anti-RV properties (Montenegro-Landívar et al. 2021). Flavonoids, even at lower concentration, have been reported to exert antiviral properties by inhibiting enzymes, namely reverse transcriptase and proteases (Shahwar et al. 2013). As per the report published by Cecílio et al. (2016), ethanolic extract from the leaves of M. urundeuva plant exhibited antiviral activity. Further, fractionation of the extracted revealed the presence of flavonoids. They concluded that flavonoids might be responsible for inhibiting viral amplification in the early phases of infection. Epigallocatechin gallate (EGCG), a flavonoid found in green tea, and proanthocyanidins (PACs) from berry and grape species have been known to possess antiviral activity. Lipson et al. (2017) reported synergistic effect of both the flavonoids against rotavirus strain SA-11(RTV). They revealed that RV titers were reduced by 32% after treatment with the flavonoids. Flavonoids can interfere with RV replication, for instance, a flavanone called norkurarinol inhibits RV virus replication by hindering Toll-like receptor (TLR 3) signaling and stimulating the release of pro-inflammatory cytokines. Furthermore, *Sophora flavescens* extract (SFE) and stevioside (SV) from *Stevia rebaudiana* were used to investigate their therapeutic effect on porcine rotavirus G5 [P7] strain. The SFE extract is known to contain more than 20 alkaloids and 50 flavonoids. All these components, in combination with stevioside, exerted synergistic antiviral effects on RV by inhibiting its replication (Alfajaro et al. 2014). Furthermore, *Achyrocline bogotensis*, a medicinal plant widely distributed in Colombia, was studied for its antidiarrheal properties or antiviral activities by Téllez Alfonso et al. (2015). Qualitative analysis of the fractions confirmed the presence of various bioactive compound, including flavonoids. The interaction of these compounds with structural viral proteins was held responsible for inhibiting its infection, as viral proteins are directly responsible for adhesion to the host cell (Téllez Alfonso et al. 2015). According to a report published by Kwon et al. (2010), six polyphenols isolated from *Glycyrrhiza uralensis* roots showed growth inhibitory effects against G5P[7] and G8P[7]. RT-PCR revealed that these polyphenols were responsible for suppressing viral RNA synthesis. In addition to this, there was a significant inhibition of both viral absorption and replication. According to a recent report, resveratrol reduced the incidence of RV-mediated diarrhea in a neonatal mouse model (Yu et al. 2018). In a recent study on Gegen Qinlian Decoction, a Chinese herbal medicine, predicted active ingredients. It comprises various flavonoids, namely formononetin, acacetin, wogonin, and quercetin. It was reported that these compounds might have the potential to damage the integrity of RV structure and protein synthesis of RV (Zhong et al.2020).

### 13.4.2 Alkaloids

Alkaloids are amino acids derivatives of containing heterocyclic ring with nitrogen. Most commonly, amino acids such as L-ornithine, L-lysine, L-phenylalanine/L-tyrosine, L-tryptophan, and L-histidine act as precursors of alkaloid. The subclass of the alkaloid depends upon the type of substrate. For instance, tyrosine is the precursor of tetrahydroisoquinoline alkaloids, indole-containing alkaloids are synthesized from tryptophan, and ornithine is the substrate amino acid for pyrrolizidine- and tropane-type alkaloids (O'Connor 2010). Natural alkaloids have been effectively transformed into chemotherapeutic agents, such as camptothecin (CPT), a well-recognized topoisomerase I inhibitor and vinblastine, a potential anticancer agent. Alkaloids have been reported to be the antiviral agents. They are known to efficiently damage the cell membrane and bind to the viral glycoprotein. The amphiphilic structure of alkaloids inhibit viral attachment to the cell. Alkaloids have the ability to interfere with replication, transcription, and translation (Abookleesh et al. 2022).

Till date, various reports focusing on the role of alkaloids as antirotaviral drug have been published. As discussed earlier, Gegen Qinlian Decoction was found to include 130 active ingredients, such as flavonoids, alkaloids, phenyl ester, and fatty acids. Zhong et al. (2020) predicted anti-RV properties of these compounds using Network Pharmacology-Based Strategy. Gupta (2016) reported antidiarrheal activity of Indian traditional medicinal plant *Tamarindus indica* L. seed extract, including alkaloids. *Thalictrum* plant belonging to the family Ranunculaceae have anti-inflammation and antidote properties. Alkaloids are considered as the major bioactive compounds in this plant (Jin et al. 2018). Two new benzazapine alkaloids isolated from this plant species were tested for antirotaviral properties by Wu et al. (2021). These compounds exhibited significant anti-rotavirus activity. The legume family (Fabaceae) member *Cassia* Linn is a source of novel alkaloids, namely cassiarin A, a tricyclic alkaloid fistulain A, dimeric chromone alkaloid containing a chromone and cassiarin A and cassiarin F, a novel hybrid alkaloid derived from cassiarin A and a biphenyl unit with an acetonyl moiety. A new chromone, alatachromone A (4), a new dimeric chromone-indole alkaloid, alataindolein D (5), and three new indole alkaloids, alataindoleins A–C (1–3) isolated from *Cassia* plant showed potent inhibition with therapeutic index (TI) values in the range of 9.75–15.3.against human rotavirus Wa (Yang et al. 2022).

### 13.4.3 Terpenoids

Terpenoids belong to the terpene class with varied functional groups and flexible oxidized methyl groups. Terpenoids can be classified into further subgroups based upon the number of carbon units. Saponins, also known as triterpene glycosides, possess antirotaviral properties. *Quillaja* saponins extracted from Chilean soapbark tree have demonstrated to be potent antiviral activity. Tam and Roner (2011) investigated the effect of these saponin against rhesus rotavirus (RRV) using a mouse model. The extract at concentration of 0.015 mg significantly reduced the incidence of RRV induced diarrhea from 79% to 11%. It was due to hindrance between virus-host attachment by degradation of cellular membrane proteins and/or virus receptors. *D. viscose* has been known to exert antibacterial, antifungal, antidiabetic, anti-inflammatory, antiulcer, antioxidant, and antidiarrheal. These properties are reflected by the bioactive components present in the leaves, namely terpenoids, p-coumarin acid ester, sterols, and volatile oil. Shaheen et al. (2015) investigated *Dodonaea viscosa* leaves extract on RV SA-11. Terpenoids present in the extract possessed antirotaviral activity by binding to the capsid and masking the receptors to inhibit virus entry to the cell. Recently, a role of novel triterpenoid, namely ursolic acid (UA), isolated from plant source has been investigated against RV. They noticed that this compound interferes during early stages by inhibiting structural and non-structural proteins. Also, no cytopathic effect of UA has been reported during late stages of replication.

## 13.5 CONCLUSION

The fact that the majority of people in the world receive their healthcare from plant-derived products illustrates the significance of medicinal plants. According to several scientific studies, plants might be a viable source to produce compounds with pharmaceutical-grade (Ali et al. 2021; Gandhi et al.). Medicinal plants synthesize various secondary metabolites, which are used as an alternative to conventional medicines to fight viral infections. The bioactivities of these compounds include antioxidant potential to prevent oxidative stress-induced cellular damage and lower the severity of disease. Although, antiviral mechanisms are yet to be properly studied, the present chapter discusses the uses of plant extracts from different plant species as antirotaviral medicines. Various studies have established that plant extracts are useful due to their synergistic effects, namely antioxidant, antiviral, and anti-inflammatory. However, to combat the severity of this disease and to use the medicinal extracts in the vaccine development, more extensive studies on the mechanism of action of these extracts should be considered.

## REFERENCES


Abdulfatai, A. & Ayotunde, O. O. (2022). Lawsonia inermis Linn; review of plant with both industrial and medicinal properties. *Media Kedokteran Hewan*, 33(2), 96–121.

Abookleesh, F. L., Al-Anzi, B. S., & Ullah, A. (2022). Potential antiviral action of alkaloids. *Molecules*, 27(3), 903.

Abou Baker, Doha H., Ryszard Amarowicz, Ahmed Kandeil, Mohamed A. Ali, and Eman A. Ibrahim. (2021). Antiviral activity of Lavandula angustifolia L. and Salvia officinalis L. essential oils against avian influenza H5N1 virus. *Journal of Agriculture and Food Research*, 4, 100135.

Adhikari, B., Marasini, B. P., Rayamajhee, B., Bhattarai, B. R., Lamichhane, G., Khadayat, K., Adhikari, A., Khanal, S., & Parajuli, N. (2021). Potential roles of medicinal plants for the treatment of viral diseases focusing on COVID-19: A review. *Phytotherapy Research: PTR*, 35(3), 1298–1312. https://doi.org/10.1002/ptr.6893

Adiguna, S. P., Panggabean, J. A., Atikana, A., Untari, F., Izzati, F., Bayu, A., Rosyidah, A., Rahmawati, S. I., & Putra, M. Y. (2021). Antiviral activities of andrographolide and its derivatives: mechanism of action and delivery system. *Pharmaceuticals (Basel, Switzerland)*, 14(11), 1102. https://doi.org/10.3390/ph14111102

AL-Ballawi, Z. F., Redhwan, N. A., & Ali, M. (2017). In vitro studies of some medicinal plants extracts for antiviral activity against rotavirus. *IOSR Journal of Pharmacy and Biological Sciences*, 12(2), 53–58.

Alakwaa, F. M. (2020). Repurposing didanosine as a potential treatment for COVID-19 using single-cell RNA sequencing data. *Msystems*, 5(2), e00297–20.

Albulescu, L., Bigay, J., Biswas, B., Weber-Boyvat, M., Dorobantu, C., Delang, L. et al. (2017). Uncovering oxysterol-binding protein (OSBP) as a target of the anti-enteroviral compound TTP-8307. *Antiviral Research*, 140, 37–44. https://doi.org/10.1016/j.antiviral.2017.01.008

Alfajaro, M. M., Kim, H., Park, J., Ryu, E., Kim, J., Jeong, Y. et al. (2012). Anti-rotaviral effects of *Glycyrrhiza uralensis* extract in piglets with rotavirus diarrhoea. *Virology Journal*, 9(1), 1–10.

Alfajaro, M. M., Rho, M. C., Kim, H. J., Park, J. G., Kim, D. S., Hosmillo, M., … Cho, K. O. (2014). Anti-rotavirus effects by combination therapy of stevioside and Sophora flavescens extract. *Research in Veterinary Science*, 96(3), 567–575.

Alhazmi, H. A., Najmi, A., Javed, S. A., Sultana, S., Al Bratty, M., Makeen, H. A., … Khalid, A. (2021). Medicinal plants and isolated molecules demonstrating immunomodulation activity as potential alternative therapies for viral diseases including COVID-19. *Frontiers in Immunology*, 1721. https://doi.org/10.3389/fimmu.2021.637553

Ali, S. I., Sheikh, W. M., Rather, M. A., Venkatesalu, V., Muzamil Bashir, S., & Nabi, S. U. (2021). Medicinal plants: treasure for antiviral drug discovery. *Phytotherapy Research*, 35(7), 3447–3483.

Álvarez, D., Castillo, E., Duarte, L., Arriagada, J., Corrales, N., Farías, M. et al. (2020). Current antivirals and novel botanical molecules interfering with herpes simplex virus infection. *Frontiers in Microbiology*, 11. https://doi.org/10.3389/fmicb.2020.00139

Amen, Y., Elsbaey, M., Othman, A., Sallam, M., & Shimizu, K. (2021). Naturally occurring chromone glycosides: sources, bioactivities, and spectroscopic features. *Molecules*, 26(24), 7646.

Anand, U., Jacobo-Herrera, N., Altemimi, A., & Lakhssassi, N. (2019). A comprehensive review on medicinal plants as antimicrobial therapeutics: potential avenues of biocompatible drug discovery. *Metabolites*, 9(11), 258.

Andersson, S. (2010). General polyhedra, virus structure and mutation. *Zeitschrift für Kristallographie-Crystalline Materials*, 225, 309–312.

Appel, N., Schaller, T., Penin, F., & Bartenschlager, R. (2006). From structure to function: new insights into hepatitis C virus RNA replication. *The Journal of Biological Chemistry*, 281, 9833–9836.

Arena, A., Bisignano, G., Pavone, B., Tomaino, A., Bonina, F. P., Saija, A., Cristani, M., D'Arrigo, M., & Trombetta, D. (2008). Antiviral and immunomodulatory effect of a lyophilized extract of Capparis spinosa L. buds. *Phytotherapy Research: PTR*, 22(3), 313–317. https://doi.org/10.1002/ptr.2313

Asrie, A. B., Abdelwuhab, M., Shewamene, Z., Gelayee, D. A., Adinew, G. M., & Birru, E. M. (2016). Antidiarrheal activity of methanolic extract of the root bark of Cordia africana. *Journal of Experimental Pharmacology*, 8, 53.

Atherly, D., Dreibelbis, R., Parashar, U. D., Levin, C., Wecker, J., & Rheingans, R. D. (2009). Rotavirus vaccination: cost-effectiveness and impact on child mortality in developing countries. *The Journal of Infectious Diseases*, 1, 200.

Aucoin, M., Cooley, K., Saunders, P. R., Carè, J., Anheyer, D., Medina, D. N., … Garber, A. (2020). The effect of Echinacea spp. on the prevention or treatment of COVID-19 and other respiratory tract infections in humans: a rapid review. *Advances in Integrative Medicine*, 7(4), 203–217.

Babji, S. & Kang, G. (2012). Rotavirus vaccination in developing countries. *Current Opinion in Virology*, 2, 443–448. https://doi.org/10.1016/j.coviro.2012.05.005.

Bae, E. A., Han, M. J., Lee, M., & Kim, D. H. (2000). In vitro inhibitory effect of some flavonoids on rotavirus infectivity. *Biological and Pharmaceutical Bulletin*, 23(9), 1122–1124.

Balkrishna, A., Pokhrel, S., Singh, H., Joshi, M., Mulay, V. P., Haldar, S., & Varshney, A. (2021). Withanone from *Withania somnifera* attenuates SARS-CoV-2 RBD and host ACE2 interactions to rescue spike protein induced pathologies in humanized zebrafish model. *Drug design, development and therapy*, 15, 1111–1133.

Ben-Shabat, S., Yarmolinsky, L., Porat, D., & Dahan, A. (2020). Antiviral effect of phytochemicals from medicinal plants: applications and drug delivery strategies. *Drug Delivery and Translational Research*, 10(2), 354–367. https://doi.org/10.1007/s13346-019-00691-6

Bennett, B. C. & Husby, C. E. (2008). Patterns of medicinal plant use: an examination of the Ecuadorian Shuar medicinal flora using contingency table and binomial analyses. *Journal of Ethnopharmacology*, 116, 422–430. https://doi.org/10.1016/j.jep.2007.12.006

Benzekri, R., Bouslama, L., Papetti, A., Hammami, M., Smaoui, A., & Limam, F. (2018). Anti HSV-2 activity of Peganum harmala (L.) and isolation of the active compound. *Microbial Pathogenesis*, 114, 291–298. https://doi.org/10.1016/j.micpath.2017.12.017

Byrne, R., Carey, I., & Agarwal, K. (2018). Tenofovir alafenamide in the treatment of chronic hepatitis B virus infection: rationale and clinical trial evidence. *Therapeutic advances in gastroenterology*, 11, 1756284818786108.

Cahn, P., Madero, J. S., Arribas, J. R., Antinori, A., Ortiz, R., Clarke, A. E., ... & Smith, K. Y. (2020). Durable efficacy of dolutegravir plus lamivudine in antiretroviral treatment–naive adults with HIV-1 infection: 96-week results from the GEMINI-1 and GEMINI-2 randomized clinical trials. *Journal of acquired immune deficiency syndromes*, 83(3), 310–318.

Casanova, V., Sousa, F. H., Stevens, C., & Barlow, P. G. (2018). Antiviral therapeutic approaches for human rhinovirus infections. *Future virology*, 13(07), 505–518.

Cecílio, A. B., Oliveira, P. D. C., Caldas, S., Campana, P. R., Francisco, F. L., Duarte, M. G. R., … Almeida, V. L. D. (2016). Antiviral activity of Myracrodruon urundeuva against rotavirus. *Revista Brasileira de Farmacognosia*, 26, 197–202.

Chassagne, F., Butaud, J. F., Torrente, F., Conte, E., Ho, R., & Raharivelomanana, P. (2022). Polynesian medicine used to treat diarrhea and ciguatera: an ethnobotanical survey in six islands from French Polynesia. *Journal of Ethnopharmacology*, 292, 115186.

Chen, S., Wang, Y., Li, P., Yin, Y., Bijvelds, M. J., de Jonge, H. R., Peppelenbosch, M. P., Kainov, D. E., & Pan, Q. (2020). Drug screening identifies gemcitabine inhibiting rotavirus through alteration of pyrimidine nucleotide synthesis pathway. *Antiviral Research*, 180, 104823. DOI:10.1016/j.antiviral.2020.104823.

Chingwaru, W., Majinda, R. T., Yeboah, S. O., Jackson, J. C., Kapewangolo, P. T., Kandawa-Schulz, M., & Cencic, A. (2011). Tylosema esculentum (Marama) tuber and bean extracts are strong antiviral agents against rotavirus infection. *Evidence-Based Complementary and Alternative Medicine*, 2011, 284795.

Christensen, M. (1989). Human viral gastroenteritis. *Clinical Microbiology Reviews*, 2, 51–89.

Cotter, T. G., & Jensen, D. M. (2019). Glecaprevir/pibrentasvir for the treatment of chronic hepatitis C: design, development, and place in therapy. *Drug design, development and therapy*, 13, 2565.

Deeks, E. D. (2018). Doravirine: first global approval. *Drugs*, 78(15), 1643–1650.

Ding, S., Yu, B., & van Vuuren, A. (2021). An Emerging Application of Cholesterol-lowering Therapy in Inhibition of Rotavirus Infection.

Dou, X., Li, Y., Han, J., Zarlenga, D.S., Zhu, W., Ren, X., Dong, N., Li, X., & Li, G. (2018). Cholesterol of lipid rafts is a key determinant for entry and post-entry control of porcine rotavirus infection. *BMC Veterinary Research*, 14, 1–12.

Elawar, F., Oraby, A. K., Kieser, Q., Jensen, L. D., Culp, T., West, F. G., & Marchant, D. J. (2021). Pharmacological targets and emerging treatments for respiratory syncytial virus bronchiolitis. *Pharmacology & Therapeutics*, 220, 107712.

El-Saadony, M. T., Zabermawi, N. M., Zabermawi, N. M., Burollus, M. A., Shafi, M. E., Alagawany, M., … Abd El-Hack, M. E. (2021). Nutritional aspects and health benefits of bioactive plant compounds against infectious diseases: a review. *Food Reviews International*, 39(4), 2138–2160.

Gandhi, G. R., Santos, V. S., Denadai, M., da Silva Calisto, V. K., & Quintans, J. D. S. S., e Silva, A. M. D. O., … Gurgel, R. Q. (2017). Cytokines in the management of rotavirus infection: a systematic review of in vivo studies. *Cytokine*, 96, 152–160.

Garcia-Candela, E., Pariona-Velarde, C., Mondrag6n-Martínez, A., & Chumpitaz-Cerrate, V. (2021). Antiviral activity of the sea cucumber tegument extract (Pattalus mollis) on human rotavirus A (RVA). *Natural Product Research*, 35(6), 1014–1018.

Glass, R. I., Parashar, U., Patel, M., Gentsch, J., & Jiang, B. (2014). Rotavirus vaccines: Successes and challenges. *Journal of Infection*, 68. DOI:10.1016/j.jinf.2013.09.010.

Ghosh, R., Chakraborty, A., Biswas, A., & Chowdhuri, S. (2021). Identification of alkaloids from Justicia adhatoda as potent SARS CoV-2 main protease inhibitors: an in silico perspective. *Journal of Molecular Structure*, 1229, 129489.

Giovinazzo, G., Gerardi, C., Uberti-Foppa, C., & Lopalco, L. (2020). Can natural polyphenols help in reducing cytokine storm in COVID-19 patients? *Molecules*, 25, 5888.

Guerriero, G., Berni, R., Muñoz-Sanchez, J. A., Apone, F., Abdel-Salam, E. M., Qahtan, A. A., ... Faisal, M. (2018). Production of plant secondary metabolites: examples, tips and suggestions for biotechnologists. *Genes*, 9(6), 309.

Gupta, R. (2016). Investigation of antidiarrhoeal activity of ethanolic extract of Tamarindus indica L. seeds in Albino Wistar rats. *Asian Journal of Pharmaceutics*, 10(4). DOI:10.22377/ajp.v10i04.883.

Hardy, M. E., Hendricks, J. M., Paulson, J. M., & Faunce, N. R. (2012). 18 $\beta$-glycyrrhetinic acid inhibits rotavirus replication in culture. *Virology Journal*, 9(1), 1–7.

Hong, I. S., Lee, H. Y., & Kim, H. P. (2014). Anti-oxidative effects of Rooibos tea (Aspalathus linearis) on immobilization-induced oxidative stress in rat brain. *PLoS One*, 9(1), e87061.

Honkimaa, A., Sioofy-Khojine, A. B., Oikarinen, S., Bertin, A., Hober, D., & Hyöty, H. (2020). Eradication of persistent coxsackievirus B infection from a pancreatic cell line with clinically used antiviral drugs. *Journal of Clinical Virology*, 128, 104334.

Hossain, M. U., Hashem, A., Keya, C. A., & Salimullah, M. (2016). Therapeutics insight with inclusive immunopharmacology explication of human rotavirus A for the treatment of diarrhea. *Frontiers in Pharmacology*, 7, 153.

Hu, Q. F., Luo, D., Lv, N., Zhu, Y. N., Liu, L., Wu, F., ... Yang, G. Y. (2020). Three new benzazepine alkaloids from thalictrum cirrhosum and their anti-rotavirus activity. *Heterocycles: An International Journal for Reviews and Communications in Heterocyclic Chemistry*, 100(8), 1260–1267.

Im, K., Kim, J., & Min, H. (2016). Ginseng, the natural effectual antiviral: protective effects of Korean Red Ginseng against viral infection. *Journal of Ginseng Research*, 40(4), 309–314. https://doi.org/10.1016/j.jgr.2015.09.002

Jacob, A. & Thomas, J. (2019) Therapeutic potential of dietary flavonoids against viral-borne infections: a review. *Drug Invent Today*, 11(2), 469–479.

Jafarzadeh, A., Jafarzadeh, S., & Nemati, M. (2021). Therapeutic potential of ginger against COVID-19: Is there enough evidence? *Journal of Traditional Chinese Medical Sciences*, 8(4), 267–279.

Jiang, V., Jiang, B., Tate, J., Parashar, U. D., & Patel, M. M. (2010). Performance of rotavirus vaccines in developed and developing countries. *Human Vaccines*, 6(7), 532–542.

Jiang, C., Yang, H., Chen, X., Qiu, S., Wu, C., Zhang, B., & Jin, L. (2020). Macleaya cordata extracts exert antiviral effects in newborn mice with rotavirusinduced diarrhea via inhibiting the JAK2/STAT3 signaling pathway. *Experimental and Therapeutic Medicine*, 20(2), 1137–1144.

Jin, Q., Yang, D., Dai, Z., Khan, A., Wang, B., Wei, X., Sun, Y., Zhao, Y. L., Wang, Y. F., Liu, Y. P., Zhao, X. D., & Luo, X. D. (2018). Antitumor aporphine alkaloids from *Thalictrum wangii*. *Fitoterapia*, 128, 204–212.

Joshi, B., Panda, S. K., Jouneghani, R. S., Liu, M., Parajuli, N., Leyssen, P., ... Luyten, W. (2020). Antibacterial, antifungal, antiviral, and anthelmintic activities of medicinal plants of Nepal selected based on ethnobotanical evidence. *Evidence-Based Complementary and Alternative Medicine*, 2020.

Julsrigival, J., Sirisa-Ard, P., Julsrigival, S., & Akarchariya, N. (2021). Antiviral medicinal plants found in Lanna traditional medicine. *Chinese Herbal Medicines*, 13(4), 494–501.

Kaushik, S., Kaushik, S., Sharma, V., & Yadav, J. (2018). Antiviral and therapeutic uses of medicinal plants and their derivatives against dengue viruses. *Pharmacognosy Reviews*, 12(24), 177. DOI:10.4103/phrev.phrev_2_18.

Kawahara, T., Tomono, T., Hamauzu, Y., Tanaka, K., & Yasui, H. (2014). Inhibitory effect of a hot-water extract of leaves of Japanese big-leaf Magnolia (Magnolia obovata) on rotavirus-induced diarrhea in mouse pups. *Evidence-Based Complementary and Alternative Medicine*, 456–468.

Kew, Royal Botanic Gardens. (2017). *State of the World's Plants 2017.* Royal Botanic Gardens, Kew. Available from https://stateoftheworldsplants.org/

Khan, S. A. & Al-Balushi, K. (2021). Combating COVID-19: the role of drug repurposing and medicinal plants. *Journal of Infection and Public Health*, 14(4), 495–503.

Kim, J. H., Kim, K., & Kim, W. (2020). Genipin inhibits rotavirus-induced diarrhea by suppressing viral replication and regulating inflammatory responses. *Scientific Reports*, 10(1), 1–11.

King, C. (2008). Exhibits tubercin activity; immune enhancers such as glycyrrhizin (GR), glycyrrhetic acid, and Chinese herbal extracts. Google Patents; 2008.

Knipping, K., Garssen, J., & van't Land, B. (2012). An evaluation of the inhibitory effects against rotavirus infection of edible plant extracts. *Virology Journal*, 9(1), 1–8.

Komape, N. P., Bagla, V. P., Kabongo-Kayoka, P., & Masoko, P. (2017). Anti-mycobacteria potential and synergistic effects of combined crude extracts of selected medicinal plants used by Bapedi traditional healers to treat tuberculosis related symptoms in Limpopo Province, South Africa. *BMC Complementary and Alternative Medicine*, 17(1). DOI:10.1186/s12906-016-1521-2.

Kryvtsova, M. V., Fedkiv, O. K., Hrytsyna, M. R., & Salamon, I. (2020). Anti-microbial, anti-biofilm-forming properties of Origanum Vulgare l. Essential oils on Staphylococcus Aureus and its antioxidant action. *Studia Biologica*, 14(2), 27–38.

Kwon, H. J., Kim, H. H., Ryu, Y. B., Kim, J. H., Jeong, H. J., Lee, S. W., … Lee, W. S. (2010). In vitro anti-rotavirus activity of polyphenol compounds isolated from the roots of Glycyrrhiza uralensis. *Bioorganic & Medicinal Chemistry*, 18(21), 7668–7674.

Frazia, S., Ciucci, A., Arnoldi, F., Coira, M., Gianferretti, P., Angelini, M., … Santoro, M. G. (2013). Thiazolides, a new class of antiviral agents effective against rotavirus infection, target viral morphogenesis, inhibiting viroplasm formation. *Journal of virology*, 87(20), 11096–11106.

Lalani, S., Anasir, M., & Poh, C. (2020). Antiviral activity of silymarin in comparison with baicalein against EV-A71. *BMC Complementary Medicine and Therapies*, 20(1). https://doi.org/10.1186/s12906-020-2880-2

Lecher, J. C., Diep, N., Krug, P. W., & Hilliard, J. K. (2019). Genistein has antiviral activity against herpes b virus and acts synergistically with antiviral treatments to reduce effective dose. *Viruses*, 11(6), 499. doi: 10.3390/v11060499.

Lee, L. J., Loe, M. W., Lee, R. C., & Chu, J. J. (2019). Antiviral activity of pinocembrin against Zika virus replication. *Antiviral Research*, 167, 13–24. doi: 10.1016/j.antiviral.2019.04.003

Levine, A. J. & Enquist, L. W. (2007). History of virology. In B. N. Fields, D. M. Knipe & P. M. Howley (eds.) *Fields Virology*, pp. 565–604, 5th edn. Philadelphia, PA: Lippincott Williams & Wilkins.

Lin, L., Hsu, W., & Lin, C. (2014). Antiviral natural products and herbal medicines. *Journal of Traditional and Complementary Medicine*, 4(1), 24–35. https://doi.org/10.4103/2225–4110.124335

Lipson, S. M., Karalis, G., Karthikeyan, L., Ozen, F. S., Gordon, R. E., Ponnala, S., … Wolfe, E. (2017). Mechanism of anti-rotavirus synergistic activity by epigallocatechin gallate and a proanthocyanidin-containing nutraceutical. *Food and Environmental Virology*, 9(4), 434–443.

Liu, C., Grillner, L., Jonsson, K., Linde, A., Shen, K., Lindell, A. T., et al. (2006). Identification of viral agents associated with diarrhea in young children during a winter season in Beijing, China. *Journal of Clinical Virology*, 35, 69–72.

Mair, C. E., Liu, R., Atanasov, A. G., Schmidtke, M., Dirsch, V. M., & Rollinger, J. M. (2016). Antiviral and anti-proliferative in vitro activities of piperamides from black pepper. *Planta Medica*, 82(S 01), P807.

Mandal, U., Jyotirmayee, B., & Mahalik, G. (2022). Traditional plants utilized for the viral disease treatment. *Plant Science Today*, 9(2), 386–398.

Maryam, M., Te, K. K., Wong, F. C., Chai, T. T., Gary, K. K., & Gan, S. C. (2020). Antiviral activity of traditional Chinese medicinal plants Dryopteris crassirhizoma and Morus alba against dengue virus. *Journal of Integrative Agriculture*, 19(4), 1085–1096.

Mendes, É. A., de Pilger, D. R. B., Nastri, A. C. D. S. S., de Mello Malta, F., dos Santos Pascoalino, B., D'Albuquerque, L. A. C., ... & Pinho, J. R. R. (2019). Sofosbuvir inhibits yellow fever virus in vitro and in patients with acute liver failure. *Annals of Hepatology*, 18(6), 816–824.

Miao, M., Jing, X., De Clercq, E., & Li, G. (2020). Danoprevir for the treatment of hepatitis C virus infection: design, development, and place in therapy. *Drug Design, Development and Therapy*, 14, 2759.

Moerman, D. E. (1996). An analysis of the food plants and drug plants of native North America. *Journal of Ethnopharmacology*, 52, 1–22.

Montenegro-Landívar, M. F., Tapia-Quirós, P., Vecino, X., Reig, M., Valderrama, C., Granados, M., … Saurina, J. (2021). Polyphenols and their potential role to fight viral diseases: an overview. *Science of the Total Environment*, 801, 149719.

Moradi, M. T., Karimi, A., & Lorigooini, Z. (2018). Alkaloids as the natural anti-influenza virus agents: a systematic review. *Toxin Reviews*, 37(1), 11–18. doi: 10.1080/15569543.2017.1323338

Moradi, M. T., Karimi, A., Rafieian-Kopaei, M., Rabiei-Faradonbeh, M., & Momtaz, H. (2020). Pomegranate peel extract inhibits internalization and replication of the influenza virus: an in vitro study. *Avicenna Journal of Phytomedicine*, 10(2), 143–151.

More-Adate, P., Lokhande, K. B., Swamy, K. V., Nagar, S., & Baheti, A. (2022). GC-MS profiling of Bauhinia variegata major phytoconstituents with computational identification of potential lead inhibitors of SARS-CoV-2 Mpro. *Computers in Biology and Medicine*, 2022, 105679.

Mullaicharam, A. R. & Halligudi, N. (2019). St John's wort (Hypericum perforatum L.): a review of its chemistry, pharmacology and clinical properties. *International Journal of Research in Phytochemical and Pharmacological Sciences*, 1(1), 5–11.

Musarra-Pizzo, M., Ginestra, G., Smeriglio, A., Pennisi, R., Sciortino, M. T., & Mandalari, G. (2019). The antimicrobial and antiviral activity of polyphenols from almond (Prunus dulcis L.) skin. *Nutrients*, 11(10), 2355.

Musarra-Pizzo, M., Pennisi, R., Ben-Amor, I., Mandalari, G., & Sciortino, M. T. (2021). Antiviral activity exerted by natural products against human viruses. *Viruses*, 13, 828.

Naseer, S., Hussain, S., Naeem, N., Pervaiz, M., & Rahman, M. (2018). The phytochemistry and medicinal value of Psidium guajava (guava). *Clinical Phytoscience*, 4(1), 1–8.

Naithani, R., Huma, L., Holland, L., Shukla, D., McCormick, D., Mehta, R., & Moriarty, R. (2008). Antiviral activity of phytochemicals: A comprehensive review. *Mini-Reviews in Medicinal Chemistry*, 8(11), 1106–1133. DOI:10.2174/138955708785909943.

O'Connor, S. E. (2010) Alkaloids in: comprehensive natural products II. *Chemistry and Biology*, 1, 977–1007.

Omotayo, A. O. & Aremu, A. O. (2021). Marama bean [Tylosema esculentum (Burch.) A. Schreib.]: an indigenous plant with potential for food, nutrition, and economic sustainability. *Food & Function*, 12(6), 2389–2403.

Patel, M. M., Steele, D., Gentsch, J. R., Wecker, J., Glass, R. I., & Parashar, U. D. (2011). Real-world impact of rotavirus vaccination. *The Pediatric Infectious Disease Journal*, 30(Suppl. 1), S1e5.

Payne, D., Wikswo, M., & Parashar, U. (2011). Rotavirus. In Roush, Sandra W & Baldy, Linda M (Eds.) *Manual for the Surveillance of Vaccine Preventable Diseases*. National Center for Immunization and Respiratory Disease, USA. 5th ed. p. 11.

Phumthum, M., Balslev, H., & Barfod, A. S. (2019). Important medicinal plant families in Thailand. *Frontiers in Pharmacology*, 10, 1125.

Rates, S. M. (2001). Plants as source of drugs. *Toxicon*, 39(5), 603–613.

Ramesh, D., Vijayakumar, B. G., & Kannan, T. (2021). Advances in Nucleoside and Nucleotide Analogues in Tackling Human Immunodeficiency Virus and Hepatitis Virus Infections. *ChemMedChem*, 16(9), 1403–1419.

Rheingans, R. D., Heylen, J., & Giaquinto, C. (2012). Economics of rotavirus gastroenteritis and vaccination in Europe: what makes sense? *The Pediatric Infectious Disease Journal*, 25, S48–S55. doi: 10.1097/01.inf.0000197566.47750.3d

Rodrigo, C., Salman, N., Tatochenko, V., Mészner, Z., & Giaquinto, C. (2010). Recommendations for rotavirus vaccination: a worldwide perspective. *Vaccine*, 28, 5100–5108. doi: 10.1016/j.vaccine.2010.04.108

Rouf, R., Uddin, S. J., Sarker, D. K., Islam, M. T., Ali, E. S., Shilpi, J. A., Nahar, L., Tiralongo, E., & Sarker, S. D. (2020). Antiviral potential of garlic (Allium sativum) and its organosulfur compounds: a systematic update of pre-clinical and clinical data. *Trends in Food Science & Technology*, 104, 219–234. https://doi.org/10.1016/j.tifs.2020.08.006

Roy, P., Pramanick, K., Das, A., Majumdar, M., Roy, A., & Sarkar, U. (2022). Review of some herbal agents having antiviral activity. *International Journal of Current Science Research and Review*, 5(3), 695–706.

Sabo, V. A. & Knezevic, P. (2019). Antimicrobial activity of Eucalyptus camaldulensis Dehn. plant extracts and essential oils: a review. *Industrial Crops and Products*, 132, 413–429.

Sadowski, L. A., Upadhyay, R., Greeley, Z. W., & Margulies, B. J. (2021). Current drugs to treat infections with herpes simplex viruses-1 and-2. *Viruses*, 13(7), 1228.

Septembre-Malaterre, A., Boumendjel, A., Seteyen, A. L. S., Boina, C., Gasque, P., Guiraud, P., & Sélambarom, J. (2022). Focus on the high therapeutic potentials of quercetin and its derivatives. *Phytomedicine Plus*, 2022, 100220.

Shaheen, M., Borsanyiova, M., Mostafa, S., Chawla-Sarkar, M., Bopegamage, S., & El-Esnawy, N. (2015). In vitro effect of Dodonaea viscosa extracts on the replication of coxackievirus B3 (Nancy) and rotavirus (SA-11). *Journal of Microbiology and Antimicrobial Agents*, 1(2), 47–54.

Shahwar, D., Raza, M. A., Atta-Ur-Rahman. (2013). Identification of flavonoids withtrypsin inhibitory activity extracted from orange peel and green tea leaves. *Journal of the Science of Food and Agriculture*, 93, 1420–1426

Sharma, V., Kaushik, S., Pandit, P., Dhull, D., Yadav, J. P., & Kaushik, S. (2019). Green synthesis of silver nanoparticles from medicinal plants and evaluation of their antiviral potential against chikungunya virus. *Applied Microbiology and Biotechnology*, 103(2), 881–891.

Shiraki, K., & Daikoku, T. (2020). Favipiravir, an anti-influenza drug against life-threatening RNA virus infections. *Pharmacology & Therapeutics*, 209, 107512. doi: 10.1016/j.pharmthera.2020.107512

Shen, J., Li, P., He, C. N., Liu, H. T., Liu, Y. Z., Sun, X. B., ... & Xiao, P. G. (2019). Simultaneous determination of 15 flavonoids from different parts of Scutellaria baicalensis and its chemometrics analysis. *Chinese Herbal Medicines*, 11(1), 20–27.

Siegel, R. D. (2018). Classification *of* human viruses. In S. S. Long, C. G. Prober & M. Fischer (eds.) *Principles and Practice of Pediatric Infectious Diseases*, 1044.

Sornpet, B., Potha, T., Tragoolpua, Y., & Pringproa, K. (2017). Antiviral activity of five Asian medicinal pant crude extracts against highly pathogenic H5N1 avian influenza virus. *Asian Pacific Journal of Tropical Medicine*, 10(9), 871–876. https://doi.org/10.1016/j.apjtm.2017.08.010

Sun, Y., Gong, X., Tan, J. Y., Kang, L., Li, D., Yang, J., & Du, G. (2016). In vitro antiviral activity of Rubia cordifolia aerial part extract against rotavirus. *Frontiers in Pharmacology*, 7, 308.

Taherkhani, R. (2015). Epidemiology of Hepatitis E virus in Iran. *World Journal of Gastroenterology*, 22 (22), 5143. DOI:10.3748/wjg.v22.i22.5143.

Taherkhani, R., Farshadpour, F., & Makvandi, M. (2013). In vitro anti-rotaviral activity of Achillea kellalensis. *Jundishapur Journal of Natural Pharmaceutical Products*, 8(3), 138.

Tam, K. I. & Roner, M. R. (2011). Characterization of in vivo anti-rotavirus activities of saponin extracts from Quillaja saponaria Molina. *Antiviral Research*, 90(3), 231–241.

Tarcsai, K. R., Corolciuc, O., Tordai, A., & Ongrádi, J. (2022). SARS-CoV-2 infection in HIV-infected patients: potential role in the high mutational load of the Omicron variant emerging in South Africa. *GeroScience*, 1–9.

Tate, J. E., Haynes, A., Payne, D. C., Cortese, M. M., Lopman, B. A., Patel, M. M., & Parashar, U. D. (2013). Trends in national rotavirus activity before and after introduction of rotavirus vaccine into the national immunization program in the United States, 2000 to 2012. *The Pediatric Infectious Disease Journal*, 32, 741–744.

Téllez Alfonso, A. N., Vélez, F., Ulloa, J. C., & Téllez-Duarte, M. A. (2015). In vitro antiviral activity against rotavirus and astrovirus infection exerted by substances obtained from Achyrocline bogotensis (Kunth) DC. (Compositae). *BMC Complementary and Alternative Medicine*, 15, 10.

Tiwari, G., Kumar, A., Dwivedi, K. K., & Sharma, D. (2022). In Silico Investigation of Electronic Structure, Binding Patterns and Molecular Docking of Nevirapine: An anti-HIV Type-1 Drug. *Polycyclic Aromatic Compounds*, 42(5), 2789–2804.

Troost, B., & Smit, J. M. (2020). Recent advances in antiviral drug development towards dengue virus. *Current opinion in virology*, 43, 9–21.

Varghese, F. S., Meutiawati, F., Teppor, M., Merits, A., & Rij, R. P. V. (2019). Posaconazole is a novel inhibitor for alphavirus viral entry. *Access Microbiology*, 1(10). https://doi.org/10.1099/acmi.imav2019.po0027

Vesikari, T., Karvonen, A., Korhonen, T., Espo, M., Lebacq, E., Forster, J., et al. (2004). Safety and immunogenicity of RIX4414 live attenuated human rotavirus vaccine in adults, toddlers and previously uninfected infants. *Vaccine*, 22, 283642.

Vesikari, T., Clark, H. F., Offit, P. A., Dallas, M. J., DiStefano, D. J., Goveia, M. G., et al. (2006). Effects of the potency and composition of the multivalent human-bovine (WC3) reassortant rotavirus vaccine on efficacy, safety and immunogenicity in healthy infants. *Vaccine*, 24, 48219.

Voukeng, I. K., Beng, V. P., & Kuete, V. (2017). Multidrug resistant bacteria are sensitive to Euphorbia prostrata and six others Cameroonian medicinal plants extracts. *BMC Research Notes*, 10(1). DOI:10.1186/s13104-017-2665-y.

Walther, C., Döring, K., & Schmidtke, M. (2020). Comparative in vitro analysis of inhibition of rhinovirus and influenza virus replication by mucoactive secretolytic agents and plant extracts. *BMC Complementary Medicine and Therapies*, 20(1), 1–12.

Wang, P., Sun, L., Tan, J., Xu, J., Guo, F., & Li, Y. (2012). Two new glycosidated coumaramides from Clerodendron cyrtophyllum. *Fitoterapia*, 83(8), 1494–1499.

Wang, S. J., Huang, C. F., & Yu, M. L. (2021). Elbasvir and grazoprevir for the treatment of hepatitis C. *Expert Review of Anti-infective Therapy*, 19(9), 1071–1081.

Wang, Y.-C. (2014). Medicinal plant activity on *Helicobacter Pylori* related diseases. *World Journal of Gastroenterology*, 20(30), 10368. DOI:10.3748/wjg.v20.i30.10368.

Weckerle, C. S., Cabras, S., Castellanos, M. E., Leonti, M. (2011). Quantitative methods in ethnobotany and ethnopharmacology: considering the overall flora—hypothesis testing for over- and underused plant families with the Bayesian approach. *Journal of Ethnopharmacology*, 137(1), 837–843.

Wei, L., Wang, F. S., Zhang, M. X., Jia, J. D., Yakovlev, A. A., Xie, W., ... & Deng, J. (2018). Daclatasvir plus asunaprevir in treatment-naïve patients with hepatitis C virus genotype 1b infection. *World journal of gastroenterology*, 24(12), 1361.

Woolhouse, M., Scott, F., Hudson, Z., Howey, R., & Chase-Topping, M. (2012). Human viruses: discovery and emergence. *Philosophical Transactions of the Royal Society B: Biological Sciences*, 367(1604), 2864–2871.

Wu, F., Liu, L., Zhu, Y. N., Lv, N., Luo, D., Zhu, L. J., … Li, Z. G. (2021). Two new benzazepine alkaloids from Thalictrum wangii and their anti-rotavirus activity. *Chemistry of Natural Compounds*, 57(2), 339–342.

Yang, P. S., Dai, J. M., Gu, X. J., Xiong, W., Huang, D. Q., Qiu, S. Y., … Zhou, M. (2022). Indole alkaloids and chromones from the stem bark of cassia alata and their antiviral activities. *Molecules, 27*(10), 3129.

Yin, Y., Dang, W., Zhou, X., Xu, L., Wang, W., Cao, W., Chen, S., Su, J., Cai, X., Xiao, S., Peppelenbosch, M. P., & Pan, Q. (2017). PI3K-Akt-mTOR axis sustains rotavirus infection via the 4E-BP1 mediated autophagy pathway and represents an antiviral target. *Virulence*, 9, 83–98.

Yu, B., Jiang, Y., Zhang, B., Yang, H., & Ma, T. (2018). Resveratrol dimer trans-ε-viniferin prevents rotaviral diarrhea in mice by inhibition of the intestinal calcium-activated chloride channel. *Pharmacological Research*, 129, 453–461.

Zambon, M. C. (2001). The pathogenesis of influenza in humans. *Reviews in Medical Virology*, 11, 227–241

Zhang, X. L., Guo, Y. S., Wang, C. H., Li, G. Q., Xu, J. J., Chung, H. Y., et al. (2014). Phenolic compounds from origanum vulgare and their antioxidant and antiviral activities. *Food Chemistry*, 152, 300–3006. doi: 10.1016/j.foodchem.2013.11.153

Zhong, P., Song, L., Gao, M., Wang, X., Tan, W., Lu, H., … Zhao, W. (2020). Network pharmacology-based strategy for predicting active ingredients and potential targets of Gegen Qinlian decoction for rotavirus enteritis. *Evidence-Based Complementary and Alternative Medicine*, 2020.

Zhou, M., Zhou, K., Gao, X. M., Jiang, Z. Y., Lv, J. J., Liu, Z. H., Yang, G. Y., Miao, M. M., Che, C. T., Hu, Q. F. (2015). Fistulains A and B, new bischromones from the bark of Cassia fistula, and their activities. *Organic Letters*, 17, 2638–2641.

# 14 Human Papillomavirus and Medicinal Plants/Herbs

## *A Perspective*

*Swati Sharma and Monika Thakur*
Shoolini Institute of Life Science and Business Management

## 14.1 INTRODUCTION

A number of viral infections are well-documented to increase the risks of cancers. In approximately 15%–20% of total human cancers, viral infections play an important role in multiple stages of cancer development. Human papillomavirus (HPV) is one such virus that is well documented to enhance the risk of cervical cancer, which is the fourth most common malignancy among women worldwide, with around 0.5 million women diagnosed with it each year (Sung et al., 2021; Venturas and Umeh, 2017). Studies as early as 1976 had suggested link between HPV infection and cervical cancer, and the first genital HPV types were identified in 1978 (zur Hausen, 1976; Orth et al., 1978). Detection of HPV-DNA in cervical neoplasia was reported by 1981, and by 1989, the ability of E6 and E7 viral proteins to transform human keratinocytes was also reported (zur Hausen et al., 1981; Hawley-Nelson et al., 1989). By mid-1990s, the virus was regarded as the "primary causal agent" in the development of the malignancy and its precursor lesions (Park et al., 1995). However, the association between human papillomavirus infection and cervical cancer was officially established in November 1991 in a workshop convened by the International Agency for Research on Cancer (IARC) and the World Health Organization (WHO) (Bosch et al., 1992). It established that infection with human papilloma virus has causal relationship with development of cervical cancer, thereby emphasising the need of global awareness about likely risks of HPV infection on women's health.

Different studies show the proportion of populations in different countries infected by HPV. Approximately 5% of European, 26% of Sub-Saharan African countries, 16.1% of Latin American, and 21.4% of Eastern European populations are infected by HPV (Clifford et al., 2005; Bruni et al., 2010). As far as India is concerned, almost 6.3% of population is infected by HPV (Sankaranarayanan et al., 2004). Side effects such as a non- specific toxicity of drugs and radiations are associated with different approaches used in cancer therapy. Furthermore, resistance to treatment also presents a great problem in management of HPV infection-associated cervical cancer (National cancer Institute). However, natural compounds with anticancer potential are a good alternative with minimum side effects and potential to overcome resistance. Plants and plant-based products are popularly used in traditional medicine systems and therefore more popularly used, especially in developing nations. Over 50% drugs that are being tested in clinical trials for their anticancer potential are obtained from plants (Newman and Cragg, 2007). A number of recent studies have also reported the discovery of new compounds with antitumoral activities from plants being used in traditional medicine systems (Hidayatullah et al., 2023; Salaria et al., 2022; Song et al., 2020; Kharisma et al., 2020; Ghanbari et al., 2019).


DOI: 10.1201/9781003329169-17

## 14.2 HUMAN PAPILLOMA VIRUS (HPV)

HPV is a non-enveloped double-stranded DNA virus belonging to the *Papovaviridae* family. The diameter of virus is 55 nm, and it has an icosahedral capsid comprising of 72 capsomeres, with each of them having two capsid proteins, namely L1 and L2. HPV can infect humans, where it may be symptomatic or asymptomatic, as well as several animals in a host specific manner. There are other cancers, such as anogenital, vulvar, vaginal, penile, head, and neck, linked to HPV infections (Kerkar et al., 2011; Petrelli et al., 2016).

More than 120 HPV genotypes that infect human skin are known today, and they can be classified into categories depending upon their potential to induce malignant transformation in infected cells (Clavel et al., 1999). High-risk types have high oncogenic potential and are the most detectable in carcinoma and dysplasias, while intermediate risk types are less ejected in carcinomas and are commonly linked with mild or severe dysplastic lesions. High-risk HPV include type 16, 18, 31, 33, 35, 39, 45, 50, 51,53, 55, 56, 58, 59, 64, and 68, and HPVs 31, 33, 35, 51, and 52 are intermediate risk types (Lorincz et al., 1992); HPV 16 and HPV18 are most commonly associated with cervical carcinoma, with HPV16 and HPV18 being detected in more than 50% of squamous cell carcinomas and more than 50% of adenocarcinomas, respectively (Milde-Langosch et al., 1993). Low-risk types, such as HPV-11, usually cause benign hyper proliferative lesions like genital warts and papillomatosis.

### 14.2.1 HPV Genome

HPV is a DNA virus with a genome organised into three regions: (i) El to E7 early genes region, (ii) L1 and L2 late gene regions, and (iii) the upper regulatory region (URR). Both the early and late gene regions encode proteins, whereas the URR does not encode any proteins but has binding sites for many transcriptions regulatory factors. Therefore, it may have some role to play in determining the hosts infected by a particular type of HPV (Turek, 1994).

Early genes 1 and 2 code for proteins that facilitate extrachromosomal replication of the viral genome and completion of the life cycle. E2 can encode different proteins that can either inhibit or increase the transcription of early region (Ward et al., 1989). Expression of E2 is lost in HPV-associated cervical cancer cells (Thierry et al., 2004). The expression of E4 gene occurs in the later stages during the assembly and maturation of virus. Its product also induces collapse of cytokeratin network in cytoplasm of infected keratinocytes, thereby facilitating release of viruses (Doorbar et al., 1991). Protein expressed from E5 gene can interact with receptors of epidermal growth factor, platelet-derived growth factor β, and colony stimulating factor. It is not required for the maintenance of cancerous state of host cells as suggested by its often deletion in cervical carcinoma cells (Hwang et al., 1995). However, in HPV16 infected cells, the E5 protein has been found to have little transforming activity (Crusius et al., 1997).

It is the E6 and E7 regions that play the most important role in cancerous transformation of infected cells. They code for oncoproteins that have functional role in viral replication immortalisation of host cells tumorigenic (Bedell et al., 1987; Phelps et al., 1988; Spitkovsky et al., 1996). The E6 protein has been reported to bind to and inactivate tumour suppressor p53 and E7 binds to and degrades tumour suppressor pRb. This resulted in E6-p53 and E7-pRb model, of how HPV infection leads to transformation of infected cells (Storey et al., 1988; Werness et al., 1990). Late genes L1 and L2 code for capsid proteins that are used during assembly stage. While the L1 protein is highly conserved in different papillomaviruses, the L2 protein has sequence variations and can be used to detect specific HPV using appropriate antibodies (Park et al., 1995).

### 14.2.2 HPV Infection-Associated Carcinogenesis

HPV-linked carcinogenesis primarily depends on interference of viral oncoproteins E6 and E7 with tumour suppressor proteins -p53 and retinoblastoma. Furthermore, factors such as dysregulated cell adhesion and cell signalling, immunomodulation in host, and downregulation of autophagy also contribute in the development of cancer.

In cells, the p53 protein can detect any cell injury or insult and, in turn, activate the transcription of genes like p21 that can delay the entry of cell into get phase of cell cycle until the damage is repaired. However, in case the repair is not possible, p53 may induce apoptosis by activating genes like bax1 and fas receptor gene (Zilfou and Lowe, 2009). The viral oncoprotein E6 forms a ternary complex with p53 and with the ubiquitin E3 ligase E6-associated protein (E6AP), resulting in the ubiquitination of p53. Ubiquitinated p53 is consequently degraded by the proteasome system. In the absence of p53, cells can evade cell cycle arrest and apoptosis leading to accumulation of cellular mutations and chromosomal instability that can eventually results in development of cancer. It must be noted that only high-risk HPVs can cause p53 degradation (Scheffner et al., 1993). pRb binds to E2F, thereby resulting in negative regulation of cell cycle and quiescent state of cell. Hyper phosphorylation of pRb by cyclin-CDK complex can release E2F, which in its free and active state can transcribe genes involved in cell cycle progression (Sherr and McCormick, 2002). However, the HPV oncoprotein E7 interacts with pRb, induces its degradation through the proteasome system, and consequently facilitates the release of E2F, leading to uncontrolled cell cycle progression even in the absence of appropriate cyclin-CDK complexes (Münger et al., 1989). Apart from these, the HPV oncoprotein E5 is involved in the activation and increase of the EGFR pathway. EGFR is a proto-oncogene with its role in cellular processes such as proliferation, migration, survival, and angiogenesis. Activation of EGFR pathway by E5 can result in intracellular signalling that activates many protooncogenes (Ilahi and Bhatti, 2020). Another signalling deregulated by viral oncoproteins E6 and E7 is the PI3K/Akt/mTOR cascade that promotes cell survival, growth, proliferation, migration, and energy metabolism. Furthermore, HPV E6 can also induce phosphorylation of JNK, resulting in activation of c-jun that promotes proliferation and expression of viral genes and can induce activation of ERK signalling (Morgan et al., 2021; Liu et al., 2016).

In addition to increasing the regulation of pathways that result in cell division and growth, HPV proteins also inhibit apoptosis. HPV16 E5 down-regulates the mRNA expression of autophagic genes, such as ATG4a, ATG5, LC3, ULK1, ULK2, Beclin 1, and ATG7, thus immortalising the cells. Interestingly, HPV-16 can also downregulate the expression of E-cadherin in epithelial cells resulting in Epithelial Mesenchymal Transition (EMT) that favours formation of cancer (Laurson et al., 2010; Jiang et al., 2019). HPV E6/E7 oncoproteins can also modulate the immune response by downregulation of NF-κB through a mutated IκB, thus facilitating survival and thriving of virus (Byg et al., 2012). The expression of E6/E7 oncoproteins is fundamentally dependent on binding of AP-1 to its consensus sequence present on HPV promoter to initiate transcription. AP-1 is a dimeric transcription factor involved in regulation of different pathways such as differentiation, proliferation, and apoptosis in cell (Steger et al., 2001).

## 14.3 THERAPEUTIC POTENTIAL OF HERBS AND MEDICINAL PLANTS AGAINST HPV

Therapies to cure HPV infections include invasive techniques like surgeries and ablation, which can be expensive, or the use of various chemotherapeutic agents. However, these therapies are not accessible and affordable, particularly in developing nations around the world. This presents a need to investigate alternative natural therapeutics that can be used to treat HPV infections. Plant-based therapies also have tremendous potential to treat cervical cancer resulting from such infections due to their ability to target multiple molecular pathways (Table 14.1).

Vegetables like broccoli, cauliflower, cabbage, and kale, when consumed, release indole-3-carbinol (I3C), which dimerises in low pH conditions of stomach to form diindolylmethane (DIM) that is further metabolised. The Brassica indoles can induce cytochrome 1A1 (CYP1A1), primarily in the liver, resulting in a shift in oestrogen metabolism from16α- and 4-hydroxylation towards 2-hydroxylation, which is beneficial as 2-hydroxyestrone suppresses HPV, while 16a- and 4-hydroxyestrone promote it (Newfield et al., 1993). The chemopreventive potential of *Ficus religiosa* has also been reported in cervical cancer. Aqueous extract of bark of *Ficus religiosa* has been reported to inhibit neoplastic growth in human cervical cancer cell lines SiHa and HeLa in a dose-dependent manner.

The cell cycle progression was blocked at $G_1/S$ phase in SiHa along with increased expression of p53, p21, pRb proteins. In HeLa cells, apoptosis was induced characterised by rise in intracellular levels of Ca2+, loss of mitochondrial membrane potential, release of cytochrome c, and increased expression of caspase-3. Furthermore, the treatment with plant extract also resulted in reduced migration and invasion ability as well as in downregulation of MMP-2 and Her-2 expression in cervical cancer cells. Expression of viral oncoproteins E6 and E7 was also found to be reduced in cervical cancer cell lines on treatment with extract (Choudhari et al., 2013). Fruit extract of *Phyllanthus emblica* has been reported to inhibit DNA binding activity of constitutively active activator protein-1 (AP-1), that is essential for tumorigenicity of cervical cancer cells. A dose and time-dependent inhibition of DNA binding activity of AP-1 was observed in both HPV 16 and HPV 18 positive cervical cells; however, the inhibition kinetics varied in the two. This inhibition was mediated through downregulation of c-Jun, JunB, JunD, and c-Fos, and was accompanied with suppression of viral transcription and reduced growth of cancerous cells. Moreover, the fruit extract induced apoptosis, thus indicating a possible use in treatment of HPV-induced cervical cancers (Mahata et al., 2013). Another study investigated the effect of lipid-soluble extract from *Pinellia pedatisecta* on HPV E6 gene expression in HPV 16 positive CaSki cells and HPV 18 positive HeLa cells. The extract was found to inhibit the growth and induce apoptosis in both the cell lines in a time and dose-dependent manner (Li et al., 2010).

Oh and Lim (2007) investigated the apoptotic effect of *Solanum nigrum* glycoprotein (SNL) on HeLa cells. 50 μg/mL of SNL affected activity of apoptotic related proteins: Release of cytochrome C into cytoplasm, activation of caspase-8, -3 and poly (ADP-ribose) polymerase (PARP) cleavage in cervical cancer cells. Furthermore, 50 μg/mL of glycoprotein treatment for 4 hours decreased the activities of transcription factors like NF-κB and AP-1; however, there was no change in DNA binding activity between NF-κB and AP-1. Similar treatment with glycoprotein also significantly reduced the amount of nitric oxide (NO) produced in cancer cells, suggesting an inhibitory effect of SNL protein on cervical cancer cells.

Studies also suggest possible use of curcumin in the management of HPV-associated tumours (Divya and Pillai, 2006). Curcumin (diferuloyl methane), a pigment found in turmeric, has been reported to be cytotoxic to cervical cancer cells infected by HPV16 and HPV18 in concentration-dependent and time-dependent manner. Apoptosis marked by nuclear fragmentation and internucleosomal fragmentation of DNA was important in regulating the growth of tumour. Furthermore, RT-PCR and western blotting results after curcumin treatment also showed decreased expression of viral oncoproteins E6 and E7, which play an important role in carcinogenesis of HPV-infected cells. Curcumin also inhibited the activation of NFκB, downregulated one of its target gene *COX*-2 and epithelial tissue-specific gene expression of HPV.

Silymarin is a flavonolignans obtained from *Silybum marianum* (L.) Gaertn. It has been reported to possess therapeutic properties like hepatoprotective, antioxidant, antiinflammatory, anti-cancerous, and cardioprotective potential (Ramasamy and Agarwal, 2008). *In vitro* studies show silymarin to induce apoptosis and necrosis in HeLa cells, depending on the dosage amount. While doses below 80 μmol/ L resulted in apoptosis hallmarked by chromatin condensation and nuclear fragmentation, high dose of 160 μmol/L induced necrosis in cells. Apoptosis was not mediated by mitochondrial Bcl-2 family, including Bcl-2 and Bax, and pan-caspase inhibitor and caspase-3 inhibitor, cannot antagonise silymarin-induced apoptosis. However, in serum free media, silymarin induces apoptosis in cervical cancer cells through p38/JNK MAPK pathway (Huang et al., 2005).

High-throughput *in vitro* screening has revealed the potential role of *carrageenan*, a sulphur containing polysaccharide from red algae, as an inhibitor of HPV. The dose required to inhibit genital HPV was found to be 50% less than that for herpes simplex viruses and some strains of HIV. The mechanism of action of *carrageenan* involves its resemblance with heparin sulphate, an HPV cell attachment factor. It primarily inhibits binding of HPV to cells and can also block it via some post attachment effect. With its widespread use in various cosmetic and food products as a thickener, such products may be effective in preventing HPV infection (Buck et al., 2006).

**TABLE 14.1**
**Herbs, Medicinal Plants or Products Obtained from Them Having Therapeutic Potential Against HPV-Infected Cancerous Cells**

| Herbs/Natural and Herbal Derivatives | Effect | References |
|---|---|---|
| *Asarinin* found in *Zanthoxylum* spp | Predicted to Disrupt MHC I localisation in the ER, the ability of infected cells to proliferate, and the virion assembly process.<br>Expected to interact with three key regions of the HPV16 E5 protein. | Hidayatullah et al., (2023) |
| *Berberis aristata* | Stigmasterol from plant can potentially inhibit HPV18 protein. | Salaria et al. (2022) |
| *Brassica oleracea* varieties | Brassica indoles can shift oestrogen metabolism in a way that HPV is suppressed. | Newfield et al. (1993) |
| *Bryophyllum pinnatum* (life plant, air plant) leaf | Inhibition of HPV18 transcription, induction of apoptosis, and suppression of the anti-apoptotic molecules on cervical cancer cells. | Mahata et al. (2013) |
| *Camellia sinensis* (green tea) | Botanical drug sinecatechins is a combination of epigallocatechin gallate (EGCG) extracted from green tea. | Ahn et al. (2003) |
| *Carrageenan* | Acts primarily by preventing the binding of HPV virions to cells. | Buck et al. (2006) |
| *Curcumin* and curcumin-based poly-herbal formulation | Inhibition of oncogenic and viral transcription through inhibition of AP-1 and NF-λB.<br>Block viral entry in cervical cells. | Divya and Pillai (2006) |
| *Ficus religiosa* (sacred fig) | Aqueous bark extract showed anti-neoplasticism potential in human cervical cancer cell lines, SiHa and HeLa. | Choudhari et al. (2013) |
| *Ficus carica latex* | Inhibition of rapid growth and invasion in HPV-positive cancer cell lines CaSki and HeLa.<br>Decreased expression of p16 and HPV oncoproteins E6, E7. | Ghanbari et al. (2019) |
| *Genistein* (flavonoid present in soy bean) | Immune benefits in a mouse model of human papillomavirus (HPV)-associated cervical cancer. | Ghaemi et al. (2012) |
| *gingerol* | Disable binding of the E6 protein to host p53 protein. | Kumar et al. (2015) |
| *Hamamelis* virginiana (witch hazel) | Tannin-rich bark extract inhibited HPV type 16 infection. | Theisen et al. (2014) |
| *Hedyotis diffusa* | Rutin, one of the ingredients in *H. diffusa,* induces a strong specific immune response against HPV-related tumours *in vivo*. | Song et al. (2020) |
| *Inula viscosa, Retama monosperma, Ormenis eriolepis* | *In vitro* studies showed methanolic extracts from all the three plants to induce significant and dose-dependent inhibitory activities against human cervical cancer cell lines, SiHa and HeLa | Merghoub et al. (2009) |
| *Periploca sepium* (silk vine) | Negative expression of HPV-DNA in anogenital warts caused by human papillomavirus (HPV). | Cui et al. (2005) |
| *Phyllanthus emblica* (Indian gooseberry) | Fruit extract suppressed viral transcription in HPV16-positive (SiHa) and HPV18-positive (HeLa) cervical cancer cells. | Mahata et al. (2013) |
| *Pinellia pedatisecta* (fan-leaf Chinese green dragon) rhizome | Lipid-soluble extract obtained from the rhizomes resulted in downregulation of HPV E6 mRN and protein and induction of apoptosis in human cervical cancer cells. | Li et al. (2010) |
| *quercetin and quercetin-like components* | *In silico* studies show potential inhibition of human cervical cancer cell line proteins. | Muthukala et al. (2015) |
| *Rheum emodi* | Clicoemodin from plant can potentially inhibit HPV18 protein. | Salaria et al. (2022) |
| *Silymarin* | Apoptosis by no-mitochondrial pathway mediated through p38/JNK MAPKs. | Huang et al. (2005) |
| *Solanum nigrum* Phytoglycoprotein | Reduces the binding activities of NF-kB and AP-1 and declines the level of nitric oxide (NO) production. | Oh and Lim (2007) |

(*Continued*)

**TABLE 14.1** (*Continued*)
**Herbs, Medicinal Plants or Products Obtained from Them Having Therapeutic Potential Against HPV-Infected Cancerous Cells**

| Herbs/Natural and Herbal Derivatives | Effect | References |
|---|---|---|
| *Soyasaponins* | Apoptosis by mitochondrial pathway. | Xiao et al. (2007) |
| *Thiazolo* [3,2-a]benzimidazole-3(2H)-one,2-(2-fluorobenzylideno)-7,8-dimethyl found in *Myristica fragrans* | Predicted to Disrupt MHC I localisation in the ER, the ability of infected cells to proliferate, and the virion assembly process. Expected to interact with three key regions of the HPV16 E5 protein. | Hidayatullah et al., (2023) |
| *Zingiber officinale* | *In silico* analysis predicted 6-gingerol to inhibit the E6 protein. | Kharisma et al. (2020) |

Soyasaponins (SS) have also been reported to inhibit proliferation of HeLa cells through changes in cell cycle and induction of apoptosis. Loss of mitochondrial transmembrane potential, increased $Ca^{2+}$ concentration in cytosol, increase in nitric oxide content, and increased activities of nitric oxide synthase (NOS) and inducible NOS were documented suggesting mitochondrial pathway-mediated apoptosis of cancer cells (Xiao et al., 2007).

Divya and Pillai (2006) demonstrated that there are changes in the expression levels of different members of the AP-1 transcription factor family in cervical cancer cells. *In vitro* and *in vivo* studies showed that c-fos is overexpressed, and fra-1 expression is downregulated during progression to malignancy. Furthermore, there were also changes in dimerisation pattern of the AP-1 complex. Treatment with curcumin, an active component of turmeric, was demonstrated to downregulate HPV18 transcription and AP-1 binding activity in tumorigenic cells. Curcumin was also found to modulate the expression of c-fos and fra-1 in HeLa cells, reestablishing it back to normal control cells. Thus, this study suggests a possible role of anti-oxidative agents such as curcumin in controlling HPV infection and its progression into malignant state.

Variant peptide-based vaccine in combination with extract of *Hedyotis diffusa*, a plant used clinically in Asian countries, has been reported to improve antitumour activity in HPV-related tumour models. Rutin, one of the major ingredients of the extract, can induce a strong specific immune response against HPV transformed cells *in vivo*. Therefore, it can be used as a natural adjuvant in such vaccines to increase immune response against HPV (Song et al., 2020).

Different *in silico* studies have also identified molecules obtained from herbs and medicinal plants that have great potential for investigating a cure for HPV infections. Docking analysis has shown inhibitory potential of quercetin and quercetin-like components against human cervical cancer cell line proteins (Muthukala et al., 2015). Interaction studies carried out by Kumar et al. (2015) using Autodock 4.2 tool showed several natural compounds such as ferulic acid, ursolic acid, jaceosidin, curcumin, gingerol, indol-3-carbinol, silymarin, resveratrol, berberine, and artemisinin to interact with HPV18 E6 oncoprotein. This interaction can disable binding of the E6 protein to host the p53 protein, thus having a potentially anticancer effect. Recently Salaria et al. (2022) selected 119 phytoconstituents from 20 unexplored medicinal plants of north-western Himalayas. Among them, two were found to have the potential to develop anti-HPV drug. Computational screening and stimulation studies demonstrated stigmasterol from *Berberis aristata*, and clicoemodin from *Rheum emodi* can bind inside the HPV18 protein pocket, thereby inhibiting its activity.

The E5 protein from HPV16 can suppress tumour suppressors p21 and p27, resulting in downregulation of MHC I and BAP31 in the ER and promotion of infected cell into S phase of cell cycle. Docking studies, conducted by Putra (Hidayatullah et al., 2023), showed two compounds, asarinin and thiazolo, found in *Zanthoxylum* spp. and *Myristica fragrans*, respectively, bind to HPV16E5 protein and are predicted to disrupt MHC I localisation in the ER, affecting the ability of proliferation and the assembly process of virion. Additionally, 6-gingerol obtained from *Zingiber officinale* has also been predicted to bind to the E6 protein with the lowest binding energy, forming hydrogen and hydrophobic bond, resulting in its inhibition (Kharisma et al., 2020).

## 14.4 CONCLUSION

The treatment of human papillomavirus-related cervical cancer still faces many limitations. Approaches such as surgery, radiation therapy, chemotherapy, and cryotherapy can have several side effects (Cheng et al., 2020; Jørgensen et al., 2020). Therefore, there is a need to develop alternative strategies for combating HPV infections and related cancer risk. Various biological and *in silico* studies have brought forward plants used in traditional medicine systems that can be explored further for their potential to develop antiviral agents and drugs. Various vaccination programmes have been ongoing in different regions of sub-Sahara African countries to provide protection against HPV-16 and HPV-18 strains. However, these programmes lack efficiency, cost-effectiveness, and the ability to reach all those in need (Nnadozie et al., 2022). Since herbal medicines have been an important part of healthcare systems in developing nations such as Africa, adopting medicinal plant-based therapies against HPV infection has high likelihood of acceptance and success in these areas. A number of computational studies have also predicted bioactive compounds found in different medicinal plants to bind and inhibit HPV proteins such as E5 and E6. While extracts and formulations prepared from medicinal plants have been shown to inhibit the growth of transformed cells and exhibit cytotoxic potential, the *in silico* predictions can be used as a reference for designing new drugs as well. This highlights the tremendous potential of bio-therapeutics in the management of HPV infection and associated cancers.

## REFERENCES


Ahn WS, Yoo J, Huh SW, Kim CK, Lee JM, Namkoong SE, Bae SM, Lee IP. Protective effects of green tea extracts (polyphenon E and EGCG) on human cervical lesions. *European Journal of Cancer Prevention.* 2003 Oct 1:383–90.

Bedell MA, Jones KH, Laimins LA. The E6-E7 region of human papillomavirus type 18 is sufficient for transformation of NIH 3T3 and rat-1 cells. *Journal of Virology.* 1987 Nov;61(11):3635–40.

Bosch FX, Munoz N, Shah KV, Meheus A. Second international workshop on the epidemiology of cervical cancer and human papillomavirus. *International Journal of Cancer.* 1992 Sep 9;52(2):171–3.

Bruni L, Diaz M, Castellsagué M, Ferrer E, Bosch FX, de Sanjosé S. Cervical human papillomavirus prevalence in 5 continents: meta-analysis of 1 million women with normal cytological findings. *Journal of Infectious Diseases.* 2010 Dec 15;202(12):1789–99.

Buck CB, Thompson CD, Roberts JN, Müller M, Lowy DR, Schiller JT. Carrageenan is a potent inhibitor of papillomavirus infection. *PLoS Pathogens.* 2006 Jul;2(7):e69.

Byg LM, Vidlund J, Vasiljevic N, Clausen D, Forslund O, Norrild B. NF-κB signalling is attenuated by the E7 protein from cutaneous human papillomaviruses. *Virus Research.* 2012 Oct 1;169(1):48–53.

Cheng L, Wang Y, Du J. Human papillomavirus vaccines: an updated review. *Vaccines.* 2020 Jul 16;8(3):391.

Choudhari AS, Suryavanshi SA, Kaul-Ghanekar R. The aqueous extract of *Ficus religiosa* induces cell cycle arrest in human cervical cancer cell lines SiHa (HPV-16 positive) and apoptosis in HeLa (HPV-18 positive). *PLoS One.* 2013 Jul 26;8(7):e70127.

Clavel C, Masure M, Bory JP, Putaud I, Mangeonjean C, Lorenzato M, Gabriel R, Quereux C, Birembaut P. Hybrid capture II-based human papillomavirus detection, a sensitive test to detect in routine high-grade cervical lesions: a preliminary study on 1518 women. *British Journal of Cancer.* 1999 Jul;80(9):1306–11.

Clifford GM, Gallus S, Herrero R, Munoz N, Snijders PJ, Vaccarella S, Anh PT, Ferreccio C, Hieu NT, Matos E, Molano M. Worldwide distribution of human papillomavirus types in cytologically normal women in the International Agency for Research on Cancer HPV prevalence surveys: a pooled analysis. *The Lancet.* 2005 Sep 17;366(9490):991–8.

Crusius K, Auvinen E, Alonso A. Enhancement of EGF-and PMA-mediated MAP kinase activation in cells expressing the human papillomavirus type 16 E5 protein. *Oncogene.* 1997 Sep;15(12):1437–44.

Cui BN, Xu X, Zhao YS. Clinical efficacy of compound periploca liquid in treating condyloma acuminatum and its effect on human papilloma virus DNA expression in vitro [in Chinese]. *Zhongguo Zhong Xi Yi Jie He Za Zhi* 2005;25:392–96.

Divya CS, Pillai MR. Antitumor action of curcumin in human papillomavirus associated cells involves down-regulation of viral oncogenes, prevention of NFkB and AP-1 translocation, and modulation of apoptosis. *Molecular Carcinogenesis: Published in cooperation with the University of Texas MD Anderson Cancer Center.* 2006 May;45(5):320–32.

Doorbar J, Ely S, Sterling J, McLean C, Crawford L. Specific interaction between HPV-16 E1-E4 and cytokeratins results in collapse of the epithelial cell intermediate filament network. *Nature.* 1991 Aug;352(6338):824–7.

Ghaemi A, Soleimanjahi H, Razeghi S, Gorji A, Tabaraei A, Moradi A, Alizadeh A, Vakili MA. Genistein induces a protective immunomodulatory effect in a mouse model of cervical cancer. *Iranian Journal of Immunology.* 2012;9(2):119–27.

Ghanbari A, Le Gresley A, Naughton D, Kuhnert N, Sirbu D, Ashrafi GH. Biological activities of *Ficus carica* latex for potential therapeutics in Human Papillomavirus (HPV) related cervical cancers. Scientific Reports. 2019;9(1):1–11.

Hawley-Nelson P, Vousden KH, Hubbert NL, Lowy DR, Schiller JT. HPV16 E6 and E7 proteins cooperate to immortalize human foreskin keratinocytes. *The EMBO Journal.* 1989 Dec;8(12):3905–10.

Hidayatullah A, Putra WE, Rifa'i M, Heikal MF, Widiastuti D, Permatasari GW, Susanto H, Shuib AS. Molecular docking and molecular dynamic study of multiple medicinal plants' bioactive compounds as human papillomavirus type 16 E5 protein inhibitor. *Kuwait Journal of Science.* 2023; 50(1A):1–14.

Huang Q, Wu LJ, Tashiro SI, Onodera S, Li LH, Ikejima T. Silymarin augments human cervical cancer HeLa cell apoptosis via P38/JNK MAPK pathways in serum-free medium. *Journal of Asian Natural Products Research.* 2005 Oct 1;7(5):701–9.

Hwang ES, Nottoli T, Dimaio D. The HPV16 E5 protein: expression, detection, and stable complex formation with transmembrane proteins in COS cells. *Virology.* 1995 Aug 1;211(1):227–33.

Ilahi NE, Bhatti A. Impact of HPV E5 on viral life cycle via EGFR signaling. *Microbial Pathogenesis.* 2020 Feb 1;139:103923.

Jiang J, Li X, Yin X, Zhang J, Shi B. Association of low expression of E cadherin and β catenin with the progression of early stage human squamous cervical cancer. *Oncology Letters.* 2019 Jun 1;17(6):5729–39.

Jørgensen L, Gøtzsche PC, Jefferson T. Benefits and harms of the human papillomavirus (HPV) vaccines: systematic review with meta-analyses of trial data from clinical study reports. *Systematic Reviews.* 2020 Dec;9(1):1–23.

Kerkar SC, Latta S, Salvi V, Mania-Pramanik J. Human papillomavirus infection in asymptomatic population. *Sexual & Reproductive Healthcare.* 2011 Jan 1;2(1):7–11.

Kharisma VD, Ansori ANM, Nugraha AP. Computational study of ginger (*Zingiber officinale*) as E6 inhibitor in human papillomavirus type 16 (HPV-16) infection. *Biochemical and Cellular Archives.* 2020;20(1):3155–59.

Kumar S, Jena L, Sahoo M, Kakde M, Daf S, Varma AK. In silico docking to explicate interface between plant-originated inhibitors and E6 oncogenic protein of highly threatening human papillomavirus 18. *Genomics & Informatics.* 2015 Jun;13(2):60.

Laurson J, Khan S, Chung R, Cross K, Raj K. Epigenetic repression of E-cadherin by human papillomavirus 16 E7 protein. *Carcinogenesis.* 2010 May 1;31(5):918–26.

Li GL, Jiang W, Xia Q, Chen SH, Ge XR, Gui SQ, Xu CJ. HPV E6 down-regulation and apoptosis induction of human cervical cancer cells by a novel lipid-soluble extract (PE) from *Pinellia pedatisecta* Schott in vitro. *Journal of Ethnopharmacology.* 2010 Oct 28;132(1):56–64.

Liu F, Lin B, Liu X, Zhang W, Zhang E, Hu L, Ma Y, Li X, Tang X. ERK signaling pathway is involved in HPV-16 E6 but not E7 oncoprotein-induced HIF-1α protein accumulation in NSCLC cells. *Oncology Research.* 2016;23(3):109.

Lorincz AT, Reid RI, Jenson AB, Greenberg MD, Lancaster WA, Kurman RJ. Human papillomavirus infection of the cervix: relative risk associations of 15 common anogenital types. *Obstetrics and Gynecology.* 1992 Mar 1;79(3):328–37.

Mahata S, Pandey A, Shukla S, Tyagi A, Husain SA, Das BC, Bharti AC. Anticancer activity of *Phyllanthus emblica* Linn.(Indian gooseberry): inhibition of transcription factor AP-1 and HPV gene expression in cervical cancer cells. *Nutrition and Cancer.* 2013 Jan 1;65(supl):88–97.

Merghoub N, Benbacer L, Amzazi S, Morjani H, El-Mzibri M. Cytotoxic effect of some Moroccan medicinal plant extracts on human cervical cell lines. *Journal of Medicinal Plants Research.* 2009;3(12);1045–50.

Milde-Langosch K, Schreiber C, Becker G, Löning T, Stegner HE. Human papillomavirus detection in cervical adenocarcinoma by polymerase chain reaction. *Human Pathology.* 1993 Jun 1;24(6):590–4.

Morgan EL, Scarth JA, Patterson MR, Wasson CW, Hemingway GC, Barba-Moreno D, Macdonald A. E6-mediated activation of JNK drives EGFR signalling to promote proliferation and viral oncoprotein expression in cervical cancer. *Cell Death & Differentiation.* 2021 May;28(5):1669–87.

Münger K, Werness BA, Dyson N, Phelps WC, Harlow E, Howley PM. Complex formation of human papillomavirus E7 proteins with the retinoblastoma tumor suppressor gene product. *The EMBO Journal.* 1989 Dec;8(13):4099–105.

Muthukala BO, Sivakumari KA, Ashok KA. In silico docking of Quercetin compound against the Hela cell line proteins. *International Journal of Current Pharmaceutical Research.* 2015;7(1):13–6.

Newman DJ, Cragg GM. Natural products as sources of new drugs over the last 25 years. *Journal of Natural Products.* 2007;70(3):461–77.

Nnadozie FNA, Tankeu S, Nwaozuzu D. The use of native flora/herbal products in human papilloma virus (HPV) infection: a global perspective study. In Sanjeet Kumar (Ed.) *Medicinal Plants*, IntechOpen, 2022. Available from: http://dx.doi.org/10.5772/intechopen.1047

Newfield L, Goldsmith A, Bradlow HL, Auborn K. Estrogen metabolism and human papillomavirus-induced tumors of the larynx: chemo-prophylaxis with indole-3-carbinol. *Anticancer Research.* 1993 Mar 1;13(2):337–41.

Oh PS, Lim KT. HeLa cells treated with phytoglycoprotein (150 kDa) were killed by activation of caspase 3 via inhibitory activities of NF-κB and AP-1. *Journal of Biomedical Science.* 2007 Mar;14(2):223–32.

Orth G, Favre M, Jabblonska S, Brylak K, Croissant O. Viral sequences related to a human skin papillomavirus in genital warts. *Nature.* 1978 Sep;275(5678):334–6.

Park TW, Fujiwara H, Wright TC. Molecular biology of cervical cancer and its precursors. *Cancer.* 1995 Nov 15;76(S10):1902–13.

Petrelli A, Di Napoli A, Rossi PG, Rossi A, Luccini D, Di Marco I, Traoré AL, Tos AG, Trevisan M, Mirisola C, Costanzo G. Prevalence of primary HPV in Djibouti: feasibility of screening for early diagnosis of squamous intraepithelial lesions. *Journal of Lower Genital Tract Disease.* 2016 Oct 1;20(4):321–6.

Phelps WC, Yee CL, Münger K, Howley PM. The human papillomavirus type 16 E7 gene encodes transactivation and transformation functions similar to those of adenovirus E1A. *Cell.* 1988 May 20;53(4):539–47.

Ramasamy K, Agarwal, R. Multitargeted therapy of cancer by silymarin. Cancer Letters. 2008;269(2):352–62.

Salaria D, Rolta R, Mehta J, Awofisayo O, Fadare OA, Kaur B, Kumar B, Araujo da Costa R, Chandel SR, Kaushik N, Choi EH. Phytoconstituents of traditional Himalayan Herbs as potential inhibitors of Human Papillomavirus (HPV-18) for cervical cancer treatment: an In silico Approach. *PLoS One.* 2022 Mar 17;17(3):e0265420.

Sankaranarayanan R, Chatterji R, Shastri SS, Wesley RS, Basu P, Mahe C, Muwonge R, Seigneurin D, Somanathan T, Roy C, Kelkar R. Accuracy of human papillomavirus testing in primary screening of cervical neoplasia: results from a multicenter study in India. *International Journal of Cancer.* 2004 Nov 1;112(2):341–7.

Scheffner M, Huibregtse JM, Vierstra RD, Howley PM. The HPV-16 E6 and E6-AP complex functions as a ubiquitin-protein ligase in the ubiquitination of p53. *Cell.* 1993 Nov 5;75(3):495–505.

Sherr CJ, McCormick F. The RB and p53 pathways in cancer. *Cancer Cell.* 2002 Aug 1;2(2):103–12.

Song YC, Huang HC, Chang CY, Lee HJ, Liu CT, Lo HY, Ho TY, Lin WC, Yen HR. A potential herbal adjuvant combined with a peptide-based vaccine acts against HPV-related tumors through enhancing effector and memory T-cell immune responses. *Frontiers in Immunology.* 2020 Feb 20;11:62.

Spitkovsky D, Aengeneyndt F, Braspenning J, von Knebel Doeberitz M. p53-independent growth regulation of cervical cancer cells by the papillomavirus E6 oncogene. *Oncogene.* 1996 Sep 1;13(5):1027–35.

Steger G, Rehtanz M, Schnabel C. Identification of a promoter in position 56 within the long control region of human papillomavirus type 18. *Archives of Virology.* 2001 Nov;146(11):2069–84.

Storey A, Pim D, Murray A, Osborn K, Banks L, Crawford L. Comparison of the in vitro transforming activities of human papillomavirus types. *The EMBO Journal.* 1988 Jun;7(6):1815–20.

Sung H, Ferlay J, Siegel RL, Laversanne M, Soerjomataram I, Jemal A, Bray F. Global cancer statistics 2020: GLOBOCAN estimates of incidence and mortality worldwide for 36 cancers in 185 countries. *CA: A Cancer Journal for Clinicians.* 2021 May;71(3):209–49.

Thierry F, Benotmane MA, Demeret C, Mori M, Teissier S, Desaintes C. A genomic approach reveals a novel mitotic pathway in papillomavirus carcinogenesis. *Cancer Research.* 2004 Feb 1;64(3):895–903.

Theisen LL, Erdelmeier CA, Spoden GA, Boukhallouk F, Sausy A, Florin L, Muller CP. Tannins from Hamamelis virginiana bark extract: characterization and improvement of the antiviral efficacy against influenza A virus and human papillomavirus. *PLoS One.* 2014 Jan 31;9(1):e88062.

Turek LP. The structure, function, and regulation of papillomaviral genes in infection and cervical cancer. *Advances in Virus Research*. 1994 Jan 1;44:305–56.

Venturas C, Umeh K. Health professional feedback on HPV vaccination roll-out in a developing country. *Vaccine*. 2017 Apr 4;35(15):1886–91.

Ward P, Coleman DV, Malcolm AD. Vegulatory mechanisms of the papillomaviruses. *Trends in Genetics*. 1989 Jan 1;5:97–9.

Werness BA, Levine AJ, Howley PM. Association of human papillomavirus types 16 and 18 E6 proteins with p53. *Science*. 1990 Apr 6;248(4951):76–9.

Xiao JX, Huang GQ, Zhang SH. Soyasaponins inhibit the proliferation of Hela cells by inducing apoptosis. *Experimental and Toxicologic Pathology*. 2007 Sep 1;59(1):35–42.

Zilfou JT, Lowe SW. Tumor suppressive functions of p53. *Cold Spring Harbor Perspectives in Biology*. 2009 Nov 1;1(5):a001883.

zur Hausen H. Condylomata acuminata and human genital cancer. *Cancer Research*. 1976 Feb;36(2_Part_2):794.

zur Hausen H, de Villiers EM, Gissmann L. Papillomavirus infections and human genital cancer. *Gynecologic Oncology*. 1981 Oct 1;12(2):S124–8.

# 15 Medicinal Plants and Herbs in Viral Meningitis

*Lakhveer Singh, Mandeep Kumar Arora, Priyanshu Kumar, Saumaya Rustogi, and Shashank Shekher Mishra*
School of Pharmaceutical & Population Health Informatics, DIT University

*Manoj Gadewar*
School of Medical & Allied Sciences, KR Mangalam University

## 15.1 INTRODUCTION

Meningitis, an inflammatory disease of brain meninges, occurs when a virus or bacteria enters the brain following any invasive procedure or traumatic injury on head (van de Beek et al., 2021). The commonly observed viral meningitis symptoms are similar to bacterial meningitis like fever, headache, sensitivity to light, lethargy and confusion and stiffness in the neck (Kohil et al., 2021).

Due to similarity in signs and symptoms, physicians often face challenges in differentiating between encephalitis, meningitis and meningoencephalitis. Encephalitis involves inflammation in the brain parenchyma and cerebral cortex, which leads to appearance of neurological signs like altered mental status and behaviour. While meningitis involves inflammation of the meninges, resulting in fever and pleocytosis of the cerebral fluid, meningoencephalitis involves the infection of central nervous system and presents the signs and symptoms of both meningitis and encephalitis (Melhuish and McGill, 2021).

Sometimes rashes are also observed, which are quite different from the bacterial meningitis. Inflammation of brain cortex and parenchyma lead to behavioural changes, focal neurological abnormalities and impairment of consciousness, which is why viral meningitis is sometimes confused with encephalitis (Soung and Klein, 2018).

## 15.2 CAUSATIVE ORGANISMS FOR VIRAL MENINGITIS

### 15.2.1 Enteroviruses (EVs)

While a variety viruses like enteroviruses, herpesviruses and influenza viruses are generally involved in the causation of viral meningitis, enterovirus has been reported as the primary cause of most of the cases of viral meningitis (PETTE, 1960). The virus can infect the healthy individuals after transmission though the oral-faecal route and through the nasal route. The virus replicates in the epithelial cells of the GIT, nasopharynx, oropharyngeal and intestinal mucosal cell. The viral then migrates and reaches the lower gastrointestinal tract (GIT), where it binds with the specific receptors on the enterocytes of GIT and continue its replication in the Peyer's patch after crossing the intestinal lining. The virus can also transmit from mother to their offspring through breastfeeding.

To gain a thorough understanding of the pathogenesis of viral meningitis, it is essential to explore the interplay between EV-infected cells and the development of resultant neuronal disorder. Several studies have confirmed that EV can infect the glial cells and astrocytes present in the cerebrospinal fluid (CSF).


DOI: 10.1201/9781003329169-18

Studies on preclinical mice models have shown that EV-71 and Coxsackievirus B can replicate in undifferentiated neuronal progenitor cells (Solomon et al., 2010). Additionally, another virus from this family, EV-D68, can cause infections in the neuroblastoma cells lines as well as in the postnatal cortical neuronal culture cells. The virus can also infect immune cells in the blood, thus disseminating to other sites like heart and central nervous system (CNS), which can further lead to the development of meningitis and encephalitis (Sooksawasdi et al., 2021).

Coxsackievirus B, a member of Enterovirus B family, is well known for its CNS infection. The virus first infects the mononuclear cells with the potential to penetrate the BBB and then invade into the epithelium of choroid plexus, subsequently, spreading the infection to the CNS. Further, in CNS, coxsackievirus causes infection in the monocytes, T-cells and B-cells, causing further exacerbation of the inflammation in the CSF. These immune cells secrete inflammatory mediators like IL-6, TNF-α, IFN-α and MCP-1 (Garmaroudi et al., 2015). Additionally, the virus can transfer from mother to her offspring.

Another virus from the Enterovirus D family, Enterovirus D68 (EV-68), has been reported to cause aseptic meningitis and paralysis, especially in children. As EV-D68 is a respiratory virus, its pathogenesis differs. Once in the respiratory tract, it binds to the N-acetylneuraminic acid α2,6-galactose receptor present on the surface of the epithelial cells. This virus causes limited neurological complications like meningitis due to the presence of N-acetylneuraminic acid α2,6-galactose receptor only in the respiratory tract. Additionally, EV-D68 can also reach the CNS through olfactory nerves in the nasal cavity (Elrick et al., 2021).

### 15.2.2 Human Par Echovirus (HPeV)

This is another virus known for causing respiratory infections and spreading from person to person through respiratory secretions and saliva. It can cause viral meningitis in children but the exact mechanism of pathogenies in meningitis is still unknown. To date, no receptor has been identified for this virus (Renaud and Harrison, 2015).

### 15.2.3 Herpesviridae

Major members of this family that can cause viral meningitis are human herpes virus-1 and 2 (HSP-1 and HSP-2), Varicella-Zoster Virus (VZV) and Epstein-Barr virus (EBV). These viruses are well known for causing viral meningitis. These viruses have special characteristics of latent infection in neurons that can reactivate and re-establish the infection. HSP-1/2 uses the four types of glycoproteins gB, gD, gH and gL to enter the target cells, while actual cell surface binding is mediated by only gB and gC. HSP-1 needs the following three receptor for binding to target cells: (i) Herpesvirus entry mediator (HVEM), (ii) Members of the immunoglobulin superfamily and (iii) Heparin sulfate generated by certain isoforms of 3-O-sulfotransferases. It is reported that HVEM receptor is also present on the epithelial cells and neurons. HSP-1 binds to HVEM receptor and, subsequently, enters the viral meningitis after reaching the CSF (Zhu and Viejo-Borbolla, 2021).

EBV initiates its pathogenesis by entering through the oropharyngeal epithelium cells from where it migrates to the lymphoid tissue. In the lymphoid tissue, it infects the B-cells. The virus remains in a latent phase in the B-cells. During this latent phase, the virus expresses Epstein-Barr nuclear antigens (EBNAs) and latent membrane proteins (LMPs). The virus can also cause latent infection in the endothelial cells of blood vessels that supply blood to the brain. In stressful condition, the virus becomes activated and causes lysis of the endothelial cells, followed by the entry of virions into the brain meninges. Inflammatory mediators like IL-1, TNF-α, IL-12 and IL-18 are also released along with the virus, leading to the inflammation of the brain meninges (Kleines et al., 2011).

### 15.2.4 Family Orthomyxoviridae

Human influenza viruses belong to this family and mainly cause respiratory infections like pneumonia and acute respiratory distress syndrome. In some rare cases, these viruses can also cause other disorders like meningitis, encephalitis, meningoencephalitis and myelitis. Influenza virus can enter the CNS through the olfactory route or through the cranial nerves. The virus requires α2,6-lnked sialic acid and α2,3-lnked sialic acid, which are found on the epithelial cells of the respiratory tract. However, studies have also confirmed the presence of α2,6-lnked sialic acid and α2,3-lnked sialic acid on the neurons of cerebral cortex and brainstem. Therefore, the possible mechanism of influenza viruses to enter in the CNS could be through these receptor molecules, subsequently allowing the virus to entry the CSF (Mtambo et al., 2021).

### 15.2.5 Arboviruses

Arboviruses include viruses belonging to Flaviviridae (e.g., West Nile Virus), Togaviridae (e.g., Eastein equine encephalitis virus), and Peribunyaviridae (e.g., La Crosse virus). Generally, all of these are associated with CNS infections. Most of these have zoonotic origins and spread in humans through bites of mosquitoes, ticks and sandflies (Kaiser, 2008). Whenever any person is bitten by the infected Arthropoda, the transmitted virus first migrates to the regional lymph node and causes infection in the immune cells of lymph nodes. Infected cells from the lymph nodes enter the bloodstream and through the bloodstream, the virus ultimately reaches the brain. Infections caused by West Nile Virus are known as West Nile neuroinvasive disorders (WNNDs), which include aseptic meningitis, paralysis and encephalitis (Sejvar, 2016). Once the virus enters the lymph nodes, it can then spread to secondary locations through bloodstream. From the bloodstream, the virus can also cross the BBB and then cause CNS infection. In CNS, the virus binds to the Toll-like receptor 3 present on the immune cells and triggers the release of inflammatory mediators like TNF-α along with other inflammatory mediators. All these factors increase the permeability of the BBB, allowing the virions to enter the CSF, and subsequently, the patient develops meningeal infection.

## 15.3 PATHOPHYSIOLOGY OF VIRAL MENINGITIS

Though the precise mechanism of virus entry in the brain meningitis is unknown, the recent findings have considerably helped in understanding the pathogenesis of viral meningitis. Pathogenesis of viral meningitis involves the infection of the individual by respiratory tract infections or gastrointestinal infections, followed by its entry to the brain meningitis through the bloodstream. Some viruses like poli and Herpes simplex virus can also invade through the peripheral and cranial nerves. Once the virus enters the CNS, it spreads to the CSF through subarachnoid space. Infection of the virus in CSF triggers inflammatory response, leading to the development of meningitis. The virus can also enter the neurons through the infected leukocytes and glial cells. Once the firm CNS infection has been established, another inflammatory cell specifically targeting the infecting virus starts gathering in the CNS. Accumulated inflammatory cells further secrete cytokines like interleukin (IL)-1β, IL-6 along with tumour necrosis factor-1α (TNF-1α) and antibodies production by the plasma cells. Combined inflammatory responses by infiltrating cells and cytokines lead to an increase in the permeability of the blood brain barrier, further exacerbating the inflammatory response. This is one mechanism of virus entry in the meninges. Another mechanism is through blood and its components (Chadwick, 2005).

Once in the bloodstream, the virus can infect the blood cells (monocytes, lymphocytes and neutrophils) or the endothelium cells of blood vessels entering the brain meninges. Infection of both endothelium and blood vessels can result in breakdown of tight endothelium junctions and the underlying basement membrane. Viral infected cells (neutrophils, monocytes) enter the pia mater through the process of diapedesis and release inflammatory mediators. Sometimes microglial cells become infected with virion particles, and together with infiltrating immune cells, they produce chemo attractants that cause further breakdown of tight junctions of endothelium cells and influx of uninfected innate (macrophages, neutrophils) and adaptive (T and B Lymphocytes). Continuous accumulation of immune cells forms a nascent granuloma, which slowly keeps secreting cytokines, causing inflammation of the meninges. Further, rupture in the nascent granuloma, following the necrosis of infiltrating cells, leads to dissemination of viral infection in the meninges and cerebrospinal fluid (Wilkinson et al., 2017).

## 15.4 TREATMENT AND MANAGEMENT OF VIRAL MENINGITIS

Most of the patients with viral meningitis can be managed and treated at home with antipyretics, antiemetics and analgesics without being admitted in the hospitals. However, those who are severely infected with visible symptoms such as suffusing, high fever, severe headache, vomiting and seizure need to be admitted to the hospital and should be kept under medical supervision. These patients can be treated with corticosteroids to reduce the inflammatory response and mediators, but specific treatment for the causative organism is still needed to prevent the recurrent inflammation (Haydar et al., 2019).

Despite the advancement in medical science, especially in the treatment and management of infectious disease, progress in the treatment of viral meningitis is still lacking. In 1990 and 2000, Pleconaril was used for the first time to control the enteroviral neurological infections due to its special property of high accumulation in the CSF. However, it was never licensed due to its failure to control the viral infection and its higher risks, including drug interactions (Desmond et al., 2006). Although it had shown some degree of success in immunocompromised patient suffering chronic enteroviral infections, it failed to show any therapeutic benefit in patients suffering with meningitis caused by EV-17 virus (Abzug et al., 2003). Most of the patients with mild viral meningitis recover quickly and do not require any specific treatment, however, specific treatment is still needed in the severely infected patients. To discover an effective treatment, researchers are evaluating the antiviral drugs against the viral meningitis. For example, acyclovir was tested against the HSP-2 induced viral meningitis, which exhibited efficacy in reducing the symptoms. Very recently, Psoromic acid, a drug derived from bioactive lichen, has shown its efficacy against HSP-1 and HSP-2 virus (Lauinger et al., 2013). The drug acted by inhibiting protease and DNA polymerase in HSPs. Psoromic acid could be a future drug for the treatment and management of viral meningitis caused by HSP-1/2 (Šmejkal, 1946). Valacyclovir was tested in clinical trials for preventing and suppressing recurrent viral meningitis, but did not document efficacy due to which the drug was withdrawn from clinical trials. Later some vaccines were also designed to prevent the viral meningitis, but none of them showed any clinical efficacy (Šmejkal, 1946). Therefore, the need for discovery of an effective treatment for viral meningitis remains prevalent.

### 15.4.1 Drawback of Current Treatment

Conventional drugs are associated with serious toxicities and drug interactions. Moreover, cost of these drugs is very high, making them unaffordable for many patients. Therefore, alternative drugs should be designed to minimize the side effects and reduce the cost of treatment.

## 15.5 MEDICINAL PLANTS FOR VIRAL MENINGITIS

Medicinal plants and herbs have been used for treatment and management of various ailments since ancient time. Neem (*Azadirachta indica*) is well known for its antiviral activity and is still used in rural area for cleansing the wounds and ulcers. The antiviral potential of certain pure compounds like andrographolide, curcumin and glycyrrhic acid has already been proven through clinical investigations (Zhu and Viejo-Borbolla, 2021). There are several undiscovered and unexplored plants that still have not yet been investigated for their antiviral potential due to a lack of available information. Moreover, some of these plants are commonly consumed in our diet, such as onion and garlic, and are easily available in our surrounding geographical area. As people are unaware of the therapeutic potential of these plants, and they are not used commonly by people to prevent viral infections like meningitis. Table 15.1 lists the names of the commonly used plants that are likely to be effective in the treatment and management of viral meningitis. A few plant descriptions are also given in more detail.

### 15.5.1 *Allium sativum* (Garlic)

Meningitis can be successfully treated with garlic. *Allium sativum*, a member of the Liliaceae family, is the plant whose bulb constitutes garlic. The four active constituents which are present to reduce the action of viral meningitis are volatile oil, allicin, carbon dioxide and diallyl sulphide.

Garlic is used as a carminative aphrodisiac, stimulant and expectorant, as well as for treating pulmonary conditions. It is often used as condiments. Garlic oil is employed as an anti-helminthic and rubefacient. Allicin is widely used in the treatment of meningitis due to its antibacterial, antiviral and antifungal effects. Atherosclerosis and high blood pressure can both be treated with garlic oil. Additionally, using fresh garlic can prevent amoebic dysentery.

### 15.5.2 *Eucalyptus globulus*

Meningitis is treated by gargling with leaf extract having volatile oil. Housewives utilise the young leaves, especially from copice growths, as a bug repellent. The leaves can also be used as floor bedding. Since its introduction in 1898, the eucalyptus has become widespread throughout Ethiopia's mid- and high-altitude areas. Eucalyptus woods can be found all across Addis Abeba and used for house construction and as firewood. Some people have the tendency to believe that eucalyptus thrives more and grows more comfortably in Ethiopia than in its native Australia. It grows to the exclusion of all other plants since in eucalyptus forests nothing else grows. Eucalyptus globulus is the species that is most frequently cultivated (Sharma, 2021).

### 15.5.3 *Adansonia digitata* (Baobab)

In many African nations, the baobab tree (*Adansonia digitata* L. Family Bombacaceae) is a native species. Various parts of the plant, particularly the leaves, fruit pulp, seeds and bark fibres, have historically been utilised for therapeutic and dietary purposes, and some industrial businesses create standardised concoctions using the seeds, fruit pulp and leaves. In addition to different usages as anti-inflammatory, antipyretic and analgesic medicines, the therapeutic applications include treatments for intestinal and skin conditions. There have also been reports of antiviral, antibacterial and anti-trypanosome properties (Komane et al., 2023).

### 15.5.4 *Allium cepa* L. *(A. cepa)* (Onion)

*Allium cepa* L. (A. cepa), or onion species, are utilised as vegetables and as medicinal agents in traditional therapies. The perennial plant onion is grown in practically every nation, mostly in areas with mild climates. Onions come in various colours, including white, yellow, purple, red and green, with some being sweet and others not. The plant may grow up to 1 m tall, featuring hollow leaves, green stalks and tiny white or purple flowers. The underground bulb of the plant is not only used as a spice for its delightful aroma and flavour but also for medicinal purposes, including meningitis. Onions contain flavonoids, which have antiviral properties (Lengbiye et al., 2020).

**TABLE 15.1**
**List of Medicinal Plants with Antiviral Potential**

| S. No. | Plant Name and Part Used | Extract | Active Constituents | Mechanism of Action | Type of Virus | References |
|---|---|---|---|---|---|---|
| 1. | *Adansonia digitate* (baobab) – Leaves and stem bark | Water, methanol and DMSO | protein, vitamin C, antioxidants, potassium, magnesium, iron, zinc, calcium and B vitamins | Increases in IL-8 secretion | Influenza, HSV, RV, RSV | Babangida et al. (n.d.); Purushothaman et al. (2018) |
| 2. | *Allium cepa* (onion) – Bulb | Aqueous extract | Polyphenolic substance and flavonoids | Blocking and destroying viral protein and nucleic acid synthesis | Herpes simplex virus type I, rabies virus, poliovirus, meningitis virus, rabies pseudovirus, sindbis virus and para influenza virus type 3 herpes simplex virus type I, rabies virus, poliovirus, meningitis virus, rabies pseudovirus, sindbis virus and para influenza virus type 3 herpes simplex virus type I, rabies virus, poliovirus, meningitis virus, rabies pseudovirus, sindbis virus and para influenza virus type 3 Herpes virus, Rabies virus, poliovirus, meningitis virus, and influenza virus | Rahul et al. (2015); Beigoli et al. (2021) |
| 3. | *Allium sativum* (garlic) – Whole plants | Diallyl disulphide (DADS) | Volatil oil, allicin, diallyl sulphide and carbon dioxide | Blocking viral entry and fusion into host cells, inhibiting viral RNA polymerase, reverse transcriptase, and viral replication, as well as enhancing host immune response | DENV-2, Huh-7 cells, Adenoviridae, Arteriviridae, Coronaviridae, Flaviviridae,Herpesviridae, Orthomyxoviridae, Picornaviridae, Paramyxoviridae, *Poxvirus*, Rhabdoviridae and Retroviridae family | Lengbiye et al. (2020); Rouf et al. (2020) |
| 4. | *Aloe vera* (L.) Burm.f. – Whole plants | Glycerine, 2% DMSO, | Vitamins, minerals, enzymes, polysaccharides, polyphenols, sterols, indoles, phenolic and organic acids | Stimulation of immune system against some viral complications | HSV-1,2, poliovirus, cytomegalovirus | Rajeswari et al. (2012) |
| 5. | *Ananas comosus* (L.) Merr. (pineapple) – Bark | 80% methanol | Vitamins A and C, flavonoids, tannins, and other polyphenolic compounds, organic acids, and soluble monosaccharides and disaccharides | Inhibition of pro-inflammatory mediators | HIV, SARS-CoV, | Rezazadeh et al. (2016) |

*(Continued)*

**TABLE 15.1 (*Continued*)**
**List of Medicinal Plants with Antiviral Potential**

| S. No. | Plant Name and Part Used | Extract | Active Constituents | Mechanism of Action | Type of Virus | References |
|---|---|---|---|---|---|---|
| 6. | *Azadirachta indica* A. Juss. (Neem) – Leaves and bark | Methanol, DMSO | Azadirachtin, nimbolinin, nimbin, nimbidin, nimbidol, sodium nimbinate, gedunin, salannin, and quercetin | Inhibiting viral entry and spread | HSV-1, | Sarkar et al. (2020) |
| 7. | *Carica papaya* L (Gwanda) - Leaves | Aqueousextract, Busulfan | Alkaloids, terpenoids, phenols, tannins, flavonoids, saponins and glycosides, Carotenoids and polyphenols | Increase the expression of ALOX 12 and inhibit destructive effects on platelets | Dengue virus, | Ahmad et al. (2011) |
| 8. | *Cassia occidentalis* L. (*Sanga sanga*) – Leaves | Acetone, dichloromethane, methane, hexane | Achrosin, aloe-emodin, emodin, anthraquinones, anthrones, apigenin, aurantiobtusin, campesterol, cassiollin, chryso-obtusin, chrysophanic acid, chrysarobin, chrysophanol, chrysoeriol, emodin, essential oils, funiculosin, galactopyranosyl, helminthosporine, islandicine, kaempferol, lignoceric acid, linoleic acid, linolenic acid, mannitol, mannopyranosyl, matteucinol, obtusifolin, obtusin, oleic acid, physcion, quercetin, rhamnosides, rhein, rubrofusarin, sitosterols, tannins, and xanthorine | Mast cell degranulation inhibition and human red blood Cell (HRBC) membrane stabilization | HSV, Flaviviruses | Egharevba et al. (2010); Yadav et al. (2010) |
| 9. | *Cassia tora* L (Sickle senna) – Leaves | Hexane, ethylacetate and methanol | Carbohydrates, saponins, sterols, flavonoids, resins, alkaloids, terpenes, anthraquinones, glycoside and balsam | Inhibiting viral entry and spread | HSV, mumps virus | Kumar et al. (2013) |

(*Continued*)

**TABLE 15.1 (*Continued*)**
**List of Medicinal Plants with Antiviral Potential**

| S. No. | Plant Name and Part Used | Extract | Active Constituents | Mechanism of Action | Type of Virus | References |
|---|---|---|---|---|---|---|
| 10. | *Combretum glutinosum* Perr. (Tarauniya) – Stem bark | Aqueous and ethanolic extract | Triterpenoid molecules, betulonic acid and cabraleone | Biosynthesis of secondary metabolites | HSV | N'Diaye et al. (2017) |
| 11. | *Detarium senegalense* J. F. Gmel. (Tallow tree) – Stem bark | Ethanol | 2-Methoxyamine 3,4,5,7-tetrahydroxy anthocyanidines | Prevent viral entry and spread | Influenza, HSV, RV, RSV | Van De Beek et al. (2021); Desmond et al. (2006) |
| 12. | *Diospyros mespiliformis* (Jackalberry) – Stem bark | Aqueous, ethanolic and methanolic extract | Alkaloids, saponins, triterpenes, tanins, flavonoids and cyanogenic glucosides | Arrested virus multiplication | NDV, FPV and IBDV, HIV, AIDS | Van De Beek et al. (2021); Abzug et al. (2003) |
| 13. | *Ficus platyphylla* (Gamji) – Stem bark | Aqueous, methanolicand Hydroalcoholic | Saponin | Longer the haemolysis time | HSV and mumps virus | Sieniawska et al. (2022) |
| 14. | *Ficus polita* (Durumi) – Leaves and Roots | Methanol | (E)-3,5,4'-trihydroxy-stilbene-3,5-O-b-D-diglucopyranoside, betulinic acid, ursolic acid, bsitosterol, sitosterol-3-O-b-D-glucopyranoside | Prevent the growth | HIV | Van De Beek et al. (2021); Sieniawska et al. (2022) |
| 15. | *Ficus sycomorus* (Fig mullberry) – Leaves and stem bark | Chloroform, methanolic | Quercetin, gallic acid, quercetin 3-O-Lrhamnopyranosyl (1→6)-β-D-glucopyranoside (Rutin), quercetin 3-O-β-Dglucopyranoside (Isoquercitrin), quercetin 3,7-O-α-L-dirhamnoside, quercetin 3-O-β-D-galactopyranosyl(1→6)-glucopyranoside and β-sitosterol-3-β-D-glucopyranoside | Prevent the growth and entry of virus | Mumps, HSV | Van De Beek et al. (2021); Sieniawska et al. (2022) |

(*Continued*)

**TABLE 15.1 (*Continued*)**
**List of Medicinal Plants with Antiviral Potential**

| S. No. | Plant Name and Part Used | Extract | Active Constituents | Mechanism of Action | Type of Virus | References |
|---|---|---|---|---|---|---|
| 16. | *Guiera senegalensis* (Sabara) – Leaves | Petroleum ether, dichloromethane | Alkaloids (hyoscyamine and solanine) tannins, terpenoids menthol, coumarins, saponins, flavonoids (quercetin), cardiotonics and cynogenic hetrosides | Stimulating various metabolites which show antiviral properties | HSV 1 and 2 | Parvez et al. (2020) |
| 17. | *Lawsonia inermis* (Henna) – Leaves | Ethanol, ethyl acetate | carbohydrates, proteins, flavonoids, tannins and phenolic compounds, alkaloids, terpenoids, quinones, coumarins, xanthones and fatty acids | Prevent the growth and entry of virus | SFV | Purushothaman et al. (2018) |
| 18. | *Mangifera indica* (Mango) – Leaves | Mathanolic extract | Mangiferin, carotenoids and polyphenols, and omega-3 and -6 polyunsaturated fatty acids | Inhibition of iNOS and cyclooxygenase-2 expression | HSV-1 and 2, HIV, hepatitis B virus | Rahul et al. (2015); Beigoli et al. (2021) |
| 19. | *Momordica charantia* – Leaves | Aqueous and methanolic extract | glycosides, phytosterols, proteins, alkaloids, flavonoids, tannins, saponins, fats and fixed oils, gums and mucilages | Secondary metabolites contain antiviral properties | DENV-2, HIV | Sooksawasdi et al. (2021); Lengbiye et al. (2020) |
| 20. | *Moringa oleifera* – Roots | Hydroalcoholic, methanol | s flavonoids, phenolic acids, alkaloids, isothiocyanates, and thiocarbamate | Inhibit viral entry into the nucleus of a cell | HSV-1, IAV | Van De Beek et al. (2021); Sooksawasdi et al. (2021); Rouf et al. (2020) |
| 21. | *Psidium guajava* – Leaves | Aqueous extract | Quercetin, avicularin, apigenin, guaijaverin, kaempferol, hyperin, myricetin, gallic acid, catechin, epicatechin, chlorogenic acid, epigallocatechin gallate, and caffeic acid | Prevention of viral entry into host cells | Influenza virus | Rajeswari et al. (2012) |

(*Continued*)

**TABLE 15.1 (*Continued*)**
**List of Medicinal Plants with Antiviral Potential**

| S. No. | Plant Name and Part Used | Extract | Active Constituents | Mechanism of Action | Type of Virus | References |
|---|---|---|---|---|---|---|
| 22. | *Parkia biglobosa* – Stem bark | Aqueous and methanolic extract | Saponins, glycosides, tannins, phenolics and alkaloids | Inhibition of growth | | Rezazadeh et al. (2016) |
| 23. | *Piliostigma thonningii* – Leaves and stem | Ethanolic extract | Flavonoids, tannins, kaurane diterpenes, alkaloids, carbohydrates, saponins, terpenes, and volatile oils | Prevention of viral entry into host cells | HSV 1 and 2, HIV, influenza viruses and syncytial viruses | Sarkar et al. (2020) |
| 24. | *Prosopis Africana* - Leaves | Aqueous, ethanolic and methanolic extract | tannins, flavonoids, alkaloids, phenols, glycosides, steroids, saponins and terpenoid | Presence of secondary metabolites shows antimicrobial activity | HSV 1 and 2, mumps vires | Ahmad et al. (2011) |
| 25. | *Sterculia setigera* – Stem bark | Hexane and mixture of dichloromethane: ethane | Polyphenols, flavonoids and tannins | Presence of lupeol beside of high phenolic contents | HSV | Egharevba et al. (2010) |
| 26. | *Tamarindus indica* – Leaves | Ethanolic extract | Sterols/terpenes, flavonoids, tannins, alkaloids, lignins, saponins and coumarins | Inhibition of protein denaturation | Influenza virus, mumps virus | Yadav et al. (2010) |
| 27. | *Terminalia avicenoides* – Stem bark | Aqueous extract | Ellagic acid, punicalagin, flavogallonic acid and terchebulin | Inhibit the growth | HSV-2, mumps virus, rabies and influenza viruses, HIV | Kumar et al. (2013) |
| 28. | *Vitellaria paradoxa* – Stem bark | Ethanol, water and acetone extracts | Flavonoids, phenolic, Triterpenoids, steroids and volatile compounds | Inhibiting the growth | Human polio virus, astrovirus, HSV, bovine parvovirus and canine parvoviru | N'Diaye et al. (2017) |
| 29. | *Ziziphus mauritiana* – Leaves | Ethanol and methanol extract | Berberine, quercetin, kaempferol, sitosterol, stigmasterol, lanosterol and diosgenin | Inhibiting the growth | HSV | Van De Beek et al. (2021) |

(*Continued*)

**TABLE 15.1 (*Continued*)**
**List of Medicinal Plants with Antiviral Potential**

| S. No. | Plant Name and Part Used | Extract | Active Constituents | Mechanism of Action | Type of Virus | References |
|---|---|---|---|---|---|---|
| 30. | *Zingiber officinale* – Root | DMSO | Ginsenosides, phenolic compounds, terpenes, polysaccharides, lipids, organic acids, and raw fibres | Interacting with the viral envelope | HRSV, hepatitis C, Feline Calcivirus | Van De Beek et al. (2021) |
| 31. | Macaranga. Kilimandscharica | Leaf extract | Flavonoids | Inhibition of viral polymerase and binding of viral nucleic acid or viral capsid proteins | anti-measles activity, Herpex simplex type 1 | Sejvar (2016); Abzug et al. (2003) |
| 32. | *M. barteri* | DCM extract | 3,5-Dicaffeoylquinic acid, acteoside, kampferol-7-O-glucoside and bastadin-11 | Inhibition of HIV integrase, an enzyme required for integration of viral DNA into cellular DNA | Herpes simplex virus (HSV) and HIV-1 | Šmejkal (1946) |
| 33. | *Allium sativum* | Methanolic | Allicine | Inhibition of virus entry via disruption of the viral envelope and cell membrane | Herpes simplex virus- (HSV-) 1 and 2 | Sarkar et al. (2020); Ahmad et al. (2011) |
| 34. | *Glycyrrhiza glabra* | Hydroalcoholic extract of roots and stem | Glycyrrhizin | Inhibition of the plaque formation n DAN and RNA viruses | Herpes simplex virus-(HSV-) 1 and 2 | Abraham and Florentine (2021) |
| 35. | *Gymnema sylvestre* | Methanolic extract | Triterpene and saponins. Flavones, anthraquinones, hentriacontane, pentatriacontane, $\alpha$ and $\beta$-chlorophylls, resins, d-quercitol, tartaric acid, phytin, butyric acid, formic acid, lupeol, $\beta$ | Inhibit HIV-1, reverse transcriptase (RT) enzyme, HBV DNA poly | HIV-1 | Mohammed et al. (2023) |
| 36. | *Gossypium herbaceum* | Water extract | Gossypol | Prevention of replication of type 1- HIV | HIV-1, HSV-2, influenza, and parainfluenza | Jansi et al. (2021) |

(*Continued*)

**TABLE 15.1 (*Continued*)**
**List of Medicinal Plants with Antiviral Potential**

| S. No. | Plant Name and Part Used | Extract | Active Constituents | Mechanism of Action | Type of Virus | References |
|---|---|---|---|---|---|---|
| 37. | *Phyllanthus niruri* | Aqueous extracts | Tannins | Significant inhibitory effect on HIV-1 reverse transcriptase | Human immunodeficiency virus (HIV), and herpes simplex virus (HSV) | Upadhyay and Tiwari (2023); Gangal et al. (2020) |
| 38. | *Withania somnifera* | Aqueous extracts | Alkaloids (isopelletierine, anferine), steroidal lactones (withanolides, withaferins), saponins containing additional classes of acyls, and withanoloids containing glucose | Mechanism of binding of prospective herbal drug withaferin A to the structure of the herpes simplex virus's DNA polymerases | herpes simplex type-1 | Parihar (2022) |
| 39. | *Centella asiatica* | Methanolic and aqueous extracts | Flavonoids and plant sterols | Prevention of attachment | HSV-1 and 2 | Maya et al. (2021) |
| 40. | *C. longa* | | Curcumin | Inhibition of the ubiquitin proteasome system has resulted while reducing the occurrence of infectious particles | HSV-1 | Suttithumsatid et al. (2022) |
| 41. | *Phyllanthus emblica* | Hydroalcoholic | Ascorbic acid, fixed oils, phosphates, essential oils, tannins, minerals, vitamins, amino acids, and fatty acids. | It inhibits viral attachment in the early stage by blocking viral attachment and penetration<br>Viral protein synthesis processes also have been reduced | (HSV-1) and HSV-2 | Garber et al. (2021) |
| 42. | *Terminalia arjuna* | Aqueous | Terpenoids | Active inhibition of viral penetration | Terpenoids | Kumar et al. (2023) |

### 15.5.5 *Aloe vera*

*Aloe vera* is the most used medicinal plant globally and is also the oldest known medicinal plant. Aloe vera extracts are known to cure skin damage. Its antibacterial effects ease itching and skin swellings, and help in healing of burns, skin irritations, wounds and insect bites. It is well recognised for actively repairing the damaged skin cells, which can slow down the outward indications of ageing and the development of wrinkles. Aloe is an effective detoxifier, antibacterial and nervous system tonic. Additionally, it strengthens the immune system and is antiviral. According to research, *Aloe vera* aids in digestion when added to the diet as a booster for general health. It is also a beneficial source of vitamins (Garmaroudi et al., 2015).

### 15.5.6 *Ananas comosus* (L) Merrill

Ananas comosus, also known as pineapple, is a tropical fruit widely cultivated in India, Thailand, Costa Rica, Philippines and Brazil. It contains several potentially bioactive metabolites, antioxidants and alpha glucosidase properties. Pineapple peels are a well-known ethnomedical component. According to anecdotal and traditional accounts, pineapple peel can be used alone or in conjunction with other healing plants to cure gastrointestinal issues, arthritis, typhoid illnesses and malaria. In addition to its medicinal benefits, pineapple peel may be used to flavour herbal concoctions. The main bioactive compound in pineapple is known as bromelain, a therapeutic enzyme that has been shown to be effective in treating gastrointestinal issues, thrombus development, inflammation, and malignant cell proliferation. According to biological studies, pineapple peel extract has anti-arthritic, hypolipidemic, antidiabetic and neuroprotective properties. Due to its polarity and capacity to extract the phenolic content of pineapple peel, aqueous methanol has been employed as an effective extraction solvent (Desai et al., 2020).

### 15.5.7 *Azadirachta indica* A. Juss (Neem)

*Azadirachta indica* A. Juss, commonly known as Neem, is an ethnomedicinal plant belonging to class Dicotyledonous, order Fagales and family Meliaceae. It is indigenous to African and Asian folk medicine. Neem contains bioactive compounds such as gallic acid, (-) epicatechin, catechin, margolone and isomergolonone, which have been shown to have anti-tumour, antifungal, anti-dermal (Nimbidin), anti-protozoal and spermicidal activities. Neem bark extract (NBE) has similar qualities. With substantial evidence of antioxidant characteristics, NBE demonstrated potential antibacterial action against *Salmonella paratyphi* and *S. typhi*, as well as hepatoprotective efficacy against $CCl4$-induced liver damage in albino rats. It is interesting to note that NBE is also said to prevent the entry of HSV-1 (Herpes Simplex Virus). NBE has been demonstrated to reduce the effects of malaria on cerebellar Purkinje cells in *Plasmodium berghei*-infected mice. However, little is understood about the exact mechanisms of how it prevents neuroinflammation (Zhu and Viejo-Borbolla, 2021).

### 15.5.8 *Carica papaya* L (Papaya)

*Carica papaya* L or papaya is a small, sparsely branched tree, usually with a single stem growing from 5 to 10 m (16–33 ft) tall, with spirally arranged leaves confined to the top of the trunk. The lower trunk is conspicuously scarred where leaves and fruits are borne. The leaves are large, 50–70 cm (20–28 in) in diameter, deeply palmately lobed, with seven lobes. All plant parts contain latex in articulated laticifers (Kleines et al., 2011). The juice or extract of the *C. papaya* L. leaf is traditionally used to cure a variety of illnesses, including pyrexia, diabetes, gonorrhoea, syphilis, inflammation and infected wounds. A great deal of clinical research has been done on the ability of papaya leaf to induce thrombocytosis.

Papaya is consumed all across the world, either as a raw fruit, vegetable or in processed foods like jam, pickles and canned goods. Different plant components have historically been utilised to cure various ailments. Papain, an enzyme found in the white latex of papayas, is used as an abortifacient, to tenderise meat, as an antiseptic for wound dressing and for treating malignant growths, ringworm, psoriasis and dyspepsia. Traditional uses for the root infusion include the treatment of piles yaws, and genital disorders (Adel et al., 2022).

### 15.5.9 *Cassia occidentalis* L. (Caesalpiniaceae)

An essential medicinal plant used in Ayurveda is *Cassia occidentalis* L. (Caesalpiniaceae), known by a number of names, including Negro Coffee, Fetid Cassia and Coffee Senna (English). In India, it is referred to by names like Kasamarda, Kaasaari (Ayurveda), Kasaundi, Bari Kasaundi (Hindi), Kasondi (Unani), Doddaagace (Kanad), Ponnaviram, Ponnarviriam (Malyalam), Kasinda (Telgu), Paeyaavarai and Thagarai (Siddha/Tamil). *Cassia occidentalis* thrives in the tropics and subtropics, including the eastern United States from Texas to Iowa, as well as Africa, Asia, and Australia. *C. occidentalis* is a widespread weed in India that may be found from Jammu and Kashmir to Kanyakumari (up to an altitude of 1500 m). It is utilised for a various purposes in indigenous and traditional remedies. In Haryana, it grows rapidly just after the rain and begins to fade at the start of the cold season (Hassan et al., 2022).

A 0.8–1.5 m tall, upright, somewhat branched, smooth, semi-woody, foetid herb or shrub, *C. occidentalis* has a firm, robust taproot and a few lateral roots on the midsection. In locations without frost, this plant species can range from a semi-woody annual herb to a woody annual shrub or even a short-lived perennial shrub (Sooksawasdi et al., 2021; Garmaroudi et al., 2015; Elrick et al., 2021). The stem of the plant has a reddish-purple colour. Young ones have four sides; as they become older, they become rounder. Each of the 4–6 pairs of almost sessile, opposite leaflets on an alternating, even pinnately compound leaf is 4–6 cm long, 1.5–2.5 cm broad, ovate or oblong, lanceolate with a pointy tip and fine white hairs on the border. When crushed, the leaflets emit a foul odour. At the base of the rachis is a large, ovoid, shiny, dark purple gland. The 5–10 mm long stipules frequently leave an oblique scar on the skin. A combination of axillary and terminal racemes makes up an inflorescence. The flower is flawless, measuring 2 cm in length and has five yellow petals and five yellowish-green sepals with pronounced red veins. The fruit is a sickle-shaped legume (pod), dry, dehiscent, transversely partitioned, hardly recurved, laterally compressed, with a rounded tip, and carrying 25–50 seeds. Oval-shaped, 3.5–4.5 mm broad, flattened, light to dark brown, somewhat glossy, smooth, and with a round pointy tip are the characteristics of the seeds. Both animals and people frequently consume this plant. However, this plant's seeds and leaves have been shown to have certain toxicological consequences (Singh et al., 2022).

In 1993, a research conducted by Humphry et al. on two Hausa communities, *C. occidentalis* was labelled as "Edible weeds of Agriculture" or "Famine food." She claimed that almost all of the locals (93%) guard these plants and do not hoe them out. *C. occidentalis* is referred to as "Rantarota" in the Nasik area of Maharashtra. The roots of this plant species are preserved in water, along with the roots of *Azadirachta indica* A. Juss. and *Caesalpinia sepiaria* Roxb. This infusion is administered to women who have a white discharge. Using three antimalarial herbs—leaves of *C. occidentalis*, leaves of *Lippia chevalieri*, and flower heads of *Spilanthes oleraces*—a traditional recipe developed in the African nation of Mali, *C. occidentalis* is used to treat malaria. Black pepper with *C. occidentalis* root decoction is a good treatment for filarial illness. The Tanla people of Orissa's Dhenkanal district bathe newborn babies in a decoction produced from 15 leaves each of *C. occidentalis*, *Glycosmis pentaphylla*, and *Vitex negundo* in the Malyagiri hills to render them nearly resistant to skin illnesses at the end of days 7, 12 and 21. According to "Bhavaprakasam,"

Kasamarda (*C. occidentalis*) is used for constipation, while "Wealth of India" claims that the leaves, roots, and seeds are purgative (Kaiser, 2008).

### 15.5.10 *Cassia occidentalis* Linn.

*Cassia occidentalis* Linn., a member of the *Caesalpiniaceae* family (Leguminosae), has several synonyms such as *Senna occidentalis, Cassia ciliata, Cassia foetida, Cassia frutesceus, Cassia geminiflora* and *Cassia linearis.* It is known by various local names such as *Akidi agbara* (Igbo), *Abo rere* (Yoruba) and *Dora rai* (Hausa). The roots are used as a diuretic, tonic, treatment for dysmenorrhoea (menstrual issues), TB, anaemia, liver problems and as a fever reducer. In the Gambia and the Ijo regions of Nigeria, the leaf-sap is used as a febrifuge and laxative as well as a treatment for eye problems in both young and old people. The leaf is known for its anti-neuralgic, purgative properties (used to treat diarrhoea and dysentery) and vermifuge properties. The leaves are used by the Ijo people of Nigeria to alleviate bodily pains and treat malaria. The palm oil mixture is utilised in Yoruba culture to treat infant convulsions. It is said that the root has purifying powers. Infusions or decoctions are used in Gabon to purify the blood and, for inhabitants of Trinidad, to clean the body after childbirth (Sejvar, 2016).

### 15.5.11 Glycyrrhiza glabra

Commonly known as licorice, it is widely used in Chinese medicine and is used as first line agent in Egypt, Greece and Rome. It contains various secondary metabolites obtained from the root and stem of *Glycyrrhiza glabra.* The active constituents present in liquorice are triterpenoid glycoside known as glycyrrhizin. The hydroalcoholic extract obtained from root and stem of *Glycyrrhiza glabra* exhibits significant antiviral activity against DNA- and RNA-based viruses by inhibiting the plaque formation in DNA and RNA of herpes simplex virus (HSV-1 and HSV-2) (Abraham and Florentine, 2021).

### 15.5.12 *Gymnema sylvestre*

Commonly known as gymnema, it is native to Australia, tropical Africa and western India. Various phytoconstituents present in *G. sylvestre* are saponins, triterpenes, flavones, anthraquinones, hentriacontane, pentatriacontane, α and β-chlorophylls, resins, d-quercitol, tartaric acid, phytin, butyric acid, formic acid and lupeol. Methanolic and ethanolic extracts of gymnema showed potent antiviral activity against HIV virus by inhibiting reverse transcriptase of HIV-1 (Mohammed et al., 2023).

### 15.5.13 *Gossypium herbaceum*

It is commonly known as cotton plant and widely grown throughout Middle East countries and western India. It is also cultivated in Africa, Russia, Iran, Central Asia and Turkey. Various phytochemical present in the Gossypium are saponins, glycosides, phenolic compounds, steroids, tannins and flavonoids. The active moiety present in Gossypium is Gossypol which exhibits potent antiviral activity against (HIV-l) immunodeficiency virus. The aqueous extract of plant showed antiviral effect by preventing replication of type 1 HIV immunodeficiency virus (Shehabeldine et al., 2022).

### 15.5.14 *Phyllanthus niruri*

The aqueous extract of *P. niruri* showed vital antiviral activity against HIV-1 by inhibiting HIV-1 reverse transcriptase. The antiviral activity of *Phyllanthus niruri* is attributed to the presence of tannins in the extract (Upadhyay and Tiwari 2023).

### 15.5.15 *Withania somnifera*

It contains more than 35 active components and is mainly concentrated in the root portion of the plant. Biologically active chemical constituents of this plant are alkaloids (isopelletierine, anferine), steroidal lactones (withanolides, withaferins), saponins containing additional classes of acyls and withanoloids containing glucose. The aqueous extract of the plant contains withanoloids which will be responsible for producing antiviral effect against herpes simplex virus type 1 (HSV-1) (Sultana et al., 2021).

### 15.5.16 *Centella asiatica*

It is commonly referred to as Mandukparni and is found in various parts like Sri Lanka, Indonesia and India. The major phytoconstituents present in the plant are triterpenoids which include asiaticosides along with this it contains asiatic acid, asiaticoside, madecassic acid, madasiatic acid, madecassoside and isothankunic acid. The aqueous extract of *Centella asiatica* shows anti-HSV-1 and 2 inhibition abilities (Khan et al., 2021).

### 15.5.17 *Curcuma longa*

The rhizome of curcuma, also known as turmeric, contains major bioactive compounds, with curcumin being the most prominent. Curcumin as well as its derivatives show antiviral activity against herpes simplex virus type-1 (HSV-1) (Vitali et al., 2020). It produces antiviral action by inhibiting the ubiquitin proteasome system, which reduces the occurrence of infectious particles.

### 15.5.18 *Phyllanthus emblica*

The plant is widely distributed in most of the tropical and sub-tropical countries and is a member of the *Euphorbiaceae* family. The main phytoconsttuent of *Phyllanthus emblica* is ascorbic acid (vitamin C). It also contains other bioactive compounds like fixed oils, phosphates, essential oils, tannins, minerals, vitamins, amino acids and fatty acids. The fatty acids present in *P. emblica* are linolenic, linoleic, oleic, stearic, palmitic and myristic acids. D-glucose, D-fructose, D-myositol, D-galacturonic acid, D-arabinose, D-rhamnosyl, D-xylosyI, D-glucosyI, D-mannosyl and D-galactosyI residue are sugars. The major tannins identified in the plant are emblicanin A, emblicanin B, pedunculagin and punicgluconin. The hydroalcoholic extract of *P. emblica* shows *in vitro* antiviral activity against herpes simplex virus. It also exhibits the inhibition of herpes simplex virus type-1 (HSV-1) and HSV-2 (Perera et al., 2021).

### 15.5.19 *Terminalia arjuna*

It is also known as Hirdya. It contains chemical components like triterpenoids and is primarily responsible for various pharmacological actions. Tannins are responsible for anticancer and anti-herpes simplex virus (HSV) activity. Casuarinin, a hydrolyzable tannin isolated for this plant, showed significant antiviral activity on herpes simplex type 2 viruses (HSV-2) *in vitro* (Kumar et al., 2023).

## 15.6 CONCLUSION

Various medicinal plants like *Allium sativum* (Garlic), *Aloe vera* and *Adansonia digitata* (Baobab) have been used traditionally for the treatment and management of various types of viral infections. Clinical efficacy of active constituents of these plants is already reported in preclinical studies against various CNS infections caused by the viruses. Additionally, no incidence of serious toxicity and drug interaction has been reported with the use of these medicinal plants. Future work should be carried out to investigate the clinical efficacy of these plant products against viral meningitis.

## REFERENCES


Abraham J, Florentine S (2021). Licorice (Glycyrrhiza glabra) extracts-suitable pharmacological interventions for COVID-19? A review. *Plants* 10(12):2600.

Abzug MJ, Cloud G, Bradley J, Sánchez PJ, Romero J, Powell D, Lepow M, Mani C, Capparelli EV, Blount S, et al. (2003). Double blind placebo-controlled trial of pleconaril in infants with enterovirus meningitis. *Pediatr Infect Dis J* 22:335–340. doi: 10.1097/00006454-200304000-00009

Adel A, Elnaggar MS, Albohy A, Elrashedy AA, Mostafa A, Kutkat O, Rabeh MA (2022). Evaluation of antiviral activity of Carica papaya leaves against SARS-CoV-2 assisted by metabolomic profiling. *RSC Adv* 12(51):32844–32852.

Ahmad N, Fazal H, Ayaz M, Abbasi BH, Mohammad I, Fazal L (2011). Dengue fever treatment with Carica papaya leaves extracts. *Asian Pac J Trop Biomed* 1:330–333. doi: 10.1016/S2221-1691(11)60055-5

Babangida I, Sani S, Malami I. (n.d.) Elsevier has created a Monkeypox Information Center in response to the declared public health emergency of international concern, with free information in English on the monkeypox virus. The Monkeypox Information Center is hosted on Elsevier Connect, the company ' s public news and information website. Elsevier hereby grants permission to make all its monkeypox related research that is available on the Monkeypox Information Center - including this research content - immediately available in publicly funded repositories, with rights for unrestricted research re-use and analyses in any form or by any means with acknowledgement of the original source. These permissions are granted for free by Elsevier for as long as the Monkeypox Information Center remains active. Traditional medicinal plants used for treating emerging and re-emerging viral diseases in northern Nigeria.

Beigoli S, Behrouz S, Memarzia A, Ghasemi SZ, Boskabady M, Marefati N, Kianian F, Khazdair MR, El-Seedi H, Boskabady MH (2021) Effects of Allium cepa and its constituents on respiratory and allergic disorders: A comprehensive review of experimental and clinical evidence. *Evidence-Based Complement Altern Med* 2021:5554259. doi: 10.1155/2021/5554259

Chadwick DR (2005). Viral meningitis. *Br Med Bull* 75–76:1–14. doi: 10.1093/bmb/ldh057

Desai A, Desai C, Desai H, Mansuri A, Desai J (2020). Possible role of medicinal plants in COVID-19-a brief review. *Int J Sci Develop Res* 5(4):205–209.

Desmond RA, Accortt NA, Talley L, Villano SA, Soong SJ, Whitley RJ (2006). Enteroviral meningitis: Natural history and outcome of pleconaril therapy. *Antimicrob Agents Chemother* 50:2409–2414. doi: 10.1128/AAC.00227-06

Egharevba H, Okwute S, Okogun J (2010). Phytochemical Analysis and Broad Spectrum Antimicrobial Activity of *Cassia Occidentalis* L (Whole plant). Adaptogenic (Anti-Stress) Effect of Aqueous Musanga cecropioides (Urticaceae) View project Effect of Heavy Metals on Seed Germination View project. https://www.researchgate.net/publication/281082499

Elrick MJ, Pekosz A, Duggal P (2021). Enterovirus D68 molecular and cellular biology and pathogenesis. *J Biol Chem* 296:100317. doi: 10.1016/j.jbc.2021.100317

Gangal N, Nagle V, Pawar Y, Dasgupta S (2020). Reconsidering traditional medicinal plants to combat COVID-19. *AIJR Preprints* 34(1):1–6.

Garber A, Barnard L, Pickrell C (2021). Review of whole plant extracts with activity against herpes simplex viruses in vitro and in vivo. *J Evid-Based Integr Med* 26: 2515690X20978394.

Garmaroudi FS, Marchant D, Hendry R, Luo H, Yang D, Ye X, Shi J, McManus BM (2015). Coxsackievirus B3 replication and pathogenesis. *Future Microbiol* 10:629–652. doi: 10.2217/fmb.15.5

Hassan ST, Šudomová M, Mazurakova A, Kubatka P (2022). Insights into antiviral properties and molecular mechanisms of non-flavonoid polyphenols against human herpesviruses. *Int J Mol Sci* 23(22):13891.

Haydar SM, Hallit SR, Hallit RR, Salameh PR, Faddoul LJ, Chahine BA, Malaeb DN (2019). Adherence to international guidelines for the treatment of meningitis infections in Lebanon. *Saudi Med J* 40:260–265. doi: 10.15537/smj.2019.3.23965

Jansi RS, Khusro A, Agastian P, Alfarhan A, Al-Dhabi NA, Arasu MV, Al-Tamimi A (2021). Emerging paradigms of viral diseases and paramount role of natural resources as antiviral agents. *Sci Total Environ* 759:143539.

Kaiser R (2008). Tick-Borne encephalitis. *Infect Dis Clin North Am* 22:561–575. doi: 10.1016/j.idc.2008.03.013

Khan MA, Khan MAAM, Rahman S, Ahsan A, Sanjana JM, Mahmud S, Moni MA (2021). Bioinformatics approaches to investigate the phytochemicals of *Centella asiatica* against the main protease of SARS-CoV-2. Available from https://doi.org/10.21203/rs.3.rs-1099054/v1

Kleines M, Schiefer J, Stienen A, Blaum M, Ritter K, Häusler M (2011). Expanding the spectrum of neurological disease associated with Epstein-Barr virus activity. *Eur J Clin Microbiol Infect Dis* 30:1561–1569. doi: 10.1007/s10096-011-1261-7

Kohil A, Jemmieh S, Smatti MK, Yassine HM (2021). Viral meningitis: An overview. *Arch Virol* 166:335–345. doi: 10.1007/s00705-020-04891-1

Komane B, Kamatou G, Mulaudzi N, Vermaak I, Fouche G (2023). Adansonia digitata. In *The South African Herbal Pharmacopoeia* (pp. 1–39). Academic Press.

Kumar Shukla SK, Kumar A, Terrence M, Yusuf J, Pratap Singh V, Mishra M (2013). The probable medicinal usage of Cassia tora: An overview. *Online J Biol Sci* 13:13–17. doi: 10.3844/ojbssp.2013.13.17

Kumar V, Sharma N, Saini R, Mall S, Zengin G, Sourirajan A, El-Shazly M (2023). Therapeutic potential and industrial applications of *Terminalia arjuna* bark. *J Ethnopharmacol* 310:116352.

Lauinger IL, Vivas L, Perozzo R, Stairiker C, Tarun A, Zloh M, Zhang X, Xu H, Tonge PJ, Franzblau SG, et al. (2013). Potential of lichen secondary metabolites against Plasmodium liver stage parasites with FAS-II as the potential target. *J Nat Prod* 76:1064–1070. doi: 10.1021/np400083k

Lengbiye EM, Mbadiko CM, Falanga CM, Matondo A, Inkoto CL, Ngoyi EM, Kabengele CN, Bongo GN, Gbolo BZ, Kilembe JT, et al. (2020). Antiviral activity, phytochemistry and toxicology of some medically interesting allium species: A mini review. *Int J Pathog Res* 5: 64–77. doi: 10.9734/ijpr/2020/v5i430145

Maya PG, Mahayasih W, Ahmad I (2021). In silico identification of natural products from *Centella asiatica* as severe acute respiratory syndromecoronavirus 2 main protease inhibitor. *J Adv Pharm Technol Res* 12(3):261.

Melhuish A, McGill F (2021). Viral meningitis and encephalitis. *Med (United Kingdom)* 49:675–680. doi: 10.1016/j.mpmed.2021.08.004

Mohammed FS, Uysal I, Sevindik M (2023). A review on antiviral plants effective against different virus types. *Prospects Pharm Sci* 21(2):1–21.

Mtambo SE, Amoako DG, Somboro AM, Agoni C, Lawal MM, Gumede NS, Khan RB, Kumalo HM (2021). Influenza viruses: Harnessing the crucial role of the M2 ion-channel and neuraminidase toward inhibitor design. *Molecules* 26:880. doi: 10.3390/molecules26040880

N'Diaye D, Mbaye MD, Gassama A, Lavaud C, Pilard S (2017). glutinosium. Ex DC (Combretaceae). *Int J Biol Chem Sci* 11:488–498.

Parihar S (2022). A literature review on pharmacological activities of *Withania somnifera*. *Biol Sci* 2(1):147–154.

Parvez MK, Al-Dosari MS, Arbab AH, Al-Rehaily AJ, Abdelwahid MA (2020). Bioassay-guided isolation of anti-hepatitis B virus flavonoid myricetin-3-O-rhamnoside along with quercetin from *Guiera senegalensis* leaves. *Saudi Pharm J* 28(5):550–559.

Perera WPRT, Liyanage JA, Dissanayake KGC, Gunathilaka H, Weerakoon WMTDN, Wanigasekara DN, Perera BT (2021). Antiviral potential of selected medicinal herbs and their isolated natural products. *BioMed Res Int* 2021:7872406.

PETTE H (1960). Viral meningitis. *Munch Med Wochenschr* 102:1349–1352. doi: 10.1177/003693306901400704

Purushothaman B, Prasannasrinivasan R, Suganthi P, Ranganathan B, Gimbun J, Shanmugam K (2018). A comprehensive review on *Ocimum basilicum*. *J Nat Remedies* 18:71–85. doi: 10.18311/jnr/2018/21324

Rahul J, Jain MK, Singh SP, Kamal RK, Naz A, Gupta AK, Mrityunjay SK (2015). *Adansonia digitata* L. (baobab): A review of traditional information and taxonomic description. *Asian Pac J Trop Biomed* 5:79–84. doi: 10.1016/S2221-1691(15)30174-X

Rajeswari R, Umadevi M, Rahale CS, Pushpa R, Selvavenkadesh S, Kumar KS, Bhowmik D (2012). Aloe vera: The miracle plant its medicinal and traditional uses in India. *J Pharmacogn Phytochem* 1:118–124. https://www.phytojournal.com/vol1Issue4/Issue_nov_2012/17.1.pdf

Renaud C, Harrison CJ (2015). Human Parechovirus 3: The most common viral cause of meningoencephalitis in young infants. *Infect Dis Clin North Am* 29:415–428. doi: 10.1016/j.idc.2015.05.005

Rezazadeh F, Moshaverinia M, Motamedifar M, Alyaseri M (2016). Assessment of Anti HSV-1 activity of aloe vera gel extract: An in vitro study. *J Dent (Shiraz, Iran)* 17:49–54. https://www.ncbi.nlm.nih.gov/pubmed/26966709%0Ahttps://www.pubmedcentral.nih.gov/articlerender.fcgi?artid=PMC4771053

Rouf R, Uddin, Jamal S, Kumer D, Torequl M (2020). Since January 2020 Elsevier has created a COVID-19 resource centre with free information in English and Mandarin on the novel coronavirus COVID-19. The COVID-19 resource centre is hosted on Elsevier Connect, the company' s public news and information. pp. 219–234.

Sarkar L, Putchala RK, Safiriyu AA, Das Sarma J (2020). *Azadirachta indica* A. Juss ameliorates mouse hepatitis virus-induced neuroinflammatory demyelination by modulating cell-to-cell fusion in an experimental animal model of multiple sclerosis. *Front Cell Neurosci* 14:1–22. doi: 10.3389/fncel.2020.00116

Sejvar JJ (2016). West nile virus infection. *Emerg Infect* 10:175–199. doi: 10.1128/9781555819453.ch10

Sharma S (2021). A review on Eucalyptus Globulus-an authentic herb. *J Pharm Res Int* 33(53B):107–114.

Shehabeldine AM, Hashem AH, Wassel AR, Hasanin M (2022). Antimicrobial and antiviral activities of durable cotton fabrics treated with nanocomposite based on zinc oxide nanoparticles, acyclovir, nanochitosan, and clove oil. *Appl Biochem Biotechnol* 194(2):783–800.

Sieniawska E, Świątek Ł, Sinan KI, Zengin G, Boguszewska A, Polz-Dacewicz M, Mahomoodally MF (2022). Phytochemical insights into Ficus sur extracts and their biological activity. *Molecules* 27(6):1863.

Singh A, Singh B, Gautam A (2022). Herbal-based resources against exanthematous viral infections and other viral diseases. In Chukwuebuka Egbuna (Ed.) *Coronavirus Drug Discovery* (pp. 173–202). Elsevier, United States.

Šmejkal K (1946). HSV-1 DNA polymerase: Shedding light on. 1:1–12.

Solomon T, Lewthwaite P, Perera D, Cardosa MJ, McMinn P, Ooi MH (2010). Virology, epidemiology, pathogenesis, and control of enterovirus 71. *Lancet Infect Dis* 10:778–790. doi: 10.1016/S1473-3099(10)70194-8

Sooksawasdi Na Ayudhya S, Laksono BM, van Riel D (2021). The pathogenesis and virulence of enterovirus-D68 infection. *Virulence* 12:2060–2072. doi: 10.1080/21505594.2021.1960106

Soung A, Klein RS. (2018). Viral encephalitis and neurologic diseases: Focus on astrocytes. *Trends Mol Med* 24:950–962. doi: 10.1016/j.molmed.2018.09.001

Sultana T, Okla MK, Ahmed M, Akhtar N, Al-Hashimi A, Abdelgawad H, Haq IU (2021). Withaferin A: From ancient remedy to potential drug candidate. *Molecules* 26(24):7696.

Suttithumsatid W, Panichayupakaranant P (2022). Herbal drugs for the management and treatment of herpes simplex infections. In Inderbir Singh, Rakesh K. Sindhu, Atul A. Shirkhedkar, Pharkphoom Panichayupakaranant (Eds.) *Herbal Drugs for the Management of Infectious Diseases*, 359–388.

Upadhyay R, Tiwari KN (2023). The antiviral potential of Phyllanthus species: A systematic review. *Arch Virol* 168(7):177.

van de Beek D, Brouwer MC, Koedel U, Wall EC (2021). Community-acquired bacterial meningitis. *Lancet* 398:1171–1183. doi: 10.1016/S0140-6736(21)00883-7

Vitali D, Bagri P, Wessels JM, Arora M, Ganugula R, Parikh A, Kaushic C (2020). Curcumin can decrease tissue inflammation and the severity of HSV-2 infection in the female reproductive mucosa. *Int J Mol Sci* 21(1):337.

Wilkinson RJ, Rohlwink U, Misra UK, Van Crevel R, Mai NTH, Dooley KE, Caws M, Figaji A, Savic R, Solomons R, et al. (2017). Tuberculous meningitis. *Nat Rev Neurol* 13:581–598. doi: 10.1038/nrneurol.2017.120

Yadav JP, Arya V, Yadav S, Panghal M, Kumar S, Dhankhar S (2010). *Cassia occidentalis* L.: A review on its ethnobotany, phytochemical and pharmacological profile. *Fitoterapia* 81:223–230. doi: 10.1016/j.fitote.2009.09.008

Zhu S, Viejo-Borbolla A (2021). Pathogenesis and virulence of herpes simplex virus. *Virulence* 12:2670–2702. doi: 10.1080/21505594.2021.1982373

# 16 Future Directions in Antiviral Potentials of Medicinal Plants

*Lakhveer Singh, Pritha Dalapati, Puja Shah, and Shashank Shekher Mishra*
DIT University

## 16.1 INTRODUCTION

Since the ancient times, natural products have served as a major source of drugs. About 50% of today's pharmaceutical drugs are derived from natural origin (Clark et al., 1996). The growing interest in natural products as a source of new drugs can be attributed to many factors including urgent therapeutic needs, the wide range of both chemical structures and biological activities of natural secondary metabolites, the adequacy of bioactive natural products as biochemical and molecular probes, the development of recent techniques to accurately detect, isolate and structurally characterize the bioactive natural products and advances in solving the demand for supply of complex natural products (Shu et al., 1998; Vanden Berghe et al., 1986a). Historically, the majority of the natural product-based drugs including cyclosporine, paclitaxel, and camptothecin derivatives were first discovered by traditional cell-based *in vitro* assays before their real molecular biological targets were identified (Cragg et al., 1997; Hughes et al., 1977). These cellular biological responses of natural products are likely to be associated with the inherent properties of secondary metabolites for the defense of their producing organisms (Starr & Taggart, 1991).

Natural products are compounds produced by living organisms, such as plants, animals, fungi, and bacteria (Joklik et al., 1988). They can be extracted from these organisms or synthesized in the laboratory based on the structure and function of the natural product (Razvi et al., 1995). While they have been used for centuries in traditional medicine to treat a variety of illnesses and diseases (Lockshin et al., 1991), they have also served as the basis for the development of many modern drugs. For example, aspirin, a widely used pain reliever, was originally derived from salicylic acid, which is found in willow bark (Feldmann & Klenk, 1996). Often considered complex molecules with diverse biological activities, they can have antimicrobial, anticancer, anti-inflammatory, and immunomodulatory properties, among others (Hudson et al., 1990). Some natural products are also used as food additives or flavoring agents, such as vanilla and cinnamon. Natural products can be classified into several categories based on their chemical structure, including alkaloids, terpenes, flavonoids, and phenolics (Vlietinck et al., 1997). They are often studied for their potential therapeutic benefits, as well as for their ecological roles and potential use in agriculture (Safrin & Chambers, 1998). Overall, natural products represent a valuable source of biologically active compounds with a wide range of potential applications in medicine, agriculture, and industry (De Clercq, 1993).

Natural products have provided a rich source of compounds with antiviral properties, and many of these compounds have been used as the basis for the development of antiviral drugs. Some examples of natural products that have been used as antiviral agents include:

**Acyclovir:** Acyclovir is an antiviral drug used to treat herpes simplex virus infections. It was derived from a natural product called guanine, which is found in DNA (Sen& Ransohoff, 1993).

**Tamiflu:** Tamiflu is an antiviral drug used to treat influenza. It was developed based on the structure of a compound found in star anise, a plant commonly used in traditional medicine (Vanden Berghe et al., 1986b).

DOI: 10.1201/9781003329169-19 

**Artemisinin**: Artemisinin is a natural product derived from the plant Artemisia annua. It is used to treat malaria and has also been shown to have activity against several other viruses, including hepatitis B virus and HIV (Munro et al., 1987).

**Echinacea:** Echinacea is a plant that is commonly used as an herbal remedy to treat colds and flu. It contains compounds that have been shown to have antiviral activity against respiratory viruses, such as rhinovirus and influenza virus (Becker et al., 1980).

**Curcumin:** Curcumin is a compound found in turmeric; a spice commonly used in Indian cuisine. It has been shown to have antiviral activity against several viruses, including influenza virus, herpes simplex virus, and HIV (Came et al., 1982).

**Green tea:** Green tea is a beverage made from the leaves of the Camellia sinensis plant. It contains compounds called catechins that have been shown to have antiviral activity against several viruses, including influenza virus, hepatitis B virus, and HIV (Chu et al., 1992).

**Lentinan:** Lentinan is a polysaccharide derived from the mushroom Lentinula edodes (shiitake). It has been shown to have antiviral activity against several viruses, including hepatitis B virus, herpes simplex virus, and HIV (Ng et al., 1997).

Overall, natural products have provided a valuable source of antiviral compounds, and many of these compounds are currently being studied for their potential as antiviral drugs.

## 16.2 OVERVIEW OF ANTIVIRAL ACTIVITIES OF THE COMPOUNDS DERIVED FROM MEDICINAL PLANTS

Four basic approaches are conducted for plant selection for antiviral screening assays:

1. Random collection of plants followed by mass screening
2. Ethnomedical approach
3. Literature-based follow up of the existing leads
4. Chemotaxonomic approach (Vlietinck et al., 1997)

The second and third approaches are the most favored ones due to their cost-effective applicability. The selection based on folkloric use proved five times higher percentage of active leads than other approaches. The random approach usually affords more novel compounds with antiviral activity (Che, 1991). Combining ethnomedical, phytochemical, and taxonomical approaches is considered the best compromise. Natural products have provided a rich source of compounds with antiviral properties, and many of these compounds have been used as the basis for the development of antiviral drugs. Some examples of natural products that have been shown to have antiviral activity include:

**Quercetin:** Quercetin is a flavonoid found in many fruits and vegetables, including apples, onions, and broccoli. It has been shown to have antiviral activity against several viruses, including influenza virus and HIV (Rinehart et al., 1993).

**Resveratrol:** Resveratrol is a compound found in grapes, red wine, and peanuts. It has been shown to have antiviral activity against several viruses, including influenza virus and herpes simplex virus (Vanden Berghe et al., 1978).

**Tea tree oil:** Tea tree oil is an essential oil derived from the leaves of the tea tree. It has been shown to have antiviral activity against several viruses, including influenza virus and herpes simplex virus (Ieven et al., 1982)

**Ginger:** Ginger is a root that is commonly used in cooking and as a medicinal herb. It contains compounds that have been shown to have antiviral activity against several viruses, including influenza virus and respiratory syncytial virus (Montanha et al., 1995).

**Honey:** Honey is a sweet substance produced by bees from the nectar of flowers. It has been shown to have antiviral activity against several viruses, including influenza virus and herpes simplex virus (Kim et al., 1969).

**TABLE 16.1**
**Medicinal Plants Having Antiviral Activities**

| S. No. | Plant Species | General Name | Family | Part of the Plants | Individual Uses | References |
|---|---|---|---|---|---|---|
| 1 | *Aloe vera* | Aloe | Asphodelaceae | Leaves | HIV and anti-HIV drug | Tafur et al. (1976) |
| 2 | *Andrographis paniculata* (Burm.f.) Nees. | Andrographis | Acanthaceae | Leaves | Antiviral effect through immunomodulation | Hohenschutz et al. (1981) |
| 3 | *Withania somnifera* (L.) Dunal | Ashwagandha leaves | Solanaceae | Leaves | Rejuvenate the immune system of HIV-positive people | Nash et al. (1988) |
| 4 | *Astragalus membranaceus* Bunge. | Astragalus | Fabaceae | Dry root | Anti-HIV activity | Scofield et al. (1986) |
| 5 | *Atractylodes macrocephala* Koidz. | Atractylodes | Compositae | Root | HIV-positive people have used roots of Atractylodes to improve body weight, muscle strength, also reduce immunodeficiency-associated diarrhea, along with immune function improvement | Boyd et al. (1994) |
| 6 | *Uncaria tomentosa* Willd. | Cat's claw (inner bark) | Aizoaceae | Inner Bark | Antioxidant and immunomodulatory properties | McMahon et al. (1995) |
| 7 | *Allium sativum* L. | Garlic | Amaryllidaceae | Garlic | HIV infection such as fungal infections (thrush) and parasitic infections (cryptosporidium) | Kmpotic et al. (1972) |
| 8 | *Zingiber officinale* Roscoe | Ginger | Zingiberaceae | Ginger | Nausea associated with antiretroviral treatment | Kite et al. (1988) |
| 9 | *Gingko biloba* L. | Gingko (seeds) | Ginkgoaceae, | Seeds | HIV-associated memory loss | Sethi (1985) |
| 10 | *Panax ginseng* C. Meyer. (Asian); *Panax quinquefolium* L. (North American); *Eleutherococcus senticosus* (Rupr. & Maxim.) Maxim. (Siberian) | Ginseng (roots) | Araliaceae | Roots | A tonic herb which improves cell-mediated immune system usually ravaged among HIV-infected individuals. Contraindicated during pregnancy | McCommick et al. (1996) |
| 11 | *Hydrastis canadensis* L. | Goldenseal (roots) | Ranunculaceae | Roots | Has the potential of controlling diarrhea and weight loss among HIV-infected individuals | Alarcon et al. (1998) |
| 12 | *Chelidonium majus* L. | Greater Celandine (flower) | Papaveraceae | Flower | Due to anticancer properties may be beneficial for HIV-associated Kaposi sarcoma | Ghoneum et al. (1998) |

*(Continued)*

**TABLE 16.1 (*Continued*)**
**Medicinal Plants Having Antiviral Activities**

| S. No. | Plant Species | General Name | Family | Part of the Plants | Individual Uses | References |
|---|---|---|---|---|---|---|
| 13 | *Hyssopus officinalis* L. | Hyssop (leaves and flowers) | Lamiaceae | Leaves and flower | Inhibits the replication of HIV without any toxicities | Nakashima et al. (1987) |
| 14 | *Melissa officinalis* L. | Lemon balm | Lamiaceae | Leaves | Antiviral properties against HIV and HSV | Batinic et al. (1992) |
| 15 | *Glycyrrhiza glabra* L. | Licorice (roots) | Fabaceae | Roots | Immunostimulatory properties and inhibit viral production | Patel et al. (1993) |
| 16 | *Lomatium dissectum* (Nutt.) Mathias & Constance) *Lomatium suksdorfii* (Wats.) Coult. & Rose | Lomatium | Apiaceae | Roots | Inhibit HIV *in vitro* | Balzarini et al. (1991) |
| 17 | *Cannabis sativa* L. | Marijuana | Cannabaceae | Leaves | HIV people use it for preventing nausea and stimulating appetite | Hatano et al. (1988) |
| 18 | *Olea europaea* L. | Olive leaf | Oleaceae | Leaves | Antiviral (against HIV) and antioxidant properties | Kashman et al. (1992) |
| 19 | *Plantago ovata* Forssk. | Psyllium (seed and husk of plant) | Plantaginaceae | Seeds and husk of plant | Useful for diarrhea, a side effect of protease inhibitors therapy among HIV-infected individuals | Hizi et al. (1993) |
| 20 | *Sanguinaria canadensis* | Sanguinaria | Papaveraceae | Flowers, leaves, rhizomes, and roots | Used for Pneumocystis carinii pneumonia (PCP) usually encountered by HIV-infected individuals | Pengsuparp et al. (1996) |
| 21 | *Asparagus racemosus* Willd. | Shatavari | Asparagaceae | Roots, flowers, leaves | Immunostimulant | Nishibe et al. (1997) |
| 22 | *Buxus sempervirens* L. | SPV30 (boxwood extract) | Buxaceae | Leaves | Mild antiretroviral effects | Raghavan et al. (1995) |
| 23 | *Hypericum perforatum* L. | St. John's Wort | Hypericaceae | Flower | Antiretroviral effects associated with exposure to light (photosensitive antiretroviral component). Inactivate conventional antiretrovirals and as such contraindicated for HIV patients on other medications | Beladi et al. (1977) |

(*Continued*)

**TABLE 16.1 (*Continued*)**
**Medicinal Plants Having Antiviral Activities**

| S. No. | Plant Species | General Name | Family | Part of the Plants | Individual Uses | References |
|---|---|---|---|---|---|---|
| 24 | *Melaleuca alternifolia* | Tea tree oil | Myrtaceae | Leaves | Controls HIV-associated thrush (fungal infections) | Selway et al. (1986) |
| 25 | *Sambucus nigra* | Elderberry | Caprifoliaceae | Flowers | Antiherpetic activity | Vlietinck et al. (1986) |
| 26 | *Salvia miltiorrhiza* | Red sage, Chinese sage | Lamiaceae | Roots | Novel therapeutic drugs against AIDS due to their high potency and absence of cytotoxicity | Streissle et al. (1985) |
| 27 | *Salvia cedronella* | Turkish sage | Lamiaceae | Aerial part | Antiherpetic activity | Hayashi et al. (1997) |
| 28 | *Euphorbia spinidens* Bornm. | Spurge | Euphorbiaceae | Aerial part | Antiherpetic activity | Ono et al. (1989) |
| 29 | *Swertia chirata* | Chirata | Gentianaceae | Stems and leaves | Antiherpetic activity | Xu et al. (1993) |
| 30 | *Baccharis erioclada* DC. | . Baccharis | Asteraceae | Leaves | Antiherpetic activity | Pengsuparp et al. (1995) |
| 31 | *Mentha arvensis* | Wild mint | Lamiaceae | Leaves | It is effective as a virucidal agent in HSV infection | Macrae et al. (1984) |
| 32 | *Allamanda schottii* | Allamanda | Apocynaceae | Roots and flowers | Antiherpetic activity | Eich et al. (1998) |
| 33 | *Rhaphiolepis indica* L. | Indian hawthorn | Rosaceae | Leaves | Anti-HIV activity | Bedows et al. (1982) |
| 34 | *Acacia arabica* | Babul/Kikar | Fabaceae | Leaves and bark | Anti-HIV effect | McRae et al. (1988) |
| 35 | *Camellia sinensis* | Green tea | Theaceae | Leaves | Antiherpetic effect | Eberhardt et al. (1996) |
| 36 | *Ocimum basilicum* | Sweet basil | Lamiaceae | Leaves | Antiherpetic effect | Eberhardt et al. (1996) |
| 37 | *Phyllanthus urinaria* L. | Stonebreaker | Euphorbiaceae | Roots, leaves | Antiherpetic activity | Eich et al. (1996) |

*(Continued)*

**TABLE 16.1** (*Continued*)
**Medicinal Plants Having Antiviral Activities**

| S. No. | Plant Species | General Name | Family | Part of the Plants | Individual Uses | References |
|---|---|---|---|---|---|---|
| 38 | *Polygonum cuspidatum* | Japanese knotweed | Polygonaceae | Roots | Anti-HBV activity | Kernan et al. (1997) |
| 39 | *Phyllanthus amarus* Schum. | Kura Amluki | Phyllanthaceae | Leaves | Anti-HBV activity | Valsaraj et al. (1997) |
| 40 | *Olea europaea* | Olive | Oleaceae | Leaves | Anti-HIV activity | Asano et al. (1996) |
| 41 | *Hyssopus officinalis* L. | Zoofa | Lamiaceae | Leaves | Anti-HIV activity | Asano et al. (1996) |
| 42 | *Terminalia sericea* | Silver cluster-leaf | Combretaceae | Leaves | Anti-HIV activity | Lin et al. (1989) |
| 43 | *Phyllanthus emblica* I | Indian gooseberry | Phyllanthaceae | Fruit | Anti-HIV activity | Lin et al. (1989) |
| 44 | *Pinus yunnanensis* | Yunnan pine | Pinaceae | Pine cone | Anti-HIV activity | Lin et al. (1989) |
| 45 | *Cassia occidentalis* | Coffee senna | Caesalpiniaceae | Leaves | Anti-HIV activity | Royer et al. (1995) |

**Ginseng:** Ginseng is a plant used in traditional medicine for thousands of years. It contains compounds that have been shown to have antiviral activity against several viruses, including influenza virus and herpes simplex virus (Kakiuchi et al., 1987).

**Lactoferrin:** Lactoferrin is a protein found in milk and other bodily fluids. It has been shown to have antiviral activity against several viruses, including influenza virus and hepatitis C virus (Tan et al., 1991).

**Aphidicolin:** Aphidicolin is a tetracyclic diterpenoid, produced by the fungus Cephalosporium aphilicola. It has the ability to inhibit HSV-1, -2, VV, and herpetic keratitis of rabbit (Beutler et al., 1992).

Overall, natural products have provided a valuable source of antiviral compounds, and many of these compounds are currently being studied for their potential as antiviral drugs. However, it is important to note that more research is needed to determine the safety and efficacy of these compounds for use in humans.

## 16.3 CHALLENGES AND LIMITATIONS

While natural products have shown promising potential as sources of antiviral compounds, there are several challenges and limitations associated with their use:

**Standardization and quality control**: Natural products can vary widely in their chemical composition, depending on factors such as geography, climate, and cultivation practices. This variability can make it challenging to ensure consistent potency and purity of natural product extracts. Standardization and quality control measures are necessary to ensure the safety and efficacy of natural product-based drugs (Helbig et al., 1982).

**Side effects and toxicity:** Natural products can cause side effects and toxicity, just like synthetic drugs. Some natural products may also interact with other medications or exacerbate existing medical conditions. Careful screening and evaluation of natural product extracts are necessary to identify potential toxicities and interactions (Corthout et al., 1992).

**Lack of understanding of mechanisms of action:** While natural products have been shown to have antiviral activity, the mechanisms by which they exert their effects are not always well understood. This lack of understanding can make it challenging to optimize and tailor natural product-based therapies (Nishizawa et al., 1989).

**Limited availability and access:** Natural product-based therapies may be limited in availability and access due to factors such as availability of raw materials, high production costs, and regulatory hurdles (Nonaka et al., 1990).

**Resistance development:** Viruses can develop resistance to antiviral compounds, including those derived from natural products. This can limit the long-term efficacy of natural product-based therapies (Gustafson et al., 1992).

In summary, while natural products offer a potentially rich source of antiviral compounds, there are challenges and limitations associated with their use that need to be carefully considered and addressed. Careful evaluation and standardization of natural product extracts, coupled with rigorous testing and monitoring, can help to overcome some of these challenges and unlock the full potential of natural products as sources of antiviral therapies.

## 16.4 FUTURE PERSPECTIVES

The future of antiviral natural product research looks promising, as scientists continue to explore the potential of natural products in the fight against viral infections. Some of the key future perspectives in this field include the following.

**Identification of novel compounds**: Despite the wealth of natural products that have been studied for their antiviral activity, there are likely many more compounds that remain undiscovered. Advances in analytical and screening technologies, as well as the increasing availability of genetic and genomic data, are likely to facilitate the identification of new compounds with potent antiviral activity (Chan et al., 1989; Arisawa et al., 1990; Takasaki et al., 1990; Nishizawa et al., 1991).

**Understanding mechanisms of action:** As our understanding of the molecular mechanisms underlying viral infections continues to grow, it is likely that we will gain a better understanding of how natural products exert their antiviral effects. This knowledge could facilitate the development of more targeted and effective natural product-based therapies (Nishizawa et al., 1991).

**Development of combination therapies:** Combining natural products with different mechanisms of action could enhance antiviral activity and reduce the risk of resistance development. The development of effective combination therapies will require a greater understanding of the mechanisms of action of individual natural products (Dai et al., 1998).

**Improvement of bioavailability:** Many natural products have poor bioavailability, meaning that they are not readily absorbed and distributed in the body. Efforts to improve the bioavailability of natural product compounds could enhance their efficacy and reduce the risk of toxicity (Sydiskis et al., 1991; Schinazi et al., 1990).

**Integration with other therapies:** Natural product-based therapies may be particularly effective when combined with other approaches, such as immunotherapy or gene therapy. Integration of natural products with other therapeutic approaches could enhance efficacy and reduce side effects (Fehr et al., 1995).

There is an urgent need to identify novel active chemotypes as lead for effective antiviral chemotherapy. Recent years have witnessed great advances in this area. The enormity of natural products as antiviral agents started to be expressed in the area of antiviral chemotherapy. This is represented by the FDA's approval for clinical investigation of two plant-derived compounds, in addition to one compound of marine origin and one microbial-derived compound (De Closterd et al., 1993). Out of ten synthetic approved drugs between 1983 and 1994, seven were modeled on a natural product (Cragg et al., 1997). The development of recent techniques to dereplicate, accurately detect, isolate, structurally define, and automate the assay of bioactive natural products will result in more leads for antiviral agents. It has been estimated that only 5%–15% of the approximately 250,000 species of higher plants have been systematically investigated for the presence of bioactive compounds, while the potential of the marine environment has barely been tapped (Cragg et al., 1997). Consequently, natural products represent potential leading resources for imaginative discoverers in the field of antiviral research.

## 16.5 CONCLUSION

In conclusion, natural products represent a promising source of antiviral compounds. They offer a diverse range of chemical structures and mechanisms of action, and have been shown to have potent activity against a variety of viral infections (Hara et al., 1991; Kernan et al., 1998; Eich et al., 1998a). However, there are also challenges and limitations associated with their use, including variability in chemical composition, potential toxicity and side effects, limited understanding of mechanisms of action, and the risk of resistance development (Takechi & Tanaka, 1981; Takechi & Tanaka, 1982).

To fully realize the potential of natural products in the development of antiviral therapies, it will be necessary to address these challenges through careful evaluation and standardization of natural product extracts, rigorous testing and monitoring, and the development of effective combination therapies (Okuda et al., 1989; Takechi et al., 1985; Corthout et al., 1991). Advances in analytical and screening technologies, as well as the increasing availability of genetic and genomic data, are likely to facilitate the identification of new compounds with potent antiviral activity (Kakiuchi et al., 1985; Asanaka et al., 1988; Take et al., 1989).

Overall, natural product-based therapies have the potential to offer safe and effective alternatives to synthetic drugs in the treatment of viral infections, particularly in cases where conventional therapies have proven ineffective or have caused significant side effects (Gallo et al., 1987). Further research in this field is likely to yield important insights into the molecular mechanisms of viral infections and lead to the development of new, more effective treatments for these challenging and often devastating diseases (Gallo et al., 1987; John et al., 1979; Balde et al., 1990).

## REFERENCES


Alarcon-Aguilara FJ, Roman-Ramos R, Perez-Gutierrez S, Aguilar-Contreras A, Contreras-Weber CC, Flores-Saenz JL. Study of the Anti-hyperglycemic Effect of Plants Used as Antidiabetics. *J Ethnopharmacol.* 1998 Jun;61(2):101–110.

Arisawa M., Fujita A., Hayashi T., Ockiai H., Morita N. Cytotoxic and Antiherpetic Activity of Phloroglucinol Derivatives from *Mallotus japonicus* (Euphorbiaceae) *Chem. Pharm. Bull.* 1990;38:1624–1626.

Asanaka M., Kurimura T., Kobayashi R., Okuda T., Mori M., Yokoi H. Anti-HIV Activity of Ellagitannins. Fourth International Conference on Immuno Pharmacology; May; 1988. p. 47. Osaka, Japan. Abstract.

Asano J., Chiba K., Tada M., Yoshi T. Antiviral Activity of Lignans and Their Glycosides from *Justicia procumbens. Phytochemistry.* 1996;42:713–717.

Balde A.M., Van Hoof L., Pieters L.A., Van den Berghe D.A., Vlietinck A.J. Plant Antiviral Agents. VII Antiviral and Antibacterial Proanthocyanidins from The Bark of *Pavetta owariensis. Phytother. Res.* 1990;4:182–188.

Balzarini J., Schols D., Neyts J., Van Damme E., Peumans W., De Clercq E. α-(1-3) -D- Mannose-Specific Plant Lectins Are Markedly Inhibitory to Human Immunodeficiency Virus and Cytomegalovirus Infections. *In Vitro. Antimicrob. Agents Chemother.* 1991;35:410–416. [PMC free article]

Batinic D., Robey F.A. The V3 Region of The Envelope Glycoprotein of Human Immunodeficiency Type 1 Binds Sulfated Polysaccharides and CD4-Derived Synthetic Peptides. *J. Biol. Chem.* 1992;267:6664–6671.

Becker Y. Antiviral Agents from Natural Sources. *Pharmac. Ther.* 1980;10:119–159.

Bedows E., Hatfield G.M. An Investigation of the Antiviral Activity of *Podophyllum peltatum. J. Nat. Prod.* 1982;45:725–729.

Beladi I., Pusztai R., Mucsi L., Bakay M., Gabor M. Activity of Some Flavonoids against Viruses. *Ann. N. Y. Acad. Sci.* 1977;284:358–364.

Beutler J.A., Cardellina J.H. II, McMahon J.B., Boyd M.R. Anti-HIV and Cytotoxic Alkaloids from Buchenavia Capitata. *J. Nat. Prod.* 1992;55:207–213.

Boyd M.R., Hallech Y.F., Cardellina J.H. II, Manfredi K.P., Blunt J.W., McMahon J.B., Buckheit R.W. Jr., Bringman S., Schaffer M., Cragg G.M., Thomas D.W., Jato J.C. Anti-HIV Michellamines from *Ancistrocladus korupensis. J. Med. Chem.* 1994;37:1740–1745.

Came P.E., Steinberg B.A. *Handbook of Experimental Pharmacology.* Springe Verlag; Berlin: 1982. *Chemotherapy of Viral Infection.* Natural Products; pp. 479–518.

Chan J.A., Schutis E.A., Carr S.A., De Brusse C.W., Eagleton D.S., Francis T.A., Hayland L.J., Johnson W.P., Killman L.B., Staiger D.R., Westley J.W. Novel Phloroglucinols from the Plant *Melicope sessiliflora* (Rutaceae). *J. Org. Chem.* 1989;54:2098–2103.

Che C-T. Marine Products as a Source of Antiviral Leads. *Drug Dev. Res.* 1991;23:201–218.

Chu C.K., Culter H.G. *Natural Products as Antiviral Agents.* Plenium Press; Berlin: 1992.

Corthout J., Pieters L., Claeys M., Vanden Berghe D.A., Vlietinck A.J. Antiviral Ellagitannins from *Spondias mombin. Phytochemistry.* 1991;30:1129–1130.

Corthout J., Pieters L., Claeys M., Vanden Berghe D., Vlietinck A.J. Antiviral Caffeoyl Esters from Spondias mombin. *Phytochemistry.* 1992;31:1979–1981.

Clark A.M. Natural Products as a Resource for New Drugs. *Pharmaceut. Res.* 1996;13:1133–1141.

Cragg G.M., Newman D.J., Sander K.M. Natural Products in Drug Discovery and Development. *J. Nat. Prod.* 1997;60:52–60.

Dai J-R., Hallock Y.F., Cardellina J.H. II, Boyd M.R. HIV-Inhibitory and Cytotoxic Oligostilbenes from the Leaves of *Hopea malibato. J. Nat. Prod.* 1998;61:351–353.

De Clercq E. Antiviral Agents: Characteristic Activity Spectrum Depending on the Molecular Target with Which They Interact. In: Maramorosch K., Murphy F. A., Shatkin A. J., editors. *Advances in Virus Research.* Academic Press; Stamford, Connecticut: 1993. pp. 1–54.

De Closterd L.A., Parsons I.C., Gustafson K.R., Cardellina J.H. II, Mcmahon J.B., Cragg G.M., Murata Y., Pannell L.K., Steiner J.R., Clardy J., Boyd M.R. Structure, Absolute Stereochemistry and Synthesis of Concocurvone A Potent Novel HIV-Inhibitory Naphthoquinone Trimer from a *Concospermum* sp. *J. Am. Chem. Soc.* 1993;115:6673–6679.

Eberhardt F.L., Young R.A. Assesment of the Anti-HIV Activity of a Pine Cone Isolate. *Planta Med.* 1996;62:63–65.

Eich E. Secondary Metabolites from Plants as Antiretroviral Agents: Promising Lead Structure for Anti-HIV Drugs of the Future. In: Lawson L. D., Bauer R., editors. *American Chemical Society*; 1998a. pp. 83–95. (ACS Synposium Series 691: Phytomedicines of Europe Chemistry and Biological Activity).

Eich E. *Phytomedicines of Europe.* Vol. 691; San Diego, London: 1998b. Secondary Metabolites from Plants as Antiretroviral Agents: Promising Lead Structures for Anti-HIV Drugs of the Future; pp. 83–96. (ACS Symp. Ser.).

Eich E., Partz H., Kaloga M., Schulte J., Fesen M.R., Mazondu A., Pommier Y.J. (–)-Arctigenin as a Lead Structure for Inhibitors of Human Immunodeficiency Virus Type-1 Integrase. *J. Med. Chem.* 1996;39:86–95.

Fehr M.J., McCloskey M.A., Petrick J.W.J. Light-Induced Acidification by the Antiviral Agent Hypericin. *J. Am. Chem. Soc.* 1995;117:1833–1836.

Feldmann H., Klenk H-D. Marburg and Ebola Viruses. In: Maramorosch K., Murphy F. A., Shatkin A. J., editors. *Advances in Virus Research.* Academic Press; Cold Spring Harbor, New York: 1996. pp. 1–52.

Gallo R.C., Stericher H.Z., Border S. *AIDS.* Marcel Dekker; Washington DC: 1987. p. 1.

Ghoneum M. Anti-HIV Activity in Vitro of MGN-3, an Activated Arabinoxylan from Rice Bran. *Biochem. Biophys. Res. Commun.* 1998;243:25–29.

Gustafson K.R., Cardellina J.H. II, McMahon J.B., Pannel L.K., Cragg G.M., Boyd M.R. The Peltatols, Novel HIV-Inhibitory Catechol Derivatives from *Pothomorphe peltata. J. Org. Chem.* 1992;57:2809–2811.

Hara H., Maruyama N., Yamashita S., Hayashi Y., Lee K.H., Basto K., Chairful F., Marumoto R., Imakura Y. Elecanacin, a Novel New Naphthoquinone from the Bulb of *Eleutherine americana. Chem. Pharm. Bull.* 1991;45:1714–1716.

Hatano T., Yasuhara T., Miyamoto K., Okuda T. Anti-Human Immunodeficiency Virus Phenolics from Licorice. *Chem. Pharm. Bull.* 1988;36:2286–2288.

Hayashi K., Hayashi T., Otsuka H., Takeda Y. Antiviral Avtivity of 5,6,7-Chemother. Trimethoxyflavone and Its Potentiation of the Antiherpes Activity of Acyclovir. *J. Antimicrob.* 1997;39:821–824.

Helbig B., Thiel K.D. Vergleich Der Antiviralen Aktivitat Von Oxidierter Kaffusaure und Hydrokaffeesaure Gegen Herpesvirus Hominis Typ 1 und typ 2. *In Vitro. Pharmazie.* 1982;37:603–604.

Hizi A., Tal R., Shaharabary M., Currens M.J., Boyd M.R., Hughes S.H., McMahon J.B. Specific Inhibition of the Reverse Transcriptase of Human Immunodeficiency Virus Type 1 and the Chimeric Enzymes of Human Immunodeficiency Virus Type 1 and 2 by Non-nucleoside Inhibitors. *Antimicrob. Agents Chemother.* 1993;37:1037–1042. [PMC free article]

Hohenschutz L.D., Bell E.A., Jewens P.J., Leworthy D.P., Pryce R.J., Arnold E., Clardy J, Castanospermine A. 1,6,7,8-Tetrahydroxyoctahydroindolizidine Alkaloid, from Seeds of Castanospermum Anstrale. *Phytochemistry.* 1981;20:811–814.

Hudson J.D. *Antiviral Compounds from Plants.* CRC Press, Inc.; San Diego, London: 1990. pp. 43–57.

Hughes S.S. *The Virus A History of the Concept.* Heineman Educational Books; Palo Alto, CA: 1977. pp. 109–113. Science History Publications, New York.

Ieven M, Vlietinck A.J., Vanden Berghe D.A., Totte J., Dommisse R., Esmans E., Alderweireldt F. Plant Antiviral Agents. III. Isolation of Alkaloids from *Clivia miniata* Regel (Amaryllidaceae). *J. Nat. Prod.* 1982;45:564–573.

John T.J., Mukundan P. Virus Inhibition by Tea, Caffeine and Tannic Acid. *Ind J. Med. Res.* 1979;69:542–545.

Joklik W.K. *Virology.* 3rd edition. Appleton & Lange; Belmont, California: 1988. pp. 1–38.

Kakiuchi N., Hattori M., Ishii H., Namba T. Effect of Benzo[c]phenanthidine Alkaloids on Reverse Transcriptase and Their Binding Property to Nucleic Acids. *Planta Med.* 1987;55:22–27.

Kakiuchi N., Hattori M., Namba T., Nishizawa M., Yamagishi T., Okuda T. Inhibitory Effects of Tannins on Reverse Transcriptase from RNA Tumor Virus. *J. Nat. Prod.* 1985;48:614–621.

Kashman Y., Gustatson K.R., Fuller R.W., Cardellina J.H., II, McMahon J.B., Curren M.J., Buckheit R.W. Jr., Hughes S.K., Cragg G.M., Boyd M.R. The Calanolides a Novel HIV-Inhibitory Class of Coumarin Derivatives from the Tropical Rainforest Tree *Callophyllum langigerum. J. Med Chem.* 1992;55:2735–2743.

Kernan M.R., Amarquaye A., Chen J.L., Chan J., Sesin D.F., Parkinson N., Ye Z-J., Barrett M, Bales C., Stoddart C.A., Solan B., Blanc P., Limbach C., Mrisho S., Rozhon E.J. Antiviral Phenylpropanoid Glycosides from the Medicinal Plant *Markhamia lutea. J. Nat. Prod.* 1998;61:564–570.

Kernan M.R., Sendl A., Chen J.L., Jolad S.D., Blanc P., Murphy J.T., Stoddart C.A., Nanakorn W., Balick M.J., Rozhon E.J. Two New Lignans with Activity Against Infuenza Virus from the Medicinal Plant Rhinacanthus nasutus. *J. Nat. Prod.* 1997;60:635–637.

Kim H.K., Farnsworth N.R., Blomster R.N, Fong H.H.S. Biological and Phytochemical Evaluation of Plants. V. Isolation of Two Cytotoxic Alkaloids from *Chelidonium majus. J. Parm. Sci.* 1969;58:372–374.

Kite G.C., Fellows L.E., Fleet G.W.J., Liu P.S., Scofield A.M., Smith N.G. α-Homono—jirimycin [2,6-Dideoxy-2,6-imino-Dglycerol-L-heptitol] from *Omphalea diandra* L.: Isolation and Glucosidase Inhibition. *Tetrahedron Lett.* 1988;29:6483–6486.

Kmpotic E., Farnsworth N.R., Messner W.M. Cryptopleurine, An Active Antiviral Alkaloid from *Boehmeria cylinderica* (L.) SW (*Urticaceae*). *J. Pharm. Sci.* 1972;61:1508–1509.

Lin T.S., Shinazi R., Griffith B., August E.M., Eriksson B.F.H., Zhang D.K., Huang L., Prusoff W.H. Selective Inhibition of Human Immunodeficiency Virus Type-1 Replication by The (–)-But Not The (+)-Enantiomer of Gossypol. *Antimicrob. Agents Chemother.* 1989;33:2149–2151. [PMC free article] Res. 1990;4:182–188.

Lockshin R.A., Zakeri Z. Apoptosis: The Molecular Basis of Cell Death. In: Tomei L. D., Cope F. O., editors. *Cold Spring Harbor Laboratory*; San Diego, London: 1991. pp. 47–60.
Macrae W.D., Towers G.H.N. Biological Activities of Lignans. *Phytochemistry*. 1984;23:1207–1220.
McCommick J.L., McKee T.C., Cardellina J.R., Boyd M.R. HIV Inhibitory Natural Products. *Quinoline Alkaloids from Euodia oxburghiana. J. Nat. Prod.* 1996;59:469–471.
McMahon J.B.M., Currem M.J., Gulakowski R.J., Buckheit R.W. Jr., Lachmen-Smith C, Hallock Y.F., Boyd M.R, Michellamine B. A Novel Plant Alkaloid, Inhibits Human Immunodeficiency Virus-Induced Cell Killing by at Least Two Distinct Mechanisms. *Antimicrob. Ag. Chemther.* 1995;39:484–488. [PMC free article]
McRae W.D., Hudson J.B., Towers G.N.R. Studies on the Pharmacological Activities of Amazonian Euphorbaceae. *J. Ethnopharmacol.* 1988;22:143–172.
Montanha J.A., Amoros M., Bastic J., Girre L. Antiherpes Virus Activity of Aporphine Alkaloids. *Planta Med.* 1995;61:419–424.
Munro M.H.G, Luibrand R.T., Blunt J.W. *Biorganic Marine Chemistry*. Springe Verlag; San Diego, London: 1987. pp. 93–112.
Nakashima H., Kudo Y., Kobyashi N., Motoki Y., Neushul M., Yamamoto N. Purification and Characterization of an Avian Myeloblastosis and HumanImmunodeficiency Virus Reverse Transcriptase Inhibitors, Sulfated Polysaccharides Extracted from Sea Algae. *Antimicrob. Agents Chemother.* 1987;31:1524–1528. [PMC free article]
Nash R.J., Fellows I.E., Dring J.V., Stirton C.H., Carter D., Hegarty M.P., Bell E.A. Castanospermine in *Alexia* species. *Phytochemistry*. 1988;27:1403–1404.
Ng T.B., Huang B, Fong W.P., Yeung H.W. Anti-Human Immunodeficiency Virus (Anti-HIV) Natural Products with Special Emphasis on HIV Reverse Transcriptase Inhibitors. *Life Sci.* 1997;61:933–949.
Nishibe S., Ono K., Nakane H., Kawamura T., Noro Y., Tanaka T. Studies of the Plantaginis herba. 9. Inhibitory Effects of Flavonoids from Plantago Species on HIV Reverese Transcriptase Activity. *Nat. Med.* 1997;51:547–549.
Nishizawa M., Yamada H., Sano J., Ito S., Hayashi Y., Ikeda H., Chairul S.M., Tokuda H. Structure of Syzygiol: A Skin-Tumor Promotion Inhibitor. *Tetrahedron Lett.* 1991;32:211–212.
Nishizawa M., Yamagishi T., Dutschman G.E., Parker W.B., Bodner A.J., Kilkhuskic R.E., Cheng Y.C., Lee K.H. Anti-AIDS Agents 1. Isolation and Characterization of Four New Tetragalloylquinic Acids as A New Class of HIV Reverse Transcriptase Inhibitors from Tannic Acid. *J. Nat. Prod.* 1989;52:762–768.
Nonaka G.I., Nishioka I., Nishizawa M., Yamagishi T., Kashiwada Y., Dutschman G.E., Bodner A.J., Kilkushic R.E., Cheng Y.C., Lee K.H. Anti-AIDS Agents 2. Inhibitory Effects of Tannins on HIV Reverse Transcriptase and HIV Replication in H9 Lymphocyte Cells. *J. Nat. Prod.* 1990;53:587–595.
Okuda T., Yoshida T., Hatano T. Ellagitannins as Active Constituents of Medicinal Plants. *Planata Med.* 1989;55:117–122.
Ono K., Nakane H., Meng Z.H., Ose Y., Sakai Y., Mizuno M. Differential Inhibitory Effects of Various Herb Extracts on the Activities of Reverse Transcriptase and Various Deoxyribonucleic Acid (DNA) Polymerase. *Chem. Pharm. Bull.* 1989;57:1810–1812.
Patel A.D., Freyer A.J., Egglerton D.S., Hultiwonger R.C., Bean M.F., Taylor P.B., Caranfa M.J., Breen A.L., Barton H.M., Johnson R.K., Herteberg R.P., Westley D.W. The Inophyllums, Novel Inhibitors of HIV-1 Reverse Transcriptase Isolated from the Malaysian Tree, *Callophyllum inophyllum Linn. J. Med. Chem.* 1993;36:4131.
Pengsuparp F., Cai L., Constant H., Fong H.H., Lin L.Z., Kinghorn A.D., Pezutto J.M., Cordell G.A. Mechanistic Evaluation of New Plant-Derived Compounds That Inhibit HIV—Reverse Transcriptase. *J. Nat. Prod.* 1995;58:1024–1031.
Pengsuparp F., Sirit M., Hughes S.H., Soejarto D.D., Pezutto J.M. Specific Inhibition of Human Immunodeficiency Virus Type 1 Reverse Transcriptase Mediated by Soulattrolide, A Coumarin Isolated from the Latexof *Calophyllum teysmannii. J. Nat. Prod.* 1996;59:839–842.
Raghavan K., Boulamwini J.K., Fesen M.R., Pommier Y., Kohn J.W., Weinstein J.N. Three Dimensional Quantitative Structure-Activity Relationship (QSAR) of HIV Integrase Inhibitors: A Comparative Molecular Field Analysis (CoMFA) Study. *J. Med Chem.* 1995;38:890–897.
Razvi E.S., Welsh R.M. Apoptosis in Viral Infections. In: Maramorosch K., Murphy F. A., Shatkin A. J., editors. *Advances in Virus Research*. Academic Press; Norwalk, Connecticut, San Mateo: 1995. pp. 1–60.
Rinehart K.L., Shield L.S., Cohen-Parson M. Antiviral Substances. In: Attaway D. H., Zaborsky O. R., editors. *Marine Biotechnology, Pharmaceutical and Bioactive Natural Products*. Plenium Press; New York: 1993. pp. 309–342.

Royer R.E., Deck L.M., Van derJagt D.C. Synthesis and Anti-HIV Activity of 1,1′-Dideoxy—Gossypol and Related Compounds. *J. Med. Chem.* 1995;38:2427–2432.

Safrin S., Chambers H.F. *Basic & Clinical Pharmacology*. 7th edition. In: Katzung B. G., Appleton & Lang, editors. Boca Raton, Florida: 1998. pp. 788–802.

Schinazi R.F., Chu C.K., Babu J.R., Oswald B.J., Saalmann V., Cannon D.L., Eriksson B.H.F., Nasa M. Anthraquinones as a New Class of Antiviral Agents Against HumanImmunodeficiency Virus. *Antiviral Res.* 1990;13:265–272.

Scofield A.M., Fellows L.E., Nash R.J., Fleet G.W.H. Inhibition of Mammalian Digestive Disaccharides by Polyhydroxy Alkaloids. *Life Sci.* 1986;39:645–650.

Selway J.W.T. Antiviral Activity of Flavones and Flavans. In: Cody V., Middleton E., Harborne J. B., editors. *Plant Flavonoids in Biology and Medicine: Biochemical, Pharmacological and Structure-Activity Relationships*. Alan R. Liss, Inc.; New York, London: 1986. pp. 521–536.

Sen G.C., Ransohoff R.M. Interferon-Induced Antiviral Actions and Their Regulation. In: Maramorosch K., Murphy F. A., Shatkin A. J., editors. *Advances in Virus Research*. Academic Press; San Diego, London: 1993. pp. 57–102.

Sethi M.L. Enzyme Inhibition. VIII: Mode of Inhibition of Reverse TranscriptaseActivity by Analogues, Isomers and Related Alkaloids of Coralyne. *J. Pharm. Sci.* 1985;74:889–891.

Shu Y-Z. Recent Natural Products Based Drug Development: A Pharamaceutical Industry Perspective. *J. Nat. Prod.* 1998;61:1053–1071.

Starr C, Taggart R. Biology, the Unity and Diversity of Life. Wadsworth Publishing Company; London: 1991. pp. 347–350.

Streissle G., Paessens A., Oedigert H. New Antiviral Compounds. In: Maramorosch K., Murphy F. A., Shatkin A. J., editors. *Advances in Virus Research*. Academic Press; London: 1985. pp. 83–138.

Sydiskis R.J., Owen D.G., Lohr J.L., Rosier K.H., Blomster R.N. Inactivation of Enveloped Viruses by Anthraquinones Extracted from Plants. *Antimicrob. Agents Chemother.* 1991;55:2463–2466. [PMC free article]

Tafur S., Nelson J.D., Delong D.C., Svoboda G.R. Antiviral Components of *Ophiorrhiza mungo*. Isolation of Camptothecin and 10-Methoxycamptothecin. *Lloydia*. 1976;39:261–262.

Takasaki M., Konishima T., Shingu T., Tokuda H., Nishino H., Iwashima A., Kozuka M. Structure of Euglobal-Gl, -G2 and -G3 from *Eucalyptus gerandis*. Three New Inhibitors of Epstein-Barr Virus Activation. *Chem. Pharm. Bull.* 1990;38:1444–1446.

Take Y., Inoye Y., Nakamura S., Allaudeen R.S., Kubo A. Comparative Studies of the Inhibitory Properties of Antibiotics on HumanImmunodeficiency Virus and Avaian Myeloblastosis Virus Reverse Transcriptase and Cellular DNA Polymerase. *J. Antibiotics*. 1989;42:107–115.

Takechi M., Tanaka Y. Purification and Characterization of Antiviral Substances from the Bud of *Syzygium aromatica*. *Planta Med.* 1981;42:69–74.

Takechi M., Tanaka Y. Antiviral Substances from the Root of Paeonia species. *Planta Med.* 1982;45:252–253.

Takechi M., Tanaka Y., Takehara M., Nonaka G.I., Nishioka I. Structure and Antiherpetic Activity among the Tannins. *Phytochemistry*. 1985;24:2245–2250.

Tan G.T., Pezzuto J.M., Kinghorn A.D. Evaluation of Natural Products as Inhibitors of Human Immunodeficiency Virus Type-1 (HIV-1) Reverse Transcriptase. *J. Nat. Prod.* 1991;54:143–154.

Valsaraj R., Pushpangadan P., Smitt U.W., Adsersen A., Christensen S.B., Sittie A., Nyman U., Nielsen C, Olsen C.E. New Anti -HIV, Antimalarial and Antifungal Compounds from *Terminalia bellerica*. *J. Nat. Prod.* 1997;60:739–742.

Vanden Berghe D.A., Ieven M., Mertens F., Vlietinck A.J., Lammens E. Screening of Higher Plants for Biological Activities. II. Antiviral Activity. *Lloydia*. 1978;41:463–471.

Vanden Berghe D.A., Vlietinck A.J., Van Hoof L. Plant Products as Potential Antiviral Agents. *Bull. Instit. Pasteur.* 1986a;84:101–147.

Vanden Berghe D.A., Vlietinck A.J., Van Hoof L. Plant Substances as Antiviral Agents. *Bull. Inst. Pasteur.* 1986b;84:101–147.

Vlietinck A.J., De Bruyne T., Vanden Berghe D.A. Plant Substances as Antiviral Agents. *Curr. Org. Chem.* 1997;1:307–344.

Vlietinck A.J., Vanden Berghe D.A., VanHoof L.M., Vrijsen R., Boeye A. Antiviral Activity of 3-Methoxyflavones. In: Cody V., Middleton E., Harborne J. B., editors. *Plant Flavonoids in Biology and Medicine: Biochemical, Pharmacological and Structure-Activity Relationships*. Alan R. Liss, Inc.; London: 1986. pp. 537–540.

Xu H.X., Kadota S., Kurokawa M., Shiraki K., Matsumoto T., Namba T. Isolation and Structure of Woodorien a New Glucoside Having Antiviral Activity from *Woodwardia orientalis*. *Chem. Pharm. Bull.* 1993;41:1803–1806.

# 17 Clinical Antiviral Natural Products and Approved Drugs

*Amit Kumar Dubey*
Indian Council of Medical Research (ICMR)-Rajendra Memorial Research Institute of Medical Sciences (RMRIMS)
School of Pharmaceuticals and Population Health Informatics (SoPPHI), DIT University

*Jyoti Kumari, Sneha Rani, Niva Singh, and Shubhankar Kumar Singh**
Indian Council of Medical Research (ICMR)-Rajendra Memorial Research Institute of Medical Sciences (RMRIMS)

## 17.1 INTRODUCTION

Viral infections play a crucial role in human diseases, and recent increase in globalization and ease of travel has made the prevention of these infections a critical challenge to protect the public health (Torres Munguía et al., 2022). Till April 2023, variated human infectious diseases (HIV, HBV, HCV, HCMV, HSV, HPV, RSV, VZV, and influenza virus) have been treated through approved antiviral drugs, although more than 200 human viruses have been introduced (Borah et al., 2023). In June 1963, the first antiviral drug, idoxuridine, was approved, marking the beginning of a new era of antiviral drug development (Witvrouw and De Clercq, 1997). Since then, for the treatment of millions of people worldwide, many antiviral drugs have been developed (Kausar et al., 2021; De Clercq and Li, 2016). However, viruses are resistant to therapy and prophylaxis due to their characteristics. They depend on the cells they infect for their multiplication and survival. Currently, infectious viral diseases continue to be a global challenge, and there are only few drugs available for the cure of viral diseases including acyclovir resistant HSV infections (Xu et al., 2022). Therefore, herbal medicines and purified natural products hold promise as a rich resource for novel antiviral drug development.

## 17.2 EPIDEMIOLOGY

Epidemiology of viral disease is the study of the determinants, dynamics and distribution of viral diseases in populations. To improve our understanding of the overall nature of disease and in alerting the disease control activities, quantitative measurements of disease trends, epidemiology has a major role. Viruses, whether of the same or another species, can survive in nature only if they have the capability to transmit from one host to another. To initiate the transmission cycle, there is a need for the virus to enter the body, undergo replication and followed by subsequent spread to another host. The modes of virus entry into the body include the skin, respiratory tract, intestinal tract, oropharynx, urogenital tract and conjunctiva. The exit of virus from an infected host is just as important as entry in maintaining its transmission cycle. Viruses can be transmitted horizontally or vertically, but the broad majority of transmission occurs horizontally, between individuals at high risk. The transmission of virus may directly affect the infected virus virulence,

* corresponding author

DOI: 10.1201/9781003329169-20

which in turn may directly affect the probability of its transmission. The main example of this is rabbit myxomatosis. With most viruses, endemic or epidemic transmission shows to a level of immunity in the host population that affects or even interrupts further transmission. Knowledge of epidemiology and modes of transmission of infectious diseases is important to the development and implementation of prevention and control strategies (Burrell et al., 2017; Louten, 2016).

In this concise report, we give an outline of the antiviral activities from several natural products, herbal medicines, drugs and approved drug names against some notable viral pathogens, including coronavirus (CoV), Influenza virus/influenza, Coxsackievirus, Hepatitis A virus/hepatitis, hepatitis B virus (HBV), hepatitis C virus (HCV), Herpes simplex virus/genital herpes, Human immunodeficiency virus 1/acquired immunodeficiency syndrome (AIDS), Respiratory syncytial virus (RSV), Dengue viruses/dengue (DENV), enterovirus 71 (EV71) and measles virus (MV). Table 17.1 presents a complete overview of studies related to herbal medication and viral infections.

## 17.3 CLINICAL ANTIVIRAL NATURAL PRODUCTS TRIALS

### 17.3.1 Coronavirus

Novel corona virus is an RNA virus with a single-stranded ribonucleic acid (ssRNA). This virus is grouped into the family of *Coronaviridae*. This Novel CoV mainly affects upper respiratory tract. Both humans/animals and birds are affected. The serious symptoms resemble pneumonia (Van der Hoek, 2007). There are several variants of COVID virus, such as α, β, Δ and Omicron, among others. Some variants cause milder symptoms, while others lead to serious symptoms. The severe acute respiratory syndrome coronavirus is considered the 6th type of infection as categorized by WHO in the Middle East. It is associated with high mortality rates (Geller et al., 2012; WHO, 2023b).

Lastly, from 2019 to 2023, these 5 years have been critical for mankind because of this deadliest virus. Vaccines developed are found to be ineffective and proper medication not been generated so far despite extensive measures taken. Thus, safe, and effective natural medicament is a need in present time. Chang PW and team reported that a famous group of oleanane derivatives, mostly found in China (saikosaponins (A, $B_2$, C, and D)), have been found to be effective against COVID virus. These apoins highly isolate/extract from *Bupleurum* spp. Other species are *Heteromorpha* spp., and *Scrophularia* spp., which hold promise in this regard (Cheng et al., 2006a).

### 17.3.2 Influenza Virus/Influenza

The infectious disease, commonly referred to as "flu", is widely transmitted worldwide. As per clinical data, influenza virus is generally categorized as A, B, C and D. Mostly, it is written as IFA, IFB and IFC where these three types have ssRNA (single-stranded RNA). These viruses are included in family of *Orthomyxoviridae*. These flu viruses mostly target respiratory tract. The most common symptoms are high fever, cephalalgia, pharyngitis, sneezing, discomfort and pain in joints and muscles. Sometimes, most fatal condition arises when the patient shows symptoms of fatal pneumonia (Eccles, 2005; Rello and Pop-Vicas, 2009). If we delve into the details of IFA, it typically spreads as epidemic and can infect other organisms such as birds and mammals. The strain IFB infects in *Homo sapiens* mostly. The less clinical cases of IFC have been isolated from piglets and humans (Mostafa et al., 2018). It was recorded that approximately 3 million to 5 million cases of infection are caused by influenza virus (20–50 million deaths reported in H1N1 flu) (CDC, 2023; WHO, 2023b).

Available data have constantly shown that flu viruses are continuously evolving. Mostly its hemagglutinin and neuraminidase (exosialidase/glycoside hydrolase) proteins evolve in circulating strains of influenza (Yamada et al., 2006; Van der Vries et al., 2012). Importantly, the transmission of pathogen from animals to humans raised concerns (Mak et al., 2012). The drug resistance is another serious issue which also affects therapeutics. The drugs such as amantadine and rimantadine which are blockers of M2 ion-channel also have shown drug resistance (Committee on Infectious Diseases

**TABLE 17.1**
**A Complete Studies Related to Herbal Medication and Viral Infections**

| S.No. | Causative Agent/ Disease | Transmittal Route | Herbal Products(s) Assessment | Target/Strains/Mechanism | References |
|---|---|---|---|---|---|
| 1 | CoV (Corona Virus) | Droplets fluid from respiratory tract | • Triterpenoid glycoside (Saikosaponins-A)<br>• Terpenoid (Saikosaponins-B2)<br>• Saikosaponins-C (ssc)<br>• Saikosaponins-D2 (ssd)<br>• Obtained from *Bupleurum falcatum* | • Targeted against Human Corona Virus22E9 | Cheng et al. (2006b); Li et al. (2005); Lin et al. (2005); Ryu et al. (2010); Yu et al. (2012); Lau et al. (2008) |
| | | | • Crystalline alkaloid (*lycorine*),<br>• Sweet wormwood *(Artemisia annua)*, | • Targeted against Novel Corona Virus (SARS-CoV-2) | |
| | | | • East Asian fern (*Pyrrosia lingua*), active constituents/ extracts obtained from *Lycoris radiata* | • Against Novel Corona Virus (SARSCoV-2) via ACE2 binding interactions | |
| | | | • Phenolic compounds *Lindera aggregate* (L. aggregate) isolated from the *Isatis indigotica* | • Against Novel Corona Virus (SARSCoV-2) | |
| | | | • Didemethyl-ginkgetin (Amentoflavone) extracted from Nut tree "*Torreya nucifera*" | • Against Novel Corona Virus (SARSCoV-2) via $M^{pro}$ 3CLpro | |
| | | | • Naturally-occurring flavonoids (Myricetin and scutellarein), Asian *Houttuynia cordata* water extract | • Against Novel Corona Virus (SARSCoV-2) RdRp enzyme | |
| 2 | Influenza virus/ influenza (Flu) | Miniature droplets and transmitted aerosol | • The liquid extract obtained from Black elderberries *Sambucus nigra*<br>• The root elicit from *Pelargonium sidoides* | • Influenza virus inhibitors<br>• (Against Influenza strain-A)<br>• The impaired viral hemagglutination inhibited at early stages (target influenza-A) | Krawitz et al. (2011); Theisen and Muller (2012); He et al. (2011); Ma et al. (2013); Grienke et al. (2012); Dao et al. (2011); Dao et al. (2012); Jeong et al. (2012) |
| | | | • Aqueous roots elicit from dandelion (*Taraxacum officinale*) | • The viral polymerase activity inhibited (IFA) | |
| | | | • The root elicit from the *Illicium oligandrum*—the component Spirooliganone B | • Against influenza virus A (H3N2) | |
| | | | • Novel Chalcones obtained from *Glycyrrhiza inflate* (Chinese licorice), Xanthones (antioxidants property) from *Polygala karensium* (a perennial herbs), Homoisoflavonoids from heartwood *contains medicinal value(Caesalpinia sappan)*-12 compounds | • Inhibit Influenza strains (H1N1, H9N2, etc.) via neuraminidases inhibition | |

*(Continued)*

**TABLE 17.1 (*Continued*)**
**A Complete Studies Related to Herbal Medication and Viral Infections**

| S.No. | Causative Agent/ Disease | Transmittal Route | Herbal Products(s) Assessment | Target/Strains/Mechanism | References |
|---|---|---|---|---|---|
| 3 | Coxsackievirus | Contamination through droplets/ aerosol from the respiratory tract, also transmitted through infected person contact | • *Ocimum basilicum* medicinal purpose elicits such as Aqueous and ethanolic extracts, Other herbal/natural components such as linalool (terpene alcohol), flavonoids (apigenin), and pentacyclic triterpenoid (ursolic acid)<br>• Specified Raoulic acid from the *Raoulia australis* (family Asteraceae) | • Against Coxsackievirus strain-B<br>• Against Coxsackievirus strain-B | Chiang et al. (2005); Choi et al. (2009); Cheng et al. (2006c); Cheng et al. (2007) |
| 4 | Hepatitis B virus | Transmitted through bodily fluids such as blood and semen.<br>Mainly infected mode through sexual contact as well as injecting via infected syringes or from mother to foetus transmission. | • MetaboliteIsochlorogenic acid-A (CGA) from Winged-Stem **Laggera**<br>• Piperine (alkaloid) from Long piper (*Piper longum*)<br>• Quaternary amm. Alkaloid dehydrocheilanthifoline isolated (DHCH) from *Corydalis saxicola*<br>• Saikosaponins (C, D) from annual or perennial herbs (*Bupleurum* sp.)<br>• Green extract from *Reynoutria japonica* Houtt (*P. cuspidatum* Sieb.)<br>• Curcumin | • **Neuraminidase blocker**<br>• Suppress *Methyltransferase*<br>• *VP35 interferon*<br>• Nucleoside analogues<br>• Inhibits pgRNA synthesis<br>• Anti-oxidant property<br>• **Downregulation HBV** gene replication via targeting cccDNA-bound histone acetylation | Qiu and Chen (2013); Hao et al. (2012); Jiang et al. (2013); Zeng et al. (2013); Chang et al. (2007); Chiang et al. (2003); Chang et al. (2005); Rechtman et al. (2010) |

(*Continued*)

**TABLE 17.1 (*Continued*)**
**A Complete Studies Related to Herbal Medication and Viral Infections**

| S.No. | Causative Agent/ Disease | Transmittal Route | Herbal Products(s) Assessment | Target/Strains/Mechanism | References |
|---|---|---|---|---|---|
| 5 | Hepatitis C virus | The person who is infected with HBV-C, contaminate blood, Physical contacts with infected surgical etc. (CDC data) | • Herbal extracts of **milk thistle** (*Silybum marianum*) such as Flavonolignans<br>• Curcumin<br>• Epigallocatechin3gallate (EGCG), also k/as Epigallocatechin gallate<br>• Griffithsin (GRFT)<br>• Dimethoxyflavone (Ladanein)<br>• Natural product (Tellimagrandin I) from species of *Rosae rugosae* Flos.<br>• Benzopyran tannin (Chebulagic acid) and ellagitannin (punicalagin) isolated from black- or chebulic myrobalan (*Terminalia chebula* Retz.) | • **Inhibition** of NF-κB, anti-inflammatory properties<br>• **HBV** gene inhibition<br>• Inhibits HCV via spreading Cell-to-Cell, also upregulate antiviral innate immune responses<br>• binds N-linked high-mannose oligosaccharides<br>• Viral load inhibition, also inhibit viral entry in genotype independent manner<br>• Inhibition potent **HCV** invasion<br>• HCV NS3/4A inhibition | El-Serag (2012); Welsch et al. (2012); Polyak et al. (2010); Kim et al. (2010); Anggakusuma et al. (2014); Ciesek et al. (2011); Calland et al. (2012); Meuleman et al. (2011); Takebe et al. (2013); Haid et al. (2012); Tamura et al. (2010) |
| 6 | Herpes simplex virus/genital herpes | Direct physical contacts (sexual activities) | • Herbal products isolation from deciduous *Cassia javanica* tree such as entEpiafzelechin(4α→8) epiafzelechin isolated<br>• Chamber bitter, also k/as grip-weed (*Phyllanthus urinaria*) extraction contains Hippomanin A<br>• Herbal extract such as Geraniin (dehydroellagitannin) and 1,3,4,6tetraOgalloylβDglucose which is isolated from *Phyllanthus urinaria*<br>• Excoecarianin isolated from annual perennial herbal *Phyllanthus urinaria* | • Against HSV2<br>• Against HSV2<br>• Against HSV1 and HSV2 virus<br>• Against HSV2 | Cheng et al. (2008a); Cheng et al. (2008b); Yang et al. (2007a); Yang et al. (2007b); Cheng et al. (2011); Lin et al. (2011); Khan et al. (2005); Superti et al. (2008); Chen et al. (2012); Gescher et al. (2011a); Danaher et al. (2011); Gescher et al. (2011b) |

(*Continued*)

**TABLE 17.1** (*Continued*)
**A Complete Studies Related to Herbal Medication and Viral Infections**

| S.No. | Causative Agent/ Disease | Transmittal Route | Herbal Products(s) Assessment | Target/Strains/Mechanism | References |
|---|---|---|---|---|---|
| 7 | Human immunodeficiency virus 1/acquired immunodeficiency syndrome (AIDS) | Direct sexual Contacts, infected blood transfusion, mother to foetus | • Liquid form isolation from family Adoxaceae (*Sambucus nigra*) aka Elderberry<br>• Herbal root extract from (Geraniaceae family) such as *Pelargonium sidoides*<br>• Aqueous extract from dandelion from herbaceous perennial (*Taraxacum officinale*)<br>• The roots extract Spirooliganone B from toxic shrub *Illicium oligandrum*<br>• "Chalcos"/Chalcones ($C_6H_5CCH{=}CHC_6H_5$) from Chinese **licorice**, *a plant Glycyrrhiza inflate*<br>• Phenolic compounds (Xanthones) from milkworts or snakeroots (*Polygala karensium*)<br>• Phenolic components Homoisoflavonoids (3-benzylidenechroman-4-ones) from thorny tree (*Caesalpinia sappan*) | • Against IFA and IFB<br>• Against IFA<br>• Against IFA<br>• Against IFA<br>• Against IFA<br>• Against IFA, anti-oxidant property<br>• Against IFA, anti-HIV integrase activity | Kim and Karadeniz (2011); Vo and Kim (2010); Lubbe et al. (2012); Huerta-Reyes et al. (2004); Cesar et al. (2011); Kudo et al. (2013); Hood et al. (2013) |
| 8 | Respiratory syncytial virus (RSV) | Through droplets/ cough/sneeze/ dried respiratory secretions on bedclothes and similar items/ hard surfaces for several hours and on skin for shorter amounts of time. | • An anti-oxidant Chebulagic acid (benzopyran tannin) and phenolic punicalagin (ellagitannin) isolated from deciduous tree *Terminalia chebula* Retz.<br>• Uncinoside A and B (chromone glycosides) from peacock moss (*Selaginella uncinate*)<br>• Genkwanol B (**O-methylated flavone**), genkwanol C, and natural product (stelleranol) extracted from Chinese herbal medicinal Radix Wikstroemiae<br>• Flavones ($C_6H_4OC_3HO$) from the leaves of grass family (*Lophatherum gracile*)<br>• **Cimicifugoside H-1and** *Cimicifugin* (*Cimicifuga foetida* L.)<br>• Resveratrol mostly present in red grapes | • Inhibited viral glycoprotein<br>• IC50 value of 6.9 and 1.3 mg/mL,<br>• **RSV** RNA synthesis<br>• Inhibit NF-κB which activate by RSV at early stage<br>• Dose-dependent inhibition<br>• Inhibit plaque formation<br>• Against RSVinduced inflammation | Lin et al. (2013); Ma et al. (2003); Huang et al. (2010); Zang et al. (2011); Wang et al. (2012a); Wang et al. (2012b); Wang et al. (2011) |

(*Continued*)

**TABLE 17.1 (*Continued*)**
**A Complete Studies Related to Herbal Medication and Viral Infections**

| S.No. | Causative Agent/ Disease | Transmittal Route | Herbal Products(s) Assessment | Target/Strains/Mechanism | References |
|---|---|---|---|---|---|
| 9 | Dengue viruses/ dengue | Zoonotic/ arthropod-borne | • Baicalein, Quercetin, and Narasin (derivative of salinomycin)<br>• Marine seaweed extracts; *Canistrocarpus cervicornis* (**brown algae**), *Padina gymnospora* (**Brown seaweed**), *Palisada perforate*, and edible green alga (*Caulerpa racemosa*)<br>• A benzopyran tannin (Chebulagic acid) and punicalagin extracted from deciduous tree *Terminalia chebula* Retz. | • Against DENV2 strain<br>• Against DENV serotypes 1–4<br>• against DENV2 | Zandi et al. (2012); Zandi et al. (2011); Low et al. (2011); Koishi et al. (2012) |
| 10 | Enterovirus 71 (EV71) | Through respiratory droplets via cough, sneeze and person-to-person contact infected objects (CDC) | • *Ocimum basilicum* extracted components (aqueous and ethanolic) such as linalool (terpene alcohol), apigenin (dietary flavonoid, and pentacyclic triterpenoid (ursolic acid)<br>• Raoulic acid from scabweed (*Raoulia australis*)<br>• Gallic acid aka 3,4,5-trihydroxybenzoic acid from flowers of *Woodfordia fruticosa*<br>• green tea isolated product Epigallocatechin gallate (EGCG) | • Inhibit viral replication and expression of proteins<br>• Inhibition concentration' IC(50) ~0.1 μg/mL<br>• Inhibition of viral absorption<br>• Inhibits EV71 replication | Chiang et al. (2005); Choi et al. (2009) |
| 11 | Measles virus (MV) | Transmitted via coughing, contaminate air, touch, sneezing (CDC) | • Biflavonoids (flavonoid dimer) from *Rhus succedanea* (earlier c/as *T. succedaneum*) and *Garcinia multiflora* (family Clusiaceae)<br>• Calcium spirulan from the microalga *Spirulina platensis* (a filamentous BGA)<br>• Extracts of *Zanthoxylum chalybeum* (tropical plant) and Ugandan greenheart (*Warburgia ugandensis*)<br>• A perennial legume Stem and root ethanolic extract of pigeon pea (*Cajanus cajan*)<br>• Chebulagic acid and punicalagin (hydrolyzable tannins) isolated from *Terminalia chebula* Retz. (Fam. Combretaceae) | • Inhibition of Measles replication<br>• Reduce plaque formation<br>• Antiviral effect<br>• Flavonoids shows anti-oxidant activity<br>• Binding of CHLA and PUG to viral glycoproteins | Huang et al. (1997); McWhorter et al. (1996); Lin et al. (1999); Hayashi et al. (1996); Petricevich and Mendonca (2003); Olila et al. (2002); Parker et al. (2007); Nwodo et al. (2011) |

**TABLE 17.2**
**Clinical Trials on COVID-19 (ClinicalTrials.gov)**

| S.No. | Title | Interventions | Locations |
|---|---|---|---|
| 1 | Hesperidin and Diosmin for Treatment of COVID-19 | • Drug: Hesperidin and Diosmin mixture | |
| 2 | Jing-Si-Herbal-Tea Accelerates SARS-Cov-2 Load Reduction Among COVID-19 Patients | • Drug: Jing-Si-Herbal-Tea | • Taipei Tzu chi hospital, The Buddhist Medical Foundation, Taipei, Taiwan |
| 3 | Investigation on the Epidemic Prevention Effect of Using Chinese Medicine During the Severe Acute Respiratory Syndrome Coronavirus 2 (SARS-CoV-2) Outbreak | • Drug: Jing-Guan-Fang(JGF) : Chineseherbal decoction | • Taipei City Hospital, Taipei, Taiwan |
| 4 | An Observational, Ambispective Cohort Study of Azvudine in the Treatment of Patients With COVID-19 Pneumonia | • Drug: Azvudine | • Department of Respiratory, The First Affiliated Hospital of Shandong First Medical University & Shandong Provincial Qianfoshan Hospital, #16766, Jingshi Road, Jinan City, Shandong Province, China, Jinan, Shandong, China |
| 5 | Nigella Sativa in COVID-19 | • Dietary Supplement: Nigella sativa oil | • King Abdulaziz University Hospital, Jeddah, Saudi Arabia |
| 6 | The Efficacy and Safety of a Chinese Herbal Medicine for Long COVID-Associated Fatigue | • Drug: Shenlingcao Oral Liquid | • Hong Kong Baptist University Chinese Medicine Clinic, Hong Kong, China |
| 7 | Safety and Efficacy of COVIDEX™ Therapy in Management of Adult COVID-19 Patients in Uganda. | • Drug: COVIDEX | • Makerere University Department of Pharmacology and Therapeutics, Kampala, Uganda |
| 8 | Assessment of Efficacy of KAN-JANG® in Mild COVID-19 | • Drug: Kan Jang capsules<br>• Other: Placebo capsules | • The First University Clinic of Tbilisi State Medical University, Tbilisi, Georgia |
| 9 | A Phase III, Randomised, Double-blind, Placebo-controlled Study to Evaluate the Safety and Efficacy of TD0069 Capsule for Patients With Mild to Moderate COVID-19 | • Drug: TD0069 hard capsule<br>• Drug: TD0069 Placebo | • Traditional Medicine Institute in Ho Chi Minh City, Ho Chi Minh City, Ho Chi Minh, Vietnam<br>• Hanoi Hospital of Traditional Medicine, Ha Noi, Vietnam |
| 10 | Colchicine Plus Phenolic Monoterpenes to Treat COVID-19 | • Drug: Standard Treatment<br>• Drug: Oral administration of Colchicine plus Herbal Phenolic Monoterpene Fractions | • Medical Biology Research Center, Kermanshah University of Medical Sciences, Kermanshah, Iran, Kermanshah, Iran |

et al., 2012). Drug-resistance cases have also been identified for oseltamivir and zanamivir, which are neuraminidase inhibitors for influenza virus (Samson et al., 2013). Due to drug-resistance cases, H1N1, H1N1 and H7N9 mutated strains have arisen recently (WHO, 2023b). Therefore, there is an urgent need for safe and effective therapeutics.

### 17.3.3 Hepatitis B and C Virus

Hepatitis B virus belongs to the family of *Hepadnaviridae* (genus Orthohepadnavirus). This deadliest virus holds a circular DNA, i.e. double-stranded virus (dsDNA) (Liang, 2009). Their infection spreads through contaminated fluids such as blood and other body fluids. Further, when Hepatitis gets severe, it can lead to challenging conditions such as hepatocellular carcinoma and less likely, sometimes liver cirrhosis. However, vaccination programmes have been instrumental in helping developing countries (Ni and Chen, 2010). The drugs/medications such as lamivudine, HEPSERA®, pyrimidine 2′-deoxyribonucleoside (Sebivo) and IFN-alpha are not effective in few clinical cases of HBV. Therefore, herbal medication with targeted effectiveness can show a salient hope for serious cases; more than four hundred billion HBV carrier cases can be treated (Prifti et al., 2021; Kwon and Lok, 2011).

A positive-strand RNA viruses (HCV) belong to the Flaviviridae family. It contains single-stranded ribonucleic acid genomic material (Polyak et al., 2007). HCV infection spreads through blood-to-blood contact, often occurring in persons who are infected via intravenous injections, blood-to-blood transfusions and HCV-contaminated utensils. As this virus is highly mutable, no vaccination therapy is still available, resulting in 300 million carrier clinical cases worldwide. Thus, untreated HCV cases bring patients to liver cirrhosis (El-Serag, 2012). The present therapeutics include such as Peginterferon alfa-2a, and nucleoside analogue (ribavirin). The more effective boceprevir and telaprevir also been prescribed by clinicians. However, the presence of 300 million carriers of HCV worldwide is still a significant challenge. Now, it has been analysed that drug resistance, viral mutations, genotype variations etc. are the major factors which leads to untreated cases (Welsch et al., 2012). Herbal anti-HCV agents will solve this hurdle because of minimal side effects and targeted therapeutics effects.

### 17.3.4 Herpes Simplex Virus/Genital Herpes

The genital herpes (Herpes simplex virus) is a painful blister disease. It is of two types, HSV-1 and HSV-2. The herpes belongs to the *Herpesviridae* family, which contains double-stranded DNA (dsDNA). Mostly, herpes lesions originate on mucosa transitions to skin. Generally,

**TABLE 17.3**
**Clinical Trials on Influenza (ClinicalTrials.gov)**

| S. No. | Study Title | Status | Interventions |
|---|---|---|---|
| 1 | Medicinal Herbs of Chinese: H1N1 target | Completed | Drug: oseltamivir<br>Other: blank<br>Drug: Chinese medicinal herbs<br>1 more |
| 2 | Herbal medicine : novel H1N1 (Pneumonia) | Status not known | Drug: oseltamivir<br>Drug: oseltamivir and Chinese medicinal herbs |
| 3 | A pragmatic randomized control trials of Herbal products for H1N1 influenza | Status not known | Drug: Chinese herbal medicines plus western therapy<br>Drug: western therapy |
| 4 | Influenza management with Polyherbal formulation | Status not known | Drug: polyherbal formulation<br>Drug: Oseltamivir 75mg |

**TABLE 17.4**
**Clinical Trials on Hepatitis B (HBV) and HCV (ClinicalTrials.gov)**

| S. No. | Study Title | Status | Interventions |
|---|---|---|---|
| 1 | Study related to optical strategy for Acute-on-chronic Liver failure with integrative treatment | Completed | Drug: Chinese Herbal medicine |
| 2 | The study of two different Chinese traditional medicine treatment on Chronic Hepatitis B | Unknown status | • Yinchengao decoction+Ganlu detoxification pill<br>• Entecavir/Tenofovir |
| 3 | Treatment of Chronic HCV infected Egyptian patients with Electromagnetic waves and herbal therapy | Unknown status | Device: UVA<br>Dietary supplement: Herbal |
| 4 | The effect of Chinese Herbal formula (KSY) on HCV carrier with abnormal Liver function | Completed | Drug: KSY |

mucocutaneous lesions are caused by HSV-1 strain, while genital lesions are primarily caused by HSV-2. The herpes virus can stay in body for a longer time after getting infection in sensory neurons. This disease can recur multiple times due to different types of sensation or trigger activity such as sunlight, immune system deficiency and stressful life (Stoopler and Balasubramaniam, 2013). This ailment spreads from mother to foetus and via infected lesions to healthy person. A few limited therapeutics are available (antiviral therapy). The immune suppressed persons with a herpes disease suffer a risk of visually impaired medical condition, infection in conjunctiva and superficial cornea. At higher stages, brain fever (meningitis) and inflammation in brain may occur frequently (Arduino and Porter, 2008; Chentoufi and BenMohamed, 2012). Presently, there is limited drugs and no vaccination available in market. Two antiviral drugs such as acyclovir and penciclovir are prescribed nowadays. Drug resistance is another hurdle in HSV cases (Morfin and Thouvenot, 2003). No clinical trials are reported in the registry of clinical trials for herpes.

### 17.3.5 Human Immunodeficiency Virus 1/Acquired Immunodeficiency Syndrome (AIDS)

Human immunodeficiency viruses are genus of retroviruses. These viruses belong to family of *Retroviridae*. HIV viruses (single-stranded genome material) mostly target on CD4 cells of immune system. Reverse transcription is the major phenomenon or specific characteristics of HIV-1, which can interfere host genome (Sierra et al., 2005). It is a well-known fact that causative organism of AIDS is transmitted through infected transfusion of blood, contaminated surgical instruments, sexual activities with carrier, and even through breastfeeding (Shaw and Hunter, 2012). CD-4 molecules are present on T-cells and the $CD4^+$ subtype T-cells downregulation or depletion steers up the deadliest by-infection. This will enhance life-threating and mortality ratio (Yoshimura, 2017; Moss, 2013). According to global index, approximately 40.1 million of cases occur. In 2021 data, around 6.5 million deaths were reported and 1.5 million cases diagnosed with HIV-carrier (WHO, 2023b). After a discovery of HIV pathogen, in the 1980s, several drugs/therapeutics tested, trials, but still prophylactic vaccine has not been developed. The diversity in antigenic variation leads to complexity in cases. Thus, this disease poses more challenges to treat (Burton et al., 2004). The several nucleoside analogues are tested for anti-HIV treatment. In the year 2011, RK Ghosh et al. mentioned HAART therapy. The HAART denotes for highly active antiretroviral therapy. However, patient intolerance, the resistance in present therapy, and co-infection cases lead to untreated AIDs worldwide, more specifically in under-privileged countries (Tian et al., 2023; Pillay, 2023; Margolis et al., 2014). Here, Table 17.5 shows few trails going on at different location with herbal therapy.

**TABLE 17.5**
**Clinical Trials on HIV (ClinicalTrials.gov) (https://ctri.nic.in/)**

| S. No. | Study Title | Status | Interventions |
|---|---|---|---|
| 1 | Acupuncture and Herbal treatment of Chronic HIV Sinusitus | Completed | Drug: Clavulanate potassium<br>Drug: Pseudoephedrine hydrochloride<br>Drug: Amoxicillin trihydrate |
| 2 | P.kinetic interactions b/w an Herbal Medicine (African Potato) and Antiretroviral agents (Lopinavir/Retonavir) | Completed | Drug: Kaletra (lopinavir/ritonavir), African Potato (hypoxisobtusa) |
| 3 | Moringa oleifera antiretroviral pharmacokinetic drug interaction | Completed | Dietary Supplement: Moringa oleifera<br>Drug: Efavirenz 600 mg<br>Drug: Nevirapine 200 mg Oral Tablet |
| 4 | Study on efficacy of herbal drugs in reducing viral load in HIV/AIDS patients CTRI/2022/09/045548 | Intervention | MVR Ayurveda medical college, Kerala |

### 17.3.6 Respiratory Syncytial Virus (RSV)

Respiratory syncytial virus is a single-stranded ribonucleic acid (RNA) pathogen belonging to the *Paramyxoviridae* family. This virus is omnipresent and is the primary cause of most of lower respiratory tract infections, especially in children below 3–5 years of the age group (Hall, 1994). It has been reported that most of the children under 2 years old suffer from RSV (Ogra, 2004). This virus enhances the suppression in immune system. Indeed, it is not highly dangerous in adults but infants suffer from bronchiolitis or pneumonia. In later stages of life, it can be a cause of childhood-asthma (Braciale, 2005; Sigurs et al., 2005). Prolong attack of RSV in infants leads to life-long immune suppression. Thus, repeated infection may occur in body (immune-compromised) (Busack and Shorr, 2022; Nam and Ison, 2019). Moving further, only limited therapeutics are available such as palivizumab, and ribavirin. Palivizumab is a mAbs (Hybridoma technology generated monoclonal Abs) and ribavirin is analogues of nucleotides. Still, effective immunization is not available although 32 trials are running/completed (*ClinicalTrials.gov*). Herbal products-related trials data is not found on the USA database, but one study has been found in Indian clinical registry (Table 17.6).

### 17.3.7 Dengue Viruses/Dengue

The dengue virus (DENV) belongs to the *Flaviviridae* family, hence called flevi-virus. It is single-stranded virus mostly found in arthopods such as mosquitoes and ticks. Dengue is generally transmitted via *Aedes aegypti* (128) (Black et al., 2002). According to clinical data, there are four types of serotypes of dengue such as DENV1, DENV2, DENV3, and DENV4, respectively (TuiskunenBäck and Lundkvist, 2013). The most prominent symptoms include relapsing fever, headache, muscle aches and pain, skin rashes, along with life-threatening reductions in platelets count. Severe dengue haemorrhagic fever (SDHF) is a significant aspect of DENV (Wang et al., 2020). If we focused on current therapeutics, limited drugs are available. Still, no specific antiviral therapeutic is available. For vaccine development, total 119 clinical studies are reported (*clinicalTrials.gov*). Dengvaxiavaccine prescription is discussed on CDC (centre for diseases control and prevention). Dengvaxia injection is

**TABLE 17.6**
**Clinical Trials on RSV (ClinicalTrials.gov)**

| S. No. | Study Title | Status | Interventions |
|---|---|---|---|
| 1 | Role of Pippali Choorna in the management of recurrent cough and cold infections in children. | Intervention | National Institute of Ayurveda Jaipur, Raj., India |

given to the subcutaneously (3 doses) within 6 months (Salje et al., 2021). However, natural products driven therapy will offer a safe and effective treatment. Most prominently, inhibition of viral replication is a necessary factor to cure diseases. Table 17.7 shows the studies related to herbal treatment.

Among many plants, herbs, shrubs, the *Carica papaya* extracts was tested widely in most part of world due to its controllable factors of human platelets count. A study by Ahmad et al. reports that 25 mL of papaya extract can enhance neutrophils and platelets count (Ahmad et al., 2011; Asghar et al., 2020). One more study conducted in Sri-Lanka with 12 patients where the group of physicians tested two doses of 5 mL of extract isolated from papaya leaves (24 hours) post-treatment. The results showed that blood cells count increased (Hettige, 2008; Sathyapalan et al., 2020). The dengue haemorrhagic fever was down by administration of 30 mL (3 consecutive days) (Subenthiran et al., 2013). This shows that *C. papaya* extract was widely used since ancient times and in recent studies (Yunita et al., 2012; Saleh and Kamisah, 2020). The studies exhibit that *C. papaya* isolated extract alters the signalling of IFN-Y, TNF-alpha and IL-6. Thus, dengue fever can be controlled. These studies reported *Ex-Vivo, and In-Vivo* (Dhiman et al., 2022). Many bioactive are yet to be discovered for treatment of DENV.

### 17.3.8 Enterovirus 71 (EV71)

Enterovirus-71 belongs to the *Picornaviridae* family. The EV-71 virus possesses the single-stranded genomic material (ssRNA). This virus is without enveloping. Mostly, it is transmitted via oral route, as such through cough, sneezing and infected or contaminated respiratory droplets. EV-71 reported as prime factor/pathogen for hand, foot, and mouth ailment in lower age children, and this can be life-threatening (Tapparel et al., 2013). This can be a cause for neurological degradation or infections such as meningitis, encephalitis (*CDC.gov*). According to the latest data, there are no specific drugs available, however few studies have been conducted (Rattanapisit et al., 2019; Ren et al., 2019). Mostly, the >5 age group children are susceptible to EV71 disease (Huang et al., 2011). No clinical data are found (Herbal treatment) for EV71.

### 17.3.9 Measles Virus (MV)

Measles virus consists of single-stranded ribonucleic acid (RNA) as genomic material. It comprises protein envelop. MV belongs to the *Paramyxoviridae* family. The symptoms of measles, also known as rubeola, first appear after 14 days. Mostly, it attacks infants and lower age group children. The common clinical characteristics reflected in patients are running nose, fever at 104°C. The ophthalmic system gets harmed (red eyes). Macular red rashes may also be present all over the body (Huang et al., 2011). This disease is associated with respiratory disorder. Therefore, respiratory droplets are the root cause behind transmission of rubeola. The MMR vaccine is presently being used for prophylaxis (Sabella, 2010). Measles is still a life-threatening disease among children (Xia et al., 2023). No clinical study has been found related to herbal medication.

## 17.4 CONCLUSION

There is an urgent need to identify a novel active chemotypes as leads for cost-effective antiviral chemotherapy. The promise of natural products as antiviral agents began to be expressed in the area of antiviral chemotherapy. The recent developed techniques have enabled the discovery of potent antiviral agents by dereplicating, accurately detecting, isolating, structurally defining and automating the assay of the bioactive natural products. Only 5%–15% of the approximately 250,000 species of higher plants have been investigated for the presence of bioactive compounds and the potential of the marine environment has barely been tapped systematically. We assure that natural products play an important role for potential antiviral resources for imaginative discoverers and contribute significantly to antiviral drug development.

**TABLE 17.7**
**Clinical Trials on Dengue (https://ctri.nic.in/)**

| S. No. | Study Title | Status | Interventions |
|---|---|---|---|
| 1 | Salutary effects of *Carica papaya* leaf extract in dengue fever patients—a Pilot study | Completed | Two 5 mL doses at an 8 hour interval for adults and two 2.5 mL doses at an 8 hour interval for children <10 years. Sri-Lanka |
| 2 | A life saving measure for dengue patients :Carica Papaya leaf extract, CTRI/2015/05/005806 | Intervention | Carica Papaya leaf extract Dr. B.R. Ambedkar Medical college, Karnataka, King George Hospital, A.P, India, Prasanth hospital, Karnataka, RGIMS, A.P., India, Sri Venkateshwara hospital, Karnataka |
| 3 | Is papaya leaf extract useful in treating dengue patients? CTRI/2017/07/009139 | Intervention | Bangalore Baptist Hospital, Karnataka |
| 4 | Clinical trial of an Ayurvedic formulation in dengue fever, CTRI/2017/11/010586 | Intervention | Dept. of Integrative medicine, Sri Devraj Urs Academy of Higher Education and Research, Karnataka, KLEU Dr Prabhakar Kore Hospital & Medical Research centre, Karnataka, KLEU Shri BMK Ayurveda Mahavidyalya Hospital and Medical Research Centre, Karnataka, Regional medical research centre, Karnataka |
| 5 | Efficacy and safety of Carica papaya leaf extract syrup in Thrombocytopenia associated with dengue fever, in Paediatric Patients, CTRI/2018/03/012792 | Intervention | Vijayanagara institute of medical sciences, Karnataka |
| 6 | Clinical trials for a life saving measure for dengue patients: Carica papaya leaf extract CTRI/2018/07/014988 | Intervention | Pristine Hospital and Research Centre Pvt. Ltd, Karnataka |
| 7 | CARIMAX (Papaya leaf extract 1,100 mg) CTRI/2018/08/015526 | Intervention | Shetty Hospital, Karnataka |
| 8 | Effect of Carica Papaya leaf extract capsule on platelet counts in dengue fever (Carica Papaya leaf extract capsule), CTRI/2019/04/018435 | Intervention | Pondicherry institute of Medical sciences, Pondicherry |
| 9 | A study to assess the safety and tolerability of aq. Extract of *Cocculus hisutus* (AQCH tablet) (SunPharma) in Healthy adult human patients under fasting condition | Intervention | Lambda Therapeutic Research Ltd., Gujrat |

## REFERENCES


Ahmad N, Fazal H, Ayaz M, Abbasi BH, Mohammad I, Fazal L. Dengue fever treatment with *Carica papaya* leaves extracts. *Asian Pacific Journal of Tropical Biomedicine*. 2011 Aug 1;1(4):330–3.

Anggakusuma, Colpitts CC, Schang LM, Rachmawati H, Frentzen A, Pfaender S, et al. Turmeric curcumin inhibits entry of all hepatitis C virus genotypes into human liver cells. *Gut*. 2014 Jul;63:1137–49.

Arduino PG, Porter SR. Herpes Simplex Virus Type 1 infection: overview on relevant clinico-pathological features. *Journal of Oral Pathology & Medicine*. 2008 Feb;37(2):107–21.

Asghar MS, Hassan M, Rasheed U, Jawed R, Khan NA, HaiderKazmi SJ, Khan SA. A study on investigating the role of Papaya extracts in the management of acute thrombocytopenia in Dengue fever, is there any clinical significance. *Merit Research Journal*. 2020;8:485–91.

Black IV WC, Bennett KE, Gorrochótegui-Escalante N, Barillas-Mury CV, Fernández-Salas I, de Lourdes Muñoz M, Farfán-Alé JA, Olson KE, Beaty BJ. Flavivirus susceptibility in *Aedes aegypti*. *Archives of Medical Research*. 2002 Jul 1;33(4):379–88.

Borah P, Hazarika S, Morsy MA, Goyal M, Chettri A, Venugopala KN, Mohanlall V, Acharya PC, Deb PK, Mailavaram RP. Antiviral drugs and vaccines. In *Medicinal Chemistry of Chemotherapeutic Agents* 2023 Jan 1 (pp. 319–359). Academic Press.

Braciale TJ. Respiratory syncytial virus and T cells: interplay between the virus and the host adaptive immune system. *Proceedings of the American Thoracic Society*. 2005 Aug;2(2):141–6.

Burrell CJ, Howard CR, Murphy FA. Epidemiology of viral infections. *Fenner and White's Medical Virology*. 2017:185–203.

Burton DR, Desrosiers RC, Doms RW, Koff WC, Kwong PD, Moore JP, Nabel GJ, Sodroski J, Wilson IA, Wyatt RT. HIV vaccine design and the neutralizing antibody problem. *Nature Immunology*. 2004 Mar 1;5(3):233–6.

Busack B, Shorr AF. Going viral-RSV as the neglected adult respiratory virus. *Pathogens*. 2022 Nov 11;11(11):1324.

Calland N, Albecka A, Belouzard S, Wychowski C, Duverlie G, Descamps V, et al. (−)-Epigallocatechin-3-gallate is a new inhibitor of hepatitis C virus entry. *Hepatology*. 2012;55:720–9.

CDC. gov. (accessed 30 june 2023) https://www.cdc.gov/flu/pandemic-resources/basics/past-pandemics.html

Cesar GZ, Alfonso MG, Marius MM, Elizabeth EM, Angel CB, Maira HR, et al. Inhibition of HIV-1 reverse transcriptase, toxicological and chemical profile of Calophyllumbrasiliense extracts from Chiapas, Mexico. *Fitoterapia*. 2011;82:1027–34.

Chang JS, Liu HW, Wang KC, Chen MC, Chiang LC, Hua YC, et al. Ethanol extract of Polygonum cuspidatum inhibits hepatitis B virus in a stable HBV-producing cell line. *Antiviral Research*. 2005;66:29–34.

Changi JS, Wang KC, Liu HW, Chen MC, Chiang LC, Lin CC. Sho-saiko-to (Xiao-Chai-Hu-Tang) and crude saikosaponins inhibit hepatitis B virus in a stable HBV-producing cell line. *The American Journal of Chinese Medicine*. 2007;35:341–51.

Chen SD, Gao H, Zhu QC, Wang YQ, Li T, Mu ZQ, et al. Houttuynoids A-E, anti-herpes simplex virus active flavonoids with novel skeletons from Houttuynia cordata. *Organic Letters*. 2012;14:1772–5.

Cheng HY, Huang HH, Yang CM, Lin LT, Lin CC. The *in vitro* anti-herpes simplex virus type-1 and type-2 activity of Long Dan Xie Gan Tan, a prescription of traditional Chinese medicine. *Chemotherapy*. 2008a;54:77–83.

Cheng HY, Lin LT, Huang HH, Yang CM, Lin CC. Yin Chen Hao Tang, a Chinese prescription, inhibits both herpes simplex virus type-1 and type-2 infections *in vitro*. *Antiviral Research*. 2008b;77:14–9.

Cheng HY, Yang CM, Lin TC, Lin LT, Chiang LC, Lin CC. Excoecarianin, Isolated from Phyllanthus urinaria Linnea, Inhibits Herpes Simplex Virus Type 2 Infection through Inactivation of Viral Particles. *Evidence-Based Complementary and Alternative Medicine. 2011;*2011:259103.

Cheng HY, Yang CM, Lin TC, Shieh DE, Lin CC. ent-Epiafzelechin—(4alpha-->8)-epiafzelechin extracted from Cassia javanica inhibits herpes simplex virus type 2 replication. *Journal of Medical Microbiology*. 2006a;55:201–6.

Cheng PW, Chiang LC, Yen MH, Lin CC. Bupleurum kaoi inhibits Coxsackie B virus type 1 infection of CCFS-1 cells by induction of type I interferons expression. *Food and Chemical Toxicology*. 2007;45:24–31.

Cheng PW, Ng LT, Chiang LC, Lin CC. Antiviral effects of saikosaponins on human coronavirus 229E in vitro. *Clinical and Experimental Pharmacology and Physiology*. 2006b Jul;33(7):612–6.

Cheng PW, Ng LT, Lin CC. Xiao chai hu tang inhibits CVB1 virus infection of CCFS-1 cells through the induction of Type I interferon expression. *International Immunopharmacology*. 2006c;6:1003–12.

Chentoufi AA, BenMohamed L. Mucosal herpes immunity and immunopathology to ocular and genital herpes simplex virus infections. *Clinical and Developmental Immunology.* 2012 Oct;2012:149135.
Chiangi LC, Ng LT, Cheng PW, Chiang W, Lin CC. Antiviral activities of extracts and selected pure constituents of Ocimumbasilicum. *Clinical and Experimental Pharmacology and Physiology.* 2005;32:811–6.
Chiang LC, Ng LT, Liu LT, Shieh DE, Lin CC. Cytotoxicity and anti-hepatitis B virus activities of saikosaponins from Bupleurum species. *Planta Medica.* 2003;69:705–9.
Choi HJ, Lim CH, Song JH, Baek SH, Kwon DH. Antiviral activity of raoulic acid from Raoulia australis against Picornaviruses. *Phytomedicine.* 2009;16:35–9.
Ciesek S, von Hahn T, Colpitts CC, Schang LM, Friesland M, Steinmann J, et al. The green tea polyphenol, epigallocatechin-3-gallate, inhibits hepatitis C virus entry. *Hepatology.* 2011;54:1947–55.
Committee on Infectious Diseases, Brady MT, Byington CL, Davies HD, Edwards KM, Glode MP, Jackson MA, Keyserling HL, Maldonado YA, Murray DL, Orenstein WA. Recommendations for prevention and control of influenza in children, 2012-2013. *Pediatrics.* 2012 Oct 1;130(4):780–92.
Danaher RJ, Wang C, Dai J, Mumper RJ, Miller CS. Antiviral effects of blackberry extract against herpes simplex virus type 1. *Oral Surgery, Oral Medicine, Oral Pathology Oral Radiology.* 2011;112:e31–5.
Dao TT, Dang TT, Nguyen PH, Kim E, Thuong PT, Oh WK. Xanthones from Polygala karensium inhibit neuraminidases from influenza A viruses. *Bioorganic & Medicinal Chemistry Letters.* 2012;22:3688–92.
Dao TT, Nguyen PH, Lee HS, Kim E, Park J, Lim SI, et al. Chalcones as novel influenza A (H1N1) neuraminidase inhibitors from Glycyrrhiza inflata. *Bioorganic & Medicinal Chemistry Letters.* 2011;21:294–8.
De Clercq E, Li G. Approved antiviral drugs over the past 50 years. *Clinical Microbiology Reviews.* 2016 Jul;29(3):695–747.
Dhiman M, Sharma L, Dadhich A, Dhawan P, Sharma MM. Traditional knowledge to contemporary medication in the treatment of infectious disease dengue: a review. *Frontiers in Pharmacology.* 2022 Mar 14;13:750494.
Eccles R. Understanding the symptoms of the common cold and influenza. *The Lancet Infectious Diseases.* 2005 Nov 1;5(11):718–25.
El-Serag HB. Epidemiology of viral hepatitis and hepatocellular carcinoma. *Gastroenterology.* 2012 May 1;142(6):1264–73.
Geller C, Varbanov M, Duval RE. Human coronaviruses: insights into environmental resistance and its influence on the development of new antiseptic strategies. *Viruses.* 2012 Nov 12;4(11):3044–68.
Gescher K, Kuhn J, Hafezi W, Louis A, Derksen A, Deters A, et al. Inhibition of viral adsorption and penetration by an aqueous extract from *Rhododendron ferrugineum* L. as antiviral principle against herpes simplex virus type-1. *Fitoterapia.* 2011a;82:408–13.
Gescher K, Kuhn J, Lorentzen E, Hafezi W, Derksen A, Deters A, et al. Proanthocyanidin-enriched extract from MyrothamnusflabellifoliaWelw. exerts antiviral activity against herpes simplex virus type 1 by inhibition of viral adsorption and penetration. *The Journal of Ethnopharmacology.* 2011b;134:468–74.
Grienke U, Schmidtke M, von Grafenstein S, Kirchmair J, Liedl KR, Rollinger JM. Influenza neuraminidase: a druggable target for natural products. *Natural Product Reports.* 2012;29:11–36.
Haid S, Novodomska A, Gentzsch J, Grethe C, Geuenich S, Bankwitz D, et al. A plant-derived flavonoid inhibits entry of all HCV genotypes into human hepatocytes. *Gastroenterology.* 2012;143:213-22.e5.
Hall CB. Prospects for a respiratory syncytial virus vaccine. *Science.* 1994 Sep 2;265(5177):1393–4.
Hao BJ, Wu YH, Wang JG, Hu SQ, Keil DJ, Hu HJ, et al. Hepatoprotective and antiviral properties of isochlorogenic acid A from Laggeraalata against hepatitis B virus infection. *The Journal of Ethnopharmacology.* 2012;144:190–4.
Hayashi T, Hayashi K, Maeda M, Kojima I. Calcium spirulan, an inhibitor of enveloped virus replication, from a blue-green alga Spirulina platensis. *Journal of Natural Products.* 1996;59:83–7.
He W, Han H, Wang W, Gao B. Anti-influenza virus effect of aqueous extracts from dandelion. *Virology Journal.* 2011;8:538.
Hettige S. Salutary effects of *Carica papaya* leaf extract in dengue fever patients-a pilot study. *Sri Lankan Family Physician.* 2008 Sep 15;29(1):17–9.
Hood JL, Jallouk AP, Campbell N, Ratner L, Wickline SA. Cytolytic nanoparticles attenuate HIV-1 infectivity. *Antiviral Therapy.* 2013;18:95–103.
Huang SP, Shieh GJ, Lee L, Teng HJ, Kao ST, Lin JG. Inhibition effect of shengma-gegen-tang on measles virus in Vero cells and human peripheral blood mononuclear cells. *The American Journal of Chinese Medicine.* 1997;25:89–96.
Huang SW, Kiang D, Smith DJ, Wang JR. Evolution of re-emergent virus and its impact on enterovirus 71 epidemics. *Experimental Biology and Medicine.* 2011 Aug;236(8):899–908.

Huang W, Zhang X, Wang Y, Ye W, Ooi VE, Chung HY, et al. Antiviral biflavonoids from Radix Wikstroemiae (Liaogewanggen) *Chinese Medicine.* 2010;5:23.

Huerta-Reyes M, BasualdoMdel C, Abe F, Jimenez-Estrada M, Soler C, Reyes-Chilpa R. HIV-1 inhibitory compounds from Calophyllumbrasiliense leaves. *Biological and Pharmaceutical Bulletin.* 2004;27:1471–5.

Jeong HJ, Kim YM, Kim JH, Kim JY, Park JY, Park SJ, et al. Homoisoflavonoids from *Caesalpinia sappan* displaying viral neuraminidases inhibition. *Biological and Pharmaceutical Bulletin.* 2012;35:786–90.

Jiang ZY, Liu WF, Zhang XM, Luo J, Ma YB, Chen JJ. Anti-HBV active constituents from Piper longum. *Bioorganic & Medicinal Chemistry Letters.* 2013;23:2123–7.

Kausar S, Said Khan F, Ishaq Mujeeb Ur Rehman M, Akram M, Riaz M, Rasool G, Hamid Khan A, Saleem I, Shamim S, Malik A. A review: mechanism of action of antiviral drugs. *International Journal of Immunopathology and Pharmacology.* 2021 Mar;35:20587384211002621.

Khan MT, Ather A, Thompson KD, Gambari R. Extracts and molecules from medicinal plants against herpes simplex viruses. *Antiviral Research.* 2005;67:107–19.

Kim K, Kim KH, Kim HY, Cho HK, Sakamoto N, Cheong J. Curcumin inhibits hepatitis C virus replication via suppressing the Akt-SREBP-1 pathway. *FEBS Letters.* 2010;584:707–12.

Kim SK, Karadeniz F. Anti-HIV activity of extracts and compounds from marine algae. *Advances in Food and Nutrition Research.* 2011;64:255–65.

Koishi AC, Zanello PR, Bianco EM, Bordignon J, Nunes Duarte dos Santos C. Screening of Dengue virus antiviral activity of marine seaweeds by an *in situ* enzyme-linked immunosorbent assay. *PLoS One.* 2012;7:e51089.

Krawitz C, Mraheil MA, Stein M, Imirzalioglu C, Domann E, Pleschka S, et al. Inhibitory activity of a standardized elderberry liquid extract against clinically-relevant human respiratory bacterial pathogens and influenza A and B viruses. *BMC Complementary and Alternative Medicine.* 2011;11:16.

Kudo E, Taura M, Matsuda K, Shimamoto M, Kariya R, Goto H, et al. Inhibition of HIV-1 replication by a tricyclic coumarin GUT-70 in acutely and chronically infected cells. *Bioorganic & Medicinal Chemistry Letters.* 2013;23:606–9.

Kwon H, Lok AS. Hepatitis B therapy. *Nature Reviews Gastroenterology & Hepatology.* 2011 May;8(5):275–84.

Lau KM, Lee KM, Koon CM, Cheung CS, Lau CP, Ho HM, et al. Immunomodulatory and anti-SARS activities of Houttuynia cordata. *The Journal of Ethnopharmacology.* 2008;118:79–85.

Li SY, Chen C, Zhang HQ, Guo HY, Wang H, Wang L, et al. Identification of natural compounds with antiviral activities against SARS-associated coronavirus. *Antiviral Research.* 2005;67:18–23.

Liang TJ. Hepatitis B: the virus and disease. *Hepatology.* 2009 May;49(S5):S13–21.

Lin CW, Tsai FJ, Tsai CH, Lai CC, Wan L, Ho TY, et al. Anti-SARS coronavirus 3C-like protease effects of Isatisindigotica root and plant-derived phenolic compounds. *Antiviriral Research.* 2005;68:36–42.

Lin LT, Chen TY, Chung CY, Noyce RS, Grindley TB, McCormick C, et al. Hydrolyzable tannins (chebulagic acid and punicalagin) target viral glycoprotein-glycosaminoglycan interactions to inhibit herpes simplex virus 1 entry and cell-to-cell spread. *Journal of Virology.* 2011;85:4386–98.

Lin LT, Chen TY, Lin SC, Chung CY, Lin TC, Wang GH, et al. Broad-spectrum antiviral activity of chebulagic acid and punicalagin against viruses that use glycosaminoglycans for entry. *BMC Microbiology.* 2013;13:187.

Lin YM, Flavin MT, Schure R, Chen FC, Sidwell R, Barnard DL, et al. Antiviral activities of biflavonoids. *Planta Medica.* 1999;65:120–5.

Louten J. Virus transmission and epidemiology. In *Essential Human Virology.* 2016:71. Elsevier.

Low JS, Wu KX, Chen KC, Ng MM, Chu JJ. Narasin, a novel antiviral compound that blocks dengue virus protein expression. *Antiviral Therapy.* 2011;16:1203–18.

Lubbe A, Seibert I, Klimkait T, van der Kooy F. Ethnopharmacology in overdrive: the remarkable anti-HIV activity of Artemisia annua. *Journal of Ethnopharmacology.* 2012;141:854–9.

Ma LY, Ma SC, Wei F, Lin RC, But PP, Lee SH, et al. Uncinoside A and B, two new antiviral chromone glycosides from Selaginella uncinata. *Chemical and Pharmaceutical Bulletin (Tokyo).* 2003;51:1264–7.

Ma SG, Gao RM, Li YH, Jiang JD, Gong NB, Li L, et al. Antiviral spirooliganones A and B with unprecedented skeletons from the roots of *Illicium oligandrum. Organic Letters.* 2013;15:4450–3.

Mak PW, Jayawardena S, Poon LL. The evolving threat of influenza viruses of animal origin and the challenges in developing appropriate diagnostics. *Clinical Chemistry.* 2012 Nov 1;58(11):1527–33.

Margolis AM, Heverling H, Pham PA, Stolbach A. A review of the toxicity of HIV medications. *Journal of Medical Toxicology.* 2014 Mar;10:26–39.

McWhorter JH. Spicebush. A Cherokee remedy for the measles. *North Carolina Medical Journal.* 1996;57:306.

Meuleman P, Albecka A, Belouzard S, Vercauteren K, Verhoye L, Wychowski C, et al. Griffithsin has antiviral activity against hepatitis C virus. *Antimicrobial Agents and Chemotherapy.* 2011;55:5159–67.

Morfin F, Thouvenot D. Herpes simplex virus resistance to antiviral drugs. *Journal of Clinical Virology*. 2003 Jan 1;26(1):29–37.
Moss JA. HIV/AIDS review. *Radiologic Technology*. 2013 Jan 1;84(3):247–67.
Mostafa A, Abdelwhab EM, Mettenleiter TC, Pleschka S. Zoonotic potential of influenza A viruses: a comprehensive overview. *Viruses*. 2018 Sep 13;10(9):497.
Nam HH, Ison MG. Respiratory syncytial virus infection in adults. *BMJ*. 2019 Sep 10;366:15021.
Ni YH, Chen DS. Hepatitis B vaccination in children: the Taiwan experience. *Pathologie Biologie*. 2010 Aug 1;58(4):296–300.
Nwodo UU, Ngene AA, Iroegbu CU, Onyedikachi OA, Chigor VN, Okoh AI. *In vivo* evaluation of the antiviral activity of Cajanus cajan on measles virus. *Archives of Virology*. 2011;156:1551–7.
Ogra PL. Respiratory syncytial virus: the virus, the disease and the immune response. *Paediatric Respiratory Reviews*. 2004 Jan 1;5:S119–26.
Olila D, Olwa O, Opuda-Asibo J. Screening extracts of Zanthoxylum chalybeum and Warburgiaugandensis for activity against measles virus (Swartz and Edmonston strains) *in vitro*. *African Health Sciences*. 2002;2:2–10.
Parker ME, Chabot S, Ward BJ, Johns T. Traditional dietary additives of the Maasai are antiviral against the measles virus. *Journal of Ethnopharmacology*. 2007;114:146–52.
Petricevich VL, Mendonca RZ. Inhibitory potential of Crotalus durissusterrificus venom on measles virus growth. *Toxicon*. 2003;42:143–53.
Pillay N. The Global Impact of HIV/AIDS on the Realisation of Health Rights. *Infectious Diseases*. IntechOpen; 2023. Available from: http://dx.doi.org/10.5772/intechopen.109382.
Polyak SJ, Morishima C, Lohmann V, Pal S, Lee DY, Liu Y, et al. Identification of hepatoprotective flavonolignans from silymarin. *Proceedings of the National Academy of Sciences of the United States of America*. 2010;107:5995–9.
Polyak SJ, Morishima C, Shuhart MC, Wang CC, Liu Y, Lee DY. Inhibition of T-cell inflammatory cytokines, hepatocyte NF-kappaBsignaling, and HCV infection by standardized Silymarin. *Gastroenterology*. 2007;132:1925–36.
Prifti GM, Moianos D, Giannakopoulou E, Pardali V, Tavis JE, Zoidis G. Recent advances in hepatitis B treatment. *Pharmaceuticals*. 2021 May 1;14(5):417.
Qiu LP, Chen KP. Anti-HBV agents derived from botanical origin. *Fitoterapia*. 2013;84:140–57.
Rattanapisit K, Chao Z, Siriwattananon K, Huang Z, Phoolcharoen W. Plant-produced anti-enterovirus 71 (EV71) monoclonal antibody efficiently protects mice against EV71 infection. *Plants*. 2019 Dec 1;8(12):560.
Rechtman MM, Har-Noy O, Bar-Yishay I, Fishman S, Adamovich Y, Shaul Y, et al. Curcumin inhibits hepatitis B virus via down-regulation of the metabolic coactivator PGC-1alpha. *FEBS Letters*. 2010;584:2485–90.
Rello J, Pop-Vicas A. Clinical review: primary influenza viral pneumonia. *Critical Care*. 2009 Dec;13:1–6.
Ren G, Ding G, Zhang H, Wang H, Jin Z, Yang G, Han Y, Zhang X, Li G, Li W. Antiviral activity of sophoridine against enterovirus 71 in vitro. *Journal of Ethnopharmacology*. 2019 May 23;236:124–8.
Ryu YB, Jeong HJ, Kim JH, Kim YM, Park JY, Kim D, et al. Biflavonoids from Torreya nucifera displaying SARS-CoV 3CL (pro) inhibition. *Bioorganic & Medicinal Chemistry*. 2010;18:7940–7.
Sabella C. Measles: not just a childhood rash. *Cleveland Clinic Journal of Medicine*. 2010 Mar 1;77(3):207–13.
Saleh MS, Kamisah Y. Potential medicinal plants for the treatment of dengue fever and severe acute respiratory syndrome-coronavirus. *Biomolecules*. 2020 Dec 30;11(1):42.
Salje H, Alera MT, Chua MN, Hunsawong T, Ellison D, Srikiatkhachorn A, Jarman RG, Gromowski GD, Rodriguez-Barraquer I, Cauchemez S, Cummings DA. Evaluation of the extended efficacy of the Dengvaxia vaccine against symptomatic and subclinical dengue infection. *Nature Medicine*. 2021 Aug;27(8):1395–400.
Samson M, Pizzorno A, Abed Y, Boivin G. Influenza virus resistance to neuraminidase inhibitors. *Antiviral Research*. 2013 May 1;98(2):174–85.
Sathyapalan DT, Padmanabhan A, Moni M, P- Prabhu B, Prasanna P, Balachandran S, Trikkur SP, Jose S, Edathadathil F, Anilkumar JO, Jayaprasad R. Efficacy & safety of Carica papaya leaf extract (CPLE) in severe thrombocytopenia ($\leq$ 30,000/μl) in adult dengue—results of a pilot study. *PLoS One*. 2020 Feb 19;15(2):e0228699.
Shaw GM, Hunter E. HIV transmission. *Cold Spring Harb Perspect Med*. 2012 Nov 1;2(11):a006965. Cold Spring Harbor Laboratory Press. doi: 10.1101/cshperspect.a006965. PMID: 23043157; PMCID: PMC3543106.
Sierra S, Kupfer B, Kaiser R. Basics of the virology of HIV-1 and its replication. *Journal of Clinical Virology*. 2005 Dec 1;34(4):233–44.

Sigurs N, Gustafsson PM, Bjarnason R, Lundberg F, Schmidt S, Sigurbergsson F, Kjellman B. Severe respiratory syncytial virus bronchiolitis in infancy and asthma and allergy at age 13. *American Journal of Respiratory and Critical Care Medicine*. 2005 Jan 15;171(2):137–41.

Stoopler ET, Balasubramaniam R. Topical and systemic therapies for oral and perioral herpes simplex virus infections. *Journal of the California Dental Association*. 2013 Apr 1;41(4):259–62.

Subenthiran S, Choon TC, Cheong KC, Thayan R, Teck MB, Muniandy PK, Afzan A, Abdullah NR, Ismail Z. Carica papaya leaves juice significantly accelerates the rate of increase in platelet count among patients with dengue fever and dengue haemorrhagic fever. *Evidence-Based Complementary and Alternative Medicine*. 2013 Oct;2013:616737.

Superti F, Ammendolia MG, Marchetti M. New advances in anti-HSV chemotherapy. *Current Medicinal Chemistry*. 2008;15:900–11.

Takebe Y, Saucedo CJ, Lund G, Uenishi R, Hase S, Tsuchiura T, et al. Antiviral lectins from red and blue-green algae show potent *in vitro* and *in vivo* activity against hepatitis C virus. *PLoS One*. 2013;8:e64449.

Tamura S, Yang GM, Yasueda N, Matsuura Y, Komoda Y, Murakami N. Tellimagrandin I, HCV invasion inhibitor from RosaeRugosaeFlos. *Bioorganic & Medicinal Chemistry Letters*. 2010;20:1598–600.

Tapparel C, Siegrist F, Petty TJ, Kaiser L. Picornavirus and enterovirus diversity with associated human diseases. *Infection, Genetics and Evolution*. 2013 Mar 1;14:282–93.

Theisen LL, Muller CP. EPs (R) 7630 (Umckaloabo (R)), an extract from Pelargonium sidoides roots, exerts anti-influenza virus activity *in vitro* and *in vivo*. *Antiviral Research*. 2012;94:147–56.

Tian X, Chen J, Wang X, Xie Y, Zhang X, Han D, Fu H, Yin W, Wu N. Global, regional, and national HIV/AIDS disease burden levels and trends in 1990-2019: A systematic analysis for the global burden of disease 2019 study. *Frontiers in Public Health*. 2023 Feb 15;11:1068664.

Torres Munguía JA, Badarau FC, Díaz Pavez LR, Martínez-Zarzoso I, Wacker KM. A global dataset of pandemic- and epidemic-prone disease outbreaks. *Scientific Data*. 2022 Nov 10;9(1):683.

TuiskunenBäck A, Lundkvist Å. Dengue viruses-an overview. *Infection Ecology & Epidemiology*. 2013 Jan 1;3(1):19839.

Van der Hoek L. Human coronaviruses: what do they cause?. *Antiviral Therapy*. 2007 Jan;12(4_part_2):651–8.

Van der Vries E, Collins PJ, Vachieri SG, Xiong X, Liu J, Walker PA, Haire LF, Hay AJ, Schutten M, Osterhaus AD, Martin SR. H1N1 2009 pandemic influenza virus: resistance of the I223R neuraminidase mutant explained by kinetic and structural analysis. *PLoS Pathogens*. 2012;8:e1002914.

Vo TS, Kim SK. Potential anti-HIV agents from marine resources: an overview. *Marine Drugs*. 2010;8:2871–92.

Wang KC, Chang JS, Chiang LC, Lin CC. *Cimicifuga foetida* L. inhibited human respiratory syncytial virus in HEp-2 and A549 cell lines. *American Journal of Chinese Medicine*. 2012a;40:151–62.

Wang KC, Chang JS, Chiang LC, Lin CC. Sheng-Ma-Ge-Gen-Tang (Shoma-kakkon-to) inhibited cytopathic effect of human respiratory syncytial virus in cell lines of human respiratory tract. *Journal of Ethnopharmacology*. 2011;135:538–44.

Wang KC, Chang JS, Lin LT, Chiang LC, Lin CC. Antiviral effect of cimicifugin from *Cimicifuga foetida* against human respiratory syncytial virus. *American Journal of Chinese Medicine*. 2012b;40:1033–45.

Wang WH, Urbina AN, Chang MR, Assavalapsakul W, Lu PL, Chen YH, Wang SF. Dengue hemorrhagic fever-A systemic literature review of current perspectives on pathogenesis, prevention and control. *Journal of Microbiology, Immunology and Infection*. 2020 Dec 1;53(6):963–78.

Welsch C, Jesudian A, Zeuzem S, Jacobson I. New direct-acting antiviral agents for the treatment of hepatitis C virus infection and perspectives. *Gut*. 2012 May 1;61(Suppl 1):i36–46.

Witvrouw M, De Clercq E. Sulfated polysaccharides extracted from sea algae as potential antiviral drugs. *General Pharmacology: The Vascular System*. 1997 Oct 1;29(4):497–511.

World Health Organization. Weekly epidemiological update on COVID-19—29 June 2023. 2023a. [Last accessed on 2023 June 29]. Available from: https://www.who.int/publications/m/item/weekly-epidemiological-update-on-covid-19---29-june-2023.

World Health Organization. 2023b. [Last accessed on 2023 June 30]. Available from: https://www.who.int/newsroom/spotlight/influenza-are-we-ready.

World Health Organization. 2023c. [Last accessed on 2023 June 30]. Available from: https://www.who.int/news-room/fact-sheets/detail/influenza-(avian-and-other-zoonotic.

World Health Organization. HIV and AIDS. 2023d. [Last accessed on 2023 June 30]. Available from: https://www.who.int/news-room/fact-sheets/detail/hiv-aids.

Xia S, Gullickson CC, Metcalf CJ, Grenfell BT, Mina MJ. Assessing the effects of measles virus infections on childhood infectious disease mortality in Brazil. *The Journal of Infectious Diseases*. 2023 Jan 1;227(1):133–40.

Xu L, Zhong XL, Xi ZC, Li Y, Xu HX. Medicinal plants and natural compounds against acyclovir-resistant HSV infections. *Frontiers in Microbiology*. 2022 Oct 10;13:1025605.

Yamada S, Suzuki Y, Suzuki T, Le MQ, Nidom CA, Sakai-Tagawa Y, Muramoto Y, Ito M, Kiso M, Horimoto T, Shinya K. Haemagglutinin mutations responsible for the binding of H5N1 influenza A viruses to human-type receptors. *Nature*. 2006 Nov 16;444(7117):378–82.

Yang CM, Cheng HY, Lin TC, Chiang LC, Lin CC. Hippomanin A from acetone extract of Phyllanthus urinaria inhibited HSV-2 but not HSV-1 infection *in vitro*. *Phytotherapy Research*. 2007a;21:1182–6.

Yang CM, Cheng HY, Lin TC, Chiang LC, Lin CC. The *in vitro* activity of geraniin and 1,3,4,6-tetra-O-galloyl-beta-D-glucose isolated from Phyllanthus urinaria against herpes simplex virus type 1 and type 2 infection. *Journal of Ethnopharmacology*. 2007b;110:555–8.

Yoshimura K. Current status of HIV/AIDS in the ART era. *Journal of Infection and Chemotherapy*. 2017 Jan 1;23(1):12–6.

Yu MS, Lee J, Lee JM, Kim Y, Chin YW, Jee JG, et al. Identification of myricetin and scutellarein as novel chemical inhibitors of the SARS coronavirus helicase, nsP13. *Bioorganic & Medicinal Chemistry Letters*. 2012;22:4049–54.

Yunita F, Hanani E, Kristianto J. The effect of *Carica papaya* L. leaves extract capsules on platelets count and hematocrit level in dengue fever patient. *International Journal of Medicinal and Aromatic Plants*. 2012;2(4):573–8.

Zandi K, Teoh BT, Sam SS, Wong PF, Mustafa MR, Abubakar S. Antiviral activity of four types of bioflavonoid against dengue virus type-2. *Virology Journal*. 2011;8:560.

Zandi K, Teoh BT, Sam SS, Wong PF, Mustafa MR, Abubakar S. Novel antiviral activity of baicalein against dengue virus. *BMC Complementary and Alternative Medicine*. 2012;12:214.

Zang N, Xie X, Deng Y, Wu S, Wang L, Peng C, et al. Resveratrol-mediated gamma interferon reduction prevents airway inflammation and airway hyperresponsiveness in respiratory syncytial virus-infected immunocompromised mice. *Journal of Virology*. 2011;85:13061–8.

Zeng FL, Xiang YF, Liang ZR, Wang X, Huang DE, Zhu SN, et al. Anti-hepatitis B virus effects of dehydrocheilanthifoline from *Corydalis saxicola*. *American Journal of Chinese Medicine*. 2013;41:119–30.

# Index

Note: **Bold** page numbers refer to tables and *italic* page numbers refer to figures.